Science and Technology in Homeric Epics

# HISTORY OF MECHANISM AND MACHINE SCIENCE
Volume 6

*Series Editor*
MARCO CECCARELLI

*Aims and Scope of the Series*

This book series aims to establish a well defined forum for Monographs and Proceedings on the History of Mechanism and Machine Science (MMS). The series publishes works that give an overview of the historical developments, from the earliest times up to and including the recent past, of MMS in all its technical aspects.

This technical approach is an essential characteristic of the series. By discussing technical details and formulations and even reformulating those in terms of modern formalisms the possibility is created not only to track the historical technical developments but also to use past experiences in technical teaching and research today. In order to do so, the emphasis must be on technical aspects rather than a purely historical focus, although the latter has its place too.

Furthermore, the series will consider the republication of out-of-print older works with English translation and comments.

The book series is intended to collect technical views on historical developments of the broad field of MMS in a unique frame that can be seen in its totality as an Encyclopaedia of the History of MMS but with the additional purpose of archiving and teaching the History of MMS. Therefore the book series is intended not only for researchers of the History of Engineering but also for professionals and students who are interested in obtaining a clear perspective of the past for their future technical works. The books will be written in general by engineers but not only for engineers.

Prospective authors and editors can contact the series editor, Professor M. Ceccarelli, about future publications within the series at:

LARM: Laboratory of Robotics and Mechatronics
DiMSAT – University of Cassino
Via Di Biasio 43, 03043 Cassino (Fr)
Italy
E-mail: ceccarelli@unicas.it

For other titles published in this series, go to
www.springer.com/series/7481

S.A. Paipetis

Editor

# Science and Technology in Homeric Epics

Springer

S.A. Paipetis
Department of Mechanical Engineering & Aeronautics
University of Patras
Patras
Greece



ISBN-13: 978-1-4020-8783-7 e-ISBN-13: 978-1-4020-8784-4

Library of Congress Control Number: 2008937901



Printed on acid-free paper

9 8 7 6 5 4 3 2 1

springer.com

# Table of Contents

## Part 7: Medicine



## Part 8: Flora and Fauna



## Part 9: Astronomy



## Part 10: Seafaring

## Part 11: Cultural Environment



## Part 12: Geography



## Part 13: General Interest

# Preface

The present volume is based on papers presented at the international symposium "Science and Technology in Homeric Epics", SPAP Conference Centre, Ancient Olympia, 27–30 August 2006. It includes a total of 41 contributions, mostly original research papers, covering diverse fields of science and technology, in the modern sense of these words.

The use of terms coined in relatively recent times (after the 15th century) to refer to situations from times so long ago as the Mycenaean Era, may sound inappropriate. However, careful studies of the Homeric Epics by specialists in the various scientific fields may convince the reader that the knowledge contained therein reflects a deep understanding of the science of nature and an ability to apply technological achievements and structures, strongly reminiscent of modern technology in its present evolution level.

The question of knowledge contained in the Homeric Epic had, until recently, received a negative answer. The seemingly scientific knowledge and admirable technological achievements presented have always been attributed to poetic inspiration rather than to a solid scientific mind. Of course, if the latter were true, which is likely to be sometimes, it is also true that the very conception of an idea can constitute a catalyst towards scientific development. For example, Isaac Asimov notices that the first reference to robots is found in the Iliad, i.e. to the *golden girls* of Hephaestus, who, although made of soulless matter, "were like real young women, with sense and reason, voice also and strength, and all the learning of the immortals" (Il. 18.419–420). The great dream of Man, i.e. the possession of fully rational, obedient and efficient mechanical servants, is about to be substantiated nowadays. On the other hand, the technical information given in the Iliad is not sufficient to reach a justified conclusion on whether such devices really existed in the Mycenaean era.

However, there are many more explicit cases, where the description of a structure provides sufficient data, from which, on the basis of fully realistic assumptions, it is possible to reconstruct it in the form of numerical models and/or experimental specimens and, consequently, to perform a theoretical and/or experimental analysis. Examples of such structures are the shields of Achilles and Ajax, which are laminated structures, of practically modern technology, exhibiting maximum penetration resistance. Their analysis confirmed their battle behaviour, as recounted in the *Iliad*, with surprising accuracy. The analysis of Circe's instruction to Ulysses, on how to cross the fearful straits of Scylla and Charybdis safely, based on the hydrodynamic investigation of the problem, is another example of analytical approach.

It is, of course, clear that these instances have been pinpointed by science and technology specialists, whose broader interests have led them to study the Homeric Epics within the frame of their own scientific area. Thus, the conclusion is drawn that investigation of knowledge contained in the Epics cannot be but an *interdisciplinary activity*. It is further noted that many points of interest would not be recognizable a few decades ago even by competent scientists given the level of scientific and technological progress at the time. Accordingly, the next conclusion is that investigation of knowledge contained in the Epics is also a *diachronic effort*. In other words, a study of the Homeric Epics must be performed by the broadest possible circle of scientific specialists, but also on a continual basis, to account for new disciplines created as science is progressing.

The idea of organizing an international symposium on Homer's science and technology was born during the study of the monumental monograph by Constantine Zeggelis "The science of nature in Homer",[1] a 1891 publication. The initial thought was "to rewrite" Zeggelis' book by a team of experts, each one a specialist in the respective area. However, from the end of 19th century to the present day both the classification and the number of established sciences, as well as the knowledge they included, has changed dramatically. In other words, the scope was so much enlarged that only an invitation to those members of the international scientific community, who are dealing with the Homeric Epics from all possible aspects, was found appropriate to guarantee a satisfactory outcome.

This choice was fully justified by the highly enthusiastic response of the academic community, as well as of independent researchers, both in Greece and internationally, and of the mass media, and last but not least of lay people, justifying the title awarded to the Epics as "The Gospel of the Hellenic Nation". Finally, besides the fact that numerous subjects remain unexplored, several distinguished scientists sent regrets for not being able to attend. This called for the preparation of a second symposium with the same theme in the near future.

The contents of the present volume are classified in 13 parts: Out of two general lectures, one in-depth presentation of Mycenaean Technology, covering all classes of activities, was delivered by Professor Theodossios P. Tassios, Greece. In this lecture, the sometimes inexplicable tendency of certain researchers to undermine the scientific and/or technological achievements of Ancient Greeks, as based on knowledge originating from the Orient or on imported know-how, and in fact by arguments unsupported or irrelevant, is addressed. This opposes long-held tendencies, which, with equally extreme attitude, used to express probably disproportionate admiration for the Greek achievements: unprejudiced research is, of course, the answer, which is greatly facilitated by the modern scientific research means available. The second lecture by Professor Emmanuel Mikroyannakis, Greece, deals with the interpretation of the term "autagreton", as it appears in the Epics, which reaches the astonishing conclusion that it refers to technical devices not just automated but "intelligent", i.e. capable of autonomous action, making decisions after a proper assessment of external excitations.

[1] Zeggelis, C.D. (1891), *The Science of Nature in Homer*, reprinted by the University of Patras Press with an Introduction by S.A. Paipetis, Patras, 1997 [in Greek].

In Part 2, Mathematics and Physics, a variety of subjects is presented. Such as the Archimedean problem of counting Helios' cattle, as described by Homer (Chris Rorres, US), a Fluid Dynamics approach to water vortices accounted for in the Odyssey (G.H. Vatistas, Canada), the implementation of Homeric automata in the ensuing centuries (D. Kalligeropoulos and S. Vasileiadou, Greece), the creation of the world and the appearance of Man according to Homeric Cosmogony (Taha Showleh, Canada) and finally the excellent formulation of the laws of curvilinear motion by King Nestor of Pylos in the *Iliad* (S.A. Paipetis, Greece).

Part 3 deals with materials, a subject for which Homeric accounts abound. The importance of iron in the Epics is presented along with the ecological sensitivity of Homer (G. Varoufakis, Greece). Two important works come from Emmanuel Pantos (UK) and his associates, dealing with early bronze technology in the Iberian Peninsula and the production of faience-like blue glaze at low temperatures. Finally, a dyeing method based on porphyra is presented, inspired by a reference in the *Iliad* "to dye ivory-made horse cheek ornament in reddish purple" (W. Nishiyama, Japan).

Part 4 deals with the development of defensive weapons as depicted in the Homeric Epics, such as helmets of copper alloys (P. Manti et al., UK), a numerical and experimental analysis of the famous shields of Achilles and Ajax (S.A. Paipetis and V. Kostopoulos, Greece) and finally an analysis of the structural details of a Corinthian-type bronze helmet from Olympia, also by E. Pantos and his associates.

Telecommunication through "friktories", i.e. huge pyres lit at specific geographical positions to transmit important messages at great distances in relatively short time, are analyzed by the propagation theory of electromagnetic radiation (N. Uzunoglu, Greece) in Part 5.

Part 6 deals with geological and geotechnical knowledge of the Greeks of the Mycenaean period, which is covered by two works (D. Zekkos et al. and I. Mariolakos, Greece). In addition, a complete dynamic analysis of the Atreus vaulted tomb in Mycenae is given (P.K. Askouni et al., Greece), revealing the amazing anti-seismic properties of the structure.

Part 7 covers the issue of medical knowledge of Homeric times, as depicted on vases of the time. It contains two papers (S. Geroulanos et al. and S. Marketos et al., Greece).

Part 8 deals with Homeric flora and fauna, subjects covered by two contributions (C. Thanassoulopoulos and E. Voultsiadou et al., Greece).

Part 9, Homeric Astronomy, consists of "*Eneoros Minos* and the Minoan calendrical abacus" (P. Gregoriadis, Greece), "Homeric Hephaestus as a god of comets and meteors" (A. Laoupi, Greece), "A comet during the Trojan War?" (S.P. Papamarinopoulos, Greece), "Homeric calendar and Helios charioteer" (M.K. Papathanassiou, Greece), "Homer and Orosius: The key in the interpretation of Deucalion's Flood, Exodus and other stories" (E. Spedicato, Italy).

Part 10 deals with seafaring in Homeric times (M.T. Wright, UK) and also with "The redness of Ulysses' ships", an issue that may reveal advanced shipbuilding knowledge (T.Th. Katsaros, Greece).

Part 11 describes the cultural environment of the Homeric era and consists of two interesting presentations: "Ambrosia, nectar and elaion in Homeric poetry"

(D.G. Zanni, Greece) and "Dietary habits in the Homeric Epics" (S. Christopoulou, Greece).

"Geography of Trojan plane and Homeric topography" (P. Malfas, Greece) covers Part 12.

Finally, Part 13 comprises works of general and practical interest, both due to the information they provide and by being the object of vivid and fruitful discussions between scientists from theoretical and applied fields, namely the following presentations: "*Mêtis* and the artificial" (C.P. Anagnostopoulos et al., Greece), "The decorations on Achilles' shield and their interpretation" (D. Kalligeropoulos et al., Greece), "The so-called Homeric problems" (A. Tziropoulou-Efstathiou, Greece), "Atlantis in Homer and other writers prior to Plato" (S.P. Papamarinopoulos), "Did Ulysses travel to Atlantis?" (R.W. Kühne, Germany), "The Homeric reference to writing in Proitos' era" (E. Polygiannaki, Greece), "Linguistic science and script technology: The Homeric evidence" (A. Teffeteller, Canada) and finally "The miraculous Homeric metre" (S.A. Paipetis), a reference to recent research indicating that the metric recitation of Homeric Epics causes coordination of heart and respiration rates, similar to the effect obtained by religious meditation techniques, as practised by Eastern peoples.

Certainly, the above papers constitute substantial contributions towards uncovering knowledge found in the Epics, but the quest in the endless world of Homer does not stop. A few questions may have been answered, but only to reveal an enormous number of further questions, waiting to be answered by new, adventurous investigators, either specialists from practically all fields of science or even lay people who just happen to be sensitive towards beauty and, therefore, ready to partake of the hidden knowledge. The Hellenic gods would never reveal their mysteries to uninitiated ones, to those ignorant of geometry!

Prof. Emeritus S.A. Paipetis
Editor

# Acknowledgements

The Symposium would have never been possible without the generous financial assistance of the West Greece Region and its then Secretary-General Mr. Pagiotis Kavadas, to whom grateful thanks by the Organizing Committee are expressed. Gratitude is also due to GEFYRA SA (Rion-Antirrion Bridge Company) and its top executives Mr. Nikos Harikiopoulos, Mr. G. Kalogerou, Mr. S. Stavris, Mrs. A. Sotiropoulou and Mr. I. Freris for material and moral support. Also, to the Central Archaeological Council of the Ministry of Culture of Greece for making available to the Organizers the SPAP Conference Centre *gratis*.

Enormous, and from every aspect invaluable, was the contribution of Mrs. Georgia Hatzi, Head of the 7th Ephorate of Prehistoric and Classical Antiquities and Director of the Olympia Museums. Not only was she close at hand providing support at every step to the Organizing Committee, but it was her idea for the Opening Ceremony to take place in the Hall of Freezes of the New Olympia Museum. This magnificent ceremony was honored by the presence of numerous important figures of the academic, archaeological and political community. Similarly, thanks are due to Mrs. V. Vasilopoulou, Head of the Directorate General of Antiquities of the Ministry of Culture, and practically to the whole of the community of archaeologists for receiving the idea with enthusiasm and positive attitude.

Several individuals and companies, in one way or another, provided assistance. Among them is the Research Committee of the University of Patras, the Technical Chamber of Greece, the Patras Agency of the Xerox Corporation and its executives Messr. N. Arvanitidis and G. Betsos, DYNACOMP Computer Company and many others.

Special thanks are due to the members of the Scientific Committee and to the invited speakers for the exceptionally high level of their presentations and also to the Greek and foreign referees of the works presented at the Symposium, a necessary procedure to ensure high quality before having them included in the Proceedings.

The contribution and the support team is greatly appreciated, namely that of Dr. John Lukas-Lekatsas, mathematician, Scientific Associate of Applied Mechanics Laboratory of the University of Patras, Mr George Mirotsos, computer specialist, responsible for the digitization of the whole event, Mr. Babis and Ms. Dimitra Nika and also of Ms. Stella-Zaira Avloniti, distinguished philologist, MSc in Comparative Linguistics and Language Diversity, University of Patras, who provided excellent translations of a number of papers authored by English-speaking delegates into Greek. Finally, grateful thanks are due to Mrs. Teta Giannarou, distinguished journalist and public relations officer of the Symposium, for the impeccably professional way she performed her duties, both during the preparation period, by organizing numerous press conferences in Athens and in the provinces, as well as during the Symposium by means of complete and detailed press releases on a daily basis and also by regular direct contacts with the Mass Media, who managed to supply this important event with the publicity it deserved.

Prof. Emeritus S.A. Paipetis
Editor

# List of Contributors

Agelopoulou, H.A., Department of Civil Engineering, University of Patras, 26001 Patras, Greece

Anagnostopoulos, K.P., Department of Production & Management Engineering, Democritus University of Thrace, 67100 Kimmeria-Xanthi, Greece
E-mail: kanagn@civil.duth.gr

Androutsos, G.J., Athens University Medical School, Athens, Greece
E-mail: lyon48@otenet.gr

Askouni, P.K., Department of Civil Engineering, University of Patras, 26001 Patras, Greece
E-mail: askounik@otenet.gr

Athanasopoulos Zekkos, A., 1316 Bonita Ave., Apt. 7, Berkeley, CA 94709, USA
E-mail: adda@berkeley.edu

Athanasopoulos, G., Department of Civil Engineering, University of Patras, 26500 Patras, Greece
E-mail: gaa@upatras.gr

Bell, T., CCLRC, Daresbury Laboratory, Keckwick Lane, Warrington WA4 4AD, UK

Bennett, S.L., CCLRC, Daresbury Laboratory, Keckwick Lane, Warrington WA4 4AD, UK

Beskos, D.E., Department of Civil Engineering, University of Patras, 26001 Patras, Greece
E-mail: d.e.beskos@upatras.gr

Bontempi, E., Laboratorio di Chimica per le Tecnologie, Dipartimento di Ingegneria Meccanica, Università di Brescia, Brescia, Italy

Chapon, L.C., CCLRC, Rutherford-Appleton Laboratory, ISIS Neutron Spallation Source, UK

Chelidoni, S., Department of Production & Management Engineering, Democritus University of Thrace, 67100 Kimmeria-Xanthi, Greece

Christopoulou, S.P., 3 Ypsilon Alonian, 26224 Patras, Greece
E-mail: schristo@otenet.gr

Colombi, P., Laboratorio di Chimica per le Tecnologie, Dipartimento di Ingegneria Meccanica, Università di Brescia, Brescia, Italy

Comendador-Rey, B., Department of History, Art and Geography, Faculty of History, University of Vigo, Spain

Cornacchia, G., Laboratorio di Metallurgia, Dipartimento di Ingegneria Meccanica, Università di Brescia, Brescia, Italy

Davidovits, J., Institut Géopolymère, Laboratoire de Recherche sur les Nouveaux Matériaux, Saint-Quentin, France

Depero, L., Laboratorio di Chimica per le Tecnologie, Dipartimento di Ingegneria Meccanica, Università di Brescia, Brescia, Italy

Garner, R., The Manchester Museum, The University of Manchester, Manchester, UK

Gelfi, M., Laboratorio di Metallurgia, Dipartimento di Ingegneria Meccanica, Università di Brescia, Brescia, Italy

Geroulanos, S., Onasseion Cardiac Surgery Centre, 356 Sygrou Ave., 17674 Athens, Greece
E-mail: sger@hol.gr

Gregoriades, P.D., 10 Aldou Manoutiou St., 11521 Athens, Greece
E-mail: griskman@otenet.gr

Kalligeropoulos, D., Department of Automation, TEI of Piraeus, Piraeus, Greece
E-mail: heron100@otenet.gr

Katsaros, T.Th., Faculty of Humanities, University of Aegean, Rhodes, Greece
E-mail: tkatsaros@rhodes.aegean.gr

Klimis, G., 32 Kaisareias St., 11527 Athens, Greece
E-mail: g_klim@yahoo.gr

Kockelmann, W., ISIS, Rutherford-Appleton Laboratory, Didcot OX11 0QX, UK

Kostopoulos, V., Department of Mechanical Engineering & Aeronautics, University of Patras, Patras, Greece
E-mail: kostopoulos@mech.upatras.gr

Kühne, R.W., Tuckermannstr. 35, 38118 Braunschweig, Germany
E-mail: kuehne70@gmx.de

Laoupi, A., 9 Heroon Polytechneiou St., 15780 Athens, Greece
E-mail: alaoupi@otenet.gr

Lymberopoulou, E., Onasseion Cardiac Surgery Centre, 356 Sygrou Ave., 17674 Athens, Greece

Macdonald, M., CCLRC, Daresbury Laboratory, Keckwick Lane, Warrington WA4 4AD, UK

Malfas, P., 44 Lemesou St., 15669 Papagou, Greece
E-mail: hxografi@otenet.gr

Manousakis, I., Dromos Consulting, 27 Monemvasias St., 15125 Halandri, Greece
E-mail: jmanous@tee.gr

Manti, P., HISAR, Cardiff University, Humanities Building, Colum Drive, Cardiff CF10 3EU, UK
E-mail: mantip@cf.ac.uk

Mariolakos, I.D., National and Kapodistrian University of Athens, Panepistimiou-poli Zografou, 15784 Athens, Greece
E-mail: mariolakos@geol.uoa.gr

Marketos, S.G., Athens University Medical School, Athens, Greece

Mikrogiannakis, E., School of Philosophy, University of Athens, Athens, Greece
E-mail: thmicro@phil.uoa.gr

Mosselmans, J.F.W., CCLRC, Daresbury Laboratory, Keckwick Lane, Warrington WA4 4AD, UK

Nishiyama, W., 530 Ishki, Yaizou-City, Shzuoka, 425-0054 Japan
E-mail: wako-n@nifty.com

Paipetis, S.A., Department of Mechanical Engineering & Aeronautics, University of Patras, Patras, Greece
E-mail: paipetis@mech.upatras.gr

Pantos, M., CCLRC, Daresbury Laboratory, Keckwick Lane, Warrington WA4 4AD, UK
E-mail: e.pantos@dl.ac.uk

Papadopoulos, K., Onasseion Cardiac Surgery Centre, 356 Sygrou Ave., 17674 Athens, Greece

Papamarinopoulos, S.P., Department of Geology, University of Patras, 26500 Patras, Greece
E-mail: papamar1@otenet.gr

Papathanassiou, M.K., Faculty of Mathematics, National and Kapodistrian University of Athens, Panepistimioupolis, 15784 Athens, Greece
E-mail: mpapatha@math.uoa.gr

Polygiannaki, E., Athens, Greece
E-mail: efipolyg@otenet.gr

Pradell, T., Departament d'Enginyeria Quimica, EPSEVG Universitat Politècnica de Catalunya, Vilanova, Spain

Prag, A.J.N.W., The Manchester Museum, The University of Manchester, Manchester, UK

Reboreda-Morillo, S., Department of History, Art and Geography, Faculty of History, University of Vigo, Spain

Rorres, C., School of Veterinary Medicine, University of Pennsylvania, 3800 Spruce St., Philadelphia, PA 19104, USA
E-mail: rorres@vet.upenn.edu

Salvado, N., Departament de Física i Enginyeria Nuclear, Universitat Politècnica de Catalunya, Barcelona, Spain

Showleh, T., Department of Mechanical Engineering, Concordia University, Montreal, Canada H3G 1M8
E-mail: tahashowleh@gmail.com

Sfakianakis, M.G., Department of Civil Engineering, University of Patras, 26001 Patras, Greece
E-mail: mgs@upatras.gr

Spedicato, E., University of Bergamo, Italy
E-mail: spedicato@cypher.a-technet.com

Tasouli, A., Onasseion Cardiac Surgery Centre, 356 Sygrou Ave., 17674 Athens, Greece

Tassios, T.P., Professor Emeritus, National Technical University of Athens, 42 Patission St., Athens, Greece
E-mail: tassiost@central.ntua.gr

Tatolas, A., Department of Zoology, School of Biology, Aristotle University of Thessaloniki, 54124 Thessaloniki, Greece

Teffeteller, A., Linguistics, Concordia University, Montreal, Quebec, Canada H3G 1M8
E-mail: teffet@vax2.concordia.ca

Thanassoulopoulos, C.C., Department of Plant Pathology, Aristotelian University of Thessaloniki, Thessaloniki, Greece
E-mail: ktl@otenet.gr

Tobin, M.J., CCLRC, Daresbury Laboratory, Keckwick Lane, Warrington WA4 4AD, UK

Tziropoulou-Efstathiou, A., "Helleniki Agoghi", School of Ancient Greek, Athens, Greece

Uzunoglu, N., National Technical University of Athens, 15773 Athens, Greece
E-mail: mnap@otenet.gr

Varoufakis, G., Halyvourghiki SA, 3 Dragatsaniou St., Athens, Greece
E-mail: georgevaroufakis@yahoo.gr

Vasileiadou, S., Department of Automation, TEI of Piraeus, Piraeus, Greece

Vatistas, G.H., Department of Mechanical Engineering, Concordia University, Montreal, Canada H3G 1M8
E-mail: vatistas@encs.concordia.ca

Voultsiadou, E., Department of Zoology, School of Biology, Aristotle University of Thessaloniki, 54124 Thessaloniki, Greece
E-mail: elvoults@bio.auth.gr

Watkinson, D., HISAR, Cardiff University, Humanities Building, Colum Drive, Cardiff CF10 3EU, UK
E-mail: watkinson@cf.ac.uk

Wright, M.T., Centre for the History of Science, Technology & Medicine, Imperial College, London SW7 2AZ, UK
E-mail: m.t.wright@imperial.ac.uk

Zanni, D.G., Vas. Olgas 145, 54645 Thessaloniki, Greece
E-mail: zannidoukaina@hotmail.com

Zekkos, D., GeoSyntec Consultants, 475 14th Street, Suite 450, Oakland, CA 94612, USA
E-mail: zekkos@geoengineer.org

# Part 1

# GENERAL THEMES

# Mycenaean Technology

T.P. Tassios

*National Technical University of Athens, Greece*

**Abstract.** We should not be surprised at the great significance of Technology during the prehistoric period in Greece; or indeed in any other part of the world of those times: Technology was already possessed by several animals; it was a natural supplement to Nature, so to say. By analogy, it was, therefore, all too obvious for humans to consider that Technology was "donated" to them by the gods. In other words, it was expected that humans were admiring Technology as much as a fundamental means for their survival and their well-being, that have attributed it to a religious Heaven. This general attitude of humankind being observed, I will maintain that the level of technological development of the Greeks during the second millennium BCE, was expectedly high. The aim of this paper is to summarize the main technical achievements of that period in Mycenaean Greece. To this end, I will first take that the writings of Hesiod and Homer are reflecting, to a certain extent, facts and beliefs during the Mycenaean period as well.

## 1 Technology in Hesiod

It is important to note that in Hesiod, the equilateral triangle of the manifested deity is formed with Titans at one summit, and the Hekatoncheires and the Cyclopes at the other two: Thus, the Greeks recognized the synthesis of (i) the spiritual element, (ii) the natural forces and (iii) Technology (as a "natural" supplement to Nature). In fact, Cyclopes were conceived as God-smiths. The same place will be kept for Technology in the tripartite deity during the subsequent reign of the Olympians who have replaced the Titans. Besides, the victory of Zeus against the Cronides was only achieved when the "inventors-Cyclopes" offered to him the absolute weapon of lightning – another proof of the significance of Technology for the Greeks.

Thus, the fundamental importance of Technology for the early Greek tribes is clearly reflected in their basic Myths. And this will be presented in more detail in the Homeric Epics.

*S.A. Paipetis (ed.), Science and Technology in Homeric Epics*, 3–33.

## 2 Technology in Homer

After the "happy end" of Theomachy, the Greeks continued to nourish their myths with technological details reflecting to a certain extent their own technical development, as well as their technological dreams (the concept of automats, for instance). It is accepted that a considerable part of these facts and beliefs are portrayed in the Homeric Epics. Therefore, it is interesting to revisit the Iliad and the Odyssey, searching for technical stories – although very briefly.

(a) *Metals*: Besides gold and silver, bronze (copper + tin) was the basic metallic material in Iliad (Il. 15.309). Iron will furiously invade Odyssey – in the form of quenched steel as well (e.g. Od. 9.391). Metal workshops are vividly described (Il. 18.412, Od. 8.274).

*Weapons*: The significance of this basic subject of fabrication of weapons is shown by the fact that almost the entire rhapsody 18 is devoted to this technology. Besides, weaponry is a frequent subject of Homeric Epics: Arrows (Il. 4.105), shields (Il. 15.308), cuirasses (Il. 8.195), chariots (Il. 5.722) – and above all, the famous weapons of Achilles (Il. 18.144). More specifically regarding the structural (not the artistic) aspect of shields, including that of Achilles (Il. 20.260, 22.290) and Ajax (Il. 6.219), I wish to refer to the original work of Paipetis et al. [1] who have reconstructed layer by layer these defensive weapons, and subjected them to rigorous testing – both experimental and analytical, concluding that the nature, the sequence and the number of these layers were in fact optimal for absorbing the piercing energy of a spear. These findings tend to belie Morris [2] stating that "no such shield ever protected a Mycenaean ..."

(b) *Buildings*: It is worth noting that in Iliad, mainly the rich and complex palaces of gods are described (e.g. Il. 1.607, 5.167, 18.371 etc), whereas in Odyssey, building technology is landed on earth, describing human buildings (e.g. Od. 4.72). In the house (and ship) building technology, it is interesting to include also the structural miracle of the Wooden Horse (Od. 4.272, 8.493, 11.24) – that complex and solid artifact of chief-carpenter Epeios: A giant mobile work, with a body measuring something like 8 by 16 by 32 meters (in order to be able to accommodate three thousand well hidden hoplites, as it is said).

(c) *Automats*: Here we find ourselves in the summit of ancient Greeks' technophilia. Such is their confidence of their technical knowledge, that they rush to the future of Technology (see [3]).

- Moving automats: self moving tripods, entering and leaving the Palaces (Il. 18.376), automatic bellows of metallurgical kilns (Il. 18.468), gates automatically opening when hearing a whip (Il. 5.749). This is an Epic of Technology ...

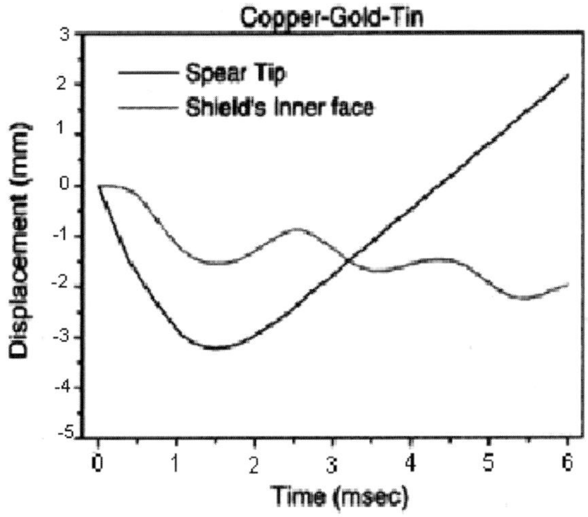


**Fig. 1** Achilles' shield: After 1.5 sec the spear's penetration is stopped, whereas after 3 sec, the spear tip moves backwards, outside of shield's inner face [1]. Courtesy Professor S.A. Paipetis.

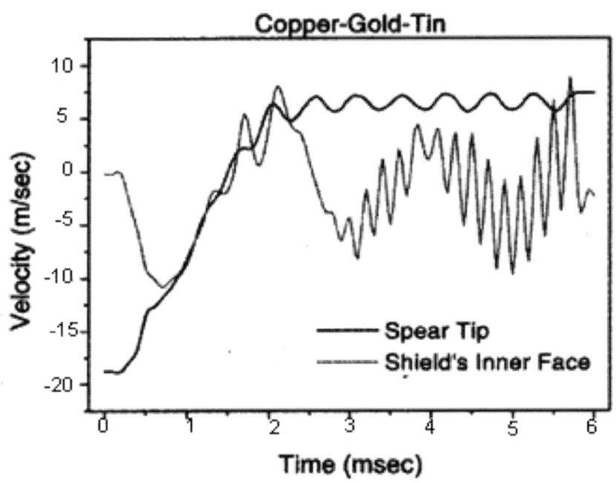


**Fig. 2** Achilles' shield: After 1 sec the shield repulses the spear, whereas after 2 sec it vibrates at velocities lower than the repulsed spear [1]. Courtesy Professor S.A. Paipetis.

- Animal and human-like automats: guardian dogs (made of gold and silver, Od. 7.91), live girls-robots ("in them is mind and wits, in them too a voice and strength", Il. 18.418), etc.
- Traps: elaborated grips, hidden above and underneath the bed, in order to catch the illegal lovers (Od. 8.274).
- Automat ships: the ships of the inhabitants of Scheria exceed the achievements of modern Automat Technology – they are "διανοούμενα" (intelligent) ships: They have no captains, but they conceive "what men have in mind", and they

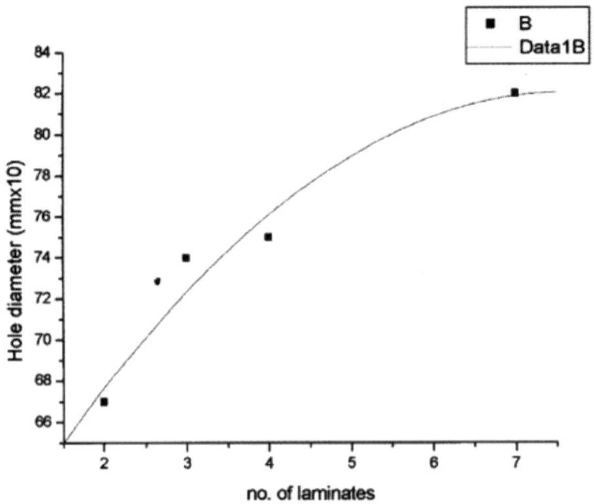


**Fig. 3** Ajax's shield: When the number of leather layers behind the number of bronze laminates is increased to seven, penetration is stopped and energy dissipation is ensured by maximization of the imprint of the bronze layer [1]. Courtesy Professor S.A. Paipetis.

travel them very rapidly across the clouds – and they are unsinkable (Od. 8.562). It is the first time that the techno-mythical thought of the Greeks dares to figure out *human* robotic achievements, whereas up to that moment automats were thought only in the service of Gods. Later on, I will reexamine this clearly humanistic attitude of the Greeks.

(d) *Artifacts-artistic objects*: in fact, the gods had initially taught humans the technical skill and crafts; Hephaestus himself "ετεφάνη"(was manifested) to humans and lived nine years with them (see Homeric Hymn to Hephaestus, 20.3), whereas, later on, Prometheus (Plato, Protagoras, 321c) will take Technology from Athēna and Hephaestus, he will transfer it to humans, and thus will save mankind. Subsequently, a more specifically human characteristic will appear on human artifacts. In the Homeric Epics we observe that every technical object was an artistic object as well; after all, in Greek language, both are called "τεχνήματα": the shield of Achilles, a perfect defensive weapon, was a densely ornamented work too (Il. 18.481) described in no less than 128 verses of Iliad![1]

## 3 The Technology of Mycenaeans

Archaeological findings, throughout Greece and in numerous areas of Mediterranean Sea (from Syria to Sardenia), have confirmed a good part of the alleged

[1] The sling of Hercules' sword was also very richly decorated (Od. xi, 609).

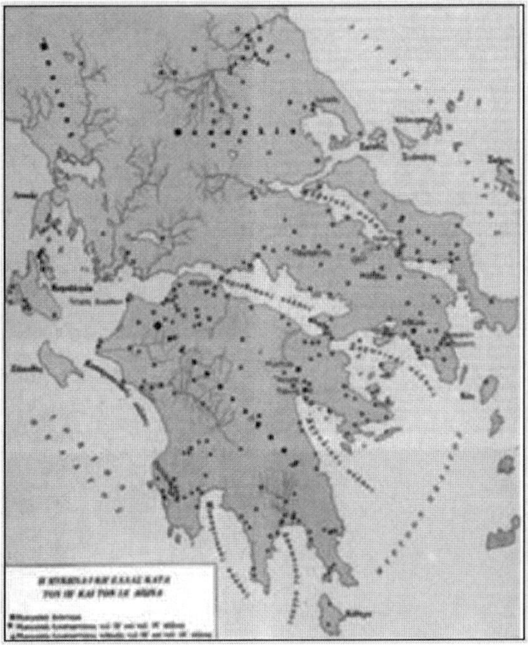

**Fig. 4** Mycenaean settlements in the mainland during 14th and 13th centuries BCE (Ekdhotikē Athinōn).

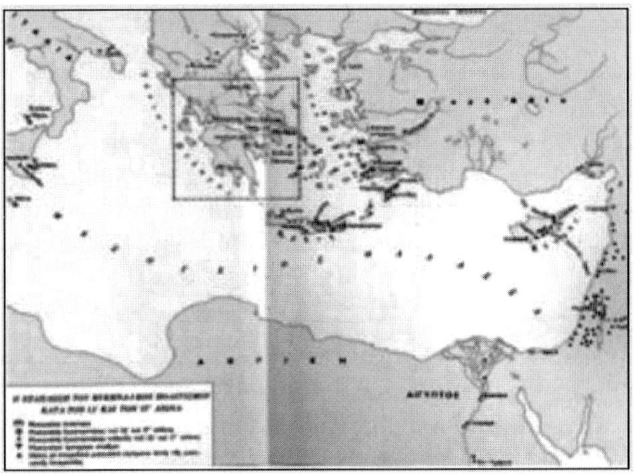

**Fig. 5** Mycenaean expansion during 14th and 13th centuries BCE (Ekdhotikē Athinōn).

technophilia of the Achaeans, i.e. the Mycenaean world. During the second half of the 2nd Millennium BCE, Mycenaean people have developed an advanced Technology, many centuries before the classical and Hellenistic times, which seem to be better studied from this point of view.

**Fig. 6** Ithaca "School of Homer", the underground well: the access tunnel, corbelling walls covered with slabs [4].

**Fig. 7** Ithaca "School of Homer", the underground well: the three-dimensional corbelled vault [4].

## *3.1 Water Supply of Cities*

First, I will briefly describe the *underground well-houses*, following the views of Knauss [4], and I will restrict this presentation only to those structures which are more characteristic to the building techniques of the Mycenaeans, i.e. the *vaulting* of access tunnels and spring chambers. The main features of these structures are the following:

- A series of retaining walls (approx. 3.00 m high) are securing (i) the stability of the sloped hillsides and (ii) a better infiltration of surface waters.
- A small entrance is built on the ground, leading to the spring chamber via an inclined tunnel, stabilized by means of a corbelled strong-vault. Slope of the tunnel: 1 to 2 (up to 4), width: 0.8 to 1.4 m.
- An underground well-house(or spring chamber) stabilized by a bidimensional or three-dimensional stone-vault. Total depth of the installation: 5 to 15 m.

In this category of water facilities belong the finds of Tiryns and Ithaca (in the area of the so called "Homer School").

Another category of underground water-supply facilities are the highly sophisticated Mycenaean structures at the Acropolis of Mycenae and the Acropolis of

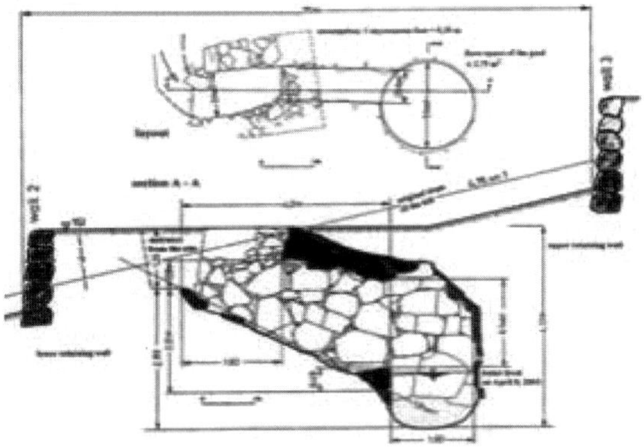

**Fig. 8** The prehistoric underground well-house in Ithaca at the so-called "School of Homer", L. Kontorli-Papadopoulou, Eranos and Corpus 2001 (supplemented by Knauss [5]). Courtesy Professor J. Knauss.

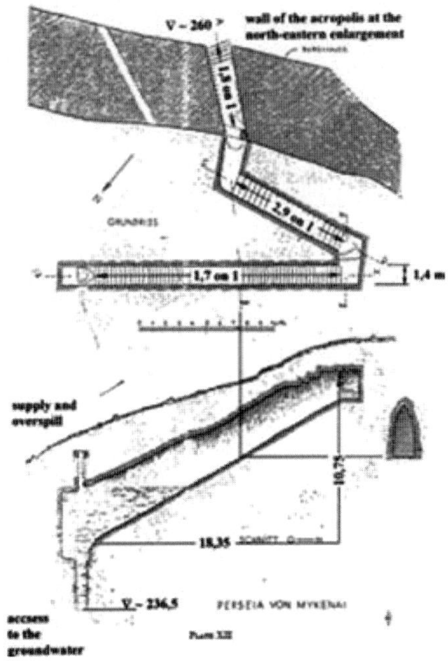


**Fig. 9** The prehistoric underground well-houses of Ithaca at Mycenae, G. Karo, AJA 38 (1934), supplemented by Knauss. Courtesy Professor J. Knauss.

Athens: The access-tunnels are hug in the rock of the hill, and lead down to a depth up to 25.0 m. In Mycenae, the lowest flight of the tunnel is plastered with a waterproof lime mortar [6], still in place.

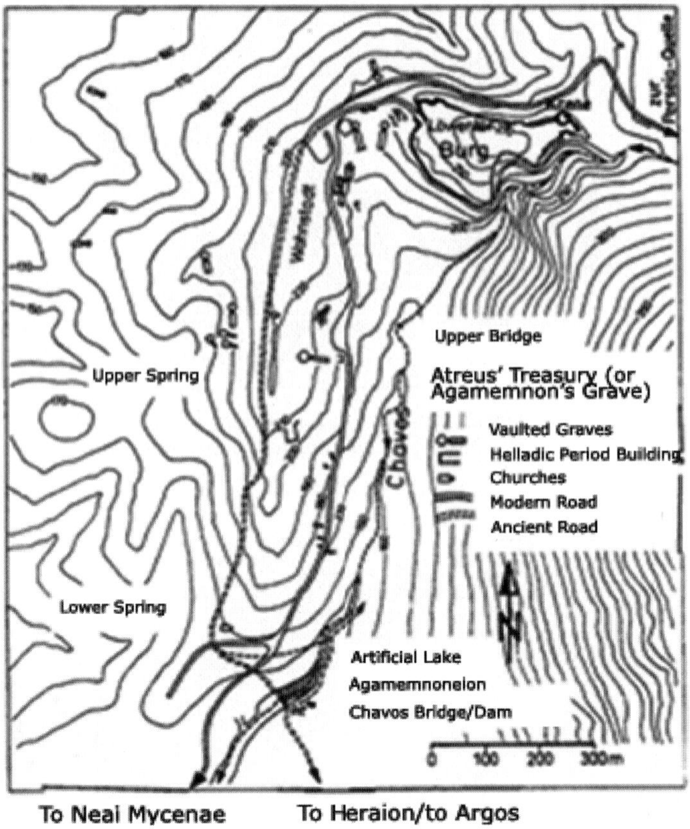


**Fig. 10** Mycenaean acropolis (upper right part of the diagram) and the masonry dam on Chavos torrent (lower central part of the diagram, from [4]).

## *3.2 Dams*

I will mention three characteristic cases of masonry dams used for three different purposes.

(a) A large artificial lake ($30 \times 100$ m) was created just outside the city walls[2] of Mycenae, by means of a masonry dam constructed at an appropriate point of Chavos torrent. Thickness 5 m, height 4.5 m, length 37 m. An efficient water-tight construction technique was used.

(b) An artificial waterfall was gradually constructed in Alyzeia (Acarnania) west-central Greece, apparently for fleece-washing in a region of intensive cattle-raising (actual Varnaka torrent). Height: initially 3.20 m, and after the natural

[2] It is worth noting that, outside the walls, the king had installed several workshops and commercial facilities.

**Fig. 11** Partial view of the Chavos dam (from [5]). Courtesy Professor J. Knauss.

**Fig. 12** Alyzeia dam (from [5]). Courtesy Professor J. Knauss.

earth filling behind it, a second height (c. 2.70), and subsequently a third and a fourth (3.00 m and 1.50 m, respectively) were added during the years, summing up a today's height of 10.50 m. Crest's length: 25.0 m.

(c) A torrent deviation (still in use today) was constructed in order to protect Tiryns from very destructive floods, which were archaeologically identified. A masonry (clay infilled) dam was erected across the deep river-bed of Tiryns-torrent. Thickness: 3.5 to 4.0 m, height up to 10.0 m, length (at the crest) 70.0 m. A 1500 m long channel was dug in order to convey the torrent waters to the riverbed of another torrent (the actual Aghios Adrianos torrent) 3.0 km away from the Mycenaean city of Tiryns.

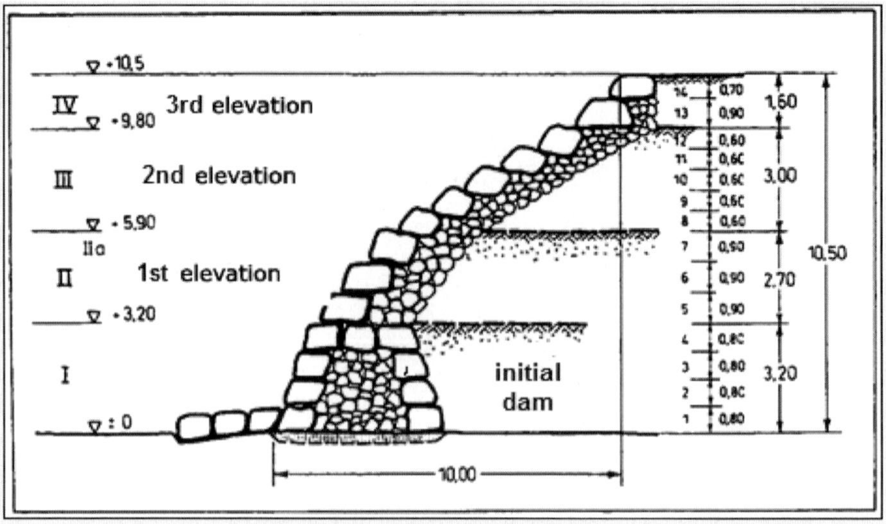


**Fig. 13** Representation of the consecutive construction stages of Alyzeia dam (from [5]). Courtesy Professor J. Knauss.

And this brings us to the other glorious chapter of Mycenaean engineering, flood control that is.

## 3.3 Flood Control and Land-Reclamation Works

During that period, agriculture was the main wealth of Greek regions – whereas only a small part of soil was cultivable. Besides, most of this cultivable soil was located in closed valleys, flooded almost every year. Flood-control and land reclamation was therefore of a paramount importance for Mycenaean peoples. The advanced hydraulic technology they developed since the middle of the 2nd millennium BCE, is very impressive by its rationality, the large scale of its applications, as well as by its efficiency. Based on the examples of such flood-control works executed in Pheneos, Tiryns, Thisbē and Kopais [6], one may describe their basic characteristics as follows.

*First solution*: The waters of the flooding torrent are contained in an artificial lake produced by means of an earth dam, covered by a protective layer of masonry. During summertime, these waters are used for irrigation through a system of small channels; this was the case e.g. of the Arcadian Orchomenos (Pausanias 8.23.2), as well as later on of Mantineia (Thukydidēs, 5.65). The pertinent dams (containing a clay core) had a height of 3.00 m and they were some hundred or occasionally some thousand meters long.

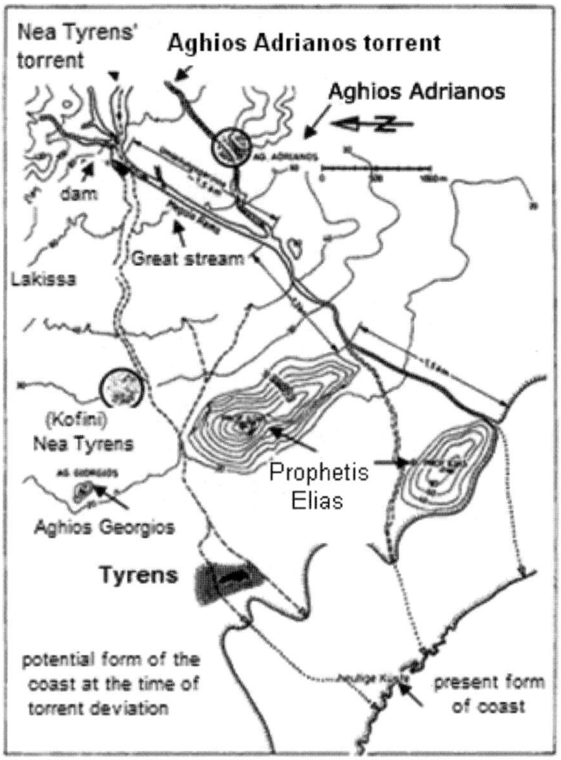


**Fig. 14** The deviation of the Tiryns torrent (upper left part of the diagram) in order to protect the acropolis of Tiryns (lower central part of the diagram, from [5]).

*Second solution*: When such an artificial lake was not sufficient (or perhaps was not feasible at all), another solution was followed. First, the torrent is deviated outside the cultivated lands by means of an appropriate large channel. Secondly, this channel is extended towards the perimeter of the valley, close to the surrounding hills, so that the outflow of water be facilitated through existing cesspits in the karstic lime stones of these hills. The internal high fill forming this channel (to the side of the valley) has a width of 20 to 30 m and is covered by an appropriate masonry, protecting the fill against erosion. In this category of solutions belong the marvelous works of the second drainage system of Kopais, during the 14th and the 13th century BCE. This giant technical achievement merits a more detailed description:

(a) By means of a 1km long dam, north to (Boeotian) Orchomenos, the waters of Melas river are collected in an artificial lake (12 km$^2$).
(b) Two kilometers northeast of Orchomenos, the waters of the Boeotian Kēphissos are deviated by means of an impressive channel, 25 km long, which conveys these waters along the north bank of the valley – in direct contact with the karstic limestone of the mountain, up to the east end of the valley. This

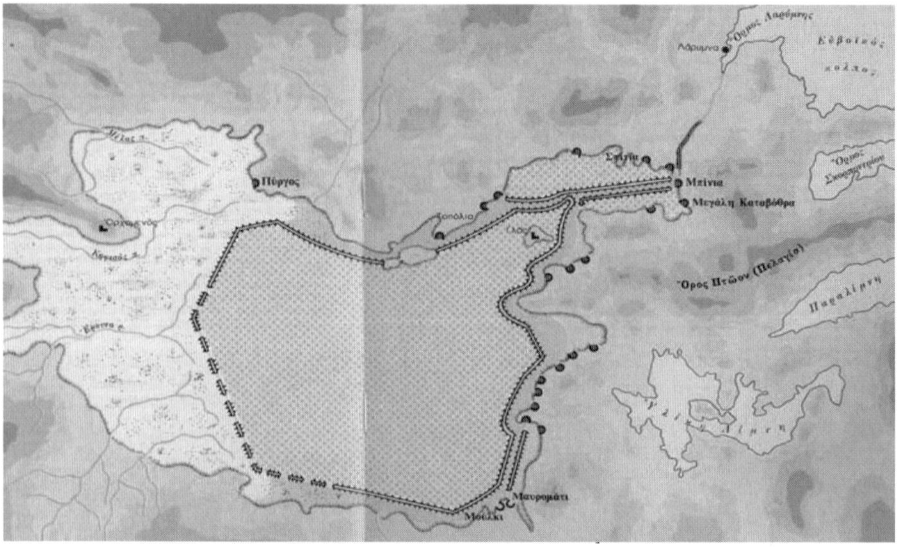


**Fig. 15** Flood control and land reclamation of Kopais lake (14th century BCE) [7].

**Fig. 16** The remnants of the Mycenaean fill, shaping the inner side of the flood control channel.

channel has a width of 40.0 m and a depth of 2,50m; it was navigable and served efficiently the transportations between the capital city of Orchomenos and the very fertile agricultural areas of Glas [7].

(c) Geomorphology of the northeast end of the valley of Kopais allows for an area of 20 km$^2$ to be drained by means of a 3 km long dam and a 5 km long ditch, heading north and meeting the large channel. In the middle of this reclaimed

**Fig. 17** Masonry cover of the Mycenaean hill shaping the inner side of the channel.

**Fig. 18** The largest cesspit receiving flood waters.

land is located the hill "Glas", headquarters of the management of this local rich agricultural enterprise.

(d) Subsequently, our large channel ends up in a large water-containment area characterized by a local system of cesspits.

(e) Yet, it seems that occasionally these cesspits were not sufficient to drain these waters to the sea. Thus, a tunnel was driven (of a cross-section roughly equal to $1.45 \times 1.75$ m and a 10% longitudinal slope), conveying these waters to Larymna bay in the Euboea gulf.[3] Incidentally, Kakavoyannis [8] has maintained that the people of Temmikes, specialists miners form Lavrion having immigrated (c. 1600 BCE) to Boeotia, were employed by the Mycenaeans to maintain cesspits or to drive land-reclamation galleries.

(f) Thus, assisted by several other peripheral fills, the reclaimed cultivated land is considerably increased. Its central area (140 $km^2$ large) is rather dry during summertime, serving only as grazing ground.

This giant-scale land-reclamation system explains the wealth of the "twelve cities of Kopais" mentioned by Homer, and it is another proof of the fundamental technophilia of the Achaeans.

In the same context, it is worth noting another category of flood-control Mycenaeans works to protect the ground of Olympia, where the equestrian Games used to take place. That piece of land was eroded both by the river Alpheios and its tributary Kladeos, near the area of their intersection. The works consisted of two major components: (i) a 500 m long regulatory wall along the east river-bank of Kladeos; the wall containing a core (made of clay and small stones) and masonry covered sides (made of 0.70 m large stones), had a cross-section of $3 \times 3$ m; and (ii) a large fill along the north river-bank of Alpheios (see also Pausanias, Ηλιακά, 20/15). The area included in the corner of these two protective works, was subsequently infilled, and is still kept intact up to our days.

## *3.4 Building Technology and Bridges*

Similar in scale and efficiency are the Mycenaean structural works. The 13th century BCE is the period of the explosive expansion of Achaeans and, at the same time, the period of the large vaulted structures in Greece, Asia Minor and Krete. The Mycenaean corbelled vault, in use earlier than 1600 BCE, is a masterpiece of Mechanics and a daring technique to cover large spans of clear openings up to 13 and 15 m. Mankind would subsequently need another 1500 years to be able to reach such large spans.

All Mycenaean cities were protected by 3 m up to 10 m thick walls; within their width, several arched passages were contained leading to the storehouses (like in the case of Tiryns) or occasionally to subterranean water houses (as in Mycenae).

On the other hand, the Mycenaean building technology was successfully following the traditional antiseismic techniques of "timber reinforced masonry".

In less monumental structures however, another traditional material was used, namely adobe, made with appropriately selected low plasticity clay, special sands and "high bond" twigers (2 mm in diameter), as fiber reinforcement; possibly, a small percentage of burnt lime was added.

[3] However, it should be noted that this tunnel has not been completely investigated up to now.

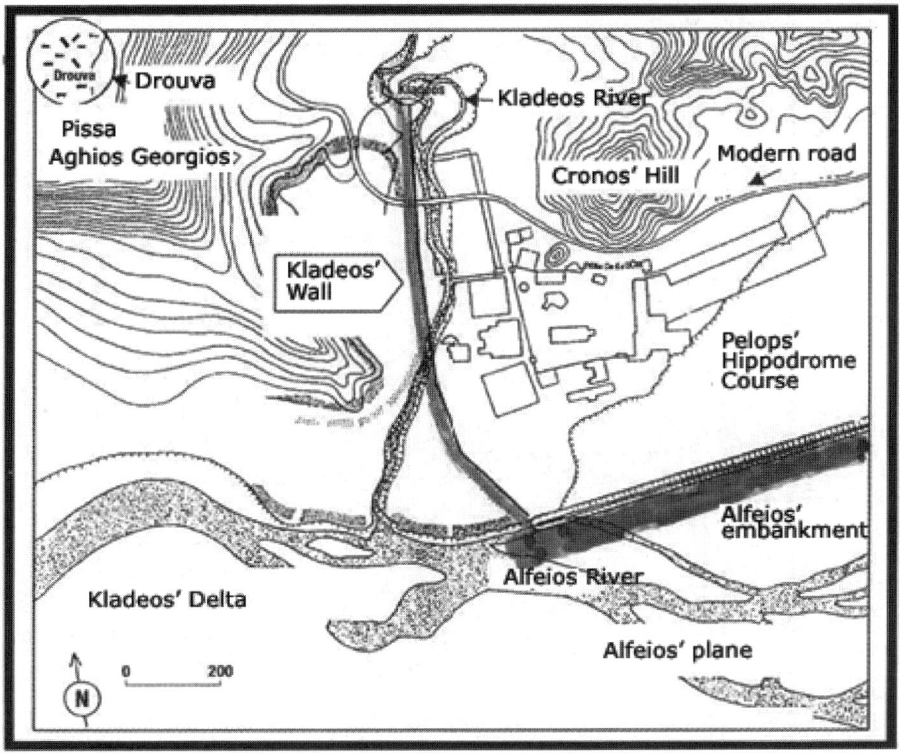


**Fig. 19** Flood protection works in Olympia (Mycenaean times) (from [5]).

This material shows a compressive strength as high as 3.0 to 5.0 MPa [9], whereas no evidence of fire or other thermal treatment was present in the examined samples. All houses of the Mycenaean period contained a complete sewage system – a feature not quite used in medieval times in Europe …

On the other hand, the extensive road network of the Mycenaean times, necessitated several bridges. Those relatively peaceful times allowed for some of these bridges to be permanent (i.e. stone bridges instead of timber). In only the area of Argolis, the remnants of 25 such bridges [10] were found; they were made with the corbel system, covering spans however lower than 2.5 m.

Finally, it is worth noting that in the Mycenaean tablets (inscriptions in linear B Greek) several technical professions are mentioned, such as δρύτομος (woodcutter), τέκτων (carpenter), ναυδόμος (shipbuilder), θρονουργός (cabinet maker), χαλκῆες (coppersmith), τοιχοδόμος (mason) and the like. And it is remarkable that, at least in one case, a woman was responsible for a team of builders: Technology and social justice are not inimical after all.

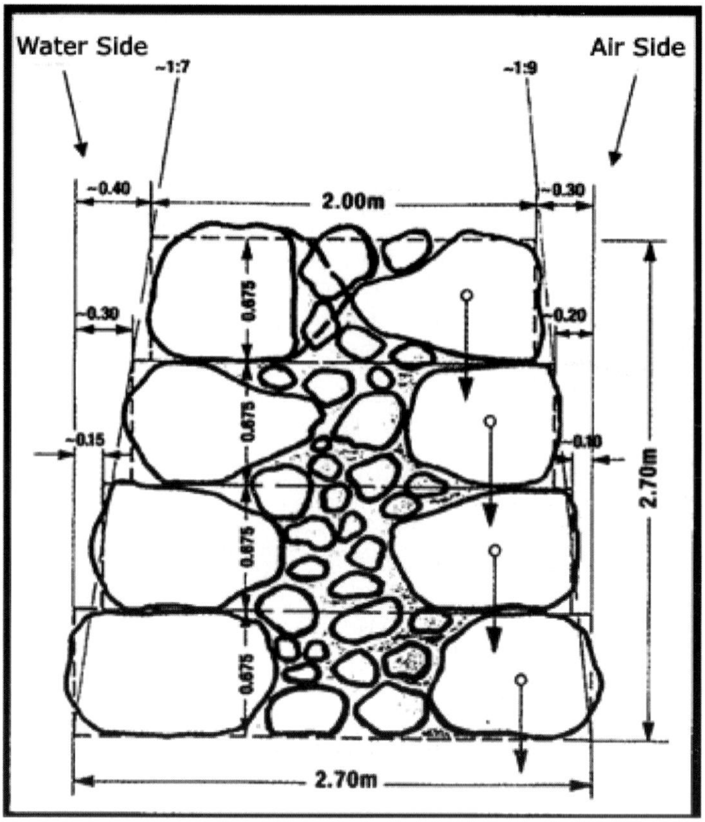


**Fig. 20** Cross-section of the wall causing deviation of ancient Kladeos torrent (from [5]).

## *3.5 Metallurgy*

Silver mining and metal-technique in Lavrion (south Attica) started around 2500 BCE [11]. "The proof of mining exploitation during the Early Bronze Age" in the Mycenaean acropolis of Velatouri (Lavrion) is given by Spitaels [12] in terms of (i) traces of working stone tools on the base of ore-bearing veins, and (ii) of archeological content of the fillings of the floor of the mining gallery, next to the ancient Mycenaean theater of Thorikos.

Such silver mining activities in Thorikos (around 2500 BCE) were accompanied by silver production activities in the broader area of South Eastern Attica, as it is clearly demonstrated by Kakavoyanni et al. [13]: a cupellation workshop was excavated at Lambrika, and *hundreds* of litharge fragments were recovered, presenting regular shapes of small shallow bowls bearing ten hemispherical cavities on their bottom surface – a proof of the process for silver separation from argentiferous lead.

**Fig. 21** Buried Mycenaean corbelled vault.

**Fig. 22** Arched passage within the width of city wall (Tiryns).

Thus, a complete evidence is now available for that very early silver-production in Lavrion; much earlier than the litharge fragments found by Servais [14], on the floor of a Mycenaean house, dated 1350 BCE or those mentioned by Konofagos [11, p. 69] in the island of Kea, dated 1500 BCE.

**Fig. 23** In Mycenaean building technology, the association of timber and masonry offered a certain level of aseismic behaviour. Pylos palace: Representation by Piet de Jong (Ekdhotikē Athinōn).

**Fig. 24** Adobe wall in Mycenae. A mud brick is also shown; its resistance is remarkably high (from [9]).

Incidentally, an "on the wing" reference of Morris [2] regarding a presence of Phoenicians in Lavrion (without any other evidence offered), does not seem relevant for metallurgical activities dating so many centuries prior to the Early Iron Period (c. 1200 BCE) accepted as the date of the expansion of Phoenician vessels in the Aegean (see *inter alia* Cooke [15]).

With such a rich mining and metallurgical background, the Mycenaeans should be in a rather favourable condition to embark in *iron* metallurgy and iron works too. Nevertheless, probably, before the 12th century BCE only meteoritic iron was used (although Varoufakis [16] has shown that non-meteoritic iron objects existed in the Mycenaean world since the 14th century BCE). Several, iron objects were also imported during the 14th century. After the defeat of the Hittites (beginning

**Fig. 25** The Mycenaean acropolis of Velatouri (Lavrion).

**Fig. 26** Red filling "B" on the rock floor of a mine gallery at Thoricos, provided proof of argentiferous lead mining exploitation during the Early Bronze Age (from [12]).

of the 12th century), their rather recent innovation of steel production[4] (which was kept secret up to that moment) was apparently propagated to the entire east Meditarranean world "where steel technology was completely established up to 1100 BCE" [17].

---

[4] Though carburization, quenching and tempering.

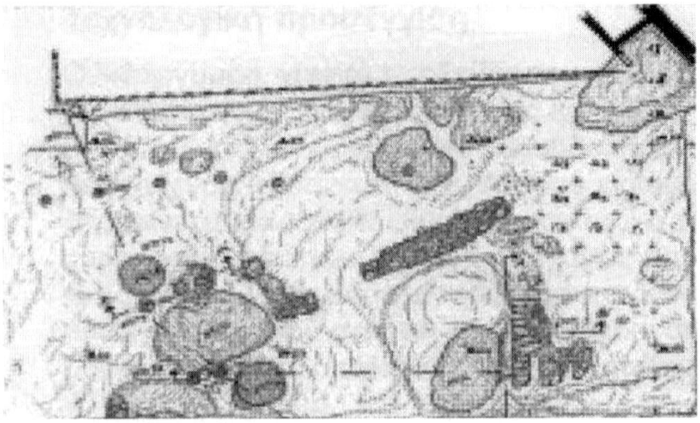

**Fig. 27** Plan of the metallurgical workshop (silver production) at Lambrika, south Attica, active during the Late Neolothic/Early Helladic periods (from [13]).

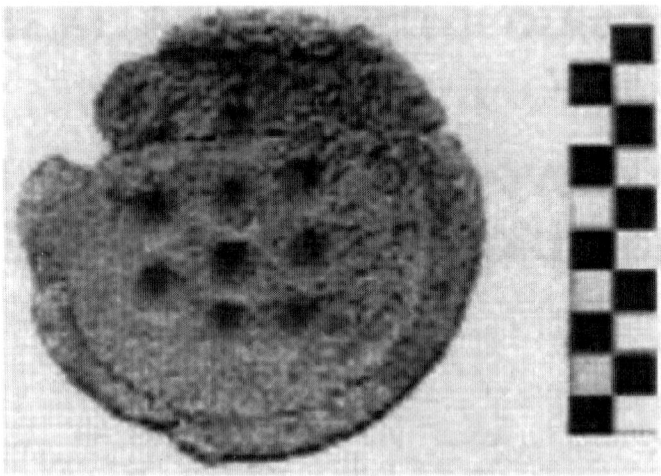

**Fig. 28** The Litharge "bowl" (approx. 1.5 × 10.00 cm) with ten hemispheric cavities (from [13]).

Vavoufakis does not exclude early smelting of *local* Mycenaean iron ores. After all, the results of recent Swedish investigations [18] in Southern Peloponnesus have shown that: (i) evidence of metal working since the Bronze Age was found in Asinē, and (ii) in the Early Iron Age bloomery iron was produced and forging took place in Asinē.

On the other hand, independently of the origin of the new materials, Mycenaeans smithworkers and metal art technicians were famous for their skills throughout Central and Eastern Mediterranean [7, p. 325].

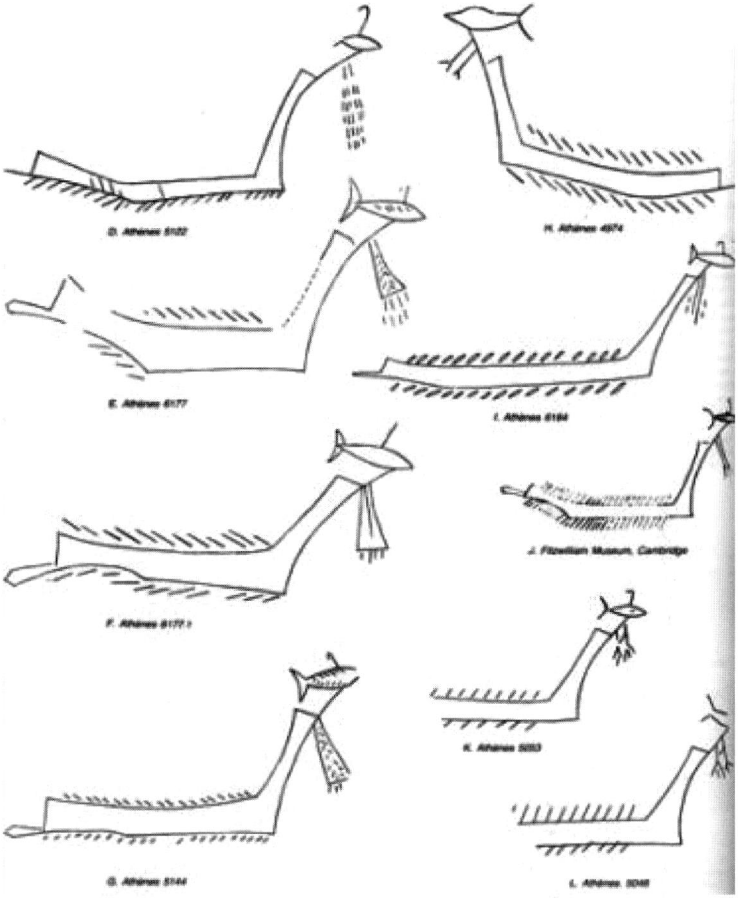

**Fig. 29** Syros-type ships. Early Bronze Aegean (from [19, p. 82]).

## *3.6 Shipbuilding*

The short presentation of this topic follows the views of Basch [19]. Eastern Mediterranean was always a region of intensive naval activity; all peoples around it had developed the necessary naval technology. After the pirogues of the Neolithic period, a considerable development was observed in the Cyclades where a sophisticated ship (the "Syros"-type ship) was used during the Early Helladic Age (2800–2200 BCE) (see Figures 29 and 30).

Its asymmetric longitudinal section and its rather angular shape insinuate distinct functions at various parts of the ship. A remarkable broad use of this ship in the Aegean Sea (from Boeotia, down to Crete) and for a long period, shows its success and popularity. A considerable development of this ship was achieved in Cyclades

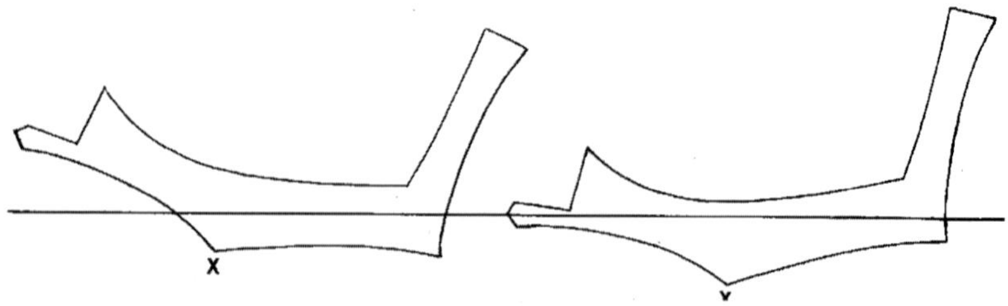


**Fig. 30** Apparent and correct position of floating line of Syros-type ships (from [19, p. 88]).

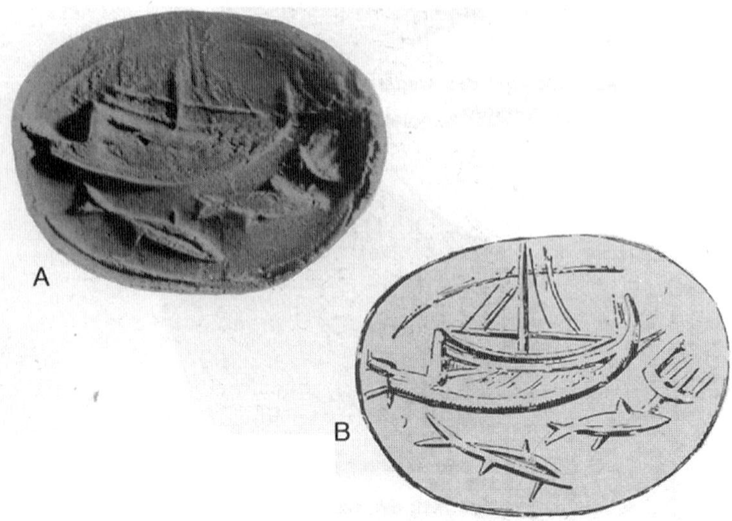


**Fig. 31** Seal ring of Platanos (Heracleion Museum, MMI) showing a development of Syros-type ship (from [19, pp. 98, 114]).

through centuries.[5] Thus, finally, up to the Mycenaean times [19, p. 148], the following important modifications were introduced to the Syros-type primitive ship:

(i) smoothening of angles,
(ii) quasi-vertical prow,
(iii) further development of the ram, as a prolongation of the keel,
(iv) high stern (with a curved stern-post),
(v) long steering oar,
(vi) considerable lengthening of the ship, to accommodate 25 rowers on each side.

[5] The specific case of the Thēra (Akrotēri) ships, the crescent-like ship, appearing in the highly artistic fresco of the west House in Akrotēri, does not seem to have much influence on the Mycenaean ships.

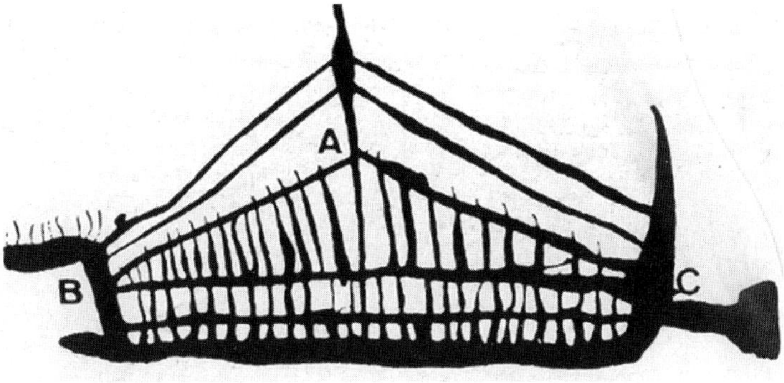


**Fig. 32** The Mycenaean ship (c. 1200 BCE) illustrated on the Gazi sarcophagi (Herakleion Museum).

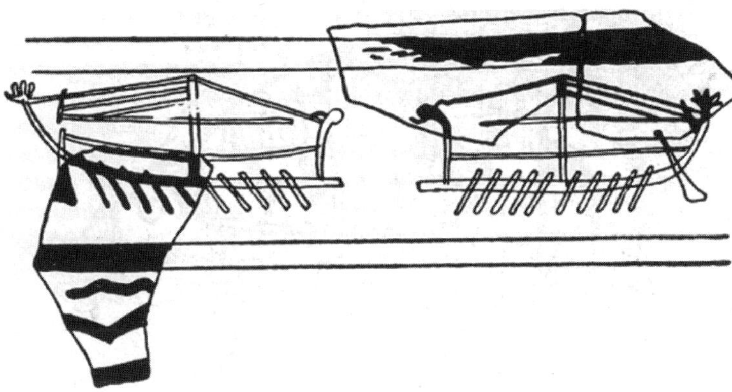

**Fig. 33** Representation of two Mycenaean ships (12th century BCE) by S. Marinatos.

**Fig. 34** Mycenaean ship depicted on a Asinē vase (12th century BCE) according to Frödin et al. (in [19, p. 147]).

This early πεντηκόντορος (pentēcontoros, i.e. a 50-rower ship) will prevail in the Mycenaean world, from Boeotia to Messēnia and to Crete, for centuries, and will be an extremely powerful means both for peace and war.

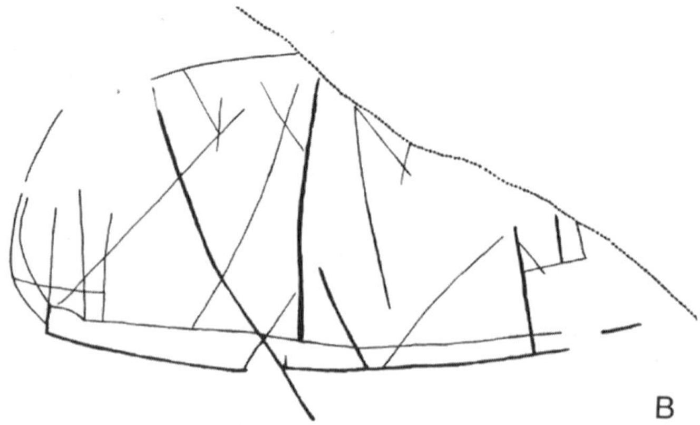


**Fig. 35** Graffito in Encomi (Cyprus), 12th cent. BCE, showing one of the Mycenaean ships, which had invaded the island (from [19, p. 148]).

**Fig. 36** One of the many Cypriot rhytons in the shape of Mycenaean ship, Lapithos, c. 1100 BCE (from [19, p. 149]).

The Mycenaeans, thanks to this novel technical achievement, dared to challenge the famous Minoan fleet, and invaded Crete around 1450 BCE. Homer may somehow be exaggerating when enumerates the 1186 Achaean ships taking part in the expedition to Troy; however, it was clear that an unprecedented naval force had appeared in Eastern Mediterranean. Finally, around 1200 BCE, short before and just after the collapse of the Hittite and Canaanit imperial systems (because of the alleged "Peoples of the Sea"), the Mycenaean ships will implement the final massive immigration of Achaeans towards Cyprus and Asia Minor (see [19, p. 149] and [7, p. 291]).

Once again, Technology is backing major historical developments …

## Appendix: How Much Coloured the Ancestors of the White Athena Were

In principle, a history of Technology should also deal with the origins of each basic technique. Although this paper was not conceived with such an ambition, I thought I should briefly address at least the "philosophy" of the subject.

1. The epoch we are dealing with, raises more intensively the issue of intercultural relationships: First, because written evidence is rather scarce during this period. Second, because "ethnic" barriers during the 2nd millennium BCE seem to be looser than during historical times: thus, intercultural osmosis might be easier.

Technology in particular, as a cultural phenomenon (and its transfer) were facilitated by commercial exchanges in the Aegean Sea. Archaeological finds, both in land and in shipwrecks (during the late Bronze Age and the beginning of Iron Age) demonstrate intensive commercial exchanges between Mycenaeans and Near East peoples, both ways.[6]

However, although a commercialized technical product "bears Technology" in it, it does not constitute per se an exchange of Technology. Such an exchange is *a posteriori* evidenced only under some other conditions, which will be the subject of the subsequent §3.

This being noted, it remains however to trace the origins of a specific technology each time on a case by case basis, rather than produce a Manichaean theory promising to explain the generation of an entire (all-inclusive) Culture. In fact, Science had suffered enough because of the "over-hellenizing" attitudes of the 19th century; nowadays, the pendulum seems to be violently pushed to the other extreme by some dedicated "orientalists". I maintain that it is our scientific duty to leave the pendulum free to oscillate by itself near its vertical position, whenever real conditions induce such a motion – on a case by case basis.

As a humble contribution to this end, the present author felt as his intellectual duty to add a warning here (inadequate and opportune as it may be) against some fashionable trends of the kind. My intention is to handle the subject, briefly though, in the following ways:

(i) I will pick some emblematic examples of orientalistic texts, in order to show their occasional epistemological flaws.
(ii) Subsequently I will submit a set of (rather elementary) rules to be followed when "hunting for the origins" of a given technology.

2. My first sampling of such (certainly non-representative) orientalistic views, refers to the book of Penglase, *Greek Myths and Mesopotamia* [20]. Here come my short comments to several proofs the author offers in support of his view that Greek myths are based on Mesopotamian ones:

[6] See *inter alia Das Schiff von Uluburun – Welthandeln vor 3000 Jahren*, Katalog der Ausstellung, Bochum, 2005.

(i) "A complex parallel [...]: Like [god] Enlil, the Greek supreme God Zeus dwells on a mountain" (p. 73).
It is however hard to find a religion without some deities living in mountains, throughout the globe ...

(ii) "Lēto settles down [in Dēlos] to give birth [to Apollo]. At this point Lēto suffers a setback for nine days. But Iris escorts from Olympus the birth goddess Eileithyia. Lēto begins to give birth" (p. 78).
"The same sequence of ideas is seen in Inanna's Descent to the Netherworld, where the goddess Inanna lies dead, and receives help [...] food and water of life [...]: a 'rebirth' which appears to be represented by a similar motif here in the Greek myth" (p. 80).
I confess I was unable to see any similarity between Lēto giving birth to Apollo and dead Inanna brought again to the upper world – otherwise a subject very frequent in many religions, anyway.

(iii) "Prometheus [offered] the gift of fire [and] taught mankind all the arts and crafts. This element recalls the Atrahasis epic" (p. 223) [...]: "Enki promised [to the senseless supreme god Enlil] to restrain mankind's numbers by natural means" (p. 219) (instead of plague, famine etc., p. 212)
No comment.

My second sampling of (certainly non-representative) orientalistic views suffering of some epistemological flaw, refers to the book by Morris, *Daidalos and the Origins of Greek Art* [2]:

(i) "Baal is 'rider of the clouds', inherited by Zeus as νεφεληγερέτης (gatherer of the clouds). Multiple deities at Ugarit manifest a collective divinity paralleled at Knossos in the linear B formula 'πᾶσι θεοῖς' (all gods)" (p. 79).
I find difficulties in these rather obvious characteristics of deities: Soma, for instance, (one of the gods of the Veda) "traverses the lights of heaven, the woolen filter". Besides, the Marut-deities "pour down rain, create darkness [...], their golden chariots gleam in the lightning" [21, p. 25]. Consequently, the dramatic statement "Zeus inherited Baal" seems to be rather cloudy ...

(ii) "The Athenian trireme [...] may have been a lesson learned from Phoenician nautical specialists" (p. 376). The following (rather irrelevant) supporting thoughts are offered by the author:

– "Rome designed its first warships from the example of a captured Carthagenian vessel.
– Disagreement reigns over the origins of the Greek trireme [...], probably because European classicists prefer it as a Greek invention, [like] democracy and other historic contributions (*sic*).
– Much of the controversy thrives on the lack of information about preclassical and non-Greek triremes" (p. 376).

Basch [19], however, after an extensive and interdisciplinary analysis, is adamantly clear: "It is certain that the Greek trireme was deeply different than the

> Phoenician trireme. And it is equally certain that these two types of ships constituted the final stage of evolution of two *independent* generations of ships" (p. 332).
>
> This fact can hardly be reduced to "a lesson learned from Phoenician ... specialists" – especially when no substantial evidence at all is offered ...

Both scholars previously commented, i.e. Penglase and Morris, have otherwise considerably contributed to their subjects, offering precious information and important analysis. What I wished to stress out here was only their disproportionate enthusiasm which occasionally led them to hasty conclusions of "general applicability" ...

Unfortunately, this syndrome seems to be somehow broader – almost the same way as the uncritical "hellenising" syndrome prevailing during the 19th and 20th centuries.

Let me finish with these examples by referring to some more recent views of Taha Showleh, contained in a paper kindly offered at the 2nd International Conference on Ancient Greek Technology entitled "The Mycenaean Engineers: Expertise and provenance" [22].

His view is that probably both the fortification walls and the tholoi (vaults) of Mycenaean cities were constructed on the basis of know-how and in presence of technicians imported from the East.

Here come my comments.

(a) The smaller tholoi [23] constructed prior to Tiryns, i.e. those at Georgiko (Late Helladic I) and Koryphasio (Middle Helladic) show that, earlier than 1600 BCE, this technique was well known. This fact does not confirm Showleh's view about a "sudden development" (p. 436).

On the other hand, tholoi were not known in Eastern Countries prior to Mycenaean times; therefore, it is rather difficult to understand Showleh's assertion that "perhaps labour would have been imported from Eastern Mediterranean" (p. 437).

Finally, the uniqum of the "cyclic houses" in Messara (Crete), is not offered at all for a theory of a "Cretan origin of Tholoi" (p. 435):

(i) As opposed to the Greek mainland where this technique was widespread (in-time and space), Messara is the unique area of hypothesized tholoi.
(ii) Besides, we have no evidence about the superstructure of such circular non-baried houses.
(iii) A corbelled tholos, without the double stabilizing effect of the upper soil layers acting both vertically and horizontally, is not safe at all.

(b) Fortification walls are the most typical example of a "universality"; almost every settlement, since neolithic times, used to defend itself by means of such walls. This is not a high-tech structure after all ...

Consequently, it is rather obvious that fortification walls were developing together with the wealth of a city and with the dangers it was facing. In the specific case of the Mycenaean fortifications, a couple of remindings are useful:

- Fortification walls 2 m thick existed already in Lerna (~2500 BCE), in Syros (~2200 BCE), in Aegina (~2100 BCE), etc. [24].
- Almost one millennium later, it is all too natural to find 3 m thick walls in Tiryns, which one hundred years later became 5 m thick – and so on, up to 7 m in Mycenae later on.

The gradual development form 2 to 3 to 5 and to 7 m, does not concur with Showleh's view of "a given absence of forerunners" (p. 434).

Besides, there is nothing technically special in increasing the thickness of a three leaf masonry: One has simply to move apart the two external masonry leaves,[7] and increase the thickness of the intermediate infill!

Consequently, it is extremely difficult to subscribe to Showleh's insinuation about a west-ward spreading of Hittite (*sic*) techniques and, possible, craftsmen, (p. 435).

My conclusion is simple: Actual archaeological and technical knowledge offers the possibility to get rid of the "perhaps-would be-matched-parallels" abuse, occasionally observed nowadays".

3. Φίλος Πλάτων, φιλτέρα Αλήθεια[8]

That is why I took the liberty to enumerate here below some elementary facts to be considered when "hunting for the origins" of a given object, technology, idea or concept.

(a) My first humble suggestion is that we should try to start from a well known human "universality".[9] Such is the case with the fundamental significance of Technology for all peoples of the world: Technology was everywhere taken as the "natural complement of Nature". Many insects, birds and animals experience the incapacity of their natural means, and they survive only thanks to additional constructions or instruments or techniques,[10] invented through their brain – in other words, thanks to a "Technology". Humans have followed the same natural/cultural pattern – and the Protagorian Prometheus[11] reflects exactly this basic feature when, in view of the four natural *incapacities* of humans, he offers to them: τεχνογνωσίαν (technical know-how) taken from the head of Athēna. It is therefore all too obvious that humans, independently of geographical area or skin colour, project this fundamental truth of life to the sky, and produce religious Myths in which Technology keeps a significant position:

---

[7] Moreover, the weight of the stones used to build these external leaves is not anymore any surprise; rudimentary techniques (soil-ramps and lever-push) were used to rise them. After all, Euripid's and Pausanias (with their disproportionate admiration of the – cyclopean walls) were not Engineers, Professor C. Palyvou points out).

[8] "Plato is dear. Truth is dearer".

[9] Such as sexual drive, defense, deeply rooted religious needs, collectivity of actions, desire for beauty, trends to face natural threats by means of artifacts (mainly by imitating Nature itself), etc.

[10] This is e.g. the case of the construction of elaborated nests of birds, of the use of webs of spiders to catch their food or of the timber splints used by chimpanzees to cure their broken leg.

[11] Plato, Protagoras, 321c.

- *Ninurta*, the Sumerian god, is responsible for the irrigation-works from the Tigris and Euphrates rivers (the fundamental technology of Mesopotamia) [20, p. 49].
- *Ilisha*, the Ugaritic carpenter-god, is responsible for the everyday structural use of timber [2, p. 91].
- *Cyclopes*, the smith-deities, were part of the pre-olympian trinity "Titans, Cyclopes, Hekatonheires" (i.e. Intellect, Technology and Natural forces) in prehistoric Greece.
- *Jehovah* himself teaches Technology to a selected human, Bezeleel, "to do all works", [Exodus 31, 2–6].

The combination of the well known "animal technicalness" and the persisting sacredness of Technology within humans, clearly demonstrates a fundamental (if not innate) tendency of most peoples to technical inventiveness.

I therefore humbly submit to specialists that *autochthonous* technical development should be the first alternative to be investigated; anthropologically speaking, Technology is by definition a "universality". Subsequently, the investigation will obviously cover the other alternatives, such as external influences, mimesis or direct importation.

(b) And my second suggestion refers precisely to the ways we investigate the "products"[12] when we try to demonstrate or belie trans-cultural influences.

- Isolated or unconnected products cannot prove significant relationships between two cultures. Multiple uncertainties,[13] accompanying the knowledge of natural or (worse) social phenomena, may jeopardise the confidence in any information potentially included in an isolated product. An appropriate number of repetitions or, best, an appropriate articulation ("system's approach") is needed for a find to be conclusive. Science is not an accumulation of speculations ...
- "Similarity" between two products belonging to different cultures, is only a first prerequisite for a potential relationship – it is not a proof.
- In fact, similar products may have been independently produced within each culture – and this is frequently the case with technological products. The probability of such a "coincidence" (as it is inappropriately labelled sometimes) is higher under the following favourable circumstances:

  - Intensive social/historical need for the product ("necessity is the mother of technology").
  - Extent and multitude of the population concerned.
  - Appropriate natural resources.

[12] With this generic term, several entities are meant, such as myths, techniques or objects.

[13] Such uncertainties are present, e.g., when an isolated object may be explained as a gift, a circumstantial commercial item, a part of looting, a property of an individual traveler or in fact a prototype of a new invention (local or remote one ... ).

- Favourable previous social attitudes or practical achievements along similar lines of development.
- A short time interval separates the similar products.
- A long distance separates (if difficult communications prevail between) the two regions of possible interaction.

When a critical mass of such favourable conditions occurs, the probability for identical or similar independent solutions is considerably higher.

In conclusion, the pleasures offered by rigorous (and patient) Science, are much subtler and longer lasting than those given by Science-fiction. Science is not a meticulous accumulation of speculations[14] ...

(c) My last humble suggestion refers to the need for a *follow-up* of the cause-to-effect relationship: When sufficient evidence suggests a foreign origin of a product, the entire investigation should now be completed by a brief comparison of the "performance level"[15] of the product in the receiving milieu, as compared to the level of the initial imported product. I maintain that this is our epistemic duty, in order to obtain a "back-lighting" on the supposed mother-to-daughter relationship: Such a feeding-back may elucidate some possibly hidden dubious points; occasionally, an extremely glorious daughter may belie her origin from an insignificant mother.

## References

1. Paipetis, S.A. et al., Defensive weapons in Homer, in *Extraordinary Machines and Structures in Antiquity*, S.A. Paipetis (Ed.), Peri Technon Publ., Patras, 2003.
2. Morris, S., *Daidalos and the Origins of Greek Art*, Princeton University Press, 1992, p. 12.
3. Kalligeropoulos, D., *Mythos Istoria tōn automatōn*, Kastaniotēs, Athens, 1999.
4. Knauss, J., Observations and considerations concerning Mycenaean underground well-houses, etc., in *Proceedings of the 2nd International Conference on Ancient Greek Technology*, EMAET, Tech. Chamber of Greece, Athens, 2006.
5. Knauss, J., *Late Helladic Hydraulic Works*, Verein zur Förderung der Hellenischen Geschichte, Weilheim/Obb., Deutschland, 2003 [in Greek].
6. Bandeka, H., Waterproof mortars of ancient Greek water-tanks, Ph.D. Thesis, National Technical University of Athens, 1974 [in Greek].
7. Iacovidēs, S., Ekistics and architecture in the 14th to the 11th centuries BCE, in *Encyclopedia of the Hellenic Nation*, Ekdotikē Athinon, Athens, 1970.
8. Kakavoyannis, E., About the Temmikes, in *3rd International Conference on Boeotian Studies*, Archaiologia, 2000, p. 121.
9. Palyvou, K., Tassios, T.P., Stournaras, K. and Geralis, K., Adobe structures at Mycenae, in *Extraordinary Machines and Structures in Antiquity*, S.A. Paipetis (Ed.), Peri Technon Publ., Patras, 2003.

[14] In any event, the reliability "$R$" of a conclusion regarding the origins of a product, seems to be inversely proportional to the square of the number "$N$" of the working hypotheses used to formulate the conclusion: $R = K/N^2$.

[15] This term here is meant as the degree of elaboration, subtleness, efficiency, etc., of the product as it appears within the "receiving" culture, compared to its predecessor product in the cultural milieu where it was supposingly conceived.

10. Boughia, P., *Ancient Greek Bridges*, EMAET/NTUA, 2002 [in Greek].
11. Konof'agos, K., *Ancient Lavrion*, Ekdotikē Ellados, Athens, 1980 [in Greek].
12. Spitaels, T., The early Helladic period in mine No 3, in *Thorikos 1972/1976*, Comité des Fouilles Belges en Grèce, Gent, Belgium, 1984, p. 162.
13. Kakavoyanni, O. et al., An attempt of technological approach of silver and lead production during Late Neolithic and Early Helladic Period in Mesogaia, in *Proceedings of 2nd International Conference on Ancient Greek Technology*, EMAET, Tech. Chamber, Athens, 2006.
14. Servais, J., in *Thorikos 1963*, Comité des Fouilles Belges en Grèce, Gent, Belgium, 1968.
15. Cooke, G.A., Phoenicia, in *Encyclopaedia Britannica*, 1965, p. 767.
16. Varoufakis, G., The origin of Mycenaean and geometric iron, in *Early Metallurgy in Cyprus*, Pieridis Foundation, Nicosia, 1982, p. 317.
17. Muhly, J.D., Texts and technology. The beginning of iron metallurgy in the Eastern Mediterranean, in *Proceedings of 2nd International Conference on Ancient Greek Technology*, EMAET, Tech. Chamber, Athens, 2006, p. 19.
18. Backe-Forsberg et al., Ancient iron sources in southern Peloponnesus, Greece, in *Proceedings of 2nd International Conference on Ancient Greek Technology*, EMAET, Tech. Chamber, Athens, 2006, p. 122.
19. Basch, L., *Le musée imaginaire de la marine antique*, Inst. Hell. pour la préservation de la tradition nautique, Athens, 1987.
20. Penglase, C., *Greek Myths and Mesopotamia*, Routledge, 1994.
21. Zaehner, R.C., *Hinduism*, Oxford University Press, 1966.
22. Showleh, T, The Mycenaean engineers: Expertise and provenance, in *Proceedings of 2nd International Conference on Ancient Greek Technology EMAET*, Tech. Chamber of Greece, Athens, 2006, p. 433.
23. Iakovidēs, S., On the form of hewn tombs at Volimidia of Messinia, in *Festschrift to A. Orlandos*, 1978, p. 110 [in Greek].
24. Palyvou, C., *Building Technology in Prehistoric Times*, Archaiologia, 2005, p. 94.

# Autagreton

Emmanuel Mikrogiannakis

*University of Athens, Greece*

**Abstract.** Homer lures, ennobles, fascinates, enriches, and stimulates. He is a large poetical encyclopaedia and inevitably a battleground for scholars. Recently, scientists believe that a modern reading of Homer will provide deep insights and even improvements in the field of technology. I think they are right.

One example, *autagreton*, a Homeric word (Od. 16,148, cf. 8.555 ff) finally expresses what is sought for today, what our final purpose is. What is the main goal of modern technology, mainly military (which is the most ambitious)? One vehicle, self-controlling mechanism according to the different situations, capable by itself to perform any mission successfully.

Is it not impressive that such a mechanism is already described in Homer? This mechanism can be called *autagreton* (already existing in the society of Phaeacia and King Alcinous). *Autagreton* is a term superior to *automaton* (self-acting), because the first is subject to the mind and brain of its possessor. Moreover, *autagreton* shares the soul of its master, especially the νοῦς (mind), which is *the governor of the soul* (according to Plato).

General remark: The technology hinted in Homer (even in mythical clothing) provides us with brand new aspects of technology and offers brilliant terms (such as *autagreton*) in the beginning of the new millennium (the *trimillennium*).

The ships of Phaeacia:

1. have outstanding knowledge of the whole earth,
2. need neither steersman, nor steering oar,
3. know what their lords have in mind and implement it,
4. they encounter successfully and handle any obstacles quickly, and
5. they are indefatigable, invisible, unassailable and invulnerable.

Homer and Olympia, an ideal combination! Two worlds with one ideal. And they challenge all those longing for distinction, for αριστεία (excellence), to throw themselves into an αγώνα (contest) with άμιλλα (competition). Homer recommends αριστεύειν (to excel) and even αιέν (always).

*S.A. Paipetis (ed.), Science and Technology in Homeric Epics,* 35–39.

He is the poet *par excellence*. He captivates. He involves us in the action. We live alongside not only Achilles and Hector, but also the minor heroes – even the swineherd Eumaeus, whose name alone indicates his good intension. Whoever is tempted by Homer's Siren song stays enchanted for life (and not any less so once the poet has been demythologised and becomes a reliable historical source).

In Homer we move on two levels. The gods too are βιωτοί (liveable). We share the θείον (divine), as the gods do the ανθρώπειον (human). There has been a democratisation of the divine here and a deification of the human. The human soul is expanded, made indivisible and statue-like.

Homer is an encyclopaedia, not in alphabetical order. Everything is woven around a catastrophic μήνις (wrath) and a πολύτροπος (much-travelled) man. In a clear way, at an analogous pace we walk alongside portraying all that the Muse αείδει (sings). The correct reading of Homer is fruitful for those who are εγκύμονες την ψυχήν (pregnant of the mind).

Here are a few examples of good studentship. Aeschylus was shaped in the grove at Marathon fighting the Persians (like a Homeric ήρως, hero) with his brother Kynaigeiros, but also by the great teacher, Homer. His tragedies (Aeschylus confesses) are bread crumbs from Homer's table. Pheidias, the greatest of the sculptors, was inspired (as one who looked on god) to create the statue of Zeus at Olympia (where we are now) by lines from the *Iliad*, where a nod from the highest of the gods shakes Olympus.

It should not surprise us that Homer's most complete αναγνώστης (reader) was Alexander the Great. This Homer was his guide, acknowledged by Alexander himself as a συστρατηγός (co-general). The association was so deep that Alexander forgot himself and modelled himself on Achilles. Virgil also showed how strong Homer's wine is in his Aeneid, the national epic of the Romans. A (displaced) Homeric hero becomes the founding father of the Romans in the work of the Homer-mad Virgil.

This, however, mainly in the sphere of culture. Homer lifts the mind (and tempts νόον [mind], μένος [wrath], κύδος [glory], κάλλος [beauty]) and excites us. But what does he offer to technology, which is an extension of the human hand and senses? Does he give us any ideas in the area of arms manufacture, of physics, medicine, astronomy? Does he offer anything to computers, the machines thinking alongside man? I shall not insist on marking the greatness of his achievement, which has been underlined many times, that he sees everything in every aspect of the subject under study with clarity and linearity. In the "Homeric Archaeology", a series of several dozen volumes, there are entries according to category on subjects such as dress, weaponry, fortifications, shipbuilding, etc.

Every science is ennobled and finalised when it is plunged into antiquity, to find its root therein, when it expresses the καινά αρχαίως (as far as the αρχαία καινώς).

There is something noteworthy that has not been noted. We hear that in large research and industrial centres attempts are being made to construct computers that will not need human hand. Perhaps not even speech. These would not have been needed to have been done, according to Homer: έπεσιν και χερσίν [with word and deed] (Il. 1.77). But how? Again a Homeric phrase: βουλή και νόω [with will and mind] (Od. 11.211, 16.374). What is that which is being sought after today and which shall be the εσόμενον (*futurum*), tomorrow? A machine, an "αρκυοστασία" (a trap with nets), which by its very nature but without έπος (speech), shall conceive and realise not just any thought but that which we wish, that which we choose and want to realise. With one word: το αιρετόν (that which is chosen).

Is there something in Homer analogous to what is hoped for today? Indeed. Telemachus (Od. 16.148), having returned from the journey to Pylos and Sparta and whilst accompanied by his faithful Eumaeus and Odysseus himself (as yet unrevealed), expresses the opinion:

ει γαρ πως είη αυτάγρετα πάντα βροτοίσιν
πρώτον κεν του πατρός ελοίμεθα νόστιμον ήμαρ.

(If people could have everything their own way,
the first thing I should choose would be the return of my father)

All the translations render this crucial αυτάγρετα along with its immediate context, either as if this were the word αιρετά or a wish, as follows: would that everything were in the hands of the mortals. Here, however, we do not have a wish or a choice, but a proposal. A simple thought is expressed. In the Attic dialect, the above would be thus:

Ει πάντα είη αυτάγρετα, ελοίμεθα αν νόστον του πατρός.

The weight falls on the αυτάγρετα (*αυτά* + αγρέω or αγρεύω). Αγρέω and αγρεύω are verbs of hunting. Hunters capture with a variety of άρκυες (nets). Αυτά*γ*ρετα (αυτάγρευτα, αυτόληπτα) are those which are caught, they are captured in some kind of mechanism, without the (physical) involvement of the hunter. Telemachus here foresees (τηλόθεν constantly from afar, I would say) the possibility that such machines could be constructed (presumably by Hephaestus, like his tripods or the κύνες [dogs] of Alcinous which have intelligence, they understand the wish and implement it).

The αυτάγρετον of Telemachus is our long-term future vision. What do humans wish for? Βουλήν (will) and the ability to realise it. The greatest will is divine. Minor β*ο*υλαί are those of humans. The αυτάγρετον that Telemachus talks of in the presence of Eumaeus and the as yet unknown stranger (his father) has already been performed and is being put into action. Telemachus does not know it, but Odysseus is already in Ithaca, is in fact right next to him. The αιρετόν (choice) of Telemachus is already a fact. The return has already taken place. How? Αυταγρέτως, I would say.

The ships of the Phaeacians carry out the αυτάγρετ*ο*ν (8.555–563). These are τιτυσκόμεναι φρεσί νήες, i.e. "intelligent" ships (to use the modern expression). These ships know the νοήματα και φρένας (thoughts and minds) of men. In other words? They conform to the spirit of their masters. They act in their interest. There are no κυβερνητήρες (steersmen) or πηδάλια (steering-oars) on these ships. All that is needed is for them to know something very simple. This is asked of the stranger on the island of the Phaeacians, that is, where does he want to go, what is his address. This is enough. The ships have stored knowledge (a type of database, navigator) and they know πόλιας και πίονας αγρούς (all the cities and fertile countries). They are in a position to deal with any difficulty, they are invisible, wrapped in clouds. The main thing: there is no concern that these ships might be hit, that they might be destroyed or lost. They cannot be hit. They sail wherever, they carry out missions perfectly and return safe.

What can we observe here? On a technological level, the society of the Phaeacians is one stage above the society of the Ithacans. That which is simply a vision or simple a thought of Telemachus has already been implemented by the Phaeacians. Indeed, the stranger who is present has already tasted the fruits of Phaeacia.

Where is our own society, at the dawn of the 21st century (the trimillennium)? Are we at the level of the Ithacans or the Phaeacians? Or perhaps somewhere in between? We aspire to build vehicles and bombs that shall carry out large missions in accordance to our νοήματα και φρένας (thoughts and minds) and which shall move safely. All the technological side, their machinery, their ability to adapt and be flexible in response to the obstacles that they shall encounter are vital elements. Our own contribution shall only be that of setting the goal.

An unavoidable remark: one objection is to what extent the technological development of the Phaeacians was a reality or a myth. From the 19th century, scholars have increasingly believed in a close relationship between Homer and history. Archaeological excavations have added to this. The conviction that generally myth, in all of its forms, conceals truths in a condensed form is becoming established. In the case of the Phaeacians the logical sequence in the recounting of the specifications of the τιτυσκομένων φρεσί νεών is impressive. The verb τιτύσκομαι, with reduplication of the present tense, indicates a repeated action. The root *tu-* (the same as in τεύχω and τυγχάνω) indicates success in every problem arising. And these ships (these vehicles of the Phaeacians) have minds similar to those of humans.

It is thus difficult to clarify whether the φρεσί (as τιτυσκόμεναι φρεσί) refers to the minds of people or to those of the ships. It appears that humans implant φρένας και νόον (mind and thought) into these ships, just as the gods implant κάλλος (beauty), κύδος (glory), μένος (wrath) and, most of all, φρένας και νουν into humans.

The word αυτάγρετον (an adjective as a noun) could end up as a term (title) for whatever is thought and aspired to by us but is executed, is captured by a mechanism that is self-controlled. Does it differ from the αυτόματον? With the αίρεσις, human will, this αίρεσις, the choice of the hunter, does not intervene. In the αυτόματον, to which the root *ma-* is subject (μάτηρ, μήτηρ and αυτομήτωρ, as αυτοπάτωρ) it is indicated that something is done by itself. If the αυτόματον and αυτάγρετον had will analogous to man, would this then lead to their rule over humans or at least over their competitors? This is a possibility that cannot be ruled out. From the perspective of humans, we aim not to lose control of the αιρείσθαι the choice (for us to be the choosers). With the αυτάγρετον the whole technological element is separated from us. We are left with the βούλησις, the will. We assign the goal to the αυτάγρετον, which is so accompanied by volition. We are left with what is perhaps most important, but setting goals cannot be done ανεμεσήτως, blame-free. We shall not examine the whole drama of the subject here.

In any case, for whatever is carried out with the αυτάγρετον, the responsible party must be sought in the person who chooses. And as Plato would have said in his *Republic*,

Αιτία ελομένου.

(The blame belongs to the one who chooses)

# Part 2

# MATHEMATICS AND PHYSICS

# Archimedes' Count of Homer's Cattle of the Sun

Chris Rorres

*University of Pennsylvania, USA*

**Abstract.** In the first few lines of *The Odyssey*, Homer foretells how Odysseus' crew "perished through their own sheer folly in eating the cattle of the Sun-god Hyperion". These Cattle of the Sun grazed near the Sicilian town of Taormina (*Tauromenion* to its ancient Greek settlers) and, although endlessly warned not to, Odysseus' crew slaughtered some of them for food. For this sacrilege Zeus tossed them from their ship to their deaths with his thunderbolts, leaving Odysseus to continue his odyssey alone. In describing the sacred cattle, Homer indirectly gives their count by writing that they comprised seven herds containing fifty cattle each (Book XII: "Of oxen fifty head in every herd feed, and their herds are seven"), leaving it to the reader to determine the total number of cattle. Centuries later this simple multiplication problem was the inspiration for Archimedes' famous "Cattle Problem", whose first line is: "If thou art diligent and wise, O stranger, compute the number of cattle of the Sun, who once upon a time grazed on the fields of the Thrinacian isle of Sicily". Archimedes, who lived in the Sicilian-Greek city-state of Syracuse, 85 kilometers south of Taormina, would have been very familiar with Homer's tale. In his problem, Archimedes challenges his colleague Eratosthenes to compute the number of the Cattle of the Sun having a larger and more complicated composition than the one described by Homer. Archimedes' problem is so complicated that the total number of cattle contains 206,545 digits. In this article I describe the origins of this problem in antiquity, its rediscovery in the eighteenth century, and the attempts since then to solve it. Its complete resolution had to await the computer age, since before then someone estimated that it would take the work of "a thousand men for a thousand years" to determine the exact solution. Attempts at its solution fueled the field of Diophantine Analysis – the analysis of problems whose solutions are restricted to whole numbers – and, in particular, the study of the so-called Pell Equation. Today a notebook computer using sophisticated algorithms can generate the number of cattle in seconds, taking more time to print out the number than to actually compute it. The amount of intellectual activity that has surrounded this problem over 23 centuries suggests the validity of Voltaire's remark, "There was more imagination in the head of Archimedes than in that of Homer".

***

> Tell me, O Muse, of that ingenious hero who travelled far and wide after he had sacked the famous town of Troy. Many cities did he visit, and many were the nations with whose manners and customs he was acquainted; moreover he suffered much by sea while trying to save his own life and bring his men safely home; but do what he might he could not save his men, for they perished through their own sheer folly in eating the *cattle of the Sun*-god

*S.A. Paipetis (ed.), Science and Technology in Homeric Epics,* 43–66.

Hyperion; so the god prevented them from ever reaching home. Tell me, too, about all these things, O daughter of Jove, from whatsoever source you may know them.

*The opening lines of Homer's Odyssey, as translated by Samuel Butler* [3]

***

There was more imagination in the head of Archimedes than in that of Homer.

*Voltaire (1694–1778)* [12]

***

The imagination in a mathematician who creates makes no less difference than in a poet who invents. ... Of all the great men of antiquity, Archimedes may be the one who most deserves to be placed beside Homer.

*Jean d'Alembert (1717–1783)*

## 1 Homer's Cattle of the Sun

A lengthy episode of Homer's *Odyssey* is devoted to the slaughter of some of the Cattle of the Sun (Βόες Ηλίου) (Figure 1) by the crew of Odysseus (Oδυσσεύς). These cattle, also called the Oxen of the Sun, belonged to the Titan god of light Hyperion (Υπερίων: "the one above") and to his son Helios ('Ηλιος: "sun"). The Greeks believed that they grazed near the Sicilian town of Taormina (Figure 2),

**Fig. 1** The Cattle of the Sun depicted on a sixth-century B.C. vase from Cerveteri, Italy, located in The Louvre, Paris.

**Fig. 2** Bronze coin of Tauromenion circa 275–216 B.C. The obverse shows a diademed head of Hercules.

which its Greek settlers called Tauromenion (Ταυρομένιον). This name is derived from tauros (ταύρος), the Greek word for bull (Diodorus Siculus [4]).

The cattle were accompanied by an equal number of sheep and the animals were guarded by Helios's two daughters: Phaethusa (Φαέθουσα: "radiance"), who looked after the cattle, and Lampetia (Λαμπέτιη: "shining"), who looked after the sheep.

Odysseus and his crew were warned by the blind seer Teiresias (Τειρεσίας) and the sorceress Circe (Κίρκη) not to harm the Cattle of the Sun. Nevertheless, after his crew exhausted their rations they chose to slaughter some of the sacred cattle and risk the wrath of the gods rather than die slowly of starvation.

The wrath of the gods came soon afterwards through Zeus's thunderbolts directed at their ship, and only Odysseus survived the resulting devastation.

The lines of Homer concerning the Cattle of the Sun that most concern us in this paper are from a lengthy passage in which Circe describes some of the dangers that lie before Odysseus (Homer [3], Book XII, lines 127–130):

> Θρινακίην δ' ἐς νῆσον ἀφίξεαι· ἔντα δὲ πολλαὶ
> βόσκοντ' Ἠελίοιο βόες καὶ ἴφια μῆλα,
> ἑπτὰ βοῶν ἀγέλαι, τόσα δ' οἰῶν πώεα καλά,
> πεντήκοντα δ' ἕκαστα.
>
> You will now come to the Thrinacian island, and here you will see many herds of cattle and flocks of sheep belonging to the sun-god – seven herds of cattle and seven flocks of sheep, with fifty head in each flock.

The word "Thrinacian" means three-cornered in Greek and refers to the triangular island of Sicily [9] (Figure 3).

It is typical of Homer not to give a total count for the Cattle of the Sun, but rather to give the smaller counts of subgroups that comprise the total count. It is left to his audience to determine the total count. He does this several times in his work when

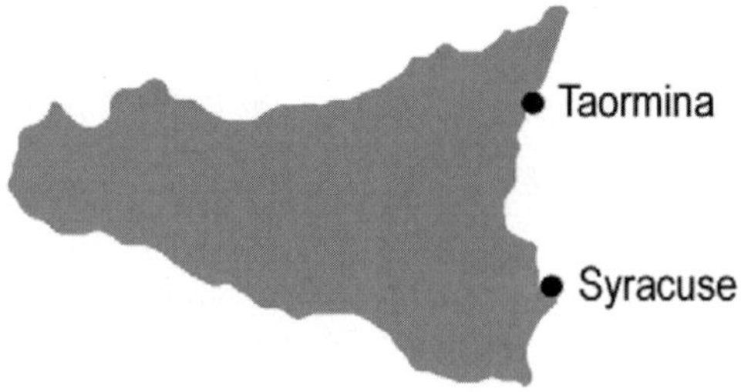


**Fig. 3** A map of Sicily showing the cities of Taormina and Syracuse.

describing the numbers of certain things. For example, in Book XVI Telemachus (Τηλέμαχος) describes to his father Odysseus the number of his mother's suitors as follows [3]:

> You shall learn their number at once. There are fifty-two chosen youths from Dulichium, and they have six servants; from Same there are twenty-four; twenty young Achaeans from Zacynthus, and twelve from Ithaca itself, all of them well born. They have with them a servant Medon, a bard, and two men who can carve at table.

While today the computation of the number of the Cattle of the Sun seems trivial ($7 \times 50 = 350$ cattle), few of Homer's audience in the early first millennium B.C. could perform even that simple multiplication. The seven herds of cattle corresponded to the seven days of the week and the 350 cattle corresponded to the 350 days in the year of the early Greek calendar. (Every other year the Greeks added an extra month of about 30 days to realign the sun with their calendar.)

## 2 Archimedes' Version of the Number of the Cattle of the Sun

Syracuse, the city in which Archimedes (c. 287–212 B.C.) lived, is only 85 kilometers south of Taormina, the mythical home of the Cattle of the Sun (Figure 3), and Archimedes would certainly have been very familiar with Homer's tale of the Cattle of the Sun. Archimedes' famous Cattle Problem written as an epigram (Πρόβλημα Βοεικόν in Greek and Problema Bovinum in Latin) would have been recognized immediately by his contemporaries as a vastly difficult version of Homer's little multiplication problem. Its introduction states that Archimedes challenged his colleague Eratosthenes of Cyrene (c. 276–194 B.C.) with the problem. There are a few oblique references to it in the ancient literature, but the only surviving text of the epigram, in Greek, was edited from a manuscript found in the Herzog August Library in Wolfenbüttel, Germany, by the famous writer Gotthold Ephraim Lessing (1729–1781) in 1773 (Figure 4) [6].

There are two parts to the problem, but no solutions to either part were found with its statement. The first part may have been amenable to solution by Archimedes,

Zur
Geschichte
und
Litteratur

Aus den Schätzen
der
Herzoglichen Bibliothek
zu
Wolfenbüttel

Zweyter Beytrag

von
Gotthold Ephraim Lessing.

Braunschweig,
im Verlage der Fürstl. Waysenhaus-Buchhandlung.
1773.

**Fig. 4** Title page of Lessing's 1773 paper on the Cattle Problem [6].

although certainly not by the method we use today. However, the second part, which many believe was a later spurious addition, was almost certainly beyond the reach of ancient mathematicians.

## 3 The First Part of the Cattle Problem

To express the problem mathematically, let

$$W = \text{number of white bulls}$$
$$B = \text{number of black bulls}$$
$$Y = \text{number of yellow bulls}$$
$$D = \text{number of dappled bulls}$$
$$w = \text{number of white cows}$$
$$b = \text{number of black cows}$$
$$y = \text{number of yellow cows}$$
$$d = \text{number of dappled cows}$$

Then the first part of the problem can be stated as the following seven equations in eight unknowns:

## A Problem

*which Archimedes solved in epigrams, and which he communicated to students of such matters at Alexandria in a letter to Eratosthenes of Cyrene.*

If thou art diligent and wise, O stranger, compute the number of cattle of the Sun, who once upon a time grazed on the fields of the Thrinacian isle of Sicily, divided into four herds of different colours, one milk white, another a glossy black, a third yellow and the last dappled. In each herd were bulls, mighty in number according to these proportions: Understand, stranger, that the white bulls were equal to a half and a third of the black together with the whole of the yellow, while the black were equal to the fourth part of the dappled and a fifth, together with, once more, the whole of the yellow. Observe further that the remaining bulls, the dappled, were equal to a sixth part of the white and a seventh, together with all of the yellow. These were the proportions of the cows: The white were precisely equal to the third part and a fourth of the whole herd of the black; while the black were equal to the fourth part once more of the dappled and with it a fifth part, when all, including the bulls, went to pasture together. Now the dappled in four parts were equal in number to a fifth part and a sixth of the yellow herd. Finally the yellow were in number equal to a sixth part and a seventh of the white herd. If thou canst accurately tell, O stranger, the number of cattle of the Sun, giving separately the number of well-fed bulls and again the number of females according to each colour, thou wouldst not be called unskilled or ignorant of numbers, but not yet shalt thou be numbered among the wise.

But come, understand also all these conditions regarding the cattle of the Sun. When the white bulls mingled their number with the black, they stood firm, equal in depth and breadth, and the plains of Thrinacia, stretching far in all ways, were filled with their multitude. Again, when the yellow and the dappled bulls were gathered into one herd they stood in such a manner that their number, beginning from one, grew slowly greater till it completed a triangular figure, there being no bulls of other colours in their midst nor none of them lacking. If thou art able, O stranger, to find out all these things and gather them together in your mind, giving all the relations, thou shalt depart crowned with glory and knowing that thou hast been adjudged perfect in this species of wisdom.

## Πρόβλημα

*ὅπερ Ἀρχιμήδης ἐν ἐπιγράμμασιν εὑρὼν τοῖς ἐν Ἀλεξανδρείᾳ περὶ ταῦτα πραγματευομένοις ζητεῖν ἀπέστειλεν ἐν τῇ πρὸς Ἐρατοσθένην τὸν Κυρηναῖον ἐπιστολῇ.*

Πληθὺν Ἠελίοιο βοῶν, ὦ ξεῖνε, μέτρησον
  φροντίδ᾽ ἐπιστήσας, εἰ μετέχεις σοφίης,
πόσση ἄρ᾽ ἐν πεδίοις Σικελῆς ποτ᾽ ἐβόσκετο νήσου
  Θρινακίης τετραχῇ στίφεα δασσαμένη
χροιὴν ἀλλάσσοντα· τὸ μὲν λευκοῖο γάλακτος,
  κυανέῳ δ᾽ ἕτερον χρώματι λαμπόμενον,
ἄλλο γε μὲν ξανθόν, τὸ δὲ ποικίλον. ἐν δὲ ἑκάστῳ
  στίφει ἔσαν ταῦροι πλήθεσι βριθόμενοι
συμμετρίης τοιῆσδε τετευχότες· ἀργότριχας μὲν
  κυανέων ταύρων ἡμίσει ἠδὲ τρίτῳ
καὶ ξανθοῖς σύμπασιν ἴσους, ὦ ξεῖνε, νόησον,
  αὐτὰρ κυανέους τῷ τετράτῳ τε μέρει
μικτοχρόων καὶ πέμπτῳ, ἔτι ξανθοῖσί τε πᾶσιν.
  τοὺς δ᾽ ὑπολειπομένους ποικιλόχρωτας ἄθρει
ἀργεννῶν ταύρων ἕκτῳ μέρει ἑβδομάτῳ τε
  καὶ ξανθοῖς αὐτοὺς πᾶσιν ἰσαζομένους.
θηλείαισι δὲ βουσὶ τάδ᾽ ἔπλετο· λευκότριχες μὲν
  ἦσαν συμπάσης κυανέης ἀγέλης
τῷ τριτάτῳ τε μέρει καὶ τετράτῳ ἀτρεκὲς ἶσαι·
  αὐτὰρ κυάνεαι τῷ τετράτῳ τε πάλιν
μικτοχρόων καὶ πέμπτῳ ὁμοῦ μέρει ἰσάζοντο
  σὺν ταύροις πάσαις εἰσ νομὸν ἐρχομέναις.
ξανθοτρίχων δ᾽ ἀγέλης πέμπτῳ μέρει ἠδὲ καὶ ἕκτῳ
  ποικίλαι ἰσάριθμον πλῆθος ἔχον τετραχῇ.
ξανθαὶ δ᾽ ἠριθμεῦντο μέρους τρίτου ἡμίσει ἶσαι
  ἀργεννῆς ἀγέλης ἑβδομάτῳ τε μέρει.
ξεῖνε, σὺ δ᾽, Ἠελίοιο βόες πόσαι, ἀτρεκὲς εἰπών,
  χωρὶς μὲν ταύρων ζατρεφέων ἀριθμόν,
χωρὶς δ᾽ αὖ, θήλειαι ὅσαι κατὰ χροιὰν ἕκασται,
  οὐκ ἄιδρις κε λέγοι᾽ οὐδ᾽ ἀριθμῶν ἀδαής,
οὐ μήν πώ γε σοφοῖς ἐναρίθμιος. ἀλλ᾽ ἴθι φράζευ
  καὶ τάδε πάντα βοῶν Ἠελίοιο πάθη.
ἀργότριχες ταῦροι μὲν ἐπεὶ μιξαίατο πληθὺν
  κυανέοις, ἵσταντ᾽ ἔμπεδον ἰσόμετροι
εἰς βάθος εἰς εὖρός τε, τὰ δ᾽ αὖ περιμήκεα πάντη
  πίμπλαντο πλήθους Θρινακίης πεδία.
ξανθοὶ δ᾽ αὖτ᾽ εἰς ἕν καὶ ποικίλοι ἀθροισθέντες
  ἵσταντ᾽ ἀμβολάδην ἐξ ἑνὸς ἀρχόμενοι
σχῆμα τελειοῦντες τὸ τρικράσπεδον οὔτε προσόντων
  ἀλλοχρόων ταύρων οὔτ᾽ ἐπιλειπομένων.
ταῦτα συνεξευρὼν καὶ ἐνὶ πραπίδεσσιν ἀθροίσας
  καὶ πληθέων ἀποδούς, ξεῖνε, τὰ πάντα μέτρα
ἔρχεο κυδιόων νικηφόρος ἴσθι τε πάντως
  κεκριμένος ταύτῃ γ᾽ ὄμπνιος ἐν σοφίῃ.

---

This English translation of Archimedes' Cattle Problem is by Ivor Thomas [10]. The Greek text was edited by Johan Ludwig Heiberg (1854-1928).

$$W = (1/2 + 1/3)B + Y \tag{1}$$

(the white bulls were equal to a half and a third of the black [bulls] together with the whole of the yellow [bulls]);

$$B = (1/4 + 1/5)D + Y \tag{2}$$

(the black [bulls] were equal to the fourth part of the dappled [bulls] and a fifth, together with, once more, the whole of the yellow [bulls]);

$$D = (1/6 + 1/7)W + Y \tag{3}$$

(the remaining bulls, the dappled, were equal to a sixth part of the white [bulls] and a seventh, together with all of the yellow [bulls]);

$$w = (1/3 + 1/4)(B + b) \tag{4}$$

(the white [cows] were precisely equal to the third part and a fourth of the whole herd of the black);

$$b = (1/4 + 1/5)(D + d) \tag{5}$$

(the black [cows] were equal to the fourth part once more of the dappled and with it a fifth part, when all, including the bulls, went to pasture together);

$$d = (1/5 + 1/6)(Y + y) \tag{6}$$

(the dappled [cows] in four parts [i.e., in totality] were equal in number to a fifth part and a sixth of the yellow herd);

$$y = (1/6 + 1/7)(W + w) \tag{7}$$

(the yellow [cows] were in number equal to a sixth part and a seventh of the white herd).

Equations (1–7) constitute a homogeneous linear system for $W$, $B$, $Y$, $D$, $w$, $b$, $y$, $d$. Using a symbolic algebra program (Maple$^{\mathrm{TM}}$, MatLab$^{\mathrm{TM}}$, Mathematica$^{\mathrm{TM}}$, etc.) it is easily determined that this system has infinitely many solutions given by

$$\begin{aligned} W &= 10{,}366{,}482k, \\ B &= 7{,}460{,}514k, \\ Y &= 4{,}149{,}387k, \\ D &= 7{,}358{,}060k, \\ w &= 7{,}206{,}360k, \\ b &= 4{,}893{,}246k, \\ y &= 5{,}439{,}213k, \\ d &= 3{,}515{,}820k, \end{aligned} \tag{8}$$

where $k$ is an arbitrary number and where the integers multiplying $k$ have no common divisor. There are thus infinitely many possible positive integer solutions, corresponding to $k = 1, 2, 3, \ldots$. The smallest positive integer solution arises when $k = 1$, and so is

$$W = 10{,}366{,}482 = \text{number of white bulls,}$$

$$B = 7{,}460{,}514 = \text{number of black bulls},$$
$$Y = 4{,}149{,}387 = \text{number of yellow bulls},$$
$$D = 7{,}358{,}060 = \text{number of dappled bulls},$$
$$w = 7{,}206{,}360 = \text{number of white cows},$$
$$b = 4{,}893{,}246 = \text{number of black cows},$$
$$y = 5{,}439{,}213 = \text{number of yellow cows},$$
$$d = 3{,}515{,}820 = \text{number of dappled cows}, \tag{9}$$

and the total number of cattle of the Sun is 50,389,082. If you could have gotten this far, then you "wouldst not be called unskilled or ignorant of numbers, but not yet shalt thou be numbered among the wise". Only by solving the second part of the problem can you be numbered among the wise.

## 4 The Second Part of the Cattle Problem

The second part of the cattle problem imposes two additional conditions that restrict the possible values of $k$ beyond $k = 1, 2, 3, \ldots$. The first additional condition states

> When the while bulls mingled their number with the black, they stood firm, equal in depth and breadth ...

The most direct interpretation of this condition is that

$$W + B = \text{a square number} \tag{10}$$

or

$$10{,}366{,}482k + 7{,}460{,}514k = \text{a square number} \tag{11}$$

or

$$17{,}826{,}996k = \text{a square number} \tag{12}$$

or

$$(2)(2)(3)(11)(29)(4657)k = \text{a square number}, \tag{13}$$

where, in the last equation, the number 17,826,996 has been expressed as a product of prime numbers. For the left-hand side of this equation to be a square number, it follows that $k$ must be of the form

$$k = (3)(11)(29)(4657)y^2 \tag{14}$$

or

$$k = 4,456,749y^2, \tag{15}$$

where $y$ is a positive integer.

The second additional condition, which further restricts the allowable value of $k$, states

> ... when the yellow and the dappled bulls were gathered into one herd they stood in such a manner that their number, beginning from one, grew slowly greater till it completed a triangular figure ...

This means that

$$Y + D = \text{a triangular number}, \tag{16}$$

where triangular numbers are numbers of the form

$$1 + 2 + 3 + 4 + 5 + \cdots + m, \tag{17}$$

where $m$ is some positive integer. By using the formula for the sum of the first $m$ integers, we can also characterize triangular numbers as those numbers of the form

$$m(m + 1)/2, \tag{18}$$

where $m$ is some positive integer. At this point we have

$$4{,}149{,}387k + 7{,}358{,}060k = m(m + 1)/2 \tag{19}$$

or

$$11{,}507{,}447k = m(m + 1)/2. \tag{20}$$

Using our previous condition for the allowable values of $k$, this becomes

$$(11{,}507{,}447)(4{,}456{,}749)y^2 = m(m + 1)/2 \tag{21}$$

or

$$(102{,}571{,}605{,}819{,}606)y^2 = m(m + 1). \tag{22}$$

The problem now is to find positive integers $m$ and $y$ that satisfy this last equation. Continuing, let us next set

$$x = 2m + 1. \tag{23}$$

Then Eq. (22) become

$$x^2 = (410{,}286{,}423{,}278{,}424)y^2 + 1. \tag{24}$$

Positive integer solutions to this equation for $x$ and $y$ then lead to positive integer values of $m$ and $y$ that satisfy Eq. (22) since $m = (x - 1)/2$. (Notice that $x$ must be an odd integer since the right-hand side of Eq. (24) is odd and only the square of an odd integer is odd.) In this way the solution of Archimedes' Cattle Problem can be reduced to finding positive integer solutions to Eq. (24).

An equation of the form

$$x^2 = dy^2 + 1, \tag{25}$$

where $d$ is a positive integer is known as a *Pell Equation* after the English mathematician John Pell (1610–1685). However, Pell himself had nothing to do with the equation that now bears his name. Equations of this form had been studied as early as the seventh century A.D. by Indian mathematicians and later by the English mathematician William Brouncker (1620–1684), whose results were mistakenly attributed to Pell by Leonhard Euler (1707–1783).

A theorem by the French mathematician Joseph Louis Lagrange (1726–1813) in 1773 states that the Pell Equation has infinitely many solutions as long as $d$ is not a perfect square (no solutions exist if $d$ is a perfect square as is fairly obvious). Since the value of $d$ in Eq. (24) is not a perfect square, we are guaranteed that Archimedes' Cattle Problem has infinite many solutions. We shall not delve further into the Pell Equation, but instead refer the reader to the excellent review article by Lenstra [5] on solution techniques for this equation.

Extending techniques previously developed for solving Pell Equations, Amthor [1] in 1880, was able to express the infinitely many solutions for the total number $T_j$ $(j = 1, 2, 3, \ldots)$ of Cattle of the Sun as

$$T_j = \frac{25194541}{184119152}(w^{4658j} - w^{-4658j})^2, \quad j = 1, 2, 3, \ldots, \tag{26}$$

where

$$w = 300426607914281713365\sqrt{609} + 84129507677858393258\sqrt{7766}. \tag{27}$$

He also found that the smallest value, for $j = 1$, results in a total number of cattle that is an integer with 206,545 digits beginning with the digits 776. Amthor's results can be thought of as solving the Cattle Problem in that they display a single computable mathematical expression for the smallest number of cattle. However, most people would consider the problem to be solved only if all 206,545 digits of the number of cattle were explicitly displayed.

Amthor's calculations were continued by an *ad hoc* group called the Hillsboro Mathematical Club (Hillsboro, Illinois, USA) in the years 1889 to 1893. The club's three members (Edmund Fish, Geo. H. Richards, and A.H. Bell) computed the first 31 digits and the last 12 digits of the smallest total number of cattle and found them to be

$$77602714064868182695302328332\underline{09}\ldots 719455081800. \tag{28}$$

However, the two underlined digits should be 13. Their results were published in 1895 by Bell [2].

From the amount of effort expended by the Hillsboro Mathematical Club to obtain a tiny fraction of the 206,545 digits of the total number of cattle, it began to appear it would be impossible to determine all of the digits. This was the sentiment expressed in the following letter to the editor of *The New York Times* (January 18, 1931, page 54):

**Fig. 5** IBM 7040 circa 1965.

> Since it has been calculated that it would take the work of a thousand men for a thousand years to determine the complete number [of cattle], it is obvious that the world will never have a complete solution.

However, such pre-computer-age thinking was considerably premature. Just 34 years later, in 1965, three researchers using an IBM 7040 computer (Figure 5) at the University of Waterloo (Waterloo, Ontario, Canada) announced a complete solution to the cattle problem [13]. Their calculations required 7 hours and 49 minutes of computing time and their printout was deposited in the Unpublished Mathematical Tables file of the above journal.

In 1981 Harry L. Nelson of the Lawrence Livermore National Laboratory (Livermore, California, USA) published the 47-page printout from a CRAY 1 computer (Figure 6) containing the 206,545 digits of the smallest possible value for the total number of cattle [7]. Nelson's computations were performed as part of the testing and validation of the laboratory's newly delivered Cray 1. The computations, together with extensive checking, took about ten minutes. In addition to the smallest solution, five additional solutions were found to further test the computer, the largest containing more than a million digits.

In 1998, Ilan Vardi of Occidental College (Los Angeles, California, USA) developed simple explicit formulas to generate solutions to the cattle problem [11]. In particular, he derived the result that the smallest value for the total number of cattle can be written as

$$\left\lceil \frac{25194541}{184119152} \left\lfloor 109931986732829734979866232821433549010880 49\right.\right.$$

**Fig. 6** Cray 1 circa 1976.

$$+\ 5054948523431503307447781973554040898634 0\sqrt{4729494}\,]^{4658}\Big\rceil, \quad (29)$$

where $\lceil x \rceil$ denotes the smallest integer greater than or equal to $x$.

Another approach to this problem was taken by Antti Nygrén of the University of Oulu (Linnanmaa, Oulu, Finland) in 2001 [8]. His approach determined the smallest total number of cattle, $T$, through the following pair of formulas:

$$\begin{bmatrix} u \\ v \end{bmatrix} = \begin{bmatrix} 109931986732829734979866232821433543901088049 & 392567302329690546856394748066206816187916440 \\ 30784636507697855142356992218944109072681060 & 109931986732829734979866232821433543901088049 \end{bmatrix}^{1164} \begin{bmatrix} 300426607914281713365 \\ 84129507677858393258 \end{bmatrix}. \quad (30)$$

$$T = \frac{48222351474}{4657}(uv)^2 \quad (31)$$

These formulas involve only integer arithmetic and can be evaluated on a personal computer in seconds (e.g., five seconds on a Pentium II notebook computer running Maple™ or Mathematica™). Below is a Maple program that implements Nygrén's algorithm and also determines the numbers of the eight different types of cattle:

```
with(linalg):

r:= array(1..2):
r[1]:= 300426607914281713365:
r[2]:=  84129507677858393258:
```

```
L:= array(1..2, 1..2):
L[1,1]:= 109931986732829734979866232821433543901088049:
L[1,2]:= 392567302329690546856394748066206816187916440:
L[2,1]:=  30784636507697855142356992218944109072681060:
L[2,2]:= L[1,1]:

L2   := multiply(L,    L):
L4   := multiply(L2,   L2):
L8   := multiply(L4,   L4):
L16  := multiply(L8,   L8):
L32  := multiply(L16,  L16):
L64  := multiply(L32,  L32):
L128 := multiply(L64,  L64):
L256 := multiply(L128,L128):
L512 := multiply(L256,L256):

r4   := multiply(L4,   r):
r12  := multiply(L8,   r4):
r140 := multiply(L128,r12):
r652 := multiply(L512,r140):
r1164:= multiply(L512,r652):

b := 3*11*29*(r1164[1]*r1164[2])^2:
c := b/4657:

whitebulls   := 2*3*7*53*b:
blackbulls   :=   2*9*89*b:
yellowbulls  :=    11*81*b:
dappledbulls:=    4*5*79*b:

whitecows    := 8*3*5*7*23*373*c:
blackcows    :=   2*9*17*15991*c:
yellowcows   :=      9*13*46489*c:
dappledcows  := 4*3*5*7*11*761*c:

totalbulls:=whitebulls+blackbulls+yellowbulls+dappledbulls:
totalcows :=whitecows +blackcows +yellowcows +dappledcows:
totalcattle := totalbulls + totalcows;
```

A contemporary mathematician once remarked that he did not know of what use a printout of all 206,545 digits of the smallest number of the Cattle of the Sun would be except to use as wallpaper. For those readers so interested, the Appendix contains that wallpaper.

# Appendix: The Smallest Number of Cattle of the Sun

77602714064868182695302328332138886664232322405923376103150619226903215930614[illegible]
55125464772584546510461602748276908192273273239624708376752171812383319307l0[illegible]
99039295240867535897511016330381995952286224898977476794934777588622737237 46[illegible]
82067522627938255320483533177453607881941951001290253537890794307708022239 04[illegible]
99773410357848770231421207023187305429219598310950037546619359116492266575 52[illegible]
83880447339843717996563382163956150672906821404597658562589543801065495963 89[illegible]
27958797951535461199202162299669681693138757515962720759360360577327122115 28[illegible]
84589167056522577623089709143598292275630551090992443744816361683463507051 78[illegible]
32101054431032936246524861777259088965968311358436670289117331111854918465 94[illegible]
14057457397137917373488794131531002376982836811331378068305363645150725055 03[illegible]
15372654601907558190199832272382845825509194548808552246369616821866714195 65[illegible]
98517305450274388722554847748705259981747565696428691310422413593026936386 51[illegible]
[illegible]

[illegible]

[illegible]

[illegible]

7309363480924247073278017070116378561400941086448279084288671110976929290665177523699495319470693188192771560841067125420395728333885097518454350057224496728178686844
4215921228276686546330981177587591368962914770164778560927268186053572811495775707587277035285440047313586267057669888848196406100450101411574122728576918473083380310
1809543573764185815297088409434904500921446831667339604657505835421272107811458861665154370103378592723916557339398320700327699414993844392031860060938233496686393986
1614209373166697405415848368548793867482069275102994469158522375374267280966019619082001798076888298914676756744598816672020126027711123062409434289610276898004770354
7104545545610286711723083671050210523104098633008838079579030510715521879831978658881908994943876091686653895533749774819722521430770877878782947545726441329340539461
0319389266375745498062915002875086759394284300465514141150026582758510587056306073444898107248803015989364326644704905368061797650775657080399548280609839298158185676
6318073143981249983173933630849732622037298543715147548862372448774368299368180928357353819664206066091083356050569135617943245114611698133203406220944302494710913814
7646022664490910038955287196497711552012429864950343434835052329877584503054676632309486180762934599955892044780799474888102917360686910929020550954460119330434509986
7993785388570814679211382451485214664684105818228226293715439526696784967961277101718343556482869129308397888114144071815566098076273987630408213613166009631191190304
9251074279850330793645838779042954420031484156964219550942409801058240597049704774078541644462639758839798413648556259287875514856769727834235681146932587601365996054
4588920231412117801087807184423502639084969965445156368386489343573165500876806423833717749949493908545084579830111353977066517273445006173647718477901511302176904085
9401769011614923418924562998200388050441081482765872461400385320771391948445537335252113288750208758758586451439166848490316944445380647223977059557893848775065281700
4209011059882095077187956040317189907322898174749825652843126331426195425317191969440425175981051788511966804763680300858844804355910308848098674621696996304450320663
4800172963085486243046347556143382390410600787514159099996924517155249729967057991709636428927046800480111071869801817438525170940919067591438765433529463150955782806 1
9729379581961405569582268483605556545372847334754568707498178339591608222887858097491977588994351086015841634737960107030625666653514323998144373846300612174033780847
5866630962306107005924062410246838132346754674186314002332327679025186745768645770993755708953606825405443268107529092825914528169426410399162477288302434326537833101
9414396110967350891424064941315479081249289754188079422394278273188659561489741354356811137496713352235127664679575240486001292811620550496417187120198500915823385948
0725164004592474003938105546226024496785433975533820461149775052441516160854895351974741630827018105667917410551252600377906997526729015118020002484590219263389484545
8926163234569520741714117690888141996243553851147432510339420983018401551608926224786905202685097626897831236585193553592718995338526125892821006294556303467257481061
8642608530379581928295494832099230060362791513859278117734025805417096076272759934072108061094702930021741821768179471177081285107518900466736853748430623201152888890
1304995141740340126820654976800229060206079945306548588687939877420319949778938367326371640340872841197615178312184353532520510071419812546839681834839328213443202784
7056195803416025092969096520668571129905979631821706593849310079180024203917015760329478067682337475550792550931240890034026798727822295587106156686075811309680928999
3746283309849708657587192501891693849182549596395393902353558370110527917121661658581546928682710341364324662869938908368242444195012537577138816916308664554867947500
2268931196510222826885714200634545108502081795737069166243629497227429660575996058645151044083411586486114628235694645836280964079634589000848960684108990468394664830
0642658267520897078663576508680559964859339618224019727756417849736149392366612784525701347412372355740183543258904415748294337270864071635979461493901722831766840414
8810980021205856274454115485099644000217827665117205722201320562444189671349188906731685068711904269862022385870489265529247494205910701191254184861618121233424234363
8459391610698194884043670636196841358906421023134979662598397218565867876542675045611154912376701435863910354690295833566615316402932527576563045640249895066098969930
9005335905173445653297619276072913580544452220888569115221210047881005162893838069832479124562584599800873096592700238726897129223987879108159087812517366022226969995
4574985181190029047874463090595353018890354421744953719840890989116295284802727674038272362556208278939078303901704693365861946157571265392431945386035344862298176507
5087436182425021103951026919609178645862967292922597790568446777252330821965114333394501054807686632950468748664976982464601728337656940341948851214701263904793615604
5016615849814825983286679192609236046553622239611588272412252284936021561781650952910706317806688341999062833041914179049235596741025467741484627786023287960537601180
1037449373149216475842303676733593386866980517955358503575938906949275917892544363199504828586621299189121168327467006567289252905361326298488139385256809872269337474
4079555451234739623476151263511813351505754415686608073731109892277449169411925908727975356021101851655476696341891872878809491228752741899130877031822840914914086173
7420977734577995391835640492636643365997241762966414834181493982609440717788991935617556904400861647242352182967049814491862527417422389059970947932902655182582662222
6049333223172923009433420657362784971287885092786359967942380169180322975498472845137930079287334981012219235057490703213707901781654977810943149417307385761632363618
9505834859785176854430128921076436941437001210224004379769286487014857869366688769275946183775065167998441617954451321455651848399158360529412147635794598916989531828
0362725762729108602055129669764341987866060859762784821456579734623308534673831041418226896438587570996993484070498530939569691545039930214737090428206495969180106731
8013662521934851729422306615090691303911516975699220526853930506845744481741726579598372022969340124862115567774691923087680405184395425715996955153534837250098251258
5111526130194300587940900659030779703611933522099026473868960031492414907565173354654571590298800365394174007435342399881868591169119703703538274943460557961137180557
8113358874974232634256827303637166503039185354590181741322364660273462784190108331821517987496769558830691062685913276164383760430458945157216941290272584238152565285
7092654521969692917375620248950010705318245745325964023014323291764548975390423635599737528653996564560543794108506389902423755827929025874506402257565559483156057910
9245143676128733235902904871820251477392340281207626635547854360523012179998044922525393686426717637377955245414086299134354976907792244988560203096670552736228385799
9197751248118520857186743447470225689774352017263081182210356639676220058586649069412253246792574070981021687735232821759963269856700967757732640119969606383537891981
7245156535852646780602659205885782703616960455528822972577061400727727872688412310914076193017320008560632337868174457774735926491615899096068342381192330091339039176
0131730624911072692772451621148410990455719064516200996905566554515302889224151857241090595403230356775720904248976981763710675527416359141268711860967166309849714753
5919173195080715831977064791653698214753961332775097402480425940583615082733331615220116470750149490824181676058549200310042516243004404856894353500626745661885351189
5872565897774570454994813278036049978417567709894811823853799429568079188292798014684179015192891412626438333443950414269376952736312159290403939353657907966782900902
5188345648961779478824772280637167801885961412845988773302668947467000445654380530268588405761385628319969129466539702526398964208353863432960828852702554834422648336
4996312214719057167141676785990100835457961019278440125338522985597372123733833398106835983624668294321521187922072646127003763554189589549996146434476961886904761645
6650357227385103671854428624084618307922909104992781531976088308492448159272817591704301624011443850720225101347034498699720875482703577146943585633501744733090919021
5867777736258952694352527754130531189663560805031090491163385962520963415088964088392345007322422890510143763872097316750978174252975733083586084925531383210427386502
3399232125747142636464417878356229683676202626232713408528697264635518348177939896034511039334701693073403121116567915542993154971168104719976137629198049047474672092 4
7371988556946711424400399505632229511746728573428147505365009457993383058968384352090441788591166147845719901992197052750235671885904498332475390284792725079647658010
2878972282552747904133534113628482585984976011752794606913159664889250838632971886854118939510013875927461048314512596160338641128624572257626460115415819239792662265
9737532543598396693620339869018632306678637662729140230790074468261445605716422161777056062357044408035684116607439713524677548295298975006238375565914859762669879836
7335257316967448064655366633962287868084784202346499700855776783264155588671571751334073652813227700958860969217468675037732074454148958999172472971147163027830242048
4246696527762614321811220610655621074494685693288060831445279637943727137582873212115611552765389157892839595716329301870140440727509999496327817818813733096051054138 7
8359801070658611742304940944621141986306613861743741080935353214841278754282791181656051737182787623603631515077212766936095938809760228259970830807355717621842510066
0057606542728878935066455695750496249206854358662657685201121265493187349187150133186906618013362495749187945337816397460125089971945270330484340971231099224822316967
5567554492222761835752562300223796476412920108386944628993422523497243288718270175080091920617334259848921363341190204819299473921654857306119489419083713595967210810
1135180872491064208300484367830245578829000049341941252579331450460750915721976950385802071438456521024942495794924194437133225349102125787625407224015520770784440898
3516880014216576670654821912961242992434220524518112037772173037832778824504723048712532753688130170524879899768736426783794774680846056518142073552538310654513346138
6846353532912548099099471905505207912101590150323418559196393501147281483825352027867904891109663979563200523517724353814264681186435188427210453720084210781577128487
4169419107069749113384469367493157382709025492105477098636888042190141922897971778810219460352202358954355756439400457635535885032947814547404318232396356543941901143
4231853294216021381397100095761961185100241753007490841167348072812982589877469762400566438033235988073329330146402544302620403691025252342789492102079924783698717810
0681356373629636921435136526523908347314719449156199474647405344508207352185707896988780960301425573100580522245188344975346059756423423929738069681514095120422727903
1601466147951077353136710287355031396200788283571430296491226284238121216948652466532986002347801052169642510365840474328769922582186635331797548459917903460475838847
1388036312608243892514896760185589074420156730606989743232221009432690060753724807417493009142002040332860290228145644851963467302670708536083623818723537593552287296
3399399700351171871682434935435455560247782237080962409180314336823801645187183956416693115622752816261937371087939957021033165770206310194425994318279199494813455309
4096939658322448307067980048204203829901337986643133770982323016606699968268296732992273644255573682453199407083511202672021162207651583087790513285004366964522492478
3121324591377938753048332350774156148465719030716479682201215375112382320889604672804766926781641565631125928535765389490815839194357829952691284977330115768490581137
4147593331382946646878538410456337421394099371551637039586414794033567687217739463990942186289740155931171049664040960294287233853784726331233458893388866126354160770
0486494548164673269460789968684879339542236298032371768466333741152233893965465096510346722177200240294019221461785055042620768442880103881250619662528507160997579142
2499764547976316851281683401152512345610710316388745584295969467724412447833694692645587494933966576242426607945635171621016049165265339313845467242552833518060424139
5746200835433349945010138687406951769867272490201025390762643333462838586328985066748033209152386646525309942231463454304906964632165539363064524638715109962552380697
7881136086355960278707837351217951147066644582459282808693915901611656726174470070237739682319866448624708478866901471817169380495248681991645946976439273295474782806
3796363195280698965208393866536077323008025228304959889424492630319957541992282636793099578970582989892023863326847010625968993215837751508647872531701723912216630684
4788881466708591279543206861030054588445092550551431933228988907438748134027070577095783284159932760283459155559247120809304173105685126216194816607818553890637702787
2177033987967046574842528248364975866316410489372756135137860166887809197163609762058178425956059255573031992167281332514142732176374774919397465596488434483910518486
7913177845638510735707746922903768551962917767105786204560282707222114097656831486323576725497373957481486110294699569434178086535339683429992661621458589023937584854
3089983173738661619721052421200268836610193609138932941022012819922248891277137173384465633667352388642519965884056361821915797104626310192168798765339316032509533511
8599633983144720312829197719161319817604014480614678586054862953153255417709218679686384909992855224958928164753033614360200418523580522417773236151911015010405216417 5
4493106032582001571680349281705125835241762914248176991757967358506724232641633327879201869923169632620469986443917201485236311810711367893040621444029710322455590596
1069728115099762198986831461811030051061747840518012974009698656089588694714824555917184983943394565596951196150084703929666001303469691349975531862213713992660079164
6936061606900329291177854449133572240134166019178456000591275121967282493316806168845542038243202741789497574660040212879210664213786374212532320372548370427533171466
6811378954534040120388393302339548658417020545580258952825245378485566525853926176094242443774267268241227823362022269990941000094946460824765642712718768037074275747
4493528483303268721114158888222542433865193505443267460784637791926268171929836230705898844594503897705870957022849849233358632809884254554745277113891297466339729987
3330075955853924503701505757903556401491735487512423393809797955919791730605200692450325645936844158195369998540892230936190078678957943481516844468446636619726171663
5566506005036198416250941312315214935316654075975143729785210564646320249969384881468074411245454874692466692898469836625451891073042131600441069375209061866049822569
3048652786797371454671347176596831807597506313576071879203558433515065466770541161865237600647825195688420586643301459547288554380976508736098117002281812365629161724
0681172840373558979029614449261973754828930269064359712996474368923145480284359367002882007378473155192973554478534123860842876633918066358511195267593851875321642232
9530484350139080896481823043835388206000677071383584876892373574602716248300446178771007342358836242372653823937355660487425343857598150352822229605711926160303263978
3601836836662647985805148741750305266026552177262498752258833980748574128465161082389796142001216487217062060557708554330826294533628639705124667631688116848965442244
2806550975683953090968749021033611304323539288978850508012146940223122476633398924358625365931959394924406591920632459108981331655659737431366860091162722320505765177
3311292377566550774158387975715234656540920914179549728143284017339858992083268719972712119587718578063568483681412656804494669199645898220822966833868179855378801 93
7103072506237655433183313490360438874195269158119314969035085824195755328528906463405006933204830712817332389275916212062462550455235349027231055828191461732999803235
4717663491832955396518747219114847079792647960758876914316262687209225031611037618672605014761434914104022172531477398687024845610361750941220120210161974294170975391
2250492409261917885936467326806793255079560329694022672765990935215978672598622393717970769863446027869870261335820494303992100126424253254821224231997944416088742037
3722660137857787743445638622684396817886145935303346926685956565144234749433674463760476778399359056106898233581736089707404584456724827242404703364823564362112916503
5826399659948905098254728481999868819680154665585273210826953468132168582771755172656734164466242070618459249249087594496154988708939310445636911158231772244760463623
3291036452564922932398381733175579185131834894279687999436205732786719961348898615703747595701064757581363320116363643788549560013456073880592604129347218549361624754
1000283280757991979648586555400698934826636509191785843799906518855778687056392722435384019978678971828058293079651885198481949204989643372270988125124430568631533437
2720048275968779080856823785135693260021822290472964158666782963431706619543305891400994438978911864888038024864092600549480245802211318031235072801071589183815654211
2189350237863908652261531120876990663698156848652351335356070395172966277068106660178148598263485605583615481397895897293816958308668207014708884731236506694435789617
1513806827654326821964737721666215964164918840549682245283520186228696123663970016383282191430515999821617661679545311371626017755539594828273761001773687547191343596
2599474196864074234796617260418421879801908184645269667804481259191543442383139991707494010970250507570955506296102530994019903286159210503320223335548311956959644375
7036985286439567411243545092940096950894342949454455388071190805387491689432901555992253557278657946229853358348330556522445627927823915471620493065939087124121370244
2770231276370745100291321092039229224674474029359608729255502745549938265802413343581534364574406474464544708377367613334074495411348754898533702270525273709539401396
1704206743049218678357419994861274784152371959835789875748951793984359708560372964454378060226182973068310589062137455480819605037983240852078170744949044163125140554
6894432830308137448975310575051669499188593654816764357147260956764295201805122932382957375152010563882283749362642016362463192295017881374344595517752025837546535367
5213460376727166063314006484331383635976489751370445597359609958135713819895448497946781379602883523921166885021029249548951202149553009914644879987090251468427411680
7869782002178917527564608031549327736492392064620942948749225135868600768394642434162647088722720623533063607807053448389957162049348664954273284123903446546947519133
0073982136236223313036259124465084124464394328132202877794333895434583837711510783535297094052725139230625772620987696808441651200898903287694619608079502599632681792
4285184156949342908196323966354873083622374120644839811461077348932233143345487595029322440839692070560895968944381957071471861729706369752991173589822607670059047824
7356154559754066205414266450539679391883371104639699263905157506104235046603428232790478033829402448819260182582984923566628095841256882859765434296296208683904445925
2669948276251989832552243397062406303014887520385595842180078875853195499173976685744541448814735713322753044021949715254116349123375336767795974847547673287160661589
3503566902718537761033757677273909480201035153834340428361840367321248497314167413631957630788759765421065696093113061593037184512931144559862125406735786157433728545
1830401487395706115981155200756977164164297415787548498167993830572109128449347921989826754854527236637201644257970465504960904435632836146332319907267212702463516874
8889343932396472967503557959987664163073889788319827028731758938800107293817265064839324579210098923806050573476613741419396405285996326580550020722292255112602581646
1431201670711890454363426595828925273451554765368236496558636867850336893008241892514009072076927712897784866616724015618057539485275558400853045796018149086414987179 9
8118777745144324242186428836068133109139243742683821093373673314376851510009307539165827533964397729810051973357618595746546133883463143991549121884684085294519910198
1925726777659179113619754576259379452848552801644235982333589367562413948612810249616453020118241899062487553587182032715282965947439730354704809474508763261983264097
8488405633138736818883935120311243291861506071784645039142322499201537690291411176936396841699944918482052619605557349596190524415897247992631676837833945141532755179
1720469924118711335743422315605006520028501623959096147574510438559038321722268599476527257493865687848763555822764924455319975289302888282438460385981581748587862173
4124989860555417080474753668703938361173625431665192168710228326324159969823958996273160380567175591484020920766605588584880646564149273350603112687764958046514925084

9716343202497087237983357202737209812732820387211588922205124899347242323484940347234534023319522455888327972898096798557245421455940035213527192343668612091014929838
0242214052686874774150591876503618916893552999396936934541934957790616666640139413978413914486325759573440320617296177742164444715635909548297738007701984501490647986
2166458059061276721842052207252254824870482131766119186413511466074804483627499575548832942809437562160592096653318117656818526273406809792580323860335807381569988256
5801155519687905823605922440102354425351530265859230269093916099829565562556501070663726577913641619323852641678092047455499060073073580752084335593668537404546801697
8377184843957776122862454647316089635781456369009866209726189436680528069668451874809056202195601061880866652047245497941484518889022513320627070770924713104654322209
3421799333824687809990620212531390263292314879255578854608541123229154680993959086987798866190773116475982199981125354035159630614387232409465427560573556234316297045
2145892217880651925001329045805383458201852621365096059215981400859368936556145084092243479419140004475514642981324766796896016052436313749314506705789271696349016538
1042071312619840322655784379945641377334294662621304323982105294727463170514770426459336757651493681252008172342841284655797022316984212722925534917940739517263065249
2752489055341684201697234188165273742045700790175214222442842046047069938931821253201724296767450897659932885921302426532235725109574077569052504178064373491994959502
9167064842299393947716586568440048265296198680791959227937667820904760455773441321129166295801829328223611215052747916244084306169021914061828094997966767100813532115
6575547729323526177750352980136679198630924262382919974157133877629743221862506622499601214845053122987699646395926726442528265485194137982164839544642407547757414995
9119061268627141363434005068592407672519774635787249529434023327087917217677047739187921224482427592707977399474503778423093711898632804830514526905497315092498661064
4601521269789614182189534193253526912717946345861129954430757068297470226099038950604760445555935879287228945333419543992242392387576273506528212621878691195016675079
2090833207062605621952865662097856638680955109794120394207264202413298042099856825877657772257976951387351876861509378595922244944404983206375691513971354293913427771
5049951278292021502570193648026844359937704055085248494777163479615983746266688253963886767442497221153551655767109760912657516229186469696059031036132379741789094170
6980211848714448329908024747106282187543803578364562787170915460145083872172275087904846064431837601409358394381921303626956117544515133353532685640222277316171148753
6542352142254491007200098646623857993735439153808590875796766554854310104691707005223676030170457365150823455333109185693795433068082167377666760982113417712640132921
1982681511011644157827429690540488050014205243650790931334748977803927372465590390027106683743587032371595790145387430007689855913736210156458474636216581956377900020
3371796281301094151616570940677023286878051269096694017070121024140263574601155283122503411253812511937202889756885785201607230660351943870593832805541977541235730104
2659418987631574948213817461186586302606295406386102050956300514965443022772734021179471811334412872860144717280291169861677668637929990755453781254852273268268644253
8539383053280254143484891789864790776544613656951811586740823396267344784408142104305187839759644880552713843996699337116897793484762215103965190312583751135486071326
1393375038290870753805564188125822711077788350751878556111919951400071837167651783958756029905620005623612293772507108387090022608234053572748625998652796166049743730
5816311381220367081222531483347455785407456042665328958666397937642748744231330082218229896541270641473471125497016224598166453881685228212727225117568404582429293281
8780950311367418724915045105995761823631193477618411235219729555140874204773288243766594654484796273764481911246847864447771629952207976280828377608192428505719517862
9525907989363252742423773851059420220911294797848952815979641484235812367964287064741320601313826347347678514466065240359479491574963033781428501435909811158038135423
7319125295382145419426072209356525332249358701213388962290523592241641194449146932159174945713052327005359229813988428923701368935017044010939106966808801502181252392
6461300784538940485578925347775417548373899332444815032079918552790001252761659819360360875927068109202093050375253498455188124726591933951761710907588701515406318767
6931883217111505938175092380941507491187904033757272580391207452269109181153153792222401806767862030945068723869602180725400947250376858678529519236743039622641328403
6886691837083754432110383842680408105927521664793191513970063305705036918185647711591607863701943669267466038710080953736479001408392853169685813376071222747826690884
8768317266489360136764107597135460010800016128858877115766877062055173700526833578721868930977223203441090642379792249498131442747690257812868854606127230823650976092
0798306498975123523261352553009953443137314232573874062519375219725612515949941493101555671120310725988077693139826448083935759427512009969438644198989131077128678065
8915868210616160036869043446898267232732595622222865815633834630953087541651695986854202233184896760193773889351412110209956642100852208646342959829784244855474692742
4031061755687250602309113919435748992832096651897311571302703276216645223309188654220908664980775524464006035182687316535462167453902528751158483730276940286381647802
4292864296919645922450462511960191824384379268101234080201206233836373259380691670933663076296002817185177545359668721296225810948134106908142815589556269778698814910
5970073075226465782954369790946748966976907018173139527016084213837370183573299217987382150329482367079880435373027689024935437005969621545782871247586565233158783148
1244353253064945058756743755086392534030996106729293498723323723726042620881470624125478543940264458563299977014864908160809224568218977264521540774614659686528402651
8162252492506835151879620736280992957153145963983998533765079232189261101734860242852105193548394536668796239556496352130522296133422394454952795017555651846153848852 6
4439921509646106328621062610499970352106932336889481305327090494805419763491408186055983091627549598085292228124093378791084655712358787571696187096939996461437659320 9
8731659223004943174161603088317407043136295813735654627779720504776884442459537804152590957540105885136817848557374955026998536404666520247864404945365148585225013355
8630079627981018998913732736967490098881605440895844359437254159106344124977668893384732037725707986780524098241781573139986061985635386615082694401569084137256224065
5531249708582222817366126365358918057770617384599999257169568091804655817928915178039109168808158922399868102637568558613465907721947708440166902114851169779874227565
0891016534497767616145160431430844101630370596710520338019792511035283174949992463521685473243361814255764972798892925331454149153778768344926398974563155638976115131
6787663879534190928819898980490519539168027409505061627942140521936403390721998851773070001927689907769213699537824138874736814419119631136817671600049176496036453797
1216748513757516199559798870321179568851832964169941537564234759700188427454354215672757767341716204161367280564625610244346250302959018791052294785029788316033151945
8227049027692508732623390805803118750171474257098722375523629664739539075221685573867903625485263140673403925790949918003543124370719711862092293361909404668851659148
2872286660170901857260953531390466318234816112607160855036117115046548770808178761700610421731633294087651549068421255130372314363128783189530420969393815538602050204
6565417705109052507817910915794121541764420586075949566353571656857399131054705092686310424483158337291110742838411962764365106097178509956722380245811897834978471429
8597768036846339920043860261724693714656214349290334401851204682026349469793728287090976720493135851887050741079037772494400551363719124348802034231507923960648522403
7925093073096172624786027966799268459139411711471847482814872871839281910796934956036054886956614049977446188674256392618422821996973447152440462904904127726039564709
0157010725153590739167076180862402841724039069635665780575274981891894056403968846683218871683522665283547473407286726592204028740621053664461758054329592146998068016
9634652745822082088777854373132854332883199122236036868462225170556711998817742761100112214922884472915659119917167266496974717972832249053083679395988462268360672523
0861380652884137523847827874080007862180448763049373200429241993299542803417347449520288769640331236907548497032369173055587130512677507046354557459587303440611658535
3944302691351417351925079209791860375494026523681436174951383613552363779675568340543034836263425861615684838738859804018616973060856000044928239410199022452531962463
4231554896027938141907218040792551994618246644397513280290643542062863525173064699895685370351118620166723655255396327191355992237604765553802633287924238455591087363
2617871251636523964519146931838620122417319810156055202301863806552439564219901415080399348700897097227898494219162345172600475850919132621686601766917705518919291677
0036866361768110241862506297324328401708610859589077446222495034186124260786653486907829774413593119664273124046591324014686076688139328507922337332379972794641971103
5278737100377894656541464325580492999864612188498332256073650951204820110934467669372673273254869840644786831962178258799805811365619773390943028043787223670917256784
1961569529167189371725998051744508462130738096467487368899492571318191655963912436830466119403034663467376608569336760248392813193123738391011348984463071070195634 68
7330553008786898663009608196239520175675098474563815696921072818908927173850344959620600339361443925563928867908839653914600477041297480175908814847330298710850755160
5922636997645635543788602253719968270991341773647668674910670931138826454560270758667520944764076322732723246506789643733490582684538238353224316067229065641157275408
9787951654280548628523442696305384438505979122243071410836052645038164161643361575354862000057816213555283188284251273359862071755258599390667607383193395080012915828
3358115523406127292817708970495329780190956700336810105977719056557921617147113519805387195722532720948139230222329412703805388281786359044813571704360381330836693182
0881603414253593787241273103331284771752416818692084678054730587740855025991369578521626084034679701638413007124742630883118826564147355081790077225311985921987724586
1708331499554610285997874824427581960726136924612644926750683780860326588670572685849822387456578758804435709811877584773039456052844210991716318110931404700349014339
2972704227912865351782480247831095003422040513802469369295358802786181314028688955776295259297301766146127921622044050686799956308096973384147041204114112699011233461
0681751694542662543467772569468404436664492049881851622658750955946247525819316793721155735022028885345015949013990705543929000442242530890985083118181181137524772681
6761405623833602129093213773793387720688400538384127078974905669418875026829361462097580664610042841676679104391169289344306301461599995552332128592639857741903817333
8576721701601082887433041504788012675826695839281659205135557341737718678001418087581689260336303429798031141705620726157645559588814921766735084093945000936997223373
7756011780877836358775814793451386234851031665092486655231872890264142200002819719406801293244124752819531275646114243155780818751002998794117976435862150771385951335
8260946686030194084924885174712525797470035413608906938970433884949940509929520605482472728482092151989687088598779844294761419148613950962979721152143281675100504622
8901566202693851783214815057218120552890235673288507346736491019763819070228351588798500540797405885752271779927542249443865749645232409878934126354313941244329061601
1611089402819833615004632394265397697204244594594227811607512454426297884530453240578109575711458564217220019488269980412733284140800367127923996093266679798440289457
9179208045368772570902843844185109915842606810621413919725253383728790536516205431938669880825916679239367728284843256934869415591796252149592353820221164611835866751
0868661778144756360196069083005132139294434810998125774832134038319115533171380254354664019739863029563063137078655402613457462102278149153550753093576579858486362307
5521103988809785209212718182193672599126283336291615930305990908776240395493645196629750119858236139547648326907977948017944239515329107555849696008378429262313181564
7995199689124911228644297506395261340583445561906634028722402662679606069484979491339312083423577257388044053899797674417523971354599254481464313906796157289974788357
3936779276708012016499608813372807033302732918693187385866055933838742244129262363495107315743980429492822441571402411473411153232890495056624466106071217572925585781
5524928165557198290912180989431404520393607177388116280536136169721488360496351865830625717535777288288360865599375651914588917724609984056565493028887222898381814107
7344152543809734211714444206381524798849479350947251228169486111369153857538926122267991055149282914328708644561647075146193771248586643621375863520142376353234060600
6843388427539827986111797420503623395516775673127780889376629299999607284547871945966901035912595617762689836276320202569846595879171082408671719567216078580156235230
8282898520591352591723803419024543994491379631278273178180721657922183302194365267074090178813327483865696568643955338880251485929247705700011051761055695969653877602
3024969563932311811236834563257258519748977022888988436898181783809060178338816535819836449348085410434688642495998922940508355704094513131677434592290805630663831827
2521674610307227609603164186641510760401858501428245422289391770969583775463429417445763809505129367563549914053117659112699461554323596556162666292944167692717829989 8
9260783200847113255588834596474098097471986515845639195559508609618231596345978409102823999934381022420241139765616980800723765218244281650793939389771091425850401540
9581194879331166158291817205402038689932134672445480258014430241316259000348366341519586168603063751618547150731590659947183200749103046892612705811356920655717867249
1797580361570181968364546635900792051566532436908196021823926235326154426989394513462573276692147100032710849528739382440330674971994376352775855607828238891786648080
2682181430164159558772504142229515067083381813586583725614820249515925835358779726881755000524413799981647109248384278501829877628992029568288532505769371835289824669
2067968914997585199819929538670719759234501559979261945863745418238980192110915130777517601974503105743344845076307064988093875494372338077018585139445639543058161786
9972846187032930808923469608331873044011841161957660588676639967868173690039409478300337427271589492209843772153998861786718131590880305823996748666726011893065994880
7815379012061344451866508681638045321003338133385846997125140520818509539719658147322305882374809007969475921002683209568812887919680132674987921548896785240986696186
4885521304304735787870182525789369663856037448717270868641639592319489776007323120132530664919573077928180255680085881246199022331843913588365728067281353822144166702
7686325074871063170331935598483054268854551642161297380097303532802949722988970029134955600070109029307334699074309084450927774726925761109304336513224272201931483619
3306306498344528661702871530559309968464797086864634435710153190618588387086031871808117085025384169329347367224113204264608443218911116846967522007327573453782653821
9570932445898730308533923157590928362943681065821180114976003744406378783862991830394035397413041453270366962179104561184969423727258630167440152296517464239118400411
4616815797938909719309037424441475160948932563254771288731198385830844084662459889702700394244149036950900402603984883621044525056400215080214311645301548033569580222
8299929291395494633463501317988099117285861301066442929124103661174898891467732746433610878691090600085643368640808386251573790612182455145940488345582221344214021750
7135295927443410818538154819013025334538822633907443644894850932271126234134486158703542674128694840671965858646052364171295042559282850627951664267666387671676922016 3
2937283605650407773087124321618916929539622781440369951163172524636242975932032131818475550123447624265272158066458537043854008483266663992212939521239696760936219360
9310105803842370396788047509510074628669379319977998600528281314858339133902829244301520587632728188526174382801268048864745126638097550518894404666916003462125115126
4799793063368621287032854362889026473985188456896437517950583305485435622621229466227649612159608957879237096251652672789004929851582025292301112258481767657228073599
2902807034709969426929169281100703751748907383012556023154948723089802924435696236852334894664340247887487554029839059395154478796644334622308409136285132461247191390
8160682333188386484550723446555829582956343933641965321140329211825541599500399256023235108886404179904243773980940506297031782999709739429434650545333284428817903553
2368370906146803732215717415208941288497530872529591857083967490999262691036699915424622863333704791160895152509610096278248359236307710213299951635734926189873629765
0804732017213475771736439158127247354269854020866490401701570219774587126169323070925564678983205975495565472601565746075116988349068042450663006735037990514195579 82
3336698393060473661631555945988794128474709267553980532634142194920153964822808340716723131929781858738919452839764790814785436561686502623427960421854620496381349761
0388052116417659381199160987663713750378825967271516821117877369705705830423792171576250641625705125907185745803174805487199064758958392389232002699711738098304591356
7041424998723947651482076716866517578620004289387620213140717813818055314444736881643859993098375555376662196316180680015599508450440019760773990505949198588116032257
4883477738750663475843107018067258805574957577174966885174186985797913779054069884694751665091549103522263685637897542239164435656166969487733963381128918063445643578
1833093512078562615586929024620741924923490385784072790067326853216016199235387763674671841135957596250465468837973061215181538114016973585363940900980400181614155724
9635968270944896359674029387112306169312920286631478268728312702614139614311383819828082305946956539233505299650797219769728074512011953407063425513948347150396250697
3684760671263171814911482012073807748074812212924671042199273837940730646417669808124136446906891515283214586780489499532639189818629141337987968402889634837966035047
2015654456822115609996385138299114776790557812080078863615019382329551981626953698278169165364189557590275205101557111413589796655241819702579708656055530842325081451
5430133342982608532263673013564682347538153808645758491774086819117355498403758301429560260016201976959764848509318666088472752697314085725964479753389279398095472359
0085130180855419613858258836854222620408448084706903295251000315279651903477808134780491596151706972422224993234807492830385385142160947469451304845383914591749936517
0592660476712187449836966130147299540202243997280277461015976793552686794138013286166394085777234787564175793112796091619968942407592088550526006869626016534019921446
7681125093093942936667509794375558735042927023327732520852231655236586648265847748075378826679418387995602372691879162015062097011591609561747518436633321777707934509
1580975065854469194177551492230661552192106069594721190084016933449029099912763554414589218924444270704327054377041423534979912505201875909638644441594144294568508490
7551843162204135317350969854976111768740310095541789533917296881817241194556197845543132720291003470431519545026463229872235165223979710840472887561057473690795454787
1705943643510815432269350136331343757894582639149342050580535603190711066902506797027584436060049392947683486334341529355702626813130394435671142069250376493057067593
7365952675213257754301824586464616978366050429302101127954600059963123932551894952962422991956763443266022298324026778336002395838487176228891824216816246542051867272
4766651780581319565687936671651468055416494169166459326334689286961214680237024935697255339928866893883463895545212831493745373483626317410936472896261102119426668434
7477625119008907277362376791885092358300092253056669359131061503198159765616050568335118002806173649697724852682521899252416915761243411452425855608644660502573416032
9295629862332799738326647577800220266717889276047514379683816341380267508892342688387284127676719769363367589994621905566684404407622792555488241877948194468012250855
5185062115618375172547346272526423880886634438245345696290423582057059069729957922585933798106255775543776703782825428246224021650417240016988246463208662639855121818
3158359819843776690329590492719746265797769803075339133114431134435840386713019230660781905213339922079876239547116994695079798680314322899055806119052392429316132995

[illegible]

[illegible]

[illegible]

6038481295884148126858001336519062825021065987613068321353053758668302946910404678286715931286645489589590473459720606612508621540329517798199826569799166995611656194
2474361290694100607190038431483563469219511463184890664489320074175023878044993373390528154853252378271608659860652176743503667710490253439260825672115624303510991172
5808184867327665759664285981065938756767154369166725595094738911471858057966829824939679331940538548552426530149540696398999852900746496089704274156775031692470598583
7707909540174494255424415110326959815080083241700700423018769759596820509689224246403745645077230735624166063367305051976285331105937927989646731652394676184645284803
6295034287571335122949946634772378645913656787488228849853554161482861974141076458017644801472414264433723660429524368055580987934871542882217581515006145065334631202
7544336136883820670741490308009437236795094583009482953037573280843737345783865962423821782155450024108134346623372794449542114432676998351245705225506698573695147420
3172034493696075115106447735385579699216565245285552051188529276299087222081790024535219507307583958876613933049135781195375853321789740259133229280864225867198293637
4047877101522681162868225583453443213728596833973584991148727376489257306847800266874835367752217792182169570166903036775862156153990340670127280837280353674314515043
5652962399467775602293749197637512172279527360693750382914208055350941842288119568159319159356110524423795310868311101641894006298907316289332780661066772062088766121
3120214702597617608977317525930970038275871884395659281064635808570049793656969080488268387800663288572030353313494939779028337539197308532441261503329926090401827867
8756269454227705702085546212758727748855046236432394161249724643276953219213145497458881150167247142015816201278813284071769913075103149007865792023657804677914347259
8276667025380750092003597191379254146442651685964937917864170022998023294178362605020542896018700555881940560340435689483254585889698574312297282420617593474917487843
7529009367785782154550396159646097204643011714286745784120977745471224113332657253629163551600595305649170660080704568841224025360409250648743340241628063729823465020
4355811890956195929446621027377540399315245622550477665685387726445062731505911963583251434394120558480920326431327313296380316625712946343012791047157858406313918610
7600386205038024085598484987809523625976099376634630981076403406420615256392170465758762167496225087347450762532714368815420377139933636305574659150373961164934469842
9927519297640759162876508843759060231167643100856938368410764424499915949052121683606296423353652720952928335819510159846331191227670018960473201173078492322602390443
8348197273754602506259099342672955815743166042145239216230463047592977847238871115414610864010405029307353725874234846108656997042941222455982735531559611855540185785
6699179141101873395060405691565628764656131305913339555841524879950051935574031143084514514427234985567204685190540543943880689060868093657878215658241293735745663708
7387807746770467629930560330094880133417152549135735213763467871701253033758303861679291683155395835246843562268965986080641736526790255864968457346304029005072094945
0856205990091170935686596387907450150940503855430450146676783778372786957368858283901065673277686467431912459362531258681514019135162314649541657575065638611495129978
2496965265562023578427889140128898682615893201039839852428942437101119877676717857127072001127693465542677363968583120501526918598781562830977712423147701383668502400
1388113205317991523743590109815240171466728480136875829051317007347043371278708716261431556097654769948483795835755033362028604698087424102574459901918753149599027491
1306553969424835633071034907519903866820513991389700652606774700274331984428343699224908730480695422895919699601706357694694264513494878470915348404764100696656729179
6578732078263128520527995428550754549237933943719396300507875359086762771435487121349428082174731995932601205752246858522921712962613071474472490579418873703572581171
0724869795799220582578885442854581968565453366776271451037428725395414633599688168840428278119795335259150848121843572735094271657640494672726452286523134459267297540
6616080027954476621228569153370374180782340585916002267285156129220357926953726980999119695717158498055597822971411628939754054413422705283589873833339724797071235497
1116597627611604068140772255711609517264939219325615040405951826336069664411465504939278823769385664531124114583150291199160159216100785832673262152453851053839500538
8953737212686175679005419492131308024962453671417353339213943650363088115187632362068995860665307721279172822261396355753614436309943087146361186557553022719257895569
2999674392721549881296485273749975538837593247113669797521656059998698981432308463695283064468233517494388946998617300345232969904253419532224404630000006665805027343
3632704280233943943865216006754475491183895388181213374716988456715389884698782486327805067707046460091751102923949634424048353560895869128093579619651554262173714461
8892635315912088071589978467328336854827684275766214571400251401165705832035763319620196582029699711088320033011483028169155937701152305859569723722836730678610704264
1329667168334238059117798401958271457070204253808760287333481120864113954515751976225967199498585956500005004481621877614649650447029042473946551899222267431340334655
4475107278515622257452973926684836766551375703223912112071952816724164202824936736413418501504765915580189621938806961993112864644304816917155545396300311642774241722
3483286055632064967596294015005301220293832627360654875287144032936783057110911877795302252357410818354837304819033385416455791626172512354095651170969694678555084709
6545049232225812241895623524144261028603276364661771073400323474217277516430903709609547926066949979160691208952264523020746771938610071295790373210687069462501448157
6806492224373137788499696609911407837456954726276202134382664598453425145303317318379610677826631902198333459544742587042776827349577610477674115595528267468987767286
8544187815752597732089108709691179959281149423762989119984671875318157264835818246623756912922922623254103987086183975114551246016330016475451741821894781710480281640
6313972205037742126131056109715136262027715270029840115654794151583607590940131936335720791987065146427490021556805953444569006915689388564287044551267498772924447793
5297014360076901409147839480961967075209188632123933119947506785049070214980546427589633404070939537599775392772669366357391467301210226633647385467443028880834551415
8590132187674971302616865955972929197927674175251707531863431753496281468605711750824398572493519109755290846940882006660982853495048773358393906076971870326775883538
2798308666983562961429421116049168735744076765949711129192100530808972903978949785682831786654317654386761980660017452046504920465402141749626784491237832541323882465
5804572924742851097004594945932063219548718644359331119029960685607426615625665204912033366386153896344416435709957604945810426799004837146594414223149575587269187536
8265298136980243756426794095697003877626417852252014439102658700203674689492669531886805282709556804857729873442963707110120869843660452757196120979874350651365640250
6194602523406059520568784182033008039925315284427375806581201294840165605410393511334546763883999840690913182190133984858151126659394917569777481192531316669819842321
3723209902564727560756767303288127929896468189845176925107232415781085196813636784056902808763661021815340323421412467001850458578400000717109810647006604683496168276
1053549497566220113710680740237618672496248467398272097106191281424575170475984803992654385799438237147277591585615121989972586544686517920489149090259373294428689659
2128252668304139793582872516199669638815889521564597782012039621089207298986986176632122175366138475533386405474866607547383750186910496244547041107069291598482145467
5965154872301683906190406429703826366948807351959443656902529535576894612806573516353411175881571446477692706488949806486748842891844712256030435199598372922381121839
9230663358730757199846728833014884296369476792333631135487050170652175813593220694250174167250994902411958383645960546520339378048615521633372813776551956246336958427
5124026425781107349596367810028538405867296705503609970495401121186513428679418825931560405898722187469338239572977018841001598852928707288811861383855896195902559306
4592652378049653314884067693417838092283369043046190711129805273052876418830537065570357977579930161018071622152933960912643109410216082305745072617851740277008000003
0765297260149458579558497390028056718071730955212178158242215254261107869338433248743758025040891441634220336722908317145738438315489000058860644408161032781090229409
4414874723732779383792387133370507794077031756596662561863171544338916665732863328634444995442897098236522964445521143677553354439484635659517730701657639400847912420
9385752756322157958841969704055596148870356337373414466956464123868880562935506891274943924969910809127418205685759418119295857273373625353891805554962317607030400171
1645552114385234484853282141726263363310464204714885599507542407762458615393510648416395505570640035708632542036914666351352571735264663872016518049270279142747697601
3682929598286509396308422975636214589871640170840153784023461593182010711321674000953522729762851744494477380911028988617723664959684797436446228109134579679347280733
7125237602648189994961816516567182096641027180972036570879864853685047862887029987049335994131244234597998034789387026384680738348250808563835508626102363192890961841
7953327507092261722359304837101041989508564599906796406269078911205399696263983721370639239520645335496845143181366100496237636264815558261212815972242051722352955 4985
0126189798050561090249088568217636265446740596949296600753024124580915348972326008731860639235149638570357969418363919544826359487932434226639999744492523742669696221
5033705649218351401751005525851679754240119522169326115522838954640323859001253812669322757844711921946161657780442519746321022421461745511685629835127768493547983850
9552935640663134229717919885563566635051919596349248072997498769431806452330628123182072478797986992067827477645696546889483359901507224344346114026410024796152270976
1133520985823016575074076605445728224801006136711418758076636460174050407897082132203389666453421283444828044245388148599790209486421067560822873922137294027921932498
5412867102840151878083274294239949479059134424739168293115901981918039435144374176248734502956257548430270883761892535156232852618875815588663500264509733229290329642
6559414212041565468782578474913836409691380414453217623395217329297974148285336864282985411668729843467827021795737234501387495649944969916210702855833476552310469704
7473792882545078398352549616825056534959849887298060815083937311704195330753379409674917683231467603523148488462139316750402009523637221544505701868832452726836372950
0521123341674363939534254364274708663573948697189095150075007853434993268214637450225328856821327701127381472014646761531035551428371888670676854663494147495334585123
9739285060685344497531349053747927447258890799614160094029952007096759062750542357183651808207519326366754576008329063125996661311962806172172373121762579030157103694
9139898781741048217353429145921021469202321123496983085365743641729199942922876736964034868023479396544036326953093993544784426792651880027513760259861430766417583009
9082054006882490326162032368977673413109090264747142554849807501006495650094334199503220828108629701414030390818171800403595335113505272843550577947327375218272062662
1175030979354510559851764609788890676849816134495708126740702849383121379310031872112538319919962271519515417400093794836842564320788433736646821380984022657864888304
9068865112031567463492894150048089670205933417368036657810149029483345130258823206180507089054012058944525485585564352018446291408853095187095267509994614544403286896
6576599702737538971497290848538731979091565859652864679615006143515674988032348301877288539294041911033562244383292638238249943113338440166727643551692365143846128853
7969045406949044572226523544114031427632550514192979449795474447370919761537378724922782285228095647188346603820730204037843206437324572596480510643068406024784250228
6128620905586250868816621500005575234211838697063276967265322807868647850977029459548233323664838487608181254864313177649046705040802122896818693259984602761358321718
4221246939829357601932841487616892588819524233750564359888696000834009880588156198592643365860721151226210119124169186917848816167493187228886677225813267994399950397
4046244492382193762384171025446101438601553617913305229951584360443683739165892589728338761714709965295189220099266312689882039694456839547941897439848671692642092108
2617536145638087464684093658883235888231356206109998377953552847197446938471717685403731107925504395733925639297724508541688924004000266420736996984142758470505732303
7688693660670449424937716369423055491360251730985249188393420300724727142962992182437273902681302889161057048579327438444163551190584599706290751213138111778802255630
6710549632995878459476827626539922977187785925010216194539829392686759137300404225322969407562083329971880503076514214891665128810327404658636447888553943776807309853
6632650762151728104519437949306579564410242084173254948994153005844020574312583483250048422642376062308934547262018435102610796872416557487761174364856061304487493907
3152318125621645925104780260008374036368595905190524226982751865849017078578988681541251646530635488907172292722974799905715150291657616258549623474440767298378453541
3370941827174960788629335289980830436577678061526822089007175381361432824258885597022749704942948543131999582680768450906862550848621132529053911928518221246568848407
3906327086999353740887357679473516067732617148150222358618135147831967953671413342686125628341740505881159470792397035993230676310072011579474972986127682318049665160
0952042083021356561783577622843942541424950486608960618060538966044827902004199560556501141942635798951524908115555692548703164446122118395737941892995939537853931293
9129389868585016348443154687862221549320172750880361395035928898503376611302728523344815390769770240467527579754110261611233752701880035298455679579804724680693016938
3686421532737885440306485815226271141481199619996273103470998150943690955213644176528997367163520035422136855761366388692735656710094079864579371802418389130454685139
3448214088941351172516123715356946242349812310686191913761467701006910094281578791915740951201492012442581805153459839510873800516257292580379124510910568321207314697
2363782739514947615093316937207430642539234862560154984162805250671045027861770460393271176979716164381395810354698167478264769561983525269892635407994205765582129737
8559376170004069741155180094016675599049523914554577598576752396489389382503715806618386629455870448827866643741083277016421144945968722429296599139503888084211988749
5111522859434271857294712754955497183307144133632735453256109346931841628443350538053845737820397947059696287844628532323926269183010811083593862225932263435872644437
5714086830368269994314879074084129693751934176293984420139796689055186317511049482371036070630914105850095967268552357199881942971094016744971004838190937882993015676
6241259329393963543866762020481406996891560381297183383225422780006765417870547830920828690483452699338843191930181178546401817896519222737873213248814292492248352405
6121322018952118501095575002556612538716825900125518305073857257763165008204434357220991999112830591436329837123901545435336578414483729393169657543733938193814572142
6820010641605556577472619738888298115916471218840084428620326120357500512809735447121999445269636412959578266419464909913730894294343989874724577342101064234601568695
2330463462416455704053064825748697398640696075257333251728628840819358786482369211679704133377485031643671586078027233501396304941155401429566187824378723345575205610
1733294598566871608190109804129190717304910326529101407065577688006821348012066161238558053625232381125643466636170364495242279863856828259210100020615576919611322477
0776906291741867791868263353846704065681669923785589029380516801431860492501075364562597738652978532586473947225454404511368789856653932731058047992476696392546038425
1783682099966404217169236477257657715352400152458520444405498866875400513003359160025963240431177209499340693935874755710849770534434156899522674488775444442744537651
9640089968247985121519450376945352391503181486611573550167525990300387596911703743881538655603914165583443303474590656418983434813506825816544739629229259194006820292
5564824121493315049929165403399055922839062949577242403680448281998312980360179633585997491771542405132253032731508548181672078139168081802072277030371495350218767484
8514932267726342498101763852059018338559963703596481136726560184380132402576913743186657603165635315959069857863553469953975516021923270858009312508815717123801773048
4752181133885340618969930284438892844327264068900889489497774423731691573345691024798396850214965401733504210389062174278330535114162792476015969462676761433456891074
3419562269049400230645891696081124188717968737338593068588448548189810957530764762677108440272985234319803974129191470421501120132088392678231305372755005279791808167
1691965327603305379324977799312258180852586507553480100222906410335572131117437648027137494896835597099576090199140222529785182579434054106170706662113794932345746391
6309829662397666072978300499070356388855348435772692907736323558513152141448167751724888537385952723750968868188887252843453704719187148402101865201479394367505291 2895
1409157015509748579820225826500671951095205662644368845458192160363224518397359117288616046477988773268097033645040268304202738485213050191043199260041200378759455752
5887354645215997748696260378056670448648685370081085710750663520502147494963310399894394711745710357960387830912569982542272779765081238315256314918706351612793174346
1825825724733892791228015579940153695805914773343618215182109202559525718612749598503158843772802768775246887004706979239291566674852431028074757069181177427994545659
7029203697769971610661674418318936178849700109217607368723291148572021302910262228256207854700318977841101027539932919201007285871014120756256884412569132052369796365 6
9269435018873061293999923859309108197160571632563054934732144682964183798382096727852543077642509441357899451537665894968486386476038097860247649124319828235680125454
9099909180039744888027319943027702307194369677457089640760135500047860081894694760433591046323309343871330130930098177571791333892795135798214070349888305040071049590
2449358139949775482776112339839260881015245291589021265037966025485771502875493759831448512234354230408785765173148703313690298271468130669477165239082591674943714934
5499592968060712521351182643666394788866608635949874572431836736936659248773751564359326712202157927711027133177000812349669073188423398557837925047600584446619487564 4
8403262251760742879086976118841169449661198799345771496903761161379945585818292795010411613926335005318453027166695115762285208631165801673770686447944464675639871505
6513042724288045141591074591076863620043667198535562465454882183919670154985638367236617466338486973729174494142525236166933467213036546602397175637465907460962874684
8165814233622430005009451602573379527303861673370591483276524005875857789809442659551666811773290635588329219904825296315872600886555418827720798468328390472249502834
2345658783554955315824129609583666197257779148498129837280768134207170928648501108320380276950889208061904736534664016603366571354957498618546305966264566434434133705
7673674179616206874658781625138196865029045509264704958936719606399417846456400299273678608094179109177428793117012336960187407085776992244787836231541880275807433007
0012702746032276727052405600124577148134121289815481942572853969039302280155449274320779641339523917876226398824974097737318815344067332501854230557590615418068647327
6615945874086238042490244927444750104693292248952709916982033631671932713388117289351930598088661286267050171612203399110283288950947449559928317835113387469470777738
5334667525693573527998304387281799502177964462591741205710067837492280129466557319149912947044253452558432006045650601749920517992422027197247251250126901098643736456
2154344225714521018311887768806863029897133785663300440680999855193917424466337493894703903752457792566996603032654356520726787288351384925616695438960481550059946301
44292500354883118973723406626719455081800

## References

1. Amthor, A. and Krumbiegel B., Das Problema bovinum des Archimedes, *Zeitschrift für Mathematik und Physik (Historisch-Literarische Abteilung)* **25**, 1880, 121–136 (by Krumbiegel alone), 153–171 (by Amthor alone).
2. Bell, A.H., The 'Cattle Problem', by Archimedies [sic] 251 B.C., *American Mathematical Monthly* **2**, 1895, 140–141.
3. Homer, *The Odyssey*, English translation by Samuel Butler, 1900.
4. Diodorus Siculus, *Library of History* (C.H. Oldfather, translator), The Loeb Classical Library, Harvard University Press, Cambridge, MA, Volume VI, 1954, sections 14.59.1–14.59.2.
5. Lenstra, H.W., Jr., Solving the Pell Equation, *Notices of the American Mathematical Society* **49**, 2002, 182–192.
6. Lessing, G.E., *Zur Geschichte und Literatur, Aus den Schatzen der Herzoglichen. Bibliothek zu Wolfenbüttel*, Zweiter Beitrag, Braunschweig, 1773.
7. Nelson, H.L., A solution to Archimedes' cattle problem, *Journal of Recreational Mathematics* **13**, 1980–1981, 162–176.
8. Nygrén, A., *A Simple Solution to Archimedes' cattle Problem*, University of Oulu Linnanmaa, Oulu, Finland, Acta Universitatis Ouluensis, Scientiae Rerum Naturalium ISBN 951-42-5932-7, March 2001.
9. Strabo, *Geography* (Horace L. Jones, translator), The Loeb Classical Library, Harvard University Press, Cambridge, MA, Volume III, 1924, section 6.2.1.
10. Thomas, Ivors, *Greek Mathematical Works*, The Loeb Classical Library, Harvard University Press, Cambridge, MA, Volume 2, 1941, pp. 203–205.
11. Vardi, I., Archimedes' cattle problem, *American Mathematical Monthly* **105**, April 1998, 305–319.
12. Voltaire, *The Philosophical Dictionary* ("Il y avait beaucoup plus d'imagination dans la tête d'Archimède que dans celle d'Homère"), 1764.
13. Williams, H.C., German, R.A. and Zarnke, C.R., Solution of the cattle problem of Archimedes, *Mathematics of Computation* **19**, 1965, 671–674.

# Vortices in Homer's Odyssey – A Scientific Approach

Georgios H. Vatistas

*Concordia University, Montreal, Canada*

*The present work is dedicated to my ailing father Charalambo Vatista, a good man who knows how to navigate life's Scyllas and Charybdis with wisdom.*

**Abstract.** The traditional approach to study Homeric epics is philological and philosophical in nature. This paper takes a forensic route to mythology elaborating on some of the accounts via the use of contemporary scientific knowledge. In particular, it will deal with the oldest clear report of two vortices found in Odyssey. First, it will examine the tidal whirlpool Charybdis [Book XII]. Based on a recent mathematical analysis of the phenomenon, we will further validate some of the exceedingly skillful navigational abilities of the ancient Greek sailors. Second, it will consider the river cisterns in the island of Phaeacians [Book VI] where princess Nausicaa and her ladies-in-waiting use the vortex action to wash clothes. The presently experimental and theoretical knowledge on cavity hydrodynamics provides a lucid picture of the four-dimensional structure of the flow, thus identifying the cause and details behind the efficacy of the method.

## 1 Introduction

A non-expert understands a vortex to represent the circular motion of water as it drains from a bathtub. To a scientist, the word vortex includes quantized whirls of super fluid helium, terrestrial tornadoes and hurricanes, the red spot in Jupiter's atmosphere, or the spiral galaxies in the heavens. Naturally occurring vortices have fascinated humans since the dawn of civilization. It is therefore no mystery to find the vortex to be one of the fundamental postulates in Anaxagoras' (499–428 BC) model of an expanding universe, in the atomic theory of Democritus (460–370 BC), or in the old Hindu sacred script *Rigveda* as the primeval bearer of the embryo. Although whirls of air or water have been widely used as the *Deus Ex Machina* in myths and legends, no other vortex has been so frequently referred to as the tidal whirlpool.

According to Greek mythology, Charybdis was the daughter of Earth and Poseidon. In order to satisfy her healthy appetite, she stole several of Hercules' cattle. As she was ready to devour them, was struck by one of Zeus' bolts. Subsequently, Charybdis was thrown into the straits of Messina and became the feared whirlpool.

*S.A. Paipetis (ed.), Science and Technology in Homeric Epics,* 67–75.

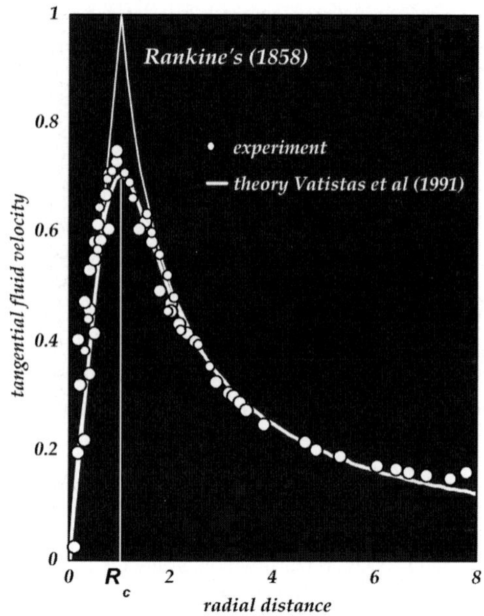


**Fig. 1** Variation of the tangential velocity inside a strong vortex.

As scientific thought replaced the primeval mystical perceptions, it was then understood that circular fluid motion is one of the most basic mechanisms to effectively transport mass, momentum, and energy in nature and technology. Today, ocean whirlpools are certainly not considered to be the effect of demonic or magical forces. Instead, these are naturally occurring phenomena that originate from the synergetic interaction of the gravitational attraction and wind-shear coupled with the morphology of the location, and probably enhanced by variations of the physical properties of the water such as differences in temperature and salinity near the site. Old mariners exaggerated the destructive power and size of these whirlpools. However, if one considers the relatively small size of crafts in antiquity, their modest thrust, the monstrous origin of the vortices along with the fear of the unknown, it is human nature to magnify their size and power.

In mythology the fact and fiction are intertwined. One however, can learn a great deal from the ancient accounts if the two are untangled via a forensic approach to mythology using contemporary scientific methods. In this article we revisit the whirlpool Charybdis dealt in a previous study. In addition, based on current work on cavity hydrodynamics, it considers the practical use of whirlpools developed in river cisterns (island of Phaeacians) to wash clothes.

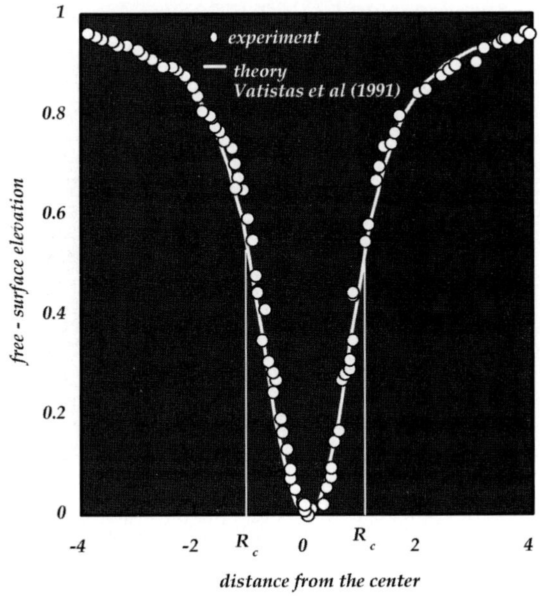


**Fig. 2** Inverted bell-like liquid free-surface profile.

## 2 Charybdis

Odysseus, the hero of Trojan expedition, had to cope with the fury of the sea god Poseidon and specifically with Charybdis. The earliest reference to this sea-born monster is found in the tale of the Western world's foremost sea-journey of the "Argonauts". The account however there, as far as science is concerned, is brief and imprecise. In this section we will deal with the oldest comprehensive report of a tidal vortex found in the epic poem "Odyssey". But first let us look closer at the general properties involved in the theory of whirlpools that will help us understand scientifically the account.

The substantial mathematical complexities involved with the details of vortex phenomena have prevented the formulation of a general analytical model. Instead, many researchers in the past developed theoretical models that are only applicable to special circumstances such as for the case of strong vortices such as for example Charybdis. A pragmatic, presently popular simulation of the whirlpool could be provided by the $n = 2$, vortex-model of Vatistas et al. [4], see Figure 1. According to the theory, which is also supported by the outcome of many experiments, the fluid velocity increases hyperbolically towards the center, reaching a maximum at a particular radius (also known as the vortex core), and then declines linearly to zero right at the center of rotation.

There are three main forces involved in the formation of the liquid surface in a whirlpool. The force of gravity is pulling the fluid particles down thus trying to keep

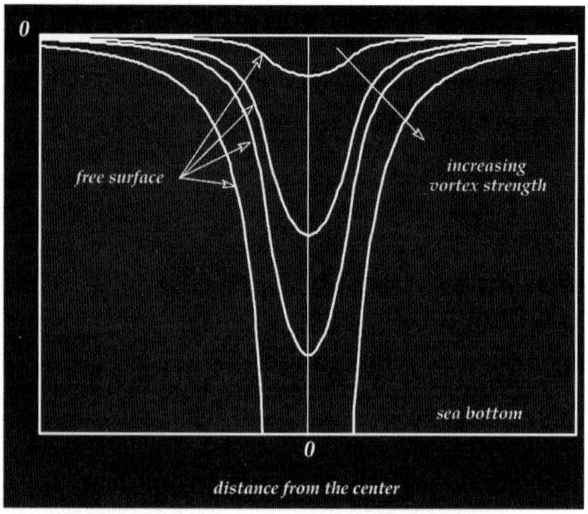


**Fig. 3** Liquid interface shape as a function of the vortex strength.

the liquid surface horizontal. The centrifugal force, due to rotation, pushes the fluid particles away from the center. The combined action of these two forces along with Archemedes' buoyancy, directs the system into a state of *permanence*, in which the free water surface attains the inverted bell-like shape of Figure 2. Under this equilibrium condition, every fluid particle within the flow domain experiences no net force. Upon arrival of the system to the state of equilibrium all fluid particles will thus remain in their last position at every subsequent time. Depending on the strength of the vortex the point of the minimum surface depression propagates further into the liquid, see Figure 3. In case of a very strong vortex, the depression can reach the impermeable bottom boundary of the physical domain.

Due to centrifugal instability, vortices are also known to host a variety of waves [3, 5]. These, generated near the axis of rotation, are convected outwards thus modulating the undisturbed free-surface profile producing a rippling effect.

The dynamics of a craft inside a whirlpool are mathematically modeled by Newton's second-law represented by a set of two non-linear ordinary equations and Archemedes' principle. Solution of these equations along with the appropriate initial conditions enables us to investigate the dynamics of floating bodies in an oceanic whirlpool. The analysis to follow is based to a large extend on parametric studies performed solving the governing equations numerically using the fourth order Runge–Kutta method.

The effect should have a cause. The ancients must have observed the vortex and the funnel-like free-surface formation during draining of a liquid like water, wine, or oil from a reservoir. They must also have noticed that matter such as pulp to be drawn into the funnel by the downwards-moving current. In the absence of any other rational explanation, the presence of the sea-born beast named *Charybdis*, was very convincing in explaining the formation of the very strong vortex. The suction and belching parts of the cycle are indeed very convincing attributes of the model since it can explain why ships with their crews are brought to the depths of the sea.

Odysseus first hears about Charybdis from the sub-goddess Circe who described her as follows:

> ... this divine Charybdis sucks down the black water. Thrice a day she belches it forth, and thrice she sucks it down terribly. Mayest thou not be there when she sucks it down, for no one could save thee from ruin. (Book XII, §105)

Circe, who probably represents the ancient navigational wisdom, provides also Odysseus instructions on how to effectively sail around the horrible monster,

> Nay, draw very close to Scylla's cliff, and drive thy ship past quickly; ... (Book XII, §110)

Homer's very clear observations regarding the whirlpool's physical manifestations are found in the following excerpt:

> We then sailed on up the narrow strait with wailing. For one side lay Scylla and on the other divine Charybdis terribly sucked down the salt water of the sea. Verily whenever she belched it forth, like a cauldron on a great fire she would seethe and bubble in utter turmoil, and high overhead the spray would fall on the tops of both the cliffs. But as often as she sucked down the salt water of the sea, within she could all be seen in utter turmoil, and round about the rock roared terribly, while beneath the earth appeared black with sand; pale fear seized my men. (Book XII, §235)

Since the phenomenon is dynamic in nature, it undergoes several phases of development. The powerful tidal bore, in conjunction with the morphological characteristics of the site, cause the generation of the vortex. In the presence of the centrifugal field the water surface begins to take the inverted "bell-like" shape. The central dip starts to propagate towards the sea bottom reaching a limiting value that depends on the vortex strength (Figure 3). For extremely strong matured vortices like the present, even the bottom of the sea may be exposed. The phrase "within she could all be seen in utter turmoil" needs closer attention. It is therefore reasonable to assume that Homer is describing the effects of instability waves that are known to accompany every vortex.

Before we analyze the ship's trajectory and the instructions of Odysseus to his crew, we should summarize briefly the most fundamental fluid and ship dynamic properties. As mentioned previously, on the surface of the tidal vortex all forces acting on any fluid particle add up to zero. Therefore, no matter where the fluid particle resides, it will stay on the surface at all times. This is however not true for the floating body. Let us now assume that a ship is riding the whirlpool's surface. Because there is always a slippage between a boat and the current, the centrifugal acceleration due to the vessel's rotation about the whirlpool's axis will be less than that required to keep it at a constant radial and axial location, see Figure 4. Accordingly, its weight will drive the ship towards the interior of the vortex. If we now take into

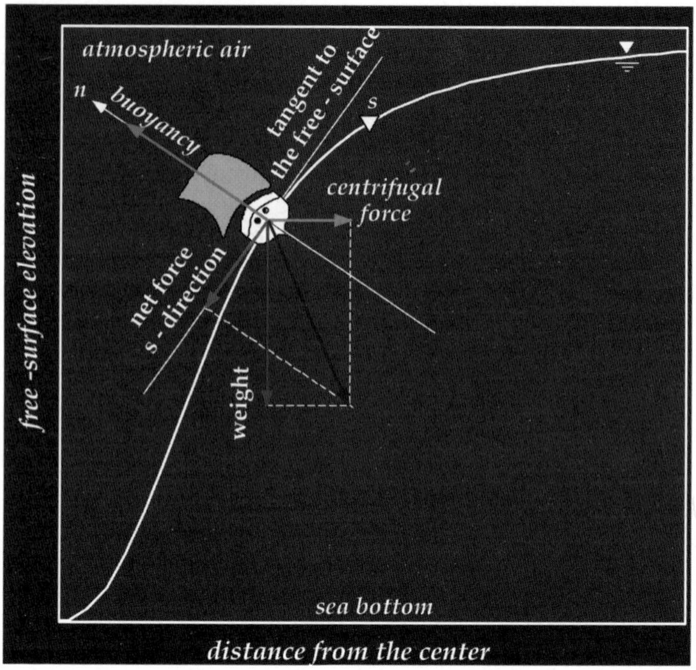


**Fig. 4** Forces acting on the floating body.

consideration the maximum available thrust generated by the oarsmen, the boat's hydrodynamic characteristics, and given its location on the surface, then we have the following three possibilities:

(i) The thrust developed is insufficient to balance all the forces.
(ii) The thrust of the boat is enough to balance the forces,
(iii) The thrust velocity of the approaching craft is large enough to produce a resultant tangent to the free surface force that acts in a direction away from the center of rotation.

In case i, the boat will be "attracted" by the whirlpool. In the second scenario, the ship will be put in orbit around the whirlpool. The third case is the rescuing possibility since the resultant force will slingshot the vessel outwards.

Navigation through the fully developed Charybdis can be presented as a classical mini-max problem with a constraint. Given that they had to go through a fully developed Charybdis, Odysseus' navigational strategy must be drawn in such a way as to maximize their chances for survival. Steering the ship through the sea opposite to Scylla's shore would have presented a maximum resistance to the boat since it had to confront a head-on current (Figure 5). This was not a viable option. Odysseus had to sail along the whirlpool's current. The ship's net attractive force parallel to the free-surface diminish as the radial location of the vessel increases. Odysseus must thus maximize the radial distance of the ship from the center of Charybdis and

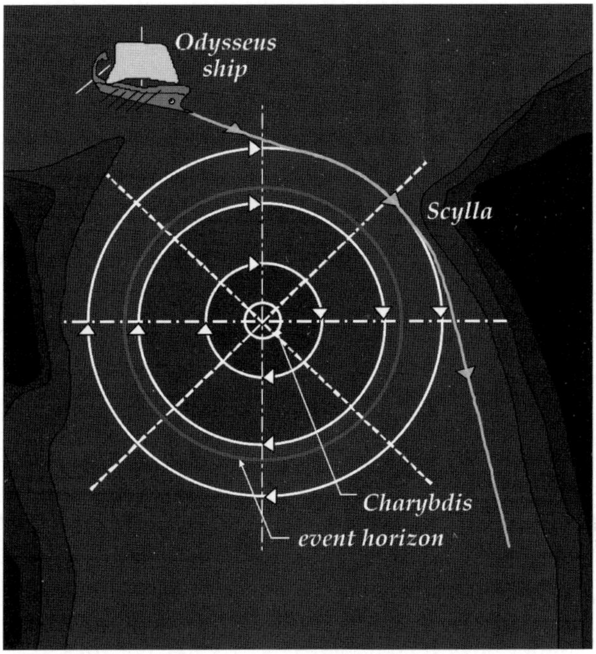


**Fig. 5** Navigational map of the trajectory for Odysseus' ship.

maximize the speed of his ship by maximizing the thrust developed by the oarsmen. The only route available was thus to the right, i.e., steering the ship close to Scyllas' reef, "and drive thy ship past quickly". The first thing that Odysseus needed was to assure an uninterrupted maximum thrust. Therefore, he gives the following orders to the oarsmen:

> Do you keep your seats on the benches and smite with your oars the deep surf of the sea, in the hope that Zeus may grant us to escape and avoid this death. (Book XII, §215)

> ... But of Scylla I went not on to speak, a curreless bane, lest haply my comrades, seized with fear, should cease from rowing and huddle together in the hold. (Book XII, §225)

There is one more matter that required his attention. Because the rotational speed of the water varies with the radius coupled with the presence of the instability waves, it is very possible that the ship could be rolled or swayed towards Scylla's reef or even towards the center of Charybdis, if the steer-man is not very attentive. Based on the previous properties, Odysseus gives the following orders to the helmsman:

> And to thee, steers-man, I give this command, and do thou lay it to heart, since thou wieldest the steering oar of the hollow ship. From this smoke and

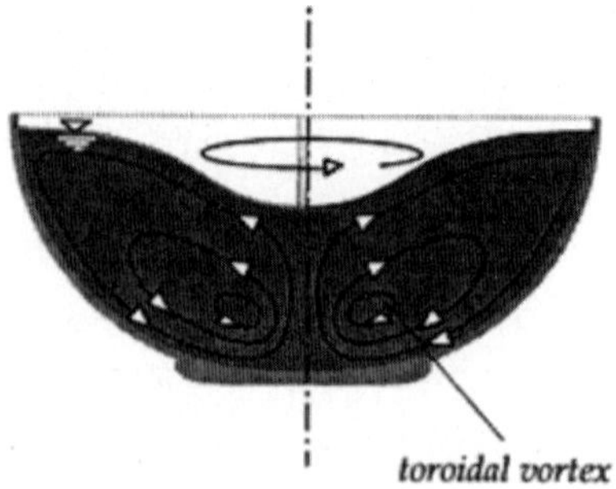


**Fig. 6** The toroidal secondary vortex flow.

> surf keep the ship well away and hug the cliff, lest, ere thou know it, the ship swerve off to the other side and thou cast us into destruction. (Book XII, §215)

The ancient Greeks were accustomed to the three forces involved, i.e., gravity, centrifugal, and buoyancy. There is also ample evidence that they had a qualitative knowledge of the main whirlpool properties, such as the formation of the dip at the center of a liquid vortex by observing the evolution of the free surface while stirring wine. Did they however, hug Scylla's rock like one tries to avoid falling off a cliff by walking by instinct near the innermost side of a path at a fast pace, or did they act in this way by using the empirical love (attraction) and strife (repulsion) properties of the whirlpool? Whatever the reasons behind Odysseus' strategy he acted wisely. Giving the correct orders as described in Homer's Odyssey indicates that the ancient mariners had empirically developed their navigational skills through oceanic whirlpools.

## 3 The Phaeacian Whirlpools

In this section we will examine technically the vortex flow structure of the river cisterns in the island of Phaeacians where princess Nausicaa and her ladies-in-waiting use the vortex action to wash clothes:

> In due course they reached the noble river with its never failing pools, in which there was enough clear water always bubbling up and swirling by to clean the dirtiest clothes. [Book VI, §10]

Hydrodynamically, vortices developed in the river's pools must bear a resemblance to the lid-driven cavity flow [1, 2]. Due to the presence of Ekman's boundary layer on the bottom of the reservoir, the fluid on a plane perpendicular to the horizontal generates a toroidal secondary vortex structure, which is alike to that of a teacup vortex of Figure 6. Due to centrifugal instability these flow patterns are also known to be unsteady. Pressure depressions inside the cavity produce vapor bubbles that are carried away by the stream and buoyancy. Present day scientific evidence indicates that the velocity field varies drastically in the three space dimensions as well as in time. The last gives rise to a shear stress field, which also varies considerably in space and time.

The swirling action in the river cisterns is clearly described by Homer:

> In due course they reached the noble river with its never failing pools, in which there was enough clear water always ... swirling by ...

However there is more information of the phenomenon. As mentioned earlier, the vapor bubbles (due to cavitating fluid) are brought into the water surface by the combined action of buoyancy and the rising stream: "... always bubbling up ...". The high shear stress variation in the three space dimensions and time oscillations produces a strong rubbing action between water to cloth and cloth to cloth, able: "... to clean the dirtiest clothes".

## 4 Conclusions

It is evident from the descriptions of the phenomenon by Homer that the ancients sailors knew a great deal about oceanic whirlpools and were able to navigate them with considerable skill. It was also shown that the ancients were also well acquainted with the practical use of whirlpools and exploited them to wash clothes.

## References

1. Lugt, H.J., *Vortex Flow in Nature and Technology*, John Wiley & Sons, New York, 1993.
2. Matyka, M., Solution to two-dimensional incompressible Navier–Stokes equations with SIMPLE, SIMPLER and vorticity-stream function approaches. Driven-lid cavity problem: Solution and visualization, Report in Computational Physics Section of Theoretical Physics, University of Wroclaw, Poland, Department of Physics and Astronomy, May 2003.
3. Thomson, W. (Lord Kelvin), Vibrations of a vortex column, *Philos. Mag.* **10**, 1880, 155.
4. Vatistas, G.H., Kozel, V. and Mih, C.W., A simpler model for concentrated vortices, *Experiments in Fluids* **11**, 1991, 73–76.
5. Vatistas, G.H., A note on liquid vortex sloshing and Kelvin's equilibria, *Journal of Fluid Mechanics* **217**, 1990, 241–248.
6. Vatistas, G.H., Floating body dynamics inside whirlpools as described in mythology and literature, in *Proceedings of 2nd World Congress, Ancient Greece and the Modern World*, University of Patras, Ancient Olympia, Greece, July 12–17, 2002.

## Bibliography

The quotations were taken from: Homer's *The Odyssey*, translated by Murray, A.T., Harvard University Press, Cambridge, Massachusetts, 1966 (Book XII), and by Rieu, E.V., Penguin Books, 1972 (Book VI).

# The Homeric Automata and Their Implementation

D. Kalligeropoulos and S. Vasileiadou

*TEI of Peiraeus, Greece*

**Abstract.** Homer, from the viewpoint of Technology, provides an abundance of technological findings and inventions, either primitive or even modern as well as fictitious future developments. These inventions include descriptions of automata, i.e., machines moving on their own, by means of internal energy, like live beings. Among them are the automatic tripods, the adaptive bellows, the female robots of Hephaestus, and the miraculous ships of Phaeacians equipped with artificial intelligence. In addition to the term 'automata', Homer suggests the evolution of machines to automobile machines, to machines with 'life'. Such a suggestion constitutes a leap in technology. In this way, Homer challenges the Presocratic philosophers to discover the natural 'roots' of self-motion, e.g. fundamental elements possessing the necessary energy, so as to be able to move by themselves. He also motivates the classical philosophers to investigate the concepts of system, control, and feedback. Finally Homer addresses to the engineers of the Hellenistic period, in order that his automata be transformed into science and applied technology – the science of Automatopoietice.

## 1 Introduction

The word 'automata' is a Homeric word. It appears frequently in the Iliad and the Odyssey, in order to describe machines moving on their own by means of internal energy, like live beings. We cannot be sure whether era such automobile machines really existed in the Homeric or it was poetic imagination daring to foresee them, to express the need for their existence and to ascribe their construction to the great Olympian craftsman, Hephaestus. "… self-bidden (Homer uses the word αυτόματα/automatically, i.e. by themselves) groaned upon their hinges the gates of heaven which the Hours had in their keeping …", writes the poet in the Iliad, Book V, 749) [7].

It was Hera who ordered the gates to open automatically. Imagination? Probably. Nevertheless, it remains an advanced formulation of the term 'automatically' and of the technological vision: it could be possible such gates to be realized. A vision that soon found its implementation.

*S.A. Paipetis (ed.), Science and Technology in Homeric Epics,* 77–84.

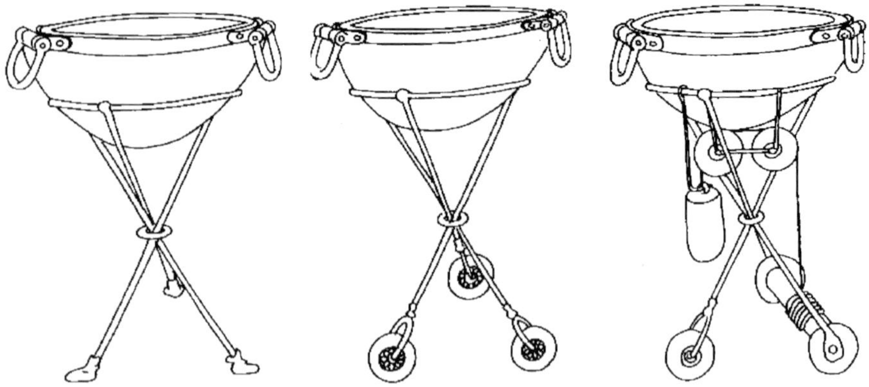

**Fig. 1** Stable, mobile and self-moving tripods.

## 2 The Automata in the Iliad

In Book XVIII of the Iliad, which is called *Oplopoiea* (= weapon manufacture), Hephaestus was working in his bronze mansion, where Thetis [7]:

> found him sweating with toil as he moved to and fro about his bellows in eager haste; for he was fashioning tripods, twenty in all, to stand around the wall of his well-built hall, and golden wheels had he set beneath the base of each that of themselves they might enter the gathering of the gods [αυτόματα by Homer] at his wish and again return to his house, a wonder to behold. (Book XVIII, 372–377)

Here Homer does not restrict himself in the formulation: – It could be possible automatic tripods to exist. He goes further: – It could be possible automatic tripods to be constructed by a competent craftsman as Hephaestus. And they could be useful in practice. They could be in the service of the Olympian gods. A vision soon implemented as well.

Homer extends his daring thought even further. The one who is able to construct automatic machines, he is also able to create something similar for his workroom. The new vision: The automata could be part of a production process. In other words, automatic workrooms could be created [7].

> He [Hephaestus] went unto his bellows, and he turned these toward the fire and bade them work. And the bellows, twenty in all, blew upon the melting-vats, sending forth a ready blast of every force, now to further him as he laboured hard, and again in whatsoever way Hephaestus might wish and his work go on. (Book XVIII, 468–473)

This description refers to a real automatic workroom, where Hephaestus commands twenty bellows to work automatically so that metals would melt. Moreover, these bellows were adaptive; he only needed to instruct them to start and they started operating automatically, faster or slower, as the work required. An ingenious conception: Automatic production was possible, by which a single person would give the initial command, and the machines would to operate on their own, regulating their operation according to the conditions and the needs of work.

The technical vision is completed when the poetic imagination arrives at its last stage: could not it be possible for the god-technologist to construct manlike self-moving machines, possessing skills and knowledge? (see also [7]):

> He [Hephaestus] spoke, and from the anvil rose, a huge, panting bulk, halting the while, but beneath him his slender legs moved nimbly ... but there moved swiftly to support their lord handmaidens wrought of gold in the semblance of living maids. In them is understanding in their hearts, and in them speech and strength, and they know cunning handiwork by gift of the immortal gods. (Book XVIII, 410–420)

Here they are: two mythical robots, two self-moving manlike machines, having sense, speech and strength. Innovative technological visions: The strength, i.e. the feature that transforms low-power commands into powerful mechanical movements, the speech, i.e. the construction of machines producing sounds to communicate, and the sense, i.e. the particular inner structure that results in skillful, learning machines.

These references are of particular interest since they introduce new concepts in technology and express technological intentions even though their realization is ascribed to gods.

## 3 The Automata in the Odyssey

In the Odyssey, the "peaceful" Homeric epic, the manufacture of automata is ascribed not only to gods, but also to men. The poet asserts that there are people – the Phaeacians, of the mythical country of Scheria, capable of manufacturing intelligent ships. Their king, Alcinous, tells Odysseus [6]:

> And tell me your country, your people, and your city, that our ships may convey you thither, discerning the course by their wits. For the Phaeacians have no pilots, nor steering-oars such as other ships have, but their ships of themselves understand the thoughts and minds of men, and they know the cities and rich fields of all peoples, and most swiftly do they cross over the gulf of the sea, hidden in mist and cloud, nor ever have they fear of harm or ruin. (Book VIII, 555–563)

A new technological vision appears here: The constructed thought, the artificial intelligence, the ability of programming, the development of a technology capable of controlling the route of a ship, using navigation organs far beyond the conventional, which find their way with the help of the stars.

## 4 The Problem of Internal Energy of the Automata

The innovative ideas of Homer paved the way to a new era: from the mythical visions to the physical interpretation of the world, as well as to the physical interpretation of the automata. In order that the vision of automatic machines be realized, two major problems had to be solved, both in theory and in practice:

- Where can the necessary energy for the self-motion of the automata be found?
- How can the operation of the automata be controlled, to produce the desired output?

The Presocratic philosophers were first to try an answer to such questions. They wondered "what is the world made of?" And "what are the properties of the world's elements – the so-called "roots" of the world?"

Thales of Miletus (ca. 624–546 BC), the first Greek philosopher and founder of scientific thought, considers *water* as the fundamental substance of which everything is made and consists of. He perceives water not as dead matter but as an active, energy-bearing element. This energy, inherently connected to the concept of motion, he calls 'soul'. The 'soul' is of a perpetually moving and self-moving nature.

One century later, Empedocles of Acragas (495–435 BC) sets four qualitatively different primary elements as the origin of the world, the so-called four roots, of which all substances known by experience are conceived to be composed.

In the same period, Anaxagoras of Clazomenae (ca. 500–428 BC) ascribes to the primary elements opposite features, which are under the control of a main controller, the so-called *Nous* (Mind).

These philosophical speculations, especially in the Hellenistic period, resulted in the solution of the problem of internal energy of the automata.

Where can man find this internal energy? It can be found in the 'soul', of the primary elements.

> Because by the composition of air, fire, water and earth, as well as by the combination of the three of even the four of them, the different functions of the automata are realized. Some of these functions are used for life needs and others cause astonishment and admiration. (Heron, Pneumatics, Introduction)

The earth: "All the mobile automata have as motive power, as the initial cause of their motion, the power of a falling lead weight" (Heron, Automatopoietice, 2.6, Figure 2a).

The water: By taking into account the properties of water, Heron constructs hydraulic automata, such as the automatic fountain (Pneumatics, A16, Figure 2b).

The air: Heron takes advantage of the expansion of heated air in the self-opening temple doors (Pneumatics, A38, Figure 2c).

The fire: Heron takes advantage of the transformation of water into high pressure steam in the sphere of Aeolus (Pneumatics B11, Figure 2d).

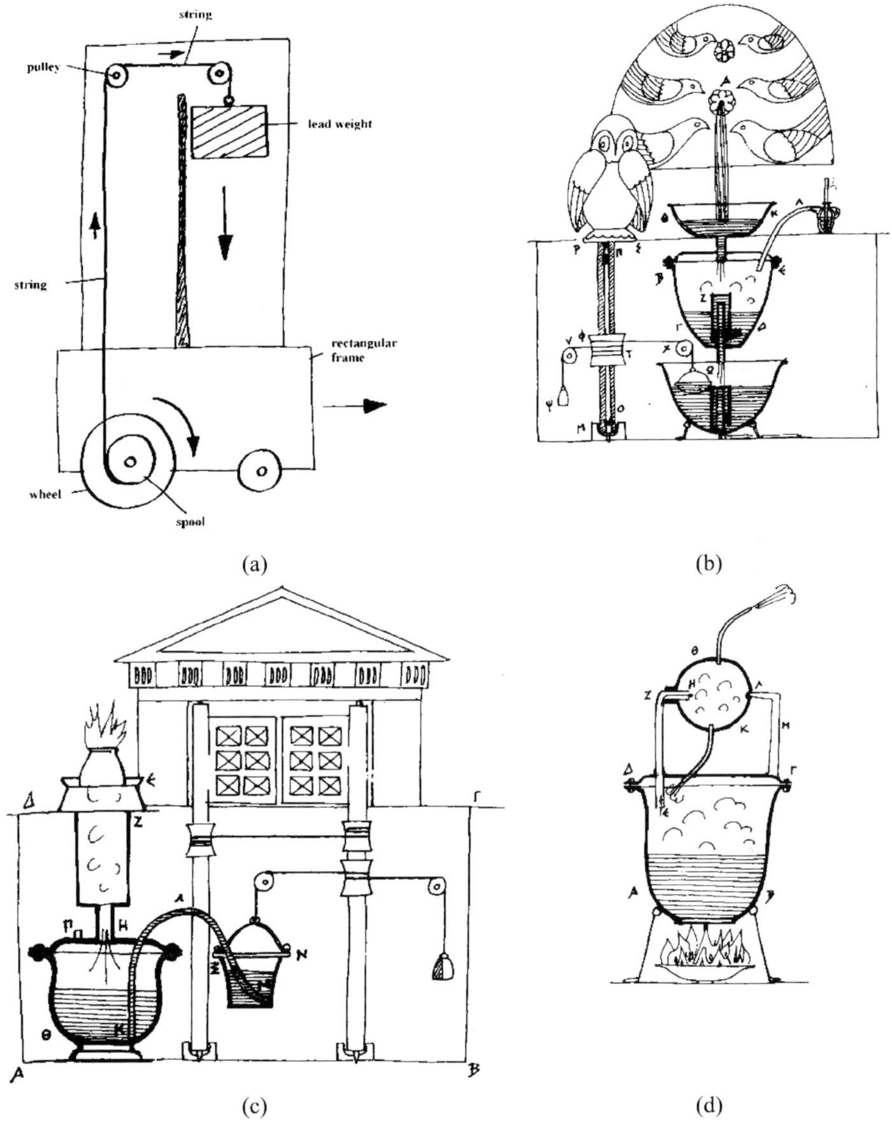


**Fig. 2** The internal motive power of earth, water, air and fire for the motion of the automata.

## 5 The Problem of Automatic Control

The difficult problem of control now remains. The Presocratic philosophers laid the foundations for the solution. The control process is a contradictory process. The simple, straight logic that goes from the cause to the result does not suffice. In the control process the desired result – output determines the cause – input, which results in this output.

Heracletus (544–484 BC) was the first to introduce, from a theoretical perspective, the concept of contradiction and the interaction of opposing elements. Plato and Aristotle undertake the scientific solution of the control problem. The aphorism of Socrates: "I only know one thing: that I do know nothing", in some way introduces the concept of feedback. By means of his maieutic method, Socrates controls his interlocutor so as to extract from him the desired answer, by asking the proper question.

Plato defines the concept of *Cybernetics*, as the art of the governor, the controller of a ship. Aristotle analyses the process of controlling the route of a ship and distinguishes the control tools into lifeless, such as the rudder, and live, such as the boatswain – the lookout man, who observes the sea, localizes the route of the ship and compares it with the desired route. Aristotle ends up to the application of automatic machines either in the manufacture or in the daily human life (*Politics*, 1253b 20):

> ...if every tool could perform its own work when ordered (κελευσθέν/by external command), or by seeing what to do in advance (προαισθανόμενον/by internal programming, having a predetermined internal function, a presentiment), like the statues of Daedalus in the story, or the tripods of Hephaestus which the poet says 'enter self-moved the company divine,' if thus shuttles wove and quills played harps of themselves (automatically), master-craftsmen would have no need of assistants and masters no need of slaves.[1] (Aristotle, Politics, A2, 4)

## 6 The Implementation of the Homeric Visions in the Hellenistic Period

The scientific investigations and the philosophical ideas of the pre-classical and the classical years lead to the methodical consideration and construction of the automata during the Classical and in particular the Hellenistic era.

The great school of the Alexandrian engineers, represented by:

- Ktesibios (∼300 BC)
- Philon of Byzantium (∼250 BC)
- Heron of Alexandria (∼100 BC)

paved the way to the implementation of the Homeric visions.

The utilization of the internal energy of natural elements, especially of water and air, is achieved by the development of the science of Pneumatics. The problem of

[1] This text is based on *Aristotle* in 23 volumes, Vol. 21, translated by H. Rackham, Harvard University Press, Cambridge, MA; William Heinemann Ltd., London, 1944.

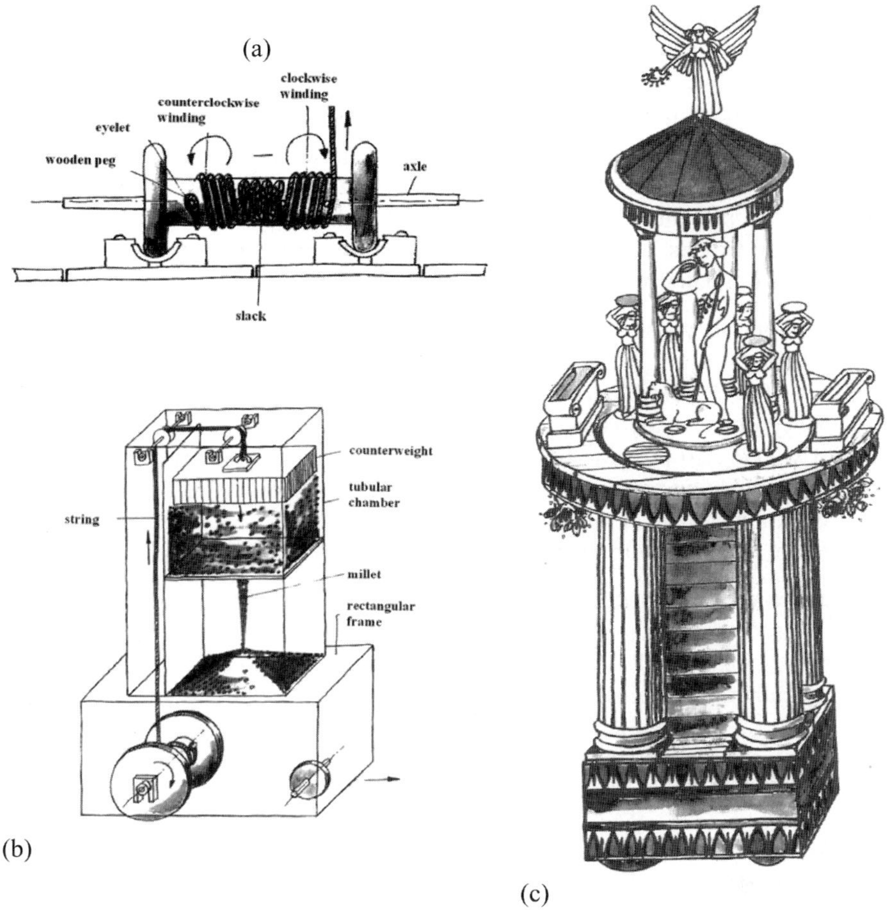


**Fig. 3** The thread-windings around the wheel axle (a), the mechanism of motion (b) and the total look (c) of Heron's mobile automaton.

systems control is solved by means of programming the movements and the operation of automata. This programming is achieved by using three different kinds of thread-windings around the wheel axle, such as in the mobile and the stable automatic theaters.

The feedback and the regulation of water flow and level into 'clever' vessels are obtained either by hydraulic, pneumatic or mechanical way. The science of Automatopoietice is established and taught in the Hellenistic Alexandria.

The mobile automatic theater of Heron with its astonishing movements, such as the straight or cyclic motion and return to initial position, lighting of fires, production of sounds, flow of milk or wine, and automatic wreathing of altar with flowers, is just great expression of gratitude to Homer, who, through his automatic tripods, showed the way to Alexandrian engineers.

## 7 Conclusions

- The Iliad and the Odyssey include not only catalogs of the races and gods of the Greeks, but also catalogs of their technical achievements and inventions. The automata possess a prominent position among these inventions.
- Homer introduces the term 'automaton' in order to describe self-moving machines. He presents Hephaestus, the god of technology, as the manufacturer of automatic tripods, adaptive bellows and gold woman-like robots. In this way, Homer formulates modern technological visions, such as:
  - It would be possible to manufacture automobile machines to replace human laborers.
  - It would be possible for these machines to operate automatically for production purposes and to have their operation regulated as needed for work at hand.
  - It would also possible be to manufacture man-like machines (robots), programmable by means of proper software, capable of obeying commands and communicating verbally.
  - Homer ends up with the Phaeacians' ships, possessing artificial intelligence.
- The problem of the necessary internal energy of the automata is solved, at first, by the Presocratic philosophers, who consider the fundamental elements of nature as possessing a 'soul', i.e., energy. The practical utilization of this energy occurs in the Hellenistic period.
- The control problem is also examined firstly from a theoretical perspective by the Presocratics, who introduce the concepts of contradiction and feedback. The practical application of control problem is completed later by the Alexandrian engineers.
- Homer paved the way to the history of automata.

## References

1. Kalligeropoulos, D. and Vasileiadou, S., *History of Technology and of Automata*, Synchroni Ekdotiki, Athens, 2005 [in Greek].
2. Kalligeropoulos, D., *Myth and History of Ancient Greek Technology and of Automata*, Vol. A, Technology in Ancient Greek Myths – Mythological Automata, Kastaniotis, Athens, 1999 [in Greek].
3. Vasileiadou, S., Evolution of system modelling and control concepts in ancient Greece, Ph.D. Thesis, City University London, 2002.
4. Kalligeropoulos, D., *Automatopoietice of Heron Alexandrinus, The Art of Manufacturing Automata*, Athens, 1996 [in Greek].
5. *Heron Alexandrinus*, Opera Vol. 1-5, Schmidt W., Teubner, Leipnig, 1899, Stuttgart, 1976.
6. Murray, A.T., *Homer: The Odyssey*, with an English translation, William Heinemann Ltd., London, in two volumes, 1919.
7. Murray, A.T., *Homer: The Iliad*, with an English translation, William Heinemann Ltd., London, in two volumes, 1924.

# The River Ocean: Homer's Cosmogony

Taha Showleh

*Concordia University, Montreal, Canada*

'Ocean is the genesis of all', Homer says. And Ocean is a river at the ends of the earth. 'I am going to see the limits of fertile earth', Hera says to Zeus, 'Okeanos begetter of gods and mother Tethys' (Il. 14.301-2)[1] – the River Okeanos, the origin of all things (Il. 14.245-6).

When Circe gives Odysseus directions to the land of the dead, she says:

> But when in your ship you have now crossed the stream of Ocean, where there is a level shore and the groves of Persephone – tall poplars, and willows that shed their fruit – there beach your ship by the deep eddying Ocean, but go yourself to the dank house of Hades. (Od. 10.508-512)[2]

And Odysseus follows her instructions; his ship, he says,

> came to deep-flowing Ocean, that bounds the earth, where is the land and city of the Cimmerians, wrapped in mist and cloud. Never does the bright sun look down on them with his rays either when he mounts the starry heaven or when he turns again to earth from heaven, but instead horrid night is spread over wretched mortals. There we came and beached our ship, and took out the sheep, and ourselves went along beside the stream of Ocean until we came to the place of which Circe had told us. (Od. 11.13-22)

And when Odysseus has returned to his home in Ithaca and killed the men who were wasting his wealth and trying to usurp his kingdom and marry his wife, Hermes calls forth the ghosts of the suitors:

> He held in his hands his wand, a beautiful wand of gold, with which he lulls to sleep the eyes of whom he will, while others again he wakens out of slumber as well; with this wand he roused and led the ghosts, and they followed

[1] Translation from Kirk and Raven.

[2] Unless otherwise noted, translations from the *Iliad* and *Odyssey* are from the Loeb editions (with occasional adjustments).

*S.A. Paipetis (ed.), Science and Technology in Homeric Epics,* 85–91.

> gibbering. And as in the innermost recess of an eerie cave bats flit about gibbering, when one has fallen off the rock from the chain in which they cling to one another, so these went with him gibbering, and Hermes, the Helper, led them down the dank ways. Past the streams of Ocean they went, past the rock Leucas, past the gates of the sun and the land of dreams, and quickly came to the meadow of asphodel, where the ghosts dwell, phantoms of men who have done with toils. (Od. 24.1-14)

The concept of Ocean as the boundary of the earth is clear in Hera's reference in the Dios Apate or Deception of Zeus. She asks Aphrodite to give her 'love' and 'desire':

> For I am going to visit the limits of the all-nurturing earth, and Ocean, from whom the gods are sprung, and mother Tethys, those who lovingly nursed and cherished me in their halls, when they had taken me from Rhea, at that time when Zeus, whose voice is borne afar, thrust Cronos down to dwell beneath earth and the unresting sea. Them am I going to visit, and I will loose for them their endless strife, since now for a long time they have held aloof from the marriage-bed and from love, because wrath has come upon their hearts. If by words I might persuade the hearts of these two, and bring them back to be joined together in love, ever should I be called dear by them and worthy of reverence. (Il. 14.198-210)

Martin West says that 'Behind this Olympian gossip there may lie a cosmogonic myth, for the separation of primeval parents who were originally united is a familiar cosmogonic motif' [14, p. 120].

The two aspects of the Homeric conception of Ocean, then, the geographical and the mythological, are intertwined in the narrative: Ocean as a river that encircles the world and Ocean as a primeval god from whom all other gods came and from which life itself arose.

The geographical conception has subsequently been proved wrong: we now know that the earth is not flat and that it is not bounded by a river-ocean. But this was the common view in early antiquity, and in spite of beliefs and observations to the contrary from at least the 1st century BC, the notion of a flat earth, floating on or surrounded by water, continued to be the commonly-held view throughout the Middle Ages and into the Renaissance, persisting in some quarters even today.

The conception of Ocean as a river encircling the earth and a body of water separate from the broad expanse of the sea is made clear in the opening lines of Book 12 of the *Odyssey*, as Odysseus and his men, having left the Underworld, return to Circe's island in the broad *thalassa*:

> Now after our ship had left the stream of the river Ocean and had come to the swell of the broad sea [the thalassa], and the Aeaean island, where is the dwelling of early Dawn and her dancing places, and the risings of the sun, there on our coming we beached our ship on the sands, and ourselves disembarked upon the shore of the sea; there we fell asleep, and waited for the bright Dawn. (Od. 12.1-7)

The liminal aspect of Ocean is manifest in the passages quoted above, where Ocean is the boundary between the world of the living and the world of the dead, at 'the limits of the fertile earth', and in several other passages of the Odyssey, which speak of the sun rising from 'softly-gliding, deep-flowing Ocean', and 'golden-throned Dawn' coming up from 'the streams of Ocean', yoking her 'swift-footed horses that bring light to men'. This is where, for Hesiod, the Isles of the Blessed are located, 'beside deepswirling Ocean' (*Works and Days* 171).

The association of rivers with snakes or serpents is important here. As Martin West records [13, p. 41]:

> Babylonian boundary-stones of the late second and early first millennium often contain, besides inscriptions, pictorial representations of religious and cosmic significance, not all of which can now be understood, but in which the sun, the moon, and Venus are often clearly identifiable in the uppermost register. Associated with them is a huge serpent who rises up from far below, his body sometimes coiling round a good part of the circumference of the stone in a horizontal direction [perhaps associated with the zodiac]. ... Here is possible evidence for a Babylonian conception of a serpent encircling the world, now rising towards the gods, now turned down again to the surrounding ocean.

The association of an encircling serpent with encircling waters is found in many traditions, perhaps most prominently in the Vedas where frequent reference is made to the great serpent Vrtra, the 'encompasser', the son of Danu, 'stream' or 'waters of heaven' [13, p. 48]:

> [Vrtra] lay upon a lofty summit, encompassed the water, or the rivers, and prevented them from flowing. But the great god Indra cast him down from his heights, destroying his ninety-nine fortresses, and uncovering the prison of the waters. The escaping waters overflowed the serpent, and he lies enveloped by them at the bottom of the lower air. Indra then produced the sun and set it in the sky. Sometimes the conflict is put in the past, sometimes it is treated as a seasonal event which is repeated constantly

as is the parallel dragon-slayer myth among the Hittites, the battle of the Stormgod with the dragon Illuyankas, recited every year at the spring festival of Purulli.

It is the conception of the cosmic serpent that we see in the boundary-stone of Gula-Eresh, from the 12th century BC (Figure 1) and in the Phoenician silver bowl from Praeneste, dated to the 8th or 7th century BC, showing various activities of mankind, encompassed by the encircling serpent, the *ouroboros*, with his tail in (or in this case near) his mouth. With these we may compare the Babylonian map of the world from the Neobabylonian period, showing Mesopotamia surrounded by a circular ocean (Figure 2). In the accompanying text the encircling ocean is called the Bitter River. We are reminded of the Shield of Achilles, made by Hephaestus, showing the range of activities of humankind: war and peace, weddings and lawcourts, sowing, reaping, vinting, dancing:

> And all around the outermost rim of the shield the god who made it set the great stream of the River Ocean, the river that is at once the frontier of the

**Fig. 1** Boundary stone of Gula-Eresh. British Museum.

**Fig. 2** Babylonian map of the world. British Museum.

> known and imagined worlds and the barrier between the quick and the dead. [5, p. 63]

The conception of Ocean encompassing the world is an iconic representation of Ocean begetting the world: Ocean is the genesis of all (Il. 14.246). Even if this passage is to be understood as 'the genesis of all gods', it comes to the same thing: the birth of the gods is the birth of the world; in the mythic world of the *Iliad* theogony is cosmogony. As the Vedic hymn (RV 1.32) says of Indra, he encompasses all things, as the rim of a wheel encompasses the spokes.

Water is the ultimate source of all life in the stories of many cultures. The Homeric Okeanos and his wife Tethys closely reflect and are almost certainly modeled on the Babylonian water-gods Apsû and Tiâmat, who are also the first parents and the creators of all life. As Lambert writes [6, p. 1829]:

> The ancients, like modern scientists, assumed that everything now in existence went back to a simple element. In recent decades a gas has been assumed by some scientists to have been the prime element in the universe. In the ancient Near East, there were various candidates – earth, water, or endless time, which can appear either alone or in combination. Only very rarely was the question put as to how the prime element arose, but it did occur and one answer is supplied in a bilingual Sumero-Babylonian incantation:
>
> Heaven was created of its own accord.
> Earth was created of its own accord.
> Heaven was abyss, earth was abyss.
>
> Here spontaneous generation of heaven and earth (namely, the universe) is proclaimed, but then we are told that there was in fact no heaven or earth but only a body of water, which is the implication of the third line quoted.

The opening lines of the Babylonian *Epic of Creation*, the *Enuma Elish*, show these primeval forces engaging in procreation to bring about the universe [6, p. 1830]:

> When above the heavens were not named,
> Below, the earth was not called a name ...
> There was Apsû, the first in order, their begetter,
> And Demiurge Tiâmat, who gave birth to them all;
> They had mingled their waters together
> Before meadowland had coalesced and reed-bed
> was to be found –
> When not one of the gods had been formed
> Or had come into being, when no destinies had been decreed,
> The gods were created within them.

Apsû is the primeval body of fresh water, lying under the earth and giving rise to all springs and rivers; Tiâmat is the sea.

'Creation', as Lambert writes, 'thus began with Apsû's and and Tiâmat's mixing their waters together – primeval sexual intercourse – and so the next generation came about' including Heaven and Earth. 'Mythologically this made great sense, since if plant life is considered a paradigm of existence, then father heaven sends down his fertilizing rain into the bosom of mother earth and plant life can flourish: cosmic sexual intercourse. Thus the theogony offered in this opening passage proceeds very rapidly from the prime element, water, to the known universe consisting of heaven and earth' [6, p. 1830].

The early Greek philosopher Thales is famous for his claim that water is the source of all life. According to Aristotle, Thales 'perhaps came to acquire this belief from seeing that the nourishment of everything is moist and that all hot things come from water and live by water (for that from which anything comes into being is its principle) – he came to his belief both for this reason and because the seeds of all things have a moist nature, and water is the natural principle of moist things' (*Metaphysics* 983b6-11, 17-27).

Jonathan Barnes [1, pp. xxii–xxiii] says that Thales' view that everything, including sand, is nothing but water 'is false in fact; but it is not foolish – on the contrary, it is thoroughly scientific in spirit'. The 'Chaos' from which all life came in Hesiod's *Theogony* was said by later philosophers to be water, which created mud, and from the mud the earth was formed.

Like the Babylonian and other early accounts of the origin of life and the world, the Greek poets' accounts are mythological stories. But while Homer's geographical conception of Ocean has been proved wrong, paradoxically the mythological/cosmic conception has been proved right. It is a fundamental tenet of modern science that life cannot exist without water. And while modern science will not make such a simple and straightforward claim as that 'water is the first principle of all things', it is interesting to note that:[3]

> Since the discovery of submarine hotsprings along mid-ocean ridges, many scientists have believed that life on Earth could have originated where hot volcanic waters well up and mingle with the sea. UK earth scientists Professor Mike Russell and Dr. Allan Hall of Glasgow University now think they have discovered a natural chemical mechanism for how this might have happened.

Russell and Hall summarize their view as follows:[4]

> ... we consider that a major geological process, the cooling by seawater of rocks under the floor of the ocean, played an important role in the origin of life. Such a process might seem remote from our everyday knowledge of life but it has now been known for more than twenty years that genetically primitive microorganisms are to be found living at warm springs on the ocean floor.
>
> ...
>
> We consider that the first living cells formed on the floor of an ocean on the earth thousands of millions of years ago. Life 'emerged' at the sites of warm submarine springs where chemical energy was focused and the mixing of spring water with seawater could lead to the precipitation of chemicals. ... In summary, we see life as having resulted from the interaction of warm sulfurous springs on the ocean floor of a young Earth.

The similarity of this current model to the account of Homer and of the Babylonians before him is striking. While we now have modern scientific models, knowledge of chemical and physical reactions, and terminology in which to express them, Homer already had the fundamental concept.

As Barnes says of Hesiod's cosmogony,

---

[3] www.resa.net/nasa/origins_life.htm

[4] www.gla.ac.uk/Project/originoflife/html/2001/laymans_abstract.htm; see also [10].

> This is not science; but it is, as it were, a scientific story: Many of Hesiod's gods and goddesses are, as their names indicate, personifications of natural features or phenomena, and in telling the birth of 'the gods' Hesiod is telling, in picturesque form, the origins of the universe.

The same is true of Homer. His account of the primary cosmogonic function of water is not science as we know it, but it is unquestionably a scientific story and a story that is now fundamental to our science.

## References

1. Barnes, J., *Early Greek Philosophy*, 2nd edn., Penguin Classics, London, 2001.
2. Burkert, W., *The Orientalizing Revolution: Near Eastern Influence on Greek Culture in the Early Archaic Age*, translated by Margaret E. Pinder and Walter Burkert, Harvard University Press, Cambridge, MA, 1992.
3. Edmunds, L., Myth in Homer, in *A New Companion to Homer*, I. Morris and B. Powell (Eds.), Brill, Leiden, 1997, pp. 415–441.
4. Kirk, G.S., Raven, J.E. and Schofield, M., *The Presocratic Philosophers*, 2nd edn., Cambridge University Press, Cambridge, 1983.
5. Knox, B., 1990. Introduction and notes, in *Homer, The Iliad*, translated by Robert Fagles, Penguin Classics, New York, 1990.
6. Lambert, W.G., Myth and mythmaking in Summer and Akkad, in *Civilizations of the Ancient Near East*, J. Sasson (Ed.), Charles Scribner's Sons, New York, 1995, pp. 1825–1835.
7. Lamberton, R., Homer in antiquity, in *A New Companion to Homer*, I. Morris and B. Powell (Eds.), Brill, Leiden, 1997, pp. 33–54.
8. Morris, I. and Powell, B. (Eds.), *A New Companion to Homer*, Brill, Leiden, 1997.
9. Morris, S., Homer and the Near East, in *A New Companion to Homer*, I. Morris and B. Powell (Eds.), Brill, Leiden, 1997, pp. 599–623.
10. Russell, M.J. and Hall, A.J., The emergence of life from iron monosulphide bubbles at a submarine hydrothermal redox and pH front, *Journal of the Geological Society of London* **154**(3), 1997, 377–402.
11. Sasson, J.M. (Ed.), *Civilizations of the Ancient Near East*, Charles Scribner's Sons, New York, 1995.
12. Schein, S.L., The *Iliad*: Structure and interpretation, in *A New Companion to Homer*, I. Morris and B. Powell (Eds.), Brill, Leiden, 1997, pp. 345–359.
13. West, M.L., *Early Greek Philosophy and the Orient*, Clarendon Press, Oxford, 1971.
14. West, M.L., *The Orphic Poems*, Clarendon Press, Oxford, 1983.
15. West, M.L., Ancient Near Eastern myths in classical Greek religious thought, in *Civilizations of the Ancient Near East*, J. Sasson (Ed.), Brill. New York, 1995, pp. 33–42.

# The Laws of Curvilinear Motion in the Iliad

S.A. Paipetis

*University of Patras, Greece*

**Abstract.** In Book 18 of the Iliad, Achilles, after the completion of the funerary rites for his deceased friend Patroclus, organizes games in his honor with precious prizes for the winners, among others, a chariot race, to which Antilochus, son of Nestor, King of Pylos, participates. Antilochus' horses appear to be inferior to those of his competitors, and Nestor instructs him how to win by knowledge rather than by the capacity of his horses. Nestor's words provide an excellent formulation of the laws governing the motion of a rigid body on a curved course. Antilochus follows his father's instructions and eventually wins the race.

## 1 The Mycenaean Chariot

A light chariot with spoked wheels was evidently developed in Syria or Northern Mesopotamia at about the beginning of the 2nd millennium BC and quickly propagated all over Middle East because of its usefulness at war. Such chariots appear in tombstones of the Mycenaean arched graves, as well as in Cretan seal rings in about 1450 BC. Besides war they were useful with hunting, even for traveling. No other form of chariots, developed for military purposes, appears in the Iliad. At around the end of the Bronze Era, roads for wheeled vehicles were constructed, especially connected to bridges of streams and rivers, some remnants of which can still be found in the area of Mycenae.

Chariots of Mycenaean and Archaic Greece used to have very light and flexible wheels, made by bending of very thin wood – of willow, elm or cypress – usually with only four spokes (Figure 1). So structured, wheel is very elastic and acts as the spring suspension of modern vehicles, allowing these chariots to trot on the rough ground of the Greek hillside, where heavier and more rigid vehicles would be useless. In fact, the wheel's hub was bending as a bow under the chariot's weight. Four-wheel chariots of similar design were a later development (Figure 2).

*S.A. Paipetis (ed.), Science and Technology in Homeric Epics,* 93–100.

**Fig. 1** Reconstruction of a Mycenaean chariot (1499–1200 BC): The Homeric chariot was constructed by bending very thin wood, therefore, very flexible.

## 2 Nestor's Instructions to Antilochos

In the Iliad, 18.306-348, Nestor, King of Pylos, gives his son Antilochos, the following instructions, in order to win the chariot race, organized by Achilles to honour his friend Patroclos, who died in a duel with Hector, the Trojan Prince:

> Antilochus, said Nestor, you are young, but Jove and Neptune have loved you well, and have made you an excellent horseman. I need not therefore say much by way of instruction. You are skilful at wheeling your horses round the post, but the horses themselves are very slow, and it is this that will, I fear, mar your chances. The other drivers know less than you do, but their horses are fleeter; therefore, my dear son, see if you cannot hit upon some artifice whereby you may insure that the prize shall not slip through your fingers. The woodman does more by skill than by brute force; by skill the pilot guides his storm-tossed barque over the sea, and so by skill one driver can beat another. If a man go wide in rounding this way and that, whereas a man who knows what he is doing may have worse horses, but he will keep them well in hand when he sees the doubling-post; he knows the precise moment at which to pull the rein, and keeps his eye well on the man in front of him. I will give you this certain token which cannot escape your notice. There is a stump of a dead tree – oak or pine as it may be – some six feet above the ground, and not yet rotted away by rain; it stands at the fork of the road; it has two white stones set one on each side, and there is a clear course all round it. It may have been a monument to some one long since dead, or it may have been used as a doubling-post in days gone by; now, however, it has been fixed on by Achilles as the mark round which the chariots shall turn; hug it as close as you can, but

**Fig. 2** A vase from around the end of Dark Ages. At the lower zone, horse-driven chariots, reflecting Mycenaean practices, are depicted at a funeral procession.

as you stand in your chariot lean over a little to the left; urge on your right-hand horse with voice and lash, and give him a loose rein, but let the left-hand horse keep so close in, that the nave of your wheel shall almost graze the post; but mind the stone, or you will wound your horses and break your chariot in pieces, which would be sport for others but confusion for yourself. Therefore, my dear son, mind well what you are about, for if you can be first to round the post there is no chance of any one giving you the goby later, not even though you had Adrastus's horse Arion[1] behind you horse which is of divine race – or those of Laomedon, which are the noblest in this country.

[1] This horse, Arion of Adrastus, was one of the favorite characters of the Thebaean Circle, still, outside the mythological Homeric circle. He was endowed with speech and reason and was related to the Arcadian cult of Poseidon (Neptune) and Demeter. In Pausanias (*Arcadian*, 25, 5), there are references to Thebais and Antimachus. According to the tradition, this mythical horse has its origin in Poseidon. It is not certain how it came to be possessed by Adrastus.

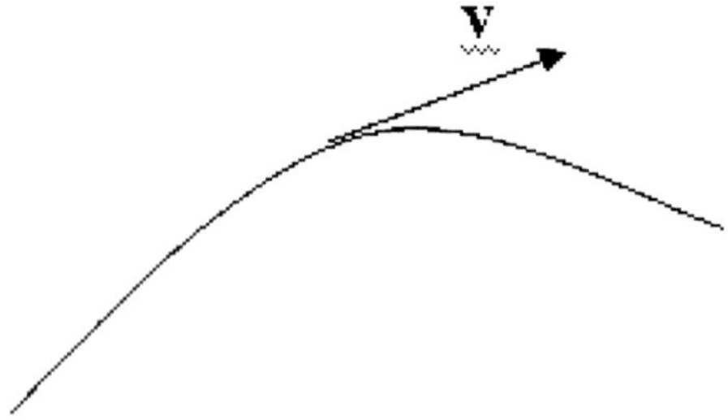


**Fig. 3** Motion on a curvilinear course.

## 3 On Curvilinear Motion

According to the First Law of Newton, pertaining to the equilibrium of a rigid body, if no force acts upon it, it either rests or it moves with uniform motion, i.e. in a constant direction at constant speed [1]. However, velocity is a *vectorial entity*, i.e. it is defined both by its meter (speed) and its direction. For example, the speed of a car appears on its speedometer, while its direction of motion is determined by the driver who is controlling the wheel. Accordingly, by speaking of constant velocity, we mean constant speed and constant direction of motion, i.e. a uniform motion along a straight line.

To change this constant velocity, a force must act on the body. The effect of this force is governed by the Second Law of Newton, expressed by a very simple equation:

$$\bar{F} = m \cdot \bar{a},$$

where $\bar{F}$ is force, $m$ the mass and $\bar{a}$ the acceleration, i.e. the rate of change of the velocity of the body with mass $m$.[2] From this Law one may conclude the following:

If, at a certain moment, a force acts on the moving body in or against its direction of motion, this will be *accelerated*, i.e. its speed in the direction of motion will increase, or will be *decelerated*, i.e. its speed decreases respectively. However, if the force acts sidewise, normally or at an angle to the direction of motion, then the latter will change as well.

For a body to move on a *curvilinear course* (Figure 3), a normal force must act continuously, since, if it stops acting, then the body will move out of its course along the tangent of the curve. In particular, for a body to remain on a circular course, a force is needed, to act constantly on it, with constant meter $F_n$ and directed towards the centre of the circular course (Figure 4).

This force is called *centripetal* and its meter is equal to:

[2] Both force and acceleration are vectorial entities, i.e. they are defined both by their meter and their direction.

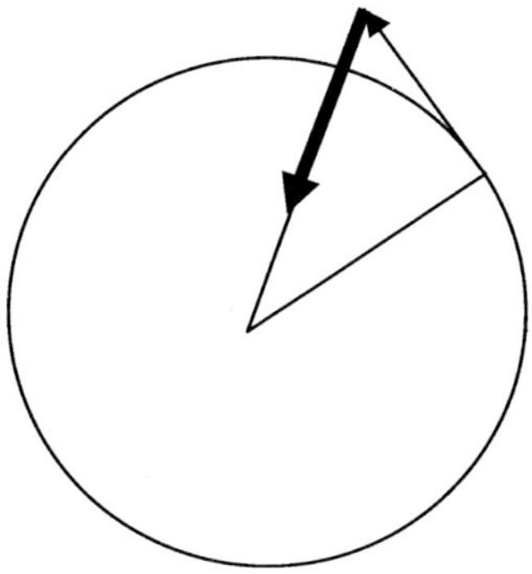

**Fig. 4** Motion on a circular course.

$$F_n = \frac{\nu^2}{R},$$

where $\nu$ is the velocity meter (or *speed*) of the body on the circular course and $R$ is the radius of the latter.

The above formula leads to the following interesting remarks:

(a) With increasing speed, the centripetal force increases by the square of the increase: if, for example, we are driving on a circular road and increase speed by 10%, the centripetal force increases by 12.1%, while with speed 15% higher, the centripetal force increases by 32.25%.
(b) The centripetal force, necessary to keep the body on the circular course, one conceives as a force of opposite direction, i.e. as a *centrifugal* force, tending to drive the body out of its course. This force does not really exist; it simply expresses the inherent tendency of the body to move rectilinearly along the tangent of its circular course.
(c) The smaller becomes the radius of curvature $R$, e.g. the more abrupt the turn, the greater the centripetal force. In fact, for very small radius of curvature, the vehicle may hardly remain on course and can jump out of the road.

This problem is especially important for air fighter pilots: A plane with an enemy plane at its tail, must execute a U-turn as quickly as possible, in order to get at the tail of the enemy and use its weapons successfully. In this case, the speed is abruptly increasing, while the radius of the course decreases: The centripetal force imposes a great acceleration towards the centre, which may assume values of the order of nine times the acceleration of gravity, e.g. the weight of the pilot increases by a ninefold.

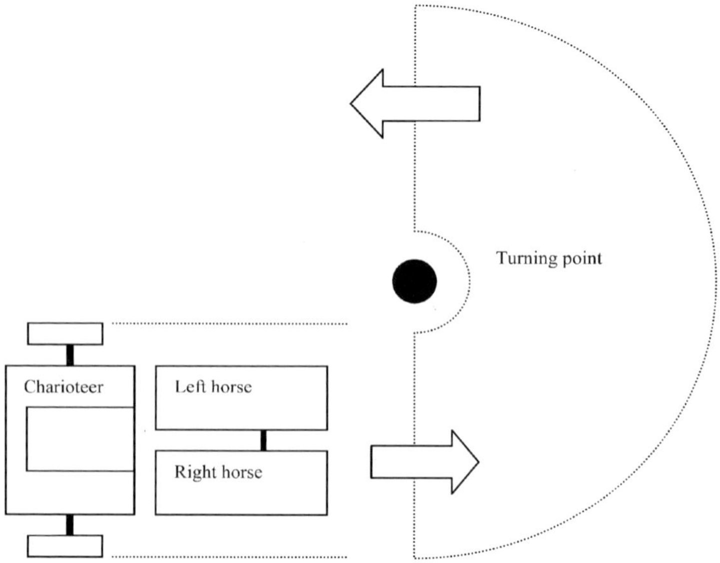


**Fig. 5** Turnaround or U-turn of a chariot.

## 4 The Chariot Race

According to the Homeric description, the race track had an oblong shape. The chariots started from one end towards the other and, when they got there, they had to execute a U-turn and get back to the starting point. This 180 degree turn (U-turn) was executed anticlockwise, i.e. in a sense opposite to the direction of motion of the hands of a clock. The end of the race track was signified by means of a wooden pole, around which the chariots should execute inversion of motion, i.e. they should move on a semicircular course, in order to move further in the opposite direction, towards the starting point. It is obvious that this moment of the race was the most difficult, since not only fast horses were needed, but also great skill from the side of the charioteer, in order to manage successfully the forces developing during the circular motion described above. One can note the following (Figures 5 and 6):

1. Nestor advises Antilochus to move as close to the pole as possible (not too close though, or the wheels of the chariot will hit on one of the two rocks supporting the pole and will be overturned). This is an obvious choice, since the length of the semicircular trajectory is proportional to its radius. In fact, if the radius is $R$, the length $S$ of the course is $S = \pi \cdot R$, where $\pi = 3{,}1416$, in other words, doubled radius corresponds to more than doubled course length. This is the well-known "internal course" of stadiums.
2. In this way though, the radius of curvature for the left horse will be very small, i.e. the centripetal force will very much increase, unless the speed of the horse

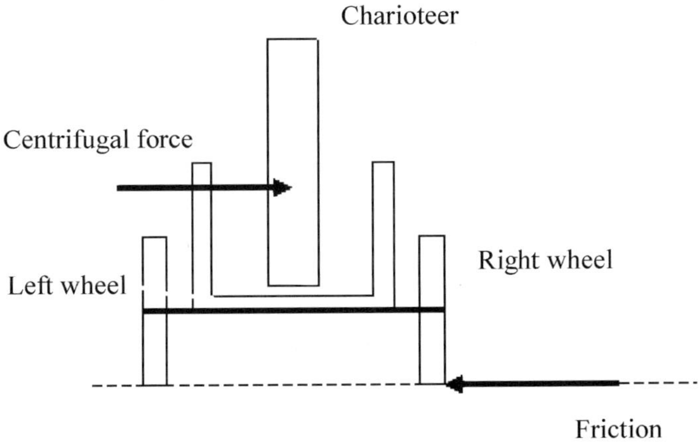


**Fig. 6** The couple of forces tending to overturn a vehicle on a curved course.

is properly reduced. If this occurs, the ratio $v^2/R$ remains within acceptable limits.

3. However, things are different for the right horse, which is at a distance from the centre greater than the left horse. Accordingly, in order that both horses be maintained on the same radius (i.e. not to be detached from the yoke), the right horse *must accelerate* its motion. This is what Nestor recommends to Antilochus to pursue, *by scream and lashes and by letting lose* the horse's reins.
4. The main requirement is that the U-turn be executed in the shortest time possible, i.e. the speed must remain the highest possible, however, with the centripetal force not exceeding a certain limit, beyond which the chariot is overturned outwards. In fact, the centripetal force is applied on the chariot through friction at the contact point of the wheels with the ground, and is directed sidewise, i.e. towards the centre of the course. On the contrary, the inertial force, expressing the resistance of the chariot to the change of direction of its rectilinear course, is manifested as a "centrifugal force", applied at the gravity centre of the chariot (more precisely, of the system "chariot/charioteer"), and is directed outwards. These two forces constitute a couple corresponding to the *turnover moment* for the chariot (Figure 7). This moment can be decreased if (a) the centre of gravity of the chariot is moving lower and (b) the charioteer, by moving his body to the left and also to a lower position, creates a moment of opposite direction than the turnover moment. This is exactly what Nestor recommends to Antilochus, by suggesting that he should *bend the elastic chariot to the left*.

The whole operation, i.e. of minimization of inversion time is very difficult indeed, and Nestor states that, if Antilochus manages to pass first, no other chariot, even the ones driven by very fast horses, will be able to overcome him: A statement in full agreement with the laws of curvilinear motion, as presented above.

## References

1. Paipetis, S.A., *Engineering Mechanics*, Vol. I, Statics, Ion Publishers, Athens, 2003.

# Part 3

# MATERIALS

# Iron in the Homeric Epics & Homer, a Sensible Ecologist

George Varoufakis

*University of Athens, Greece*

**Abstract.** The present research shows that iron metallurgy was greatly developed during the geometric period and influenced Homer to such an extent, that he refers very often to this metal either as an indication of wealth, or as an important material for hard tools and weapons. What is more, Homer was impressed by its magical property to harden when (a) transformed into steel by carbonization and (b) subject to heat treatment of the latter. No other metal possesses these unique properties. In this work, the author refers to this magical metal as appears in the Homeric epics. Finally, he states the revolutionary effect of iron on a rapid development of mankind during the 1st millennium BC onwards. In the second part of the present, Homer appears to be a sensible ecologist, blaming both Achaeans and Trojans for the enormous ecological disaster they inflict to the environment.

## 1 Introduction

Iron was a metal of great importance in the 2nd millennium BC, and much more expensive than silver. In the following text, some Mycenaean iron signet rings are presented, indicating that iron was considered as a jewel during that period. It is interesting that all of these contain nickel and, in two cases, cobalt. The author speculates whether the raw materials are of meteoric origin or of metallurgical smelting of iron-nickel ores, available in the Greek mainland.

In the *Odyssey*, Homer is so intensely influenced by the extraordinary property of iron to harden when transformed into steel, followed by the heat treatment of quenching and tempering, that he describes the blinding of Polyphemus by the wily Odysseus on the basis of the said treatment.

In a second part of the present, Homer appears to be indignant against the catastrophe of the environment provoked by both sides. They devastate large regions of woods by committing cruel arsons. The worse of all disasters is that of Scamandros river, where Achilles slaughtered hundreds of Trojans and the river got red and muddy by their blood. Homer considers this as an unacceptable ecological catastrophe.

*S.A. Paipetis (ed.), Science and Technology in Homeric Epics,* 103–111.

## 2 Iron, a Poor, Disdained and Ugly Metal

Iron is regarded as a poor, disdained and ugly metal. Actually, who amongst us would ever stop, while walking, just to collect a piece of a rusty iron metal? Certainly nobody. However, it constitutes an important metal of our everyday life. Iron dominates everywhere around. Ships, cars, motorcycles, trains, cranes, factories, buildings, skyscrapers, bridges and so many constructions are made of iron. Certainly, no one could ever imagine life without the presence of this ugly, but so necessary metal.

## 3 Iron, a Metal with Magical Properties

Iron is the only metal that has the extraordinary property to acquire high hardness, when transformed into steel, followed by heat treatment by quenching and tempering. Homer in the 8th century BC was influenced by these magical properties to such an extent, that he refers to it several times in his Epics, though the Trojan War took place in an age when copper and bronze, rather than iron, were the main metals used to manufacture tools and weapons.

In some cases, however, Homer becomes conscious that during the Trojan War iron was regarded as a valuable metal, and its possession an indication of wealth. It is worth mentioning that iron was well known in the 2nd millennium BC, but in the shape of expensive signet rings. Thus Homer, sometimes mentions iron as a precious metal and sometimes as a hard metal, i.e. steel. It is very interesting to note that tablets found at Akkad, a region near present-day Baghdad, Iraq, quote that iron was six times more expensive than silver at the time.

## 4 Minoan and Mycenaean Iron Seal-Rings of 2nd Millenneum BC

Actually, the present research on iron seal rings exhibited in the Museum of Herakleion, Crete and the National Archaeological Museum, Athens, confirms this view, and leads to some interesting conclusions.

The iron seal-ring of Figure 1 of 16th century BC shows a golden foil, covering half of an iron bezel, the other half of the golden foil being fortunately missing. Fortunately because in this way only, one could ensure that its bezel was iron and that the ring would not be just gold but a combination of the two metals. The golden foil is engraved by three beautiful eight-shaped Minoan shields. The combination of iron and gold supports the view that iron constituted an expensive metal at that remote time. This ring is exhibited in the Archaeological Museum of Heracleion, Crete.

**Fig. 1** Iron seal ring of 16th century BC. A golden foil covers half of an iron bezel. The remaining golden foil is engraved by three eight-shaped Minoan shields.

**Fig. 2** A bezel of 15th century BC found at Kakovaton Pylos, Peloponnese.

Figure 2 shows just a bezel of 15th century BC found at Kakovatos, Peloponnese.

The pretty iron seal ring of Figure 3 is very similar to the previous one, found in Crete. Again, half of the golden foil is missing.

Figure 4 shows an iron seal-ring without a bezel. It is interesting that some parts look to be shining. The reason of this interesting fact is given below.

In Figure 5, an iron seal-ring is presented, found at Dendra Medea, Peloponnese. Its bezel consists of three different layers distinctly separated as a result of corrosion. The lower layer is silver, the middle lead, and the upper one iron.

Figure 6 shows an iron seal ring found in Mycenae. It belongs to 14th–13th century BC. The remaining of its bezel is silver, the rest of the layers are missing.

Except for the Minoan iron seal ring, all others are exhibited at the National Archaeological Museum, Athens, along with numerous similar iron seal rings. Unexpectedly, all of them were found to contain nickel and, on two occasions, cobalt.

**Fig. 3** Iron seal ring. Half of its bezel is covered by a golden foil, showing a beautiful deer.

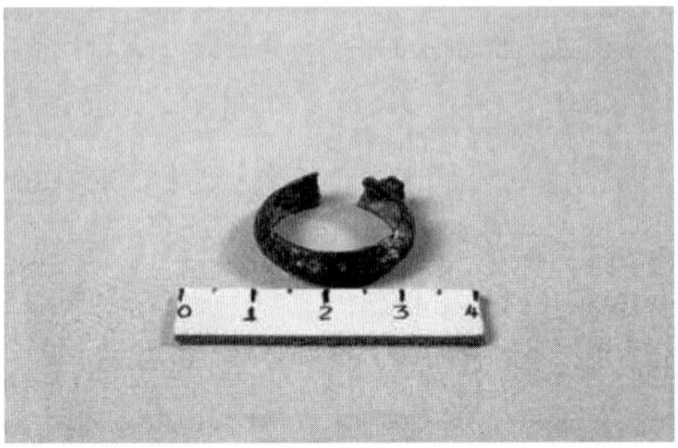

**Fig. 4** Iron seal ring without a bezel. Kakovaton, Pylos, Peloponnese.

**Fig. 5** Iron ring from Medea, Peloponnese. Its bezel consists of three distinct layers separated by corrosion. The lower is solver, the middle lead and the upper one is iron.

For this reason, one of them, appears to possess some shining parts. The question is whether the raw material used for their manufacture was derived from iron meteorites, all of which contain nickel. However, Larymna in Euboea and Skyros island possess nickeliferous iron ores, and the question is whether ancient metal workers used these ores to smelt nickeliferous iron as raw material and produce the said iron seal rings. However, the present research showed that nickel disappears from 13th century BC onwards. All iron findings contain no nickel. Metalworkers of the

**Fig. 6** A seal ring from Mycenae, 14th–13th century BC.

Geometric, Archaic, Classical, Hellenistic and Roman eras gave up smelting nickeliferous iron ores. Further research will reveal the reason for this change.

## 5 Iron Used in Exchange Commodities

As previously said, during 2nd millennium BC, iron was always considered as a valuable metal and was used in trade and exchange commodities. It is worth noting that, during the Trojan war, Achaeans obtained wine from Lemnos in exchange of copper and *shining iron* (Il. 7.472-473).

*The possession of iron as an indication of wealth*

It is interesting to note certain cases, during the Trojan War, that possession of iron was an indication of wealth. In Il. 7.44-448, is stated that, during a battle just outside the walls of Troy and after a hard pursuit, Menelaus and Agamemnon captured Adrastus, a young Trojan of a rich family, who, in despair, clasped King Menelaus' knees and beseeched him:

> Take me alive, thou son of Atreus, and accept a worthy ransom; treasures of copper, gold and hard to work iron; are stored in my father's palace, and thereof he would grant thee ransom ... should he hear that I am alive on the ships of Achaeans.

In this case, iron appears to be equally expensive as gold and copper.

The same scene is repeated (Il. 10.376-381), when Dolon was sent by Hector to spy out the Achaeans' camp, to make sure whether they would depart after the great casualties they suffered during the last fierce battle. In the same way, Odysseus and Diomedes appeared in the darkness to spy out on the intentions of Trojans. When they realized that Dolon was there for the same purpose, they followed him and, after a harsh pursuit, they captured and interrogated him about the future plans of Trojans. Dolon, pale with fear and in tears, begged of them:

> Take me alive, and I will ransom myself; for my father's palace possesses bronze, gold, and hard to work iron ... my father will grant you ransom ... should he hear that I am alive on the ships of the Achaeans.

It is obvious that, in all of these cases, iron is an indication of wealth. And this conclusion is in full harmony with archaeological findings and also with our research work on Minoan and Mycenaean iron rings.

## 6 The Use of Iron as a Weapon

Homer, influenced by his own era, refers to the use of iron in the manufacture of weapons and tools as hard steel, forgetting perhaps that weapons during the Trojan War were manufactured mainly of copper and its alloys.

1. In Book 4 of the *Iliad*, Homer refers to the use of an iron arrow-head. He describes how the Trojan Pandarus aimed at King Menelaus with his bow and an arrow with *iron head*.
2. Once again (Il. 4.485-486), during one of the bloody conflicts between Achaeans and Trojans, Ajax, son of Telamon, killed Simoeisios, a Trojan, with a *shining iron weapon*: "The man on the chariot killed him with a *cutting tool of shining iron ...*".

## 7 Odyssey

As already stated, Homer was so utterly impressed by the extraordinary property of iron to harden when transformed into steel, followed by the heat treatment of quenching and tempering, that he refers to this magical metal several times in his Epics, especially in the *Odyssey*. It is worth mentioning the blinding of Polyphemus by wily Odysseus described by Homer in Od. 9.390-394):

> ... As a blacksmith plunges a hatchet into cold water to harden it – the latter makes a great hiss, giving strength to iron- even thus did the Cyclops' eye hiss round the beam of olive wood ...

The poet is obviously deeply impressed by the discovery of this new metal. He could easily describe the terrible blinding scene by using any other metal or even a non-metal. They all hiss similarly when heated at high temperature, and then plunged into cold water. However, the power generated in iron, or, even more, in steel through quenching, stirred the poet so deeply, that he refers to this magical transformation, transmitting the message of his time: the great evolution of iron metallurgy during 9th and 8th centuries BC.

Hesiod, who lived in 8th century BC, was also deeply affected, and he attributes the misery of his own time to iron (*Works and Days*):

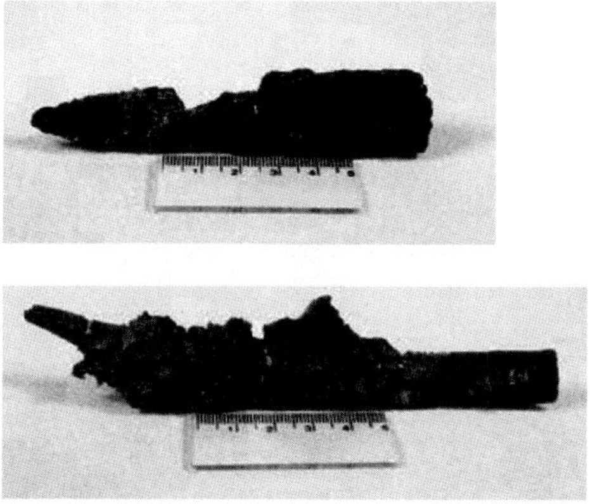

**Fig. 7** Two small iron daggers from Perati, Attica.

> ... I wish I did not live amongst people of the fifth generation, but that I had died earlier or been born later, because this generation is a generation of iron and the exhaustion and the misery of the people never ceases ...

Actually, Hesiod feared not iron itself, but the consequences it would entail as far as human relations of his time were concerned. And he was right. After 8th century, the pace of history accelerated with unimaginable rapidity and this caused social unrest and extensible bloodshed. Nevertheless, there is a positive aspect. From 8th century BC onwards, a remarkable activity in all fields of civilization and culture appears: art, poetry, philosophy, science, pottery, metallurgy, metal crafts, and technology in general. Whilst during the Bronze Age history is written in millennia, from now on a rapid evolution every two centuries occurs. Thus Geometric Period is followed by Archaic, Classical, Hellenistic and Roman periods. So many changes within only eight centuries!

It is worth mentioning some iron weapons, that the author had the opportunity to study from a metallurgical point of view some years ago. Among them, two small iron knives of 12th century BC (Figure 7), found at Perati by S. Iakovidis, Professor of Archaeology and distinguished member of Athens Academy. No similar items, belonging to this remote era, were found.

Next Figures 8 and 9 show archaic and geometric iron, or rather steel weapons. At this time many iron weapons and tools are found. Actually, iron age has appeared, and the revolutionary effect of iron on the development of mankind was very important from the 1st millennium BC up to present time, justifying the view that *the history of civilization is the history of iron*.

**Fig. 8** Geometric iron daggers.

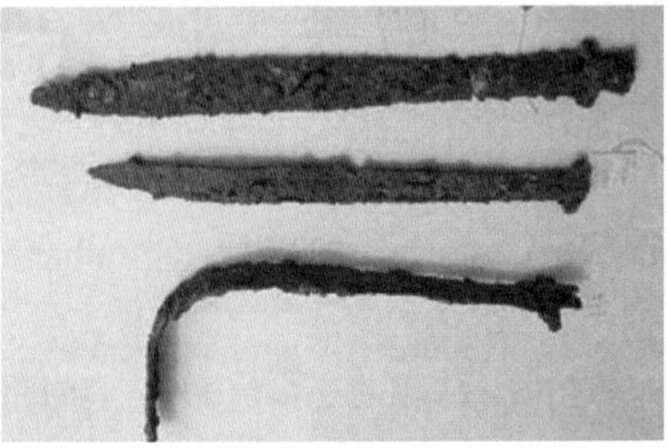

**Fig. 9** Iron daggers of the Archaic period.

## 8 Homer, a Sensible Ecologist of 8th Century BC

It is to admire Homer's great sensibility as far as environment and its protection are concerned. In Book 21 of the *Iliad*, Achilles slaughters whoever came into his sight, furious because of the tragic death of his beloved Patroclus. River Scamander, with the deep and rapid waters, got red and muddy by the blood of the Trojans slaughtered. Homer considers the fact as an unacceptable ecological catastrophe; an insult to the river god, as ancient Greeks believed that each river was an autonomous deity. Scamander, itself one of the sons of Zeus, got furious with the great disaster caused by Achilles and sought revenge, even to overwhelm him. Thus, he rushed on him with swollen waves and terrible whirlpools, "splashing out with foam, blood and dead people ...". Achilles was in complete despair. His drowning seemed almost unavoidable. As the poet recalls, he begged of Zeus to save him from the rage of Scamander. "I wish Hector had killed me (21.279)", "since a brave man would have killed another brave man (...), rather than to find a miserable death into the black waters of the wild river".

Finally, goddess Hera intervenes and with Hephaestus' help saved the hero. The god of metallurgy set fire to the nearby woods, a fire that spread out to the whole plane near the river. It burnt everything, even the bodies of deceased Trojans. The waters of the river started boiling due to heat and now Scamander was the one to find himself in a difficult position; he begged of Hera to make Hephaestus stop the fire, promising to stop tantalizing Achilles. Finally, the water of the river calmed down and Achilles was saved.

This scenery, as described by Homer, shows how sensitive he was towards environment. The great fire of the woods was certainly not caused by Hephaestus, but both by Achaeans and Trojans, who were trying to create major problems to one another. In any case, the great ecological catastrophe, inflicted upon to the river and the surrounding area seemed to have a profound effect upon Homer.

## 9 Discussion and Conclusion

In part 1 of the present, Homer emphasizes his admiration for the magical properties of iron, having, among all metals, the privilege to harden when transformed into steel and be heat-treated by quenching and tempering. However, since his epics refer to the Trojan War, taking place at around the end of 2nd millennium BC, he knows well, that iron is a very valuable metal, actually 6 times more expensive than silver, according to archaeological sources, and that is why he mentions it as an indication of wealth.

In part 2, Homer proves to be a very sensitive ecologist, objecting all environmental catastrophes caused both by Achaeans and Trojans.

# Early Bronze Technology at Land's End, North Western Iberia

Beatriz Comendador-Rey[1], Susana Reboreda-Morillo[1], Winfried Kockelmann[2], Mike Macdonald[3], Tony Bell[3] and Manolis Pantos[1]

[1]*Department of History, Art and Geography, Faculty of History, University of Vigo, Spain*
[2]*ISIS, Rutherford-Appleton Laboratory, Didcot, U.K.*
[3]*CCLRC, Daresbury Laboratory, Warrington, U.K.*

**Abstract.** The North Western Iberia metal ore wealth, especially tin ore and gold, have been proposed as the main reason for the development of intense trade routes since early prehistory. Several authors have argued the existence of interactions between the northwest of the Iberian Peninsula and other cultures of the European Occident and the Mediterranean area. Ancient sources comment on the abundance of minerals and metals in the *Cassiterides* or *Tin Islands*. These accounts must have originated from sailors who from time immemorial were trading in these coasts. The name *Cassiterides* represents the first vague knowledge of the Greeks that tin was found overseas somewhere in or off Western Europe. The word κασσιτερος was known to Homer and is mentioned ten times in the Iliad. Cape Finisterre (*Land's End* for the Romans) was proposed as the northernmost point recorded in the *Periplous* of Pytheas the Massaliot, which seems to be the basic source used by Rufus Festus Avienus. B. Cunliffe has suggested that if Cape Finisterre was the place called Oestrymnis by Avienus in *Ora Marítima*, then *Periplous* could be seen as the guide that led Greek sailors from Marseille to the northwest of Iberia to trade for the coveted Galician tin some time around 500 BC.

Recently, the study of prehistoric bronze-working places more emphasis on technological aspects as a means of detecting changes in the pattern of metal production in the archaeological record. Bronze working appears in North Western Iberia at the end of the 3rd and the beginning of the 2nd millennium BC for short-scale production, distribution and consumption, mainly as prestige goods. Metals are a major component of the prestige economy described in the Homeric Epics and Homer's accounts of Phoenician traders carrying metals hither and thither constitute the earliest literature reference. They suggest a prestige economy based in interchange of presents (most of them metals) that contribute to the perpetuation of the aristocracy, excluding ownership by the rest of the population.

Recent preliminary analyses carried out at the synchrotron and the neutron sources at the Daresbury and Rutherford Laboratories have contributed to the understanding of technological details of this very early bronze metallurgy. Archaeological evidence sustains the hypothesis of an increase of the production during the late Bronze Age. Most of these objects are produced locally, imitating foreign styles, especially in the Atlantic area, with singular features related to the alloy composition and other features. Indeed, as early as the beginning of the 8th century BC, the Phoenicians had established a trading post at Gadir. From here Phoenician ships regularly sailed north up the Atlantic coast of Iberia. We can observe differences between the Late Bronze Age and Iron Age metal production in the nortwest, the so-called *tartessic bronzes*. The technical aspects of the production of bronze during the Orientalising Period in the Iberian Peninsula favours the individualisation of different manufacturing traditions.

*S.A. Paipetis (ed.), Science and Technology in Homeric Epics,* 113–131.

# 1 Introduction

The North Western Iberia metal ore wealth, especially tin ore and gold, has been proposed as the main reason for the development of intense trade routes since early prehistory. Several authors have argued the existence of interactions between the northwest of the Iberian Peninsula and other cultures of the European Occident and the Mediterranean area. Ancient sources comment on the abundance of minerals and metals in the *Cassiterides* or *Tin Islands* [1]. These accounts must have originated from sailors who from time immemorial were trading in these coasts.

The name *Cassiterides* represents the first vague knowledge of the Greeks that tin was found overseas somewhere in or off Western Europe. The word κασσιτερος was known to Homer and is mentioned ten times in the Iliad (Table 1), very often in connection with Hephaistian metal-working, an age-old craft. Cape Finisterre (*Land's End* for the Romans) was proposed as the northernmost point recorded in the *Periplous* of Pytheas the Massaliot, which seems to be the basic source used by Rufus Festus Avienus. Cunliffe [2] has suggested that if Cape Finisterre was the place called Oestrymnis by Avienus in *Ora Marítima*, then *Periplous* could be seen as the guide that led Greek sailors from Marseille to the north-west of Iberia to trade for the coveted Galician tin some time around 500 BC.

Recently, the study of prehistoric bronze-working places more emphasis on technological aspects as a means of detecting changes in the pattern of metal production, distribution and consumption in the archaeological record. Several works [3] have highlighted the importance of physical techniques in addressing specific problems, i.e. the composition of the alloy, particularly the concentration of tin and how variations in time reflect evolution of bronze production.

Bronze-working appears in North Western Iberia in the beginning of the 2nd millennium BC for short-scale production, distribution and consumption. Traces of bronze metallurgical production have been reported from the *A Sola* (Level IIb) settlement site (Portugal) [4], and from *O Fixón-A Costa da Seixeira* (Galicia, Spain) [5] by the middle of the 2nd millennium B.C. (Figure 1). Important evidence of bronze metallurgical production was recently detected at the settlement site of *Fraga dos Corvos* (Trás-Os-Montes, North Portugal) dated to the 1700 B.C. CAL [6]. The production of flat-flanged axes has been indirectly confirmed by the presence of stone moulds (Figure 2).

Archaeological evidence from North Western Iberia supports the hypothesis of an increase in the production of bronze objects, such as axes, during the Late Bronze Age.

Most of the objects found are considered to have been produced locally, imitating foreign styles, especially of the Atlantic area, with singular features related to the alloy composition and others. Indeed, as early as the beginning of the 8th century BC, the Phoenicians had established a trading post at Gadir. From here Phoenician ships regularly sailed north up the Atlantic coast of Iberia [9]. It is interesting to observe technological differences between the so-called tartessic bronzes and the other LBA bronze traditions in Iberia. The technological aspects of bronze production during the Orientalising Period favour the development of individual manufacturing tra-

**Table 1** Verses using the word κασσιτερος in the Iliad. English translation from [7]. The full passage on Hephaistos preparing the shield for Achilles, in Il. Σ:468–617 is given in [8].

| | |
|---|---|
| Λ24 του δ ητοι δεκα οιμοι εσαν μελανος κυανοιο<br>Λ25 δωδεκα δε χρυσοιο και εικοσι κασσιτεροιο | Ten rods indeed were of dark metal, twelve of gold, and twenty of tin |
| Λ33 καλην ην περι μεν κυκλοι δεκα χαλκεοι ησαν<br>Λ34 εν δε οι ομφαλοι ησαν εεικοσι κασσιτεροιο<br>Λ35 λευκοι εν δε μεσοισιν εην μελανος κυανοιο | Round which were ten brazen orbs. Upon it were twenty white bosses of tin, and in the midst was one of dark metal. |
| Σ474 χαλκον δ εν πυρι βαλλεν ατειρεα κασσιτερον τε<br>Σ475 και χρυσον τιμηντα και αργυρον αυταρ επειτα | He cast into the fire impenetrable brass*, and tin, precious gold and silver; |
| Σ561 εν δ ετιθει σταφυλησι μεγα βριθουσαν αλωην<br>Σ562 καλην χρυσειην μελανες δ ανα βοτρυες ησαν<br>Σ563 εστηκει δε καμαξι διαμπερες αργυρεησιν<br>Σ564 αμφι δε κυανεην καπετον περι δ ερκος ελασσε<br>Σ565 κασσιτερου μια δ οιη αταρπιτος ηεν επ αυτην | On it likewise he placed, of huge size, a vine-yard, heavy with grapes, beautiful and golden ; but the clusters were black ; and they (the vines) stood against silver poles, placed in order. Round it he drew an azure ditch, about which he planted an edge of tin. |
| Σ574 αι δε βοες χρυσοιο τετευχατο κασσιτερου τε | But the oxen were made of gold and of tin |
| Σ609 αυταρ επει δη τευξε σακος μεγα τε στιβαρον τε<br>Σ610 τευξ αρα οι θωρηκα φαεινοτερον πυρος αυγης<br>Σ611 τευξε δε οι κορυθα βριαρην κροταφοις αραρυιαν<br>Σ612 καλην δαιδαλεην επι δε χρυσεον λοφον ηκε<br>Σ613 τευξε δε οι κνημιδας εανου κασσιτεροιο | But when he had completed the shield, huge and solid, he next formed for him a corselet, brighter than the splendour of fire. He also made for him a strong helmet, fitted to his temples, beautiful and variously wrought, and put upon it a golden crest; and formed greaves for him of ductile tin. |
| Υ268 ρηξε σακος χρυσος γαρ ερυκακε δωρα θεοιο<br>Υ269 αλλα δυω μεν ελασσε δια πτυχας αι δ αρ ετι τρεις<br>Υ270 ησαν επει πεντε πτυχας ηλασε κυλλοποδιων<br>Υ271 τας δυο χαλκειας δυο δ ενδοθι κασσιτεροιο<br>Υ272 την δε μιαν χρυσην τηι ρ εσχετο μειλινον εγχος | But the gold stayed it, the gifts of a god. Through two folds, however, it past, but there were yet three; since Vulcan had drawn five folds over it, two brazen, two of tin, and one golden ; where at length the brazen spear was stopped. |
| Φ592 αμφι δε μιν κνημις νεοτευκτου κασσιτεροιο<br>Φ593 σμερδαλεον κοναβησε παλιν δ απο χαλκοςορουσε<br>Φ594 βλημενου ουδ επερησε θεου δ ηρυκακε δωρα | But the greave of newly-wrought tin around it sounded horribly ; and the brazen weapon leaped back from it when struck, nor penetrated |
| Ψ503 αρματα δε χρυσωι πεπυκασμενα κασσιτερωι τε | The chariot, ornamented with gold and tin, |
| Ψ560 δωσω οι θωρηκα τον Αστεροπαιον απηυρων<br>Ψ561 χαλκεον ωι περι χευμα φαεινου κασσιτεροιο<br>Ψ562 αμφιδεδινηται πολεος δε οι αξιος εσται | I will give him the corselet which I took from Asteropaeus, brazen, round which is poured a rim of shining tin ; and it is of great value. |

*The translator uses the word *brass* instead of *bronze*. Brass (alloy of copper and zinc) was not used until much later times.

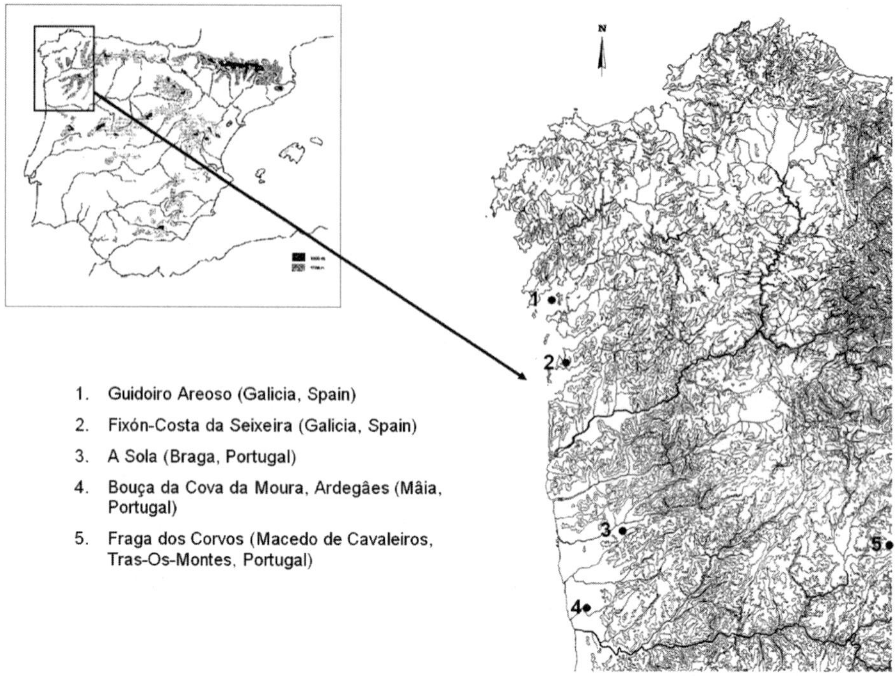


**Fig. 1** North Western Iberia. Main archaeological sites quoted.

ditions between the tartessic bronzes [10]. High-leaded bronzes are usual in North Western Iberia, whereas in South Western Iberia the absence of ternary alloys is almost total. These regional peculiarities could be related to the availability of the raw material [11]. But to know about this LBA tradition, we must understand about the origins of bronze alloy in the northwest in a technological and archaeological context. An interesting question is whether the bronze alloy is related to the tin sources. Different technological domain systems have been identified also for gold [12].

Metallurgy appears related primarily to power and not with domestic and functional use. As for cultural interaction, we talk about "imitation" or "trade", but we cannot solve this question only with discussions on typology. We should instead talk about technological transmission. The analysis of different aspects, composition, craft techniques, morphology, distribution, etc., help us distinguish between different productions.

## 2 Tin in Early Bronzes

A few of these Iberian early bronzes have been previously examined using Energy Dispersive X-ray Fluorescence (ED-XRF) and metallography [13]. Two metal

**Fig. 2** Stone Moulds: 1. *Monte das Carballas* (Guillade, Galicia); 2. *A Sola* (Braga, North Portugal); 3. *A Erosa* (Galicia); 4. Bun ingot from *Santo Ovidio* (North Portugal).

droplets and a bronze bar from the Bronze Age settlement site A Sola (Level IIb) (Braga, Portugal) dated to 1600–1500 B.C. (Figure 3), and two large-sized bronze awls [14] from the site on the islet of *Guidoiro Areoso* (Vilanova de Arousa, Pontevedra), are both considered to be amongst the earliest bronze objects from Iberia [15] (Figure 4). A new find, not previously examined, was a metal drop in two parts from the archaeological site of *Ardegães* (Bouça da Cova da Moura, Maia, Porto, Portugal) [16] (Figure 5).

The tin concentrations measured by ED-XRF are relatively high with most of the analysis showing percentages higher than 20 wt% with a small amount of trace elements that are considered to be impurities (As, Sb and other). However, it should be noted that ED-XRF is a surface characterisation technique and that the conclusions reached were based on the analysis of a limited number of objects. Moreover, bronze artefacts can have considerable variations of bulk and surface (corrosion) compositions [17]. Metallographic examination (Fig. 6) suggests that one of the examined objects was annealed, a practice which rarely occurs during the Chalcolithic and EBA period. The surface of all the objects analysed was always cleaned, look-

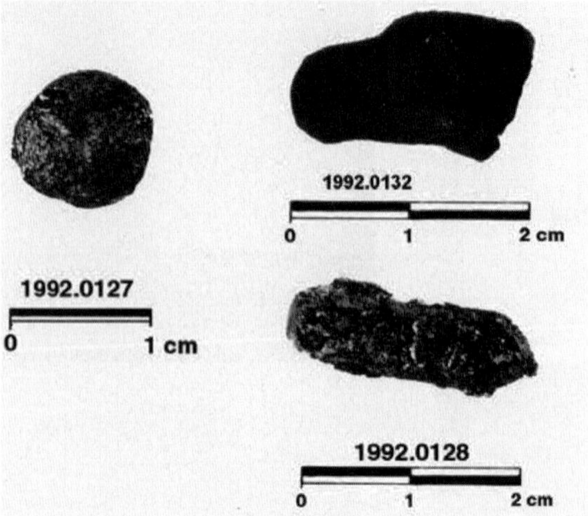


**Fig. 3** Bronze objects from *A Sola* (Level IIb) (Braga, North, Portugal).

**Fig. 4** Bronze objects (awls) from the site on the inlet of *Guidoiro Areoso* (Galicia, Spain).

ing for the bulk, but if the object is very corroded, it is impossible to remove all the corrosion products.

This paper focuses on the utilisation of the Synchrotron Radiation Source (SRS) at Daresbury Laboratory and the Neutron Spallation Source ISIS at the Rutherford-Appleton Laboratory during a Short Term Scientific Mission (STSM) in the frame of the EU COST Action G8 "Non-destructive analysis and testing of museum objects" [18]. The objective of these preliminary analyses on a few selected samples has been the understanding of technological details of this very early bronze metallurgy.

**Fig. 5** Bronze drop from the site of *Ardegâes* (Bouça da Cova da Moura, Maia, Porto, Portugal).

**Fig. 6** Metallography of the big awl of Guidoiro Areoso (Galicia, Spain), showing a metal cold-worked and annealed, a feature certainly rare at the sight of the general reckoning of Chalcolithic workshop techniques. Polished at the edge of the butt end of the object. Etched by ferric chloride and hydrochlroric acid in alcoholic solution, magnification ×250.

# 3 Neutron and Synchrotron X-Ray Diffraction Analysis

The instrumentation employed for the time-of-flight (TOF) neutron diffraction measurements on bronze and other metal objects has been described previously [19–21]. The high penetration power of neutrons for most materials allows a non-

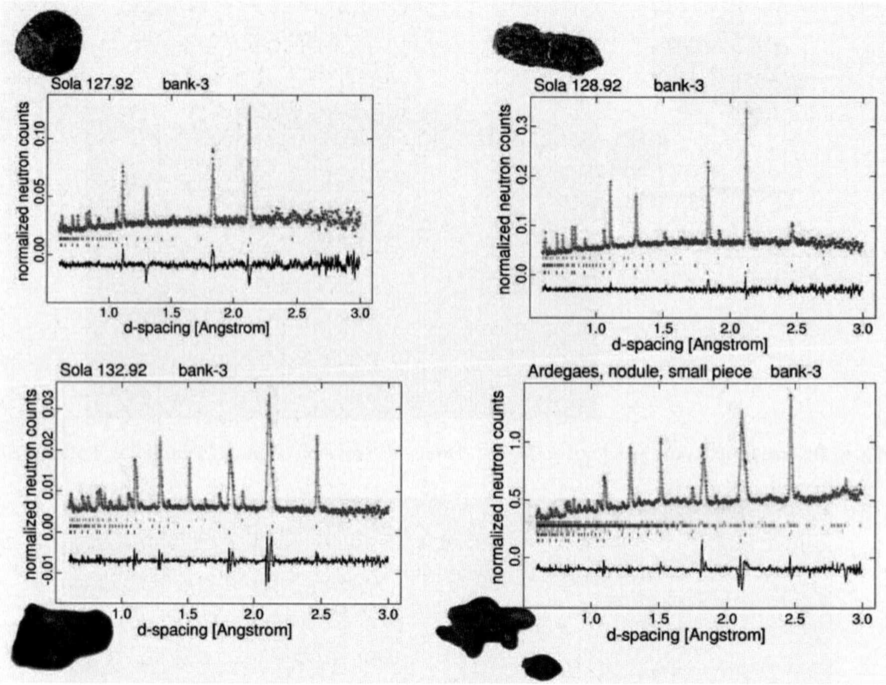


**Fig. 7** Neutron diffraction patterns were collected from bronze artefacts on ROTAX in order to determine the alloy compositions and to look for indications of microstrains as a result of cold working. Broad and narrow diffraction peaks are observed as a result of cold working and thermal annealing, respectively. (a) A Sola 127.92: 90/10 Cu/Sn bronze with a small amount of cuprite. The bronze peaks are only slightly (almost insignificantly) broadened, suggesting that the samples have been subjected to a homogenisation process. The presence of tin-rich delta and eta phases cannot be excluded. (b) A Sola 128.92: 90/10 Cu/Sn bronze with a small amount of cuprite. The bronze peak shapes are only slightly broadened, suggesting that the samples have been subjected to a homogenisation process. There are no high-tin phases present. (c) A Sola 132.92: 95/5 Cu/Sn bronze with amounts of cuprite and nantokite. The bronze peaks are very broad and "structured", as for an as-cast sample with an inhomogeneous tin distribution. The presence of tin-rich delta and eta phases cannot be excluded. (d) Ardagâes. Small piece: 95/5 Cu/Sn bronze with amounts of cuprite, nantokite, and malachite. The bronze peaks are distinctly broadened, as for an as-cast sample with an inhomogeneous tin distribution. The presence of tin-rich delta and eta phases cannot be excluded.

invasive bulk analysis of relatively large, intact objects which can be studied in situ without preparation. Neutron diffraction allows identification of crystalline phases throughout the examined area including information on the phases present in the bulk alloy and the corrosion phases on the surface of the object. For the measurements on the ROTAX diffractometer at ISIS, the samples were illuminated with a beam of cross section 20 $\times$ 20 mm. This means that a large part of the sample was immersed in the neutron beam, thus a more representative volume of the whole ob-

a b

**Fig. 8** (a) The X-ray diffractometer at the SRS station 2.3 used to study the corrosion products on the surface of a sample from *A Sola*. (b) View of the station 14.1 instrumentation showing the 2D CCD detector, the computer controlled $x$–$y$ stage and the sample cassette used to mount small corrosion flakes taken from areas on corroded objects, as indicated.

ject is sampled than with any other methods which normally probe small surface areas at a time.

Figure 7 shows the neutron diffraction patterns and the bronze objects from which the data was obtained. The concentration values of tin for these objects are between 5–10 wt%. There are also indications of tin-rich alloy phases in some of the samples. For this early production, the absence of control on the final product could result in achieving the bronze alloy by co-smelting of copper and tin ores, even the use of natural Cu-Sn ores [14]. Firm conclusions on this rather important and much discussed issue [22] cannot be reached from the small number of objects we have studied in this work. The concentration of trace elements is another important parameter which can actually be obtained using a different neutron technique, Prompt Gamma Activation Analysis (PGAA) [23], not available on the instrument we used at ISIS.

Object Sola 127.92, was also analysed using X-ray powder diffraction in flat-plate geometry at the SRS station 2.3 [24, 25] with the aim of identifying surface corrosion products in order to assess the preservation state of the object (Figure 8a). The diffraction patterns showed broad and weak Bragg reflections. Some of these reflections can be assigned to the oxides tenorite (CuO) and cassiterite ($SnO_2$). Because of the roughness of the object it was not possible to do depth-profiling of the corrosion layers, in order to establish the order in which the various corrosion phases were present. Such measurements are possible by varying the wavelength or the angle of incidence of the X-ray beam [26]. Such measurements require time and precise alignment procedures to conduct them properly, something that was not possible during the STSM beam allocation. X-ray powder diffraction of six micro-size corrosion fragments from both Guidoiro awls and other objects were measured at the SRS station 14.1 using a fast 2D CCD detector and a multi-sample cassette on a computer controlled $x$–$y$ stage [17] (Figures 8b and 9).

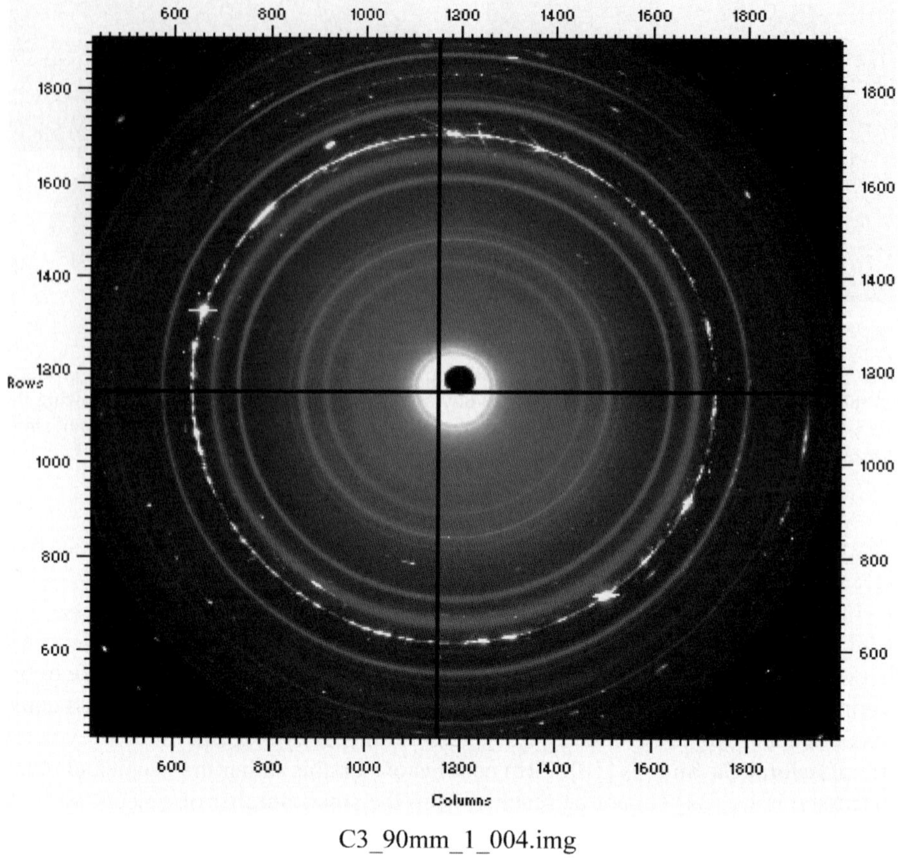


**Fig. 9** Two-dimensional X-ray diffraction pattern recorded by the CCD detector at the SRS station 14.1 from a micro-size fragment of corrosion on the big Guidoiro awl. Data collection time 30 sec.

Analysis of the diffraction patterns revealed the presence of cuprite, paratacamite and atacamite. In the case of a fragment from one of the objects (the big Guidoro awl) there is clear evidence of cassiterite. The identification of copper chlorides in particular (Figure 10) indicates the different preservation state of these objects, and allows the development of a conservation plan for the stabilisation of these objects.

The hypothesis of high values for tin in the early bronzes developed in earlier work [27] (surface analysis) is not put to doubt by the present measurements which clearly show that the tin content of the bulk alloy is below 10 wt%. The high values obtained by ED-XRF could be interpreted as tin enrichment in the surface due to corrosion/burial environment. ED-XRF analysis made on the surface of an archaeological metal artefact is able to characterise it, that is to provide important information on whether, for example, the alloy is bronze, leaded bronze or brass, but cannot identify it (that is quantify the tin content of the unexposed bulk) because of possible segregation phenomena. To determine the bulk composition of archae-

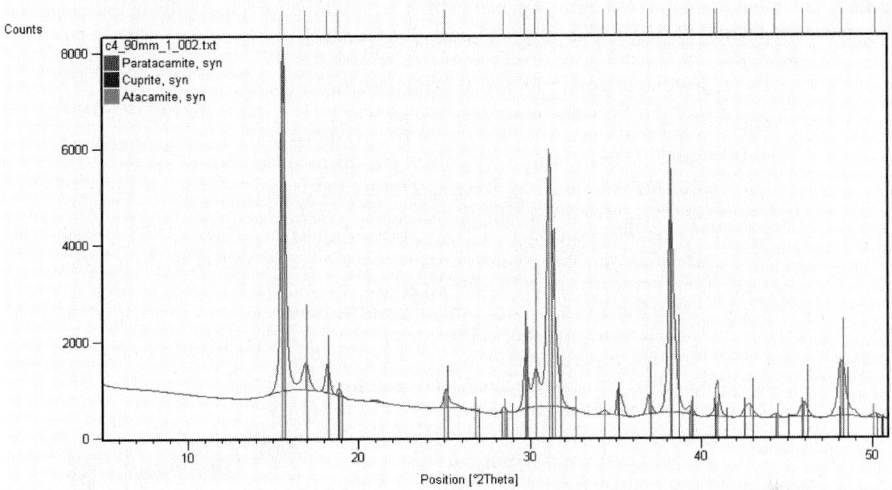


**Fig. 10** The integrated diffraction pattern corresponding to Figure 9 analysed using the X'Pert Highscore software package. The corrosion phases paratacamite, atacamite and cuprite are identified.

ological bronzes with ED-XRF one has to either study polished sections where the uncorroded interior of the object is exposed, or to use techniques such as neutron diffraction which permits precise and non-destructive determination of the bulk alloy percentages of the copper-tin phases, even in the presence of corrosion layers [18, 28].

## 4 Metals as Prestige Goods in the Homeric Epics

Let us return our attention now to non-technological matters and discuss how technology of production fitted in with customs and daily life practices in the Homeric Age which in most scholars' minds equates to the Bronze Age in Europe. Metals are a major component of the prestige economy described in the Homeric Epics. Homer's accounts of Phoenician traders carrying metals hither and thither constitute the earliest literature reference [29] to metal trading (as well as to the word Phoenician). In ancient societies, it is possible to detect two types of economy co-existing. On one hand the so-called subsistence economy, based on the exchange of necessary products, and on the other hand, the gift economy, which is based on gift exchange. It is precisely the latter aspect referred to in the Homeric poems, which we wish to comment on below.

The Homeric epics reflect a society whose political and economic power was residing in an aristocracy that was defined by the possession of land and goods [30]. In fact, the *basileus/anax* [31] (king) had to establish his position by demon-

**Table 2** Odysseus, acting on behalf of Agamemnon, offers Achilles precious gifts in exchange for his return to the battle. Il. I:260–282. English translation from [7].

παυε εα δε χολον θυμαλγεα σοι δ Αγαμεμνων
αξια δωρα διδωσι μεταλληξαντι χολοιο
ει δε συ μεν μευ ακουσον εγω δε κε τοι καταλεξω
οσσα τοι εν κλισιησιν υπεσχετο δωρ Αγαμεμνων
επτ απυρους τριποδας δεκα δε χρυσοιο ταλαντα
αιθωνας δε λεβητας εεικοσι δωδεκα δ ιππους
πηγους αθλοφορους οι αεθλια ποσσιν αροντο
ου κεν αληιος ειη ανηρ ωι τοσσα γενοιτο
ουδε κεν ακτημων εριτιμοιο χρυσοιο
οσσ Αγαμεμνονος ιπποι αεθλια ποσσιν αροντο
δωσει δ επτα γυναικας αμυμονα εργ ειδυιας
Λεσβιδας ας οτε Λεσβον ευκτιμενην ελες αυτος
εξελεθ αι τοτε καλλει ενικων φυλα γυναικων
τας μεν τοι δωσει μετα δ εσσεται ην τοτ απηυρα
κουρην Βρισηος και επι μεγαν ορκον ομειται
μη ποτε της ευνης επιβημεναι ηδε μιγηναι
η θεμις εστιν αναξ η τ ανδρων η τε γυναικων
ταυτα μεν αυτικα παντα παρεσσεται ει δε κεν αυτε
αστυ μεγα Πριαμοιο θεοι δωως αλαπαξαι
νηα αλις χρυσου και χαλκου νηησασθαι
εισελθων οτε κεν δατεωμεθα ληιδ Αχαιοι
Τρωιαδας δε γυναικας εεικοσιν αυτος ελεσθαι
αι κε μετ Αργειην Ελενην καλλισται εωσιν
.....

*"Yet even now desist, and lay aside thy disgraceful anger. Agamemnon gives gifts worthy of thee, if thou wilt, hear thou from me, and I will repeat to thee how many presents Agamemnon in his tents hath promised thee : seven tripods, which have not touched the fire, and ten talents of gold, twenty glittering goblets, and twelve strong steeds, victorious in the race, which have borne off prizes by their speed. The man need not be poor nor needy in precious gold, to whom so many prizes belong as the horses of Agamemnon have born off their speed. He will likewise give seven beautiful Lesbian women, skilful in working, whom he selected when thou didst thyself take the well-inhabited Lesbos, and who then excelled woman-kind in beauty. These will be given thee, and amongst them will be her whom he once took away, the daughter of Briseïs ; and he will moreover add a great oath, that he never ascended her bed, not embraced her, as is his custom. O king, both of men and women. All these shall immediately be thine ; and if the gods even yet grant that we take the vast city of Priam, thou mayst fill thy ships with abundance of gold and brass, entering in when we Greeks divide the spoil. Thou shalt thyself also choose twenty Trojan women, who may be fairest next to the Argive Helen. ...."*

.... and more offerings of more women, princesses, cattle and horses, lands and kudos. For Hector is about to make life difficult for the Achaeans.

strating that he was the most powerful amongst his aristocrat peers. The purpose of these goods was to be hoarded in the *keimelion* (*store of precious objects*), generally placed in a secret place of the *anaktoron* (*palace, the residence of the anax*) often in the cellar and jealously guarded by a faithful serf, to be given away as gifts again, an act that was extending the prestige of an already valuable object by enriching it with the genealogy of its previous owners.

The context and the meaning of these objects is interesting to be discussed. First of all, it is necessary to point out that access to such goods was restricted to a group

**Table 2** (Continued) Achilles refuses the gifts that wily Odysseus is offering, with contempt. Il. I:307–322 and 356–369. English translation from [7].

τον δ απαμειβομενος προσεφη ποδας ωκυς Αχιλλευς
διογενες Λαερτιαδη πολυμηχαν Οδυσσευ
χρη μεν δη τον μυθον απηλεγεως αποειπειν
ηι περ δη κρανεω τε και ως τετελεσμενον εσται
ως μη μοι τρυζητε παρημενοι αλλοθεν αλλος
εχθρος γαρ μοι κεινος ομως Αιδαο πυληισιν
ος χ ετερον μεν κευθηι ενι φρεσιν αλλο δε ειπηι
αυταρ εγων ερεω ως μοι δοκει ειναι αριστα
ουτ εμε γ Ατρειδην Αγαμεμνονα πεισεμεν οιω
ουτ αλλους Δαναους επει ουκ αρα τις χαρις ηεν
μαρνασθαι δηιοισι μετ ανδρασι νωλεμὲς αιει
ιση μοιρα μενοντι και ει μαλα τις πολεμιζοι
εν δε ιηι τιμηι ημεν κακος ηδε και εσθλος
κατθαν ομως ο τ αεργος ανηρ ο τε πολλα εοργως
ουδε τι μοι περικειται επει παθον αλγεα θυμωι
αιει εμην ψυχην παραβαλλομενος πολεμιζειν

*Whom answering, the swift-footed Achilles addressed : "Most noble son of Laertes, wily Ulysses, it behoves me undisguisedly to speak my opinion, even as I think, and as will be accomplished, that ye may not, sitting beside me, trouble me one after another. Hateful to me as the gate of hell is he who conceals one thing in his mind and utters another. But I will speak as appears to me the best ; and I think that neither Agamemnon the son of Atreus, nor the other Greeks will persuade me, since there is no satisfaction for fighting ever ceaselessly with hostile men. An equal portion is given to him who hangs back, and if one keenly fight : and the coward is in equal honour, even as the brave. The man who is useless, and he who performs many deeds, equally die ; nor does anything belong to me, because I have suffered sorrows in my soul, always exposing my life in combat."*

νυν δ επει ουκ εθελω πολεμιζεμεν Εκτορι διωι
αυριον ιρα Δτι ρεξας και πασι θεοισι
νηησας ευ νηας επην αλαδε προερυσσω
οψεαι ην εθεληισθα και αι κεν τοι τα μεμηληι
ηρι μαλ Ελλησποντον επ ιχθυοεντα πλεουσας
νηας εμας εν δ ανδρας ερεσσεμεναι μεμαωτας
ει δε κεν ευπλοιην δωηι κλυτος εννοσιγαιος
ηματι κεν τριτατωι Φθιην εριβωλον ικοιμην
εστι δε μοι μαλα πολλα τα καλλιπον ενθαδε ερρων
αλλον δ ενθενδε χρυσον και χαλκον λευκον
ηδε γυναικας ευζωνους πολιον τε σιδηρον
.....

*"But now since I choose not to combat with the noble Hector, having performed sacrifices tomorrow to Jove and all the gods, when I have dragged to the sea my well-loaded vessels, thou shalt behold, if thou wilt, and if such things be a care to thee, my ships early in the morn sailing upon the fishy Hellespond, and men within them eager for rowing ; and if the glorious Neptune grant but a prosperous voyage, on the third day I shall surely reach the fertile Pthia. I have there very many possessions, which I left, coming unauspiciously hither. And I will carry hence other gold and rudely brass, women elegantly zoned, and bright iron, which I have obtained by lot. But the reward which he gave, king Agamemnon, the son of Atreus, hath himself taken insolently from me : to whom report all, as I charge you, openly, that other Greeks also may be indignant, if he, ever clothed in impudence, still hopes to deceive any of the Greeks; not let him dare, dog though he be, to look upon my countenance. ..."*

of aristocrats who were perpetuated in this way, distinguished from other social groups. Acquisition of such goods (never by purchase) was achieved in two main ways: by war, where the goods of the defeated enemies were appropriated by force, and therefore turned into symbols of victory, or they were acquired as symbols of friendship, very often into the bosom of the bonds of hospitality that was implying obligatory exchange of prestigious goods. A third way, halfway in context between the previously quoted routes of gift acquisition, was through athletic competitions carried out in an ambience of warmth, but dominated by the competition. Again, Homer is the first available source of such societal practices. Gifts and gift exchange, at least as described in the Homeric Epics, had a common characteristic independent of origin; the possibility of changing the "contextual sphere". There exist other contexts mentioned in the Epics as gifts given as prizes in athletic games. In many cases the final destination of a prestigious object was to be used as gift (offering) to the gods, whereby its circulation stopped. Metal objects were the most common gifts, especially those made of gold, silver and bronze in that order of importance [32]. Iron, as deduced from the Homeric text (for example in the episode in the Iliad where Odysseus attempts to sweet-talk Achilles into cooperating (Table 2), seems to be a rather special case in the Homeric Epics, more precious than bronze or even gold perhaps [33].

One could group metal goods in different categories:

- The first category includes (bronze) tripods, cauldrons and craters, usual objects for water heating, or for mixing wine and water. Perhaps because of their function (suitable objects for ceremonial purposes or for social events) these objects were highly valued, possibly acquiring more value if they had never been used. In this case possession of such an object had the effect of increasing the prestige of its owner.
- The second category includes gold vessels, likely to be objects of ritual character destined, for instance, to perform libations in honour to the gods. Undoubtedly, in giving this gift there was the hidden intention that the receiver did not forget about the giver at a crucial moment.
- A third category includes weapons: swords, spears, shields, helmets and greaves for which, on the contrary to the first category, their value was increased if they were reputed to have been used previously by an illustrious warrior. The episodes accounted in the Iliad for the exquisite sceptre (material not specified) made by Hephaistos for Zeus and then given or inherited down generations by Agamemnon (Il. Β:100–107), Menelaos' gift to Telemachos (Od. δ:613–618) or King Alkinoos' farewell gifts to Odysseus (Od. ν:134–136) constitute good examples of this type of gift giving (Table 3).

In all cases mentioned above, objects were perpetuating the same reality: the social hierarchical structure of a society marked by the possession of prestige goods that were contributing much to the increase of the power of the giver, who was exhibiting publicly his aptitude to offer gifts. As of the receiver, accepting of gifts was increasing his wealth and therefore also the possibility of turning into giver. In all the cases of exchanging gifts, bonds of friendship but also of dependency were

**Table 3** Homeric accounts of prestigious goods handed down or given as gifts.

A. Hephaistos to Zeus to Hermes to Pelops to Atreus to Thyestes to Agamemnon, Il. B:100-108. English translation from [7].

παυσαμενοι κλαγγης ανα δε κρειων Αγαμεμνων
εστη σκηπτρον εχων το μεν Ηφαιστος καμε τευχων
Ηφαιστος μεν δωκε Δι Κρονιωνι ανακτι
αυταρ αρα Ζευς δωκε διακτορωι αργειφοντηι
Ερμειας δε αναξ δωκεν Πελοπι πληξιππωι
αυταρ ο αυτε Πελοψ δωκ Ατρει ποιμενι λαων
Ατρευς δε θνησκων ελιπεν πολυαρνι Θυεστηι
αυταρ ο αυτε Θυεστ Αγαμεμνονι λειπε φορηναι
πολλησιν νησοισι και Αργει παντι ανασσειν

... when king Agamemnon arose, holding in his hand the sceptre which Vulcan had toiled to form. Vulcan gave it to king Jove, the son of Saturn, and Jupiter bestowed it upon his messenger, the slayer of Argus. But king Mercury presented it to Pelops, the knight, and Pelops again gave it to Atreus, shepherd of the people. But Atreus, at his death, left it to Thyestes, rich in cattle ; and Thyestes again bequeathed it to be borne by Agamemnon, that with it he might govern many islands, and the whole of Argos.

B. Menelaos offers Telemachos a silver crater, made by Hephaistos for the king of Sidon who in turn gave it as a gift of friendship to Menelaos. Od. δ:613-618. English translation from [34].

δωρων δ οσσ εν εμωι οικωι κειμηλια κειται
δωσω ο καλλιστον και τιμηεστατον εστι
δωσω τοι κρητηρα τετυγμενον αργυρεος δε
εστιν απας χρυσωι δ επι χειλεα κεκραανται
εργον δ Ηφαιστοιο πορεν δε ε Φαιδιμος ηρως
Σιδονιων βασιλευς οθ εος δομος αμφεκαλυψε
κεισε με νοστησαντα τειν δ εθελω τοδ οπασσαι

Selected from my stores, of matchless price,
An urn shall recompense your prudent choice:
Not mean the massy mould of silver, graced
By Vulcan's art, the verge with gold enchased.
A pledge the sceptred power of Sidon gave,
When his realm I plough'd the orient wave

C. Alkinoos, the king of the Phaiakans, has sent Odysseus back to Ithaca, loaded with presents. Od. ν:135-137. English translation from [32].

κατθεσαν ειν Ιθακηι εδοσαν δε οι ασπετα δωρα
χαλκον τε χρυσον τε αλις εσθητα θ υφαντην
πολλ ος αν ουδε ποτε Τροιης εξηρατ Οδυσσευς

Of brass, rich garments, and refulgent ore ;
And bears triumphant to his native isle
A prize more worth than Ilion's noble spoil.

created, since a gift would increase in value (reciprocated) at the moment of an expected exchange.

## 5 Conclusions

The results of neutron diffraction investigation of some of the finds from excavations in North Western Iberia have been compared with ED-XRF results from earlier work which seemed to suggest rather high concentrations of tin in bronzes during the Iberian Bronze Age. The use of neutron diffraction has pointed out the limitations of the ED-XRF method, when applied on the surface of the corroded objects, in order to provide meaningful data on the bulk alloy composition. These limitations are associated with tin segregation phenomena that occur on bronzes during casting and in the burial environment. In the case of very corroded objects, as is the case of the objects studied here, they cannot be properly cleaned.

The neutron data on four of the objects show that the copper:tin ratio is lower than previously determined by ED-XRF analysis of the surface of such objects. Corrosion phases have been identified unambiguously by either neutron or X-ray diffraction and the procedures followed have demonstrated that results can be obtained non-destructively from whole objects.

We have also set the archaeological evidence from North Western Iberia in the context of the Homeric texts which refer to an era not far removed from the times of tin bronze manufacture in a land, far away from the Aegean, but still within reach of early travellers. Although Homer makes tantalising references to Bronze Age technology, details such as alloy composition, provenance or trading of tin and copper ore and the significance on the social organisation of early metal production cannot be deduced directly from the available archaeological evidence or by literary analysis of Homeric verses or accounts of ancient travellers, without resorting to hard evidence, i.e. technological facts obtained by employing materials science techniques. This is where archaeology and science meet.

It is clear to us that this pilot study at ISIS and the SRS has set the scene for more systematic studies of archaeological material from North Western Iberia, or indeed other archaeologically documented material from other Bronze Age sites in Europe, the Aegean included.

## Acknowledgemetns

Daresbury Laboratory is gratefully acknowledged for funding Beatriz Comendador's participation in the COST-G8 sponsored Master Class on SR in Cultural Heritage. The authors wish to thank their colleague Giota Manti of the Conservation Department, Cardiff University, for her valuable comments on the paper and very helpful suggestions.

## References

1. González, F.J., 2003. *Os Ártabros: Estudio Xeográfico e Etnohistórico*, Deputación Provincial, Ourense.
2. Cunliffe, B., 2002. *The Extraordinary Voyage of Pytheas the Greek: The Man Who Discovered Britain*, Penguin, London.
3. (a) Rovira, S., Montero, I. and Consuegra, S., 1997. *Las Primeras Etapas Metalúrgicas en la Península Ibérica*, Instituto Universitario Ortega y Gasset, Madrid; (b) Rovira, S. and Gómez, P., 2003. *Las Primeras Etapas Metalúrgicas en la Península Ibérica. III. Estudios metalográficos*, Instituto Universitario Ortega y Gasset, Madrid; (c) Comendador, B., 1999. *Los Inicios de la Metalurgia en el Noroeste de la Península Ibérica*, Brigantium, 11, Museo Arqueolóxico, A Coruña; (d) Comendador, B., 1999. The early development of metallurgy in the north-west of the Iberian Peninsula, in *Metals in Antiquity*, Young, Pollard, Budd and Ixer (Eds.), BAR

Int. Series No. 792, Archeopress, Oxford, pp. 63–67; (e) Comendador, B., 2005. Beyond the Cassiterides: Tin and bronze in the Bronze Age of the north-west of the Iberian Peninsula, in *Metallurgy: A Touch-Stone for Cross-Cultural Interaction*, 29–30 April 2005, London. See also http://srs.dl.ac.uk/arch/posters/beatriz-comendador.pdf (accessed October 2006).

4. Bettencourt, A., 2000. *O Povoado da Iade do Bronze da Sola, Braga, Norte de Portugal*, Cadernos de Arqueología, Universidade do Minho, Braga.
5. Suárez, X., 1995. O fixón: Una nueva perspective del bronce inicial en Galicia, in *Proceedings of XXII Congreso Nacional de Arqueología*, Vigo, 1993, pp. 56–67.
6. (a) Senna-Martínez, J.C., Ventura, J.M. and Carvalho, H.A., 2005. A Fraga dos Corvos (Macedo de Cavaleiros): Um sítio de habitat do "Mundo Carrapatas" da primeira Idade do Bronze em Trás-Os-Montes Oriental, *Cadernos Terras Quentes* **2**, 61–82. (a) Senna-Martínez, J.C., Ventura, J.M., Carvalho, H.A. and Figueredo, E., 2006. A Fraga dos Corvos (Macedo de Cavaleiros): Um sítio de habitat da primeira Idade do Bronze em Trás-Os-Montes Oriental. A Campanha 3. *Cadernos Terras Quentes* **3**, 61–85.
7. *Iliad of Homer*, translated by "A Graduate of the University of Oxford", 2nd Edn., Whitaker, London, 1825. First published in 1821.
8. Pantos, E., Davidovits, J., Pradell, T., Gelfi, M., Cornacchia, G., Bontempi, E., Colombi, P., Depero, L. (2008). Technology transfer in the Bronze Age: The case of a faience-like blue-glaze produced at breadoven temperatures, in *Proceedings of International Symposium on Science and Technology in the Homeric Epics*, Olympia, Greece, 27–30 August 2006, S. Paipetis (Ed.), Springer, Dordrecht (this volume).
9. (a) Luzón, J., 2004. Navegación antigua en el Atlántico: Arqueología y fuentes, in *Hasta el confín del mundo: Diálogos entre Santiago y el mar*, F. Singul and J. Suárez (Eds.), Galaxia, Vigo, pp. 31–37; (b) Aubet, M.E., 2003. El comercio fenicio en Homero, in *Estudios de arqueología dedicados a la profesora Ana María Muñoz Amilibia*, S.F. Ramallo (Ed.), Murcia, pp. 85–101; (c) Senna-Martínez, J.C., 2005. O outro lado do comercio orientalizante: Aspectos da produçâo metalúrgica no polo indígena, o caso das Beiras Portuguesas, *Anejos do Archivo Español de Arqueología* **XXXV**, 901–909.
10. (a) Galán, E., 2004. Noroeste y Suroeste: Dos ámbitos para el tránsito, in *Ámbitos Tecnológicos, Ámbitos de Poder. La transición Bronce Final-Hierro en la Península Ibérica*, A. Perea (Ed.), 18 March 2004. Madrid; (b) Jimenez, J., 2004. El trabajo del bronce en el Orientalizante Peninsular: algunas cuestiones referidas a la tecnología, in *Ámbitos Tecnológicos, Ámbitos de Poder. La transición Bronce Final-Hierro en la Península Ibérica*, A. Perea (Ed.), 18 March 2004. Madrid.
11. Montero, I., Rovira, S., Delibes, G., Fernández-Manzano, J., Ernández-Posse, Ma. D, Herrán, J.I., Martín, C., Maicas, R., 2003. High leaded bronze in the Late Bronze metallurgy of the Iberian Peninsula, in *Proc. International Conference Archaeometallurgy in Europe*, 24–26 September, Vol. 2, Assoziacione Italiana di Metallurgia, Milan, pp. 39–46.
12. Perea, A. (Ed.) 2004. *Proceedings of Conference Ámbitos Tecnológicos, Ámbitos de Poder. La transición Bronce Final-Hierro en la Península Ibérica*, 18 March 2004, Madrid.
13. *A Sola* PA6971, PA6972, PA6973; *Guidoiro* PA2491, PA4206. Rovira, S., Montero and I. Consuegra, S., 1997; Rovira, S. and Gómez, P., 2003. See references in [3].
14. According to http://en.wikipedia.org/wiki/Awl: Scratch awl: a tool with a long pointed spike used for marking wood. Stitching awl: a tool used by leatherworkers, such as cobblers (shoemakers), to pierce holes in leather. Bradawl: a tool for making holes in wood.
15. Rovira, S. and Montero, I., 2003. Natural tin-bronze alloy in Iberian Peninsula metallurgy: Potentiality and reality, in *Le problème de l'étain à l'origine de la métallurgie, Proceedings of the XIVth UISPP Congress*, A. Giumlia-Mair and F. Lo Schiavo (Eds.), 2/8.
16. Ribeiro, A., Tomé, Alves, L., Bacelar, Bettencourt, A.M.S., Menezes and R. Teles De (in press). Space of memory and representation: Bouça da Cova da Moura (Ardegães, Maia, Northwest of Portugal) – A case study, in *Proceedings of the Xth UISPP Congress*, Lisboa, 4–9 September, BAR International Series.
17. Figueiredo, E., Araujo, Ma.F., 2005. Influence of corrosion layers in EDXRF analysis of buried copper based prehistoric artefacts, in *European Corrosion Congress Proceedings*, Lissabon, 4–8 September.

18. Bidaud, E., Benedetti, D., Bugoi, R., Comendador, B., Goidanich, S., Gutierrez, C., Garipoli, D., Manti, P., Mifsud, T., Urbina, V., Kockelmann, W., Smith, A.D., Bahrami, F., MacDonald, M.A., Bell, A.T.M., Jones, R.L., Rizkallah, P.J. and Pantos, E., 2006. The COST-G8 master class on SR in cultural heritage at DL-SRS and RAL-ISIS, in *Proceedings of Synchrotron Radiation in Art and Archaeology, SR2A06*, Berlin, 27–30 September. See also http://srs.dl.ac.uk/arch/Posters/E.Bidaud-etal-SR2A06-Berlin-slides.htm (accessed October 2006).
19. Pantos, E., Kockelmann, W., Chapon, L.C., Lutterotti, L., Bennet, S.L., Tobin, M.J., Mosselmans, J.F.W., Pradell, T., Salvadó, N., Butí, S., Garner, R. and Prag, A.J.N.W., 2005. Neutron and X-ray characterisation of the metallurgical properties of a 7th century BC Corinthian-type bronze helmet, *Nucl. Instrum. Methods B* **239**, 16–26.
20. Kockelmann, W., Pantos, E. and Kirfel, E., 2000. Neutron and synchrotron radiation studies of archaeological objects, in *Radiation in Art and Archaeometry*, D.C. Creagh and D.A. Bradley (Eds.), Elsevier Science, pp. 347–377.
21. Siano, S., Kockelmann, W., Bafile, U., Celli, M., Iozzo, M., Miccio, M., Moze, O., Pini, R., Salimbeni, R. and Zoppi M., 2002. Quantitative multiphase analysis of archaeological bronzes by neutron diffraction, *Applied Physics A Materials Science & Processing* **74**(Suppl.), S1139–S1142.
22. (a) Charles, J.A., 1980. The coming of copper and copper base alloys and iron: A metallurgical sequence, in *The Coming of the Age of the Iron*, Yale University Press, New Haven, pp. 151–181. (b) Craddock, P.T., 1999. Paradigms of metallurgical innovation in prehistoric Europe, in *The Beginnings of Metallurgy*, A. Hauptmann (Ed.), Der Anschnitt 9, Bochum, pp. 175–192.
23. Kasztovszky, Z.S., Visser, D., Kockelmann, W., Pantos, E., Brown, A., Blaauw, M., Hallebeek, P., Veerkamp, J., Krook, W. and Stuchfield, H.M. (2007). Combined prompt gamma activation and neutron diffraction analyses of historic metal objects and limestone samples, *Nuovo Cimento C* **30**, 67–78.
24. Tang, C.C., MacLean, E.J., Roberts, M.A., Clarke, D.T., Pantos, E., Prag, A.J.N.W., 2001. The study of attic black gloss sherds using synchrotron X-ray diffraction, *J. Arch. Sci.* **28**(10), 1015–1024.
25. Collins, S.P., Cernik, R.J., Pattison, P., Bell, A.M.T. and Fitch, A.N., 1992. A two-circle powder diffractometer for synchrotron radiation on Station 2.3 at the SRS, *Reviews of Scientific Instruments* **63**(1), 1013–1014.
26. Gliozzo, E., Kirkman, I.W., Pantos, E. and Memmi-Turbanti, I., 2004, Black gloss pottery: Production sites and technology in Northern Etruria, Part II: Gloss technology, *Archaeometry* **46**(2), 227–246.
27. Comendador, B., Bettencourt, A. and Comendador, B., 2004. Los inicios de la metalurgia del bronce en el noroeste peninsular, in *Proceedings of the IV International Conference Sobre Patrimonio Geológico y Minero*, J. Mata (Ed.), September, Utrillas, Teruel.
28. Siano, S., Bartoli, L., Zoppi, M., Kockelmann, W., Daymond, M., Dann, J.A., Garagnani, G.L. and Miccio, M., 2003. Microstructural bronze characterisation by time of flight neutron diffraction, in *Proceedings International Conference Archaeometallurgy in Europe*, 24–26 September, Vol. 2, Assoziacione Italiana di Metallurgia, Milan, pp. 319–329.
29. See also http:// phoenicia.org/homer.html (accessed October 2006).
30. Reboreda, S., 1997. Los agalmata en los poemas homéricos, in *Homenaje al Prof. Dr. Fernando Gascó*, F. Presedo, J.M. Cortés, R. Urías and P. Guinea (Eds.), Sevilla, pp. 107–114.
31. Homer uses the word Anax (Wanax in Mycenean Greek – Linear B) as the adjective for important leaders like Odysseus, Achilles, Agamemnon, Menelaos, Nestor, and others. Anax is understood to mean chief amongst kings, king of kings, while Basileus is the title for a lesser aristocrat, a general or local military leader.
32. M. Helms has appointed not only the role of luxury objects on the reproduction of power in traditional, pre-industrial societies, but also the esoteric knowledge. She argues that fine artisanship and long-distance trade, both of which are more available to powerful elites than to ordinary people, are means of creating or acquiring tangible objects that embody intangible

powers that confer honour and power on their possessors. (a) Helms, M., 1993. *Craft and the Kingly Ideal: Art, Trade and Power*, University of Texas, Austin; (b) Helms, M., 1998. *Ulysses' Sail: An Ethnographic Odyssey of Power, Knowledge and Geographical Distance*, University Press,Princeton.

33. In total, there are 25 verses in the Odyssey and 21 in the Iliad where iron is mentioned. Il. 6.48 , 10.379 , 11.133 χαλκός τε χρυσός τε πολύκμητος τε σίδηρος (*copper* and *gold* and *well-wrought iron*), Od. 1.184 ες Τεμεσην μετα χαλκον αγω δ αιθωνα σιδηρον (*shining/ glittering* iron), Il. 4.485 ει πυρι uchiειpualphaς εοικε μενος δ αιθωνα σιδηρον.
34. Pope, A., 1903. *The Odyssey of Homer*, Grant Richards, London. First published in five volumes in 1725.

# Porphyra: In Search of Dyeing Methods in Ancient Greece

Wako Nishiyama

*Japan*

**Abstract.** It has been 50 years since I started studying about Yuzen dyeing, a traditional dying skill in Japan when I was 15. In the process I revived some ancient dyeing methods. Porphyra is one of these methods but it gave a great meaning to my life. I got to know how the civilization is noble especially I was deeply impressed by Greek civilization, built by noble thoughts searching for truth, who have supported me while continuing my study on porphyra for 28 years.

## 1 Introduction

Chatting with friends once, one of them said that in the Ancient Roman period people used to dye cloths using purple dye extracted from shellfish. That was news and a surprise to me. That question, made by a friend not engaged in dyeing, was my first step in the study of porphyra.

I wanted to see the purple colour dyed by shellfish. I started by finding the shellfish in illustrated books. I visited fishermen's houses one by one for many months and finally got ten Thais clavigera shellfish about the size of three centimeters. I extracted a yellow white secretion from the shellfish and dyed cloth. Then the secretion, spread only one square centimeter on the cloth, changed its yellow colour to green, blue and finally to purple in 30 minutes. I felt something mysterious in the process. However, the purple had a fatal defect. It had an offensive smell. The purple colour, emerged by the experiment made on a piece of cloth of only one square centimetre, looked divine and caught me with an indescribable impression.

As I got to understand the pigment, I had many opportunities to read literature about ancient times. I learned that ancient Greece took porphyra into her culture to sophistication. As the literature was piling thick, my yearning to visit Greece grew. Most interesting was the information which came very recently to my knowledge (2006), that since mid 2nd millennium BC the islanders of the Aegean were extracting porphyra from the murex shell and that this pigment was used on the wall paintings of Thera (Santorini).

In 1988, I had at last an opportunity to visit Greece. I still remember how much impressed I was by the beautiful landscape of the Aegean Sea. I felt it was my duty

*S.A. Paipetis (ed.), Science and Technology in Homeric Epics,* 133–138.

**Table 1** Porphyra dyeing process.

Porphyra dyeing process

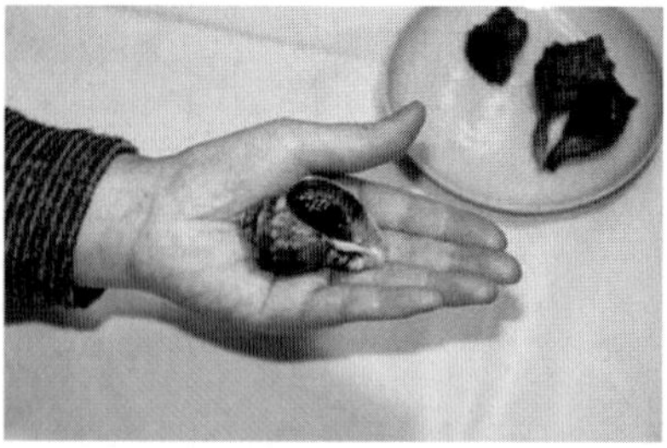

1. Rapana venosa Muroscidae

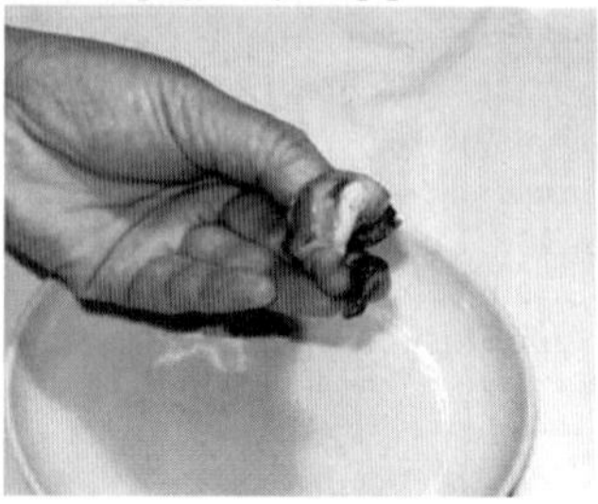

3. You remove the shell meat from the shell

4. The yellow secretion in the centre is the raw material for porphyra dyestuff

5. Purple secretion (raw material for porphyra dyestuff)

6. 1g dyestuff extracted from 10 kg shellfish and refined

as a dyer to revive porphyra, which the ancient Greek was proud of and gave a deep impression. Thus the landscape of Greece which I had seen for the first time, showed me the way I should take and filled me with enthusiasm to pursue the goal. I decided to study the actual dyeing method, not in the tube. Fortunately, fishermen helped me to collect the large quantity of shellfish required for my study.

I continued studying day after day on how to remove the offensive smell. I thought of using the characteristic of the pigment of porphyra. The dyestuff is dissolved by alkaline solution and coagulated by acid. I noticed this characteristic and tried to separate the dyestuff and viscous liquid, which seemed to send forth the offensive smell. The procedure is as follows (the various stages of porphyra dyeing appear in the photographs of Table 1):

**Table 1** (Continued)

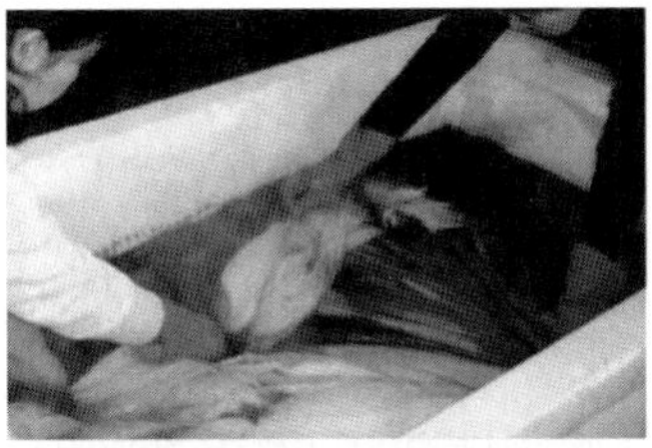

7. 90 litre solution where 5g porphyra is resolved

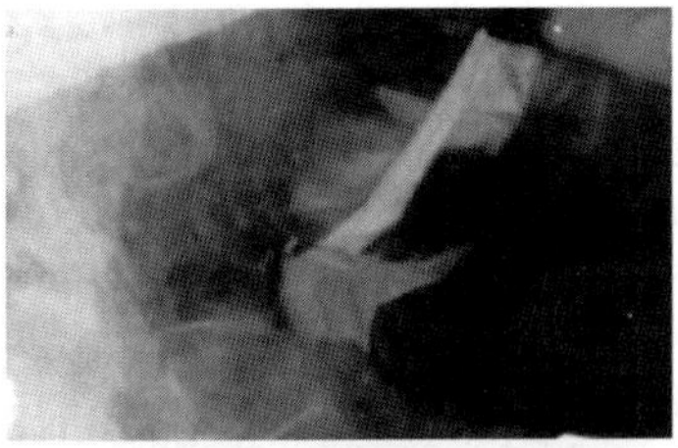

8. Yellow colour first appears on the silk cloth

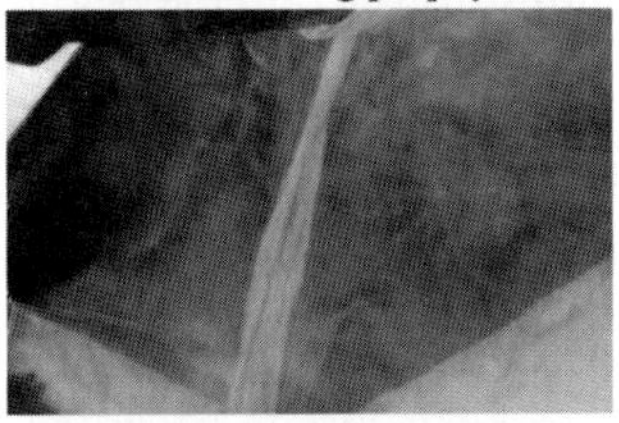

9. The yellow colour changes to green

10. The green colour changes to deep blue

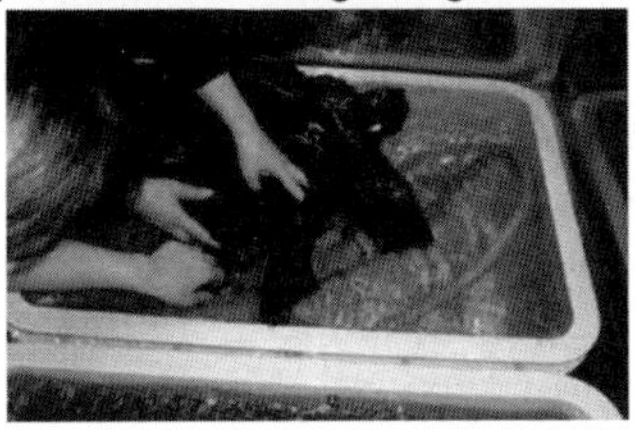

11. Porphyra purple colour finally appears

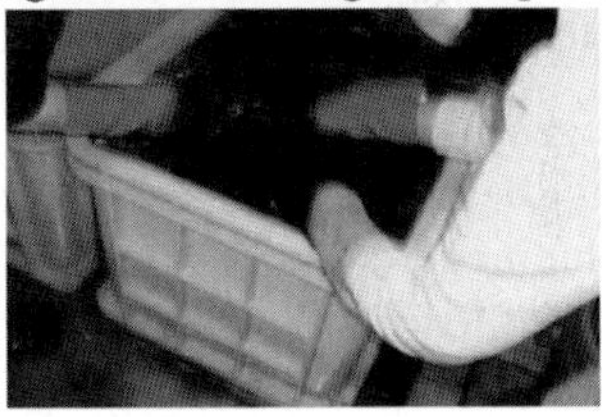

11a. Porphyra purple colour finally appears

1. Take out the purple gland (a tube form internal organs, about two by three centimetres) from a Rapana Venosa shellfish of the size of eight centimetres. Open the tube and take out the yellow viscous liquid.
2. Dissolve the viscous liquid with an alkaline solution, heat it and add acid to oxidize. Then the purple dyestuff is separated from the viscous liquid into particles. The particles are precipitated as the solution gets cold.
3. After repeating the above refining method, good quality crystals were obtained. However, the smell like a raw fish still remained slightly though the offensive smell was almost removed.

I needed 100 litres of warm water with 20 gram dyestuff in it when dyeing 1 kilogram of silk cloth. That is, the solution is diluted by 500 times, so the smell may not be defected. However, I could not help thinking that the ancient method was not fully reproduced yet, since the dyestuff extracted by this method requires some auxiliary chemical to oxidize it. That is, the dyestuff must have been yellow colour as it was in the shellfish. It has not been proved yet that such chemicals were used in ancient days. I really felt how difficult it would be to solve the mystery of porphyra. Five years had passed since I had started my study.

## 2 Extract by Corrosion

I left a purple gland with water in the test tubes for six months. The shellfish meat went rotten and dissolved into water, and the pigment of dyestuff was precipitated in the bottom of the tubes. The room was filled with a very intense smell of shellfish decomposition. Then I tried this method in a vase of 10 litres outdoors based on the following instructions:

4. Take purple gland (viscous liquid) out of about 10 kilograms of shellfish (about 200 pieces) and put them in the vase with 3 litres of water. In the beginning, stir everyday to accelerate corrosion.
5. The best season is from April to October, during which corrosion is accelerated. After six months, decomposition smell becomes intense.
6. After 12 months, passing through winter, the smell changes to a gaseous smell.
7. After passing summer twice, the smell of decomposition completely disappears and changes into a strong gaseous smell.
8. Refine the dyestuff using methods 3 and 4 above. The raw fish smell disappeared but the smell of the sea remained.

The Aegean people loved fragrance as part of their culture in 2000 BC. Accordingly, it cannot be presumed that people wore stinking cloths. The ancient Greeks must have obtained perfect porphyra as they pursued and reached the perfect thought in the field of learning, art, science, philosophy and so on. In the literature, porphyra was transported to China through the Continent of Eurasia. When? I believe that ancient porphyra was perfect and odourless, so that it was highly valued by many mainland people who did not like the smell of sea in those days, hence, odourless quality gave porphyra high value in the world of those days. I decided to search another method to reach perfect porphyra.

## 3 Aiming at Complete Removal of Smell

Though I still continued to search for the ancient method of removing smell completely, I could not find the definite method. After many experiments, I finally reached the conclusion that the only solution would be to find some liquid that would dissolve the offensive smell. I was sure that the mysterious method of refining the dyestuff must have been found in their daily life, so I tried all methods I could think of. In the end, a vague thought still remained. It was honey wine. This blend was very much used by the ancient Greeks. I thought that honey wine must provide a clue to solve the mystery of refining.

## 4 Perfect Dyestuff by Using Honey Wine

To dye cloth, I needed a large quantity of solution with dyestuff. Porphyra required 100 times the amount of solution in relation to the weight of cloth to be dyed. The method using honey wine is as follows.

I mixed purple gland extracted from *Rapana venosa* shellfish and honey to ferment. Honey and porphyra were related to gods in ancient Greece. I felt as if I had stepped into ancient mythology.

I had a reason for trying this method. An idea had been on my mind. Herodotus in his "Histories", 5th century BC, wrote that Alexander the Great, when he occupied Susa in Persia, discovered 100–120 tons of porphyra produced in Hermione (Greece) dipped into honey. Since I read this, I had thought that honey and porphyra might have some connection with honey wine, that is, a dyeing method using fermentation. Then, the method which I pursued for a long time was made clear by fermentation. Fifteen years had already passed, since I started this study. So, I tried this fermentation method:

9. Add water into a purple gland in a vase. Add one fourth of honey of the total weight.
10. Fermentation starts the next day. Stir in the morning and evening.
11. Fermentation was over after one week and reached to deoxidization. This method is not complicated as it is like A or B and no offensive smell. The room was filled around with a pleasant smell like wine or brandy.

## 5 Dyeing Ivory

The description of the Iliad (4.141–147) is as follows:

Ὡς δ᾽ ὅτε τίς τ᾽ ἐλέφαντα γυνὴ φοίνικι μιήνῃ
Μῃονὶς ἠὲ Κάειρα παρήϊον ἔμμεναι ἵππων·
κεῖται δ᾽ ἐν θαλάμῳ, πολέες τέ μιν ἠρήσαντο
ἱππῆες φορέειν· βασιλῆϊ δὲ κεῖται ἄγαλμα,
ἀμφότερον κόσμός θ᾽ ἵππῳ ἐλατῆρί τε κῦδος·
τοῖοί τοι Μενέλαε μιάνθην αἵματι μηροὶ
εὐφυέες κνῆμαί τε ἰδὲ σφυρὰ κάλ᾽ ὑπένερθε.

(As when some woman of Meonia or Caria strains purple dye on to a piece of ivory that is to be the cheek-piece of a horse, and is to be laid up in a treasure house – many a knight is fain to bear it, but the king keeps it as an ornament of which both horse and driver may be proud – even so, O Menelaus, were your shapely thighs and your legs down to your fair ankles stained with blood.)

This passage remained in my mind for a long time. I decided to reproduce the ornament some day. Since I had found no description of dyeing ivory with porphyra, I thought, it must have been a very important material in history.

## 6 Shaping and Preparation for Dyeing

12. Cut the shape of a cheek cover out of ivory a little larger than actual size.
13. Cut the corners to get a roughly round shape.
14. Shape the ivory piece with a lathe.
15. Wet a piece of sheepskin with water. Sprinkle the rough sand on the back of the sheepskin and grind the ivory carefully. Grind the ivory carefully using rough sand first, then fine sand.
16. When the ivory surface becomes bright, polish with wet linen first, then the back of a sheepskin, dried wool cloth, cashmere cloth, then sable finally for 40 hours.
17. For dyeing, choose a day with the temperature over 30° and humidity under 40%.
18. Put the ivory into warm water for one and a half hour.
19. Dry the ivory in the shade of a temperature of over 30° and humidity under 40% for three or four days.

Dyeing ivory

20. Prepare porhyra to be fermented as mentioned in D. Warm it till it becomes 60° to 65°. Add limewater or lime till it will get a pH10 to pH15.
21. Put the ivory for one hour into the water.
22. Dip the dried ivory into the solution vase. This is a work for 15 to 20 minutes.
23. Wash the dyestuff away with a thin piece of sheepskin.
24. Dry the dyed ivory in the shade with the temperature of over 30° and the humidity under 40% for two weeks.
25. Confirm that the ivory dried, then polish it as explained. Use sable for the final polish. This is a method revived from the Iliad.

## 7 Epilogue

When I started my study about porphyra, I could never have imagined that I would encounter the magnificent and age-long history of Greece. Thanks to the ideas of ancient Greeks, who pursued perfect beauty, I managed to complete my study. It is a great honour for me to have had the opportunity to present my work in the birthplace of porphyra and dyed ivory, both of which I could reproduce by studying Greek culture and literature, such as the Iliad.

# Technology Transfer in the Bronze Age: The Case of a Faience-Like Blue Glaze Produced at Bread-Oven Temperatures

E. Pantos[1], J. Davidovits[2], M. Gelfi[3], G. Cornacchia[3], E. Bontempi[4], P. Colombi[4] and L. Depero[4]

[1]*CCLRC, Daresbury Laboratory, Warrington, U.K.*
[2]*Institut GĆopolymŁre, Laboratoire de Recherche sur les Nouveaux MatĆriaux, Saint-Quentin, France*
[3]*Laboratorio di Metallurgia, Dipartimento di Ingegneria Meccanica, Università di Brescia, Brescia, Italy*
[4]*Laboratorio di Chimica per le Tecnologie, Dipartimento di Ingegneria Meccanica, Università di Brescia, Italy*

**Abstract.** How familiar were the pre-Homeric Greeks with Egyptian technology? Accounts of apparent mythological nature and archaeological evidence indicate cultural contacts. To what extent did Greek craftsmen learn their art from Egyptians?

At excavations in Djoser's pyramid (3rd dynasty, ca. 2750 BC) some 36000 glazed tiles were found, most with a turqoise-blue glaze coating but also many with other colours. The glazing process has been interpreted as a self-glazing process involving alkali salts, copper oxide (CuO) and firing temperatures of around 850°C. One investigation (Schiegel, 1988) indicated that this glaze was composed mainly of quartz, tenorite and water with a minor amount of alkalis, iron, phosphorus and chlorine. The interpretation of the data as indicative of the presence of chemically bound water is rather puzzling if such high temperatures were indeed used to produce the glaze on these tiles, and in such large quantities.

An alternative explanation has been proposed (Davidovits, 2005) according to which what is needed in order to replicate this self-glazing process is soluble silicate $SiO_2$, $K_2O$, $Na_2O$ + synthetic turquoise (*mafkat*), aluminium phosphate hydrate + copper phosphate hydrate and a firing temperature of 250°C. The resulting glaze can be described as a geopolymer-type material where the vitrified matrix is (K, Na)-Poly(Sialate-decaSiloxo), (Si-O-Al-O)-(Si-O-)10. The process can be controlled to produce blue ($K^+$) or green ($Na^+$) at 250°C or black at 350°C in the presence of Cu-phosphate, Cu-silicate and tenorite, or beige in the presence of Fe-salts (phosphate, arsenate). The process results in a vitreous matrix with the appearance of a faience-type material and with a composition corresponding to the starting materials.

We have used a number of techniques to characterise this turquoise-blue "geopolymer glaze" produced experimentally by Davidovits. Synchrotron-XRD of small flakes from the outer blue surface layer and the white inner core was used to determine the %weight of the diffracting phases as well as the amorphous component which constitutes some 87% of the scattering mass. SEM/EDS was employed to study the vitreous matrix morphology as well as the local elemental composition. Raman spectroscopy was used to correlate the other observations with the molecular composition of the areas examined.

It is an appealing thought, that ancient craftsmen were as ingenious in matters of solid state chemistry and metallurgy as present-day industrial materials scientists. Could similar skills have been used for decorating metal objects such as mentioned in the Homeric epics, particularly when referring to Hephaestian arts? Why not? What else could the Greeks have learned from the wise-old Egyptians, a mature and technologically advanced civilisation well before Homeric times? The

*S.A. Paipetis (ed.), Science and Technology in Homeric Epics,* 139–164.

great pyramids at Giza had been standing for some 1500 years before *artful Odysseus* and his *bronze-clad Achaean* warrior friends *sacked the holy citadel of Troy*.

## 1 Introduction

How familiar were the pre-Homeric Greeks with Egyptian technology? Bronze Age contact between Greece and Egypt was quite extensive and there is a huge amount of literature based on archaeological evidence and interpretation to support this (see for instance [1–3] for an overview of the evidence for relations between Egypt and Greece during the Late Bronze Age). In this article and in the spirit and context of this conference we restrict ourselves to quoting from Homer as the primary literary source and also from Herodotus. We then attempt to make a link, a rather tenuous and speculative one perhaps, between preliminary experimental data obtained from a modern product, a geopolymer vitreous material that one of us (JD) proposes to be considered as an exemplar of a self-glazing process that may have been used by the 3rd Dynasty Egyptians in the production of the glazed tiles found in Djoser's stepped pyramid, and the likelihood that Homer's decription of the making of Achiles' shield by Hephaistos contains technological information on aspects of its decorative themes which may reflect technological exchanges between Bronze Age Greece and ancient Egypt.

It should be noted at the outset, that we do not claim to offer either proof of the validity of the "Davidovits proposition" or scholarly analysis of Homer's descriptions. Our main objective is simply to stimulate discussion on these topics, even if the only positive outcome is to enlighten the co-authors and to prompt other colleagues who have a more advanced knowledge of both the archaeology and the technology of ancient Egypt to respond, hopefully with constructive criticism. Homer's accounts on Egypt, the most lengthy ones of which are found in the Odyssey (Table 1), some of apparent mythological nature, as well as accounts on Egypt by other later ancient writers, notably Herodotus [4] who devotes a whole book in his *Histories* (Book 2, Euterpe) on Egypt, Egyptian customs and achievements, indicate that close cultural contacts existed between the two civilisations over a very long period of time. Archaeological evidence supports this, although for how long, since when, and exactly to what extend during the Bronze Age is not as clear to non-specialists. Wall paintings from Thera (famous boat scene) before the eruption of 1648 BC [5] and depictions of a foreign delegation dressed in Minoan-like attire offering gifts to the Pharaoh [6] allude to Egyptian influences and diplomatic relations before the events in Troy that Homer reports. The story of the Danaids, daughters of Danaos, king of Argos and son of Poseidon and Libya, who slaughtered their husbands and married locals (Table 2) invokes speculations of Egyptian nobility marrying into the Mycenean dynastic household. Such *infusions* and *importations* have taken place several times in the long history of Greece, right up to very recent times, with the concomitant influx of customs, legal systems, art and technology from foreign lands, most notably Roman influences through the Roman occupation of Greece and the first Byzantine Emperors. All have been absorbed in time, melted down in the crucible of Greece, and *became Greeks*.

The story of the two twin brothers, Danaos and Aigyptos, offspring of Libya with their fifty daughters and sons who were forced to marry, is well embedded in Greek mythology [7]. It sounds suspiciously like a memory of a historical event of social intermixing that ended up with the Danaan Dynasty from which Agamemnon and his brother Menelaos descended. Theocaris et al. [8] reported the dating of a pyramid-like stone structure (a small one) at Hellenikon in the Argolid. A date of 2720–2630 BC was obtained using Optically Stimulated Luminescence (OSL). Pausanias [9] reports that "... walking on the road away from Argos towards Epidaurus there is a pyramid structure on our right, it is adorned with shields in the Argolic style ...". He continues to say that

> the pyramid was erected after the fratricidal battle between the twin brothers, Proetus and Akrisius, during a war of succession following the death of their father, king Avas of Argos. The battle ended in a deadlock and the pyramid was erected as a burial monument in honour of the fallen in this battle.

According to Greek mythology, Proetus was the brother of Danae; he is most noted for his poor treatment of Bellerophon [10]. The celebrated story of Bellerophon is known as the only place in the Homeric epics where the word for writing (σήματα λυγρὰ γράψας ἐν πίνακι πτυκτῶ) is mentioned (Iliad, 6:150–190). Other archaeological evidence from Knossos and Mycenae also attests relations of some kind, based on the presence of objects of Egyptian character. To what extent did Greek craftsmen learn their art from Egyptian master craftsmen? What can we glean from the Homeric verses? Is it all poetic licence or is there technological information included in them?

## 2 The Case of the 36000 Glazed Tiles from Djoser's Pyramid

At excavations in Djoser's (3rd dynasty, ca. 2750 BC) pyramid (Figure 1) some 36000 tiles were found in the funerary complex at Saqqarah [13], most coated with a turqoise-blue faience glaze [14], but also many with other colours (Figure 2). Egyptian faience is a non-clay ceramic ware made in Egypt and the Near East from about 4000 BC. The composition is based on mixtures of powdered quartz or sand, often containing a lime impurity, with sodium and potassium salts and a copper colorant [14]. The Djoser pyramid tiles are the earliest example of the innovative use of molding for a major building project using ceramic tiles. Details of shape and visual appearance are found in [13, 14] and beautiful colour photographs in [15]. The quality of the quartz source (ordinary iron-containing sand or fine crushed quartz) has a determining affect on the colour and morphology of the faience body [14].

The process of manufacture was interpreted by Kieffer and Alibert [16] in 1971 as a self-glazing process involving alkali salts and copper oxide and then firing at

**Table 1** References to Egypt in the Odyssey and the Iliad. English translation of the passages from the Odyssey from [11] and passages from the Iliad from [12].

| ΟΔΥΣΣΕΙΑ | ODYSSEY |
|---|---|
| γ:299–302<br>ἀτὰρ τὰς πέντε νέας κυανοπρῳρείους<br>Αἰγύπτῳ ἐπέλασσε φέρων ἄνεμός τε καὶ ὕδωρ.<br>ὣς ὁ μὲν ἔνθα πολὺν βίοτον καὶ χρυσὸν ἀγείρων<br>ἠλᾶτο ξὺν νηυσὶ κατ' ἀλλοθρόους ἀνθρώπους: | But five tall barks the winds and water toss'd,<br>Far from their fellows, on the Aegyptian coast.<br>There wander'd Menelaus through foreign shores<br>Amassing gold, and gathering naval stores; |
| δ:81–85<br>ἦ γὰρ πολλὰ παθὼν καὶ πόλλ' ἐπαληθεὶς<br>ἠγαγόμην ἐν νηυσὶ καὶ ὀγδοάτῳ ἔτει ἦλθον,<br>Κύπρον Φοινίκην τε καὶ Αἰγυπτίους ἐπαληθείς,<br>Αἰθίοπάς θ' ἱκόμην καὶ Σιδονίους καὶ Ἐρεμβοὺς<br>καὶ Λιβύην, ἵνα τ' ἄρνες ἄφαρ κεραοὶ τελέθουσι. | For eight slow-circling years, by tempests toss'd,<br>From Cypress to the far Phoenician coast<br>(Sidon the capital), I stretch'd my toil<br>Through regions fatten'd with the flows of Nile. |
| δ:125–131<br>Φυλὼ δ' ἀργύρεον τάλαρον φέρε, τόν οἱ ἔθηκεν<br>Ἀλκάνδρη, Πολύβοιο δάμαρ, ὃς ἔναι' ἐνὶ Θήβῃς<br>Αἰγυπτίῃς, ὅθι πλεῖστα δόμοις ἐν κτήματα κεῖται:<br>ὃς Μενελάῳ δῶκε δύ' ἀργυρέας ἀσαμίνθους,<br>δοιοὺς δὲ τρίποδας, δέκα δὲ χρυσοῖο τάλαντα.<br>χωρὶς δ' αὖθ' Ἑλένῃ ἄλοχος πόρε κάλλιμα δῶρα:<br>χρυσέην τ' ἠλακάτην τάλαρόν θ' ὑπόκυκλον ὄπασσεν<br>ἀργύρεον, χρυσῷ δ' ἐπὶ χείλεα κεκράαντο. | A silver canister, divinely wrought,<br>In her soft hands the beauteous Phylo brought;<br>To Sparta's queen of old the radiant vase<br>Alcandra gave, a pledge of royal grace;<br>For Polybus her lord (whose sovereign sway<br>The wealthy tribes of Pharian Thebes obey),<br>When to that court Atrides came, caress'd<br>With vast munificence the imperial guest:<br>Two lavers from the richest ore refined,<br>With silver tripods, the kind host assign'd;<br>And bounteous from the royal treasure told<br>Ten equal talents of refulgent gold.<br>Alcandra, consort of his high command,<br>A golden distaff gave to Helen's hand; |
| δ227–232<br>τοῖα Διὸς θυγάτηρ ἔχε φάρμακα μητιόεντα,<br>ἐσθλά, τά οἱ Πολύδαμνα πόρεν, Θῶνος παράκοιτις<br>Αἰγυπτίη, τῇ πλεῖστα φέρει ζείδωρος<br>ἄρουρα φάρμακα, πολλὰ μὲν ἐσθλὰ μεμιγμένα<br>πολλὰ δὲ λυγρά:<br>ἰητρὸς δὲ ἕκαστος ἐπιστάμενος περὶ πάντων<br>ἀνθρώπων: ἦ γὰρ Παιήονός εἰσι γενέθλης. | These drugs, so friendly to the joys of life.<br>Bright Helen learn'd from Thone's imperial wife;<br>Who sway'd the sceptre, where prolific Nile<br>With various simples clothes the fatten'd soil.<br>With wholesome herbage mix'd, the direful bane<br>Of vegetable venom taints the plain;<br>From Paeon sprung, their patron-god imparts<br>To all the Pharian race his healing arts. |
| δ:351–355<br>"Αἰγύπτῳ μ' ἔτι δεῦρο θεοὶ μεμαῶτα νέεσθαι<br>ἔσχον, ἐπεὶ οὔ σφιν ἔρεξα τεληέσσας ἑκατόμβας.<br>οἱ δ' αἰεὶ βούλοντο θεοὶ μεμνῆσθαι ἐφετμέων.<br>νῆσος ἔπειτά τις ἔστι πολυκλύστῳ ἐνὶ πόντῳ<br>Αἰγύπτου προπάροιθε, Φάρον δέ ἑ κικλήσκουσι, | "Long on the Egyptian coast by calms confined,<br>Heaven to my fleet refused a prosperous wind;<br>No vows had we preferr'd, nor victims slain!<br>For this the gods each favouring gale restrain<br>Jealous, to see their high behests obey'd;<br>Severe, if men the eternal rights evade.<br>High o'er a gulfy sea, the Pharian isle<br>Fronts the deep roar of disemboguing Nile: |
| δ:384–387<br>πωλεῖταί τις δεῦρο γέρων ἅλιος νημερτὴς<br>ἀθάνατος Πρωτεὺς Αἰγύπτιος, ὅς τε θαλάσσης<br>πάσης βένθεα οἶδε, Ποσειδάωνος ὑποδμώς:<br>τὸν δέ τ' ἐμόν φασιν πατέρ' ἔμμεναι ἠδὲ τεκέσθαι. | The oracular seer frequents the Pharian coast,<br>From whose high bed my birth divine I boast;<br>Proteus, a name tremendous o'er the main,<br>The delegate of Neptune's watery reign. |
| δ:477–483<br>πρίν γ' ὅτ' ἂν Αἰγύπτοιο, διιπετέος ποταμοῖο,<br>αὖτις ὕδωρ ἔλθῃς ῥέξῃς θ' ἱερὰς ἑκατόμβας<br>ἀθανάτοισι θεοῖσι, τοὶ οὐρανὸν εὐρὺν ἔχουσι:<br>καὶ τότε τοι δώσουσιν ὁδὸν θεοί, ἣν σὺ μενοινᾷς.'<br>"ὣς ἔφατ', αὐτὰρ ἐμοί γε κατεκλάσθη φίλον ἦτορ, | Then, steering backward from the Pharian isle,<br>We gain the stream of Jove-descended Nile;<br>There quit the ships, and on the destined shore<br>With ritual hecatombs the gods adore; |

**Table 1** (Continued)

| | |
|---|---|
| οὕνεκά μ' αὖτις ἄνωγεν ἐπ' ἠεροειδέα πόντον<br>Αἴγυπτόνδ' ἰέναι, δολιχὴν ὁδὸν ἀργαλέην τε. | |
| ξ:246–286 | |
| Αἴγυπτόνδε με θυμὸς ἀνώγει ναυτίλλεσθαι,<br>νῆας ἐΰ στείλαντα σὺν ἀντιθέοις ἑτάροισιν.<br>ἐννέα νῆας στεῖλα, θοῶς δ' ἐσαγείρατο λαός.<br>ἑξῆμαρ μὲν ἔπειτα ἐμοὶ ἐρίηρες ἑταῖροι<br>δαίνυντ': αὐτὰρ ἐγὼν ἱερήϊα πολλὰ παρεῖχον<br>θεοῖσίν τε ῥέζειν αὐτοῖσί τε δαῖτα πένεσθαι.<br>ἑβδομάτῃ δ' ἀναβάντες ἀπὸ Κρήτης εὐρείης<br>ἐπλέομεν Βορέῃ ἀνέμῳ ἀκραέϊ καλῷ<br>ῥηϊδίως, ὡς εἴ τε κατὰ ῥόον: οὐδέ τις οὖν μοι<br>νηῶν πημάνθη, ἀλλ' ἀσκηθέες καὶ ἄνουσοι<br>ἥμεθα, τὰς δ' ἄνεμός τε κυβερνῆταί τ' ἴθυνον.<br>"πεμπταῖοι δ' Αἴγυπτον ἐϋῤῥείτην ἱκόμεσθα,<br>στῆσα δ' ἐν Αἰγύπτῳ ποταμῷ νέας ἀμφιελίσσας.<br>ἔνθ' ἦ τοι μὲν ἐγὼ κελόμην ἐρίηρας ἑταίρους<br>αὐτοῦ πὰρ νήεσσι μένειν καὶ νῆας ἔρυσθαι,<br>ὀπτῆρας δὲ κατὰ σκοπιὰς ὤτρυνα νέεσθαι:<br>οἱ δ' ὕβρει εἴξαντες, ἐπισπόμενοι μένεϊ σφῷ,<br>αἶψα μάλ' Αἰγυπτίων ἀνδρῶν περικαλλέας ἀγροὺς<br>πόρθεον, ἐκ δὲ γυναῖκας ἄγον καὶ νήπια τέκνα,<br>αὐτούς τ' ἔκτεινον: τάχα δ' ἐς πόλιν ἵκετ' ἀϋτή.<br>οἱ δὲ βοῆς ἀΐοντες ἅμ' ἠοῖ φαινομένηφιν<br>ἦλθον: πλῆτο δὲ πᾶν πεδίον πεζῶν τε καὶ ἵππων<br>χαλκοῦ τε στεροπῆς: ἐν δὲ Ζεὺς τερπικέραυνος<br>φύζαν ἐμοῖς ἑτάροισι κακὴν βάλεν, οὐδέ τις ἔτλη<br>μεῖναι ἐναντίβιον: περὶ γὰρ κακὰ πάντοθεν ἔστη.<br>ἔνθ' ἡμέων πολλοὺς μὲν ἀπέκτανον ὀξέϊ χαλκῷ,<br>τοὺς δ' ἄναγον ζωούς, σφίσιν ἐργάζεσθαι ἀνάγκῃ.<br>αὐτὰρ ἐμοὶ Ζεὺς αὐτὸς ἐνὶ φρεσὶν ὧδε νόημα<br>ποίησ'--ὡς ὄφελον θανέειν καὶ πότμον ἐπισπεῖν<br>αὐτοῦ ἐν Αἰγύπτῳ: ἔτι γάρ νύ με πῆμ' ὑπέδεκτο--<br>αὐτίκ' ἀπὸ κρατὸς κυνέην εὔτυκτον ἔθηκα<br>καὶ σάκος ὤμοιϊν, δόρυ δ' ἔκβαλον ἔκτοσε χειρός:<br>αὐτὰρ ἐγὼ βασιλῆος ἐναντίον ἤλυθον ἵππων<br>καὶ κύσα γούναθ' ἑλών: ὁδ' ἐρύσατο καί μ' ἐλέησεν,<br>ἐς δίφρον δέ μ' ἕσας ἄγεν οἴκαδε δάκρυ χέοντα.<br>ἦ μέν μοι μάλα πολλοὶ ἐπήϊσσον μελίῃσιν,<br>ἱέμενοι κτεῖναι--δὴ γὰρ κεχολώατο λίην--<br>ἀλλ' ἀπὸ κεῖνος ἔρυκε, Διὸς δ' ὠπίζετο μῆνιν<br>ξεινίου, ὅς τε μάλιστα νεμεσσᾶται κακὰ ἔργα.<br>ὣ"ἔνθα μὲν ἑπτάετες μένον αὐτόθι, πολλὰ δ' ἄγειρα<br>χρήματ' ἀν' Αἰγυπτίους ἄνδρας: δίδοσαν γὰρ<br>ἅπαντες. | Nine ships I mann'd, equipp'd with ready stores,<br>Intent to voyage to the Aegyptian shores;<br>In feast and sacrifice my chosen train<br>Six days consum'd; the seventh we plough'd the main.<br>Crete's ample fields diminish to our eye;<br>Before the Boreal blast the vessels fly;<br>Safe through the level seas we sweep our way;<br>The steersman governs, and the ships obey.<br>The fifth fair morn we stem the Aegyptian tide,<br>And tilting o'er the bay the vessels ride:<br>To anchor there my fellows I command,<br>And spies commission to explore the land.<br>But, sway'd by lust of gain, and headlong will,<br>The coasts they ravage, and the natives kill.<br>The spreading clamour to their city flies,<br>And horse and foot in mingled tumult rise.<br>The reddening dawn reveals the circling fields,<br>Horrid with bristly spears, and glancing shields.<br>Jove thunder'd on their side. Our guilty head<br>We turn'd to flight; the gathering vengeance spread<br>On all parts round, and heaps on heaps lie dead.<br>I then explored my thought, what course to prove<br>(And sure the thought was dictated by Jove):<br>Oh, had he left me to that happier doom,<br>And saved a life of miseries to come!<br>The radiant helmet from my brows unlaced,<br>And low on earth my shield and javelin cast,<br>I meet the monarch with a suppliant's face,<br>Approach his chariot, and his knees embrace,<br>He heard, he saved, he placed me at his side;<br>My state he pitied, and my tears he dried,<br>Restrain'd the rage the vengeful foe express'd,<br>And turn'd the deadly weapons from my breast.<br>Pious! to guard the hospitable rite,<br>And fearing Jove, whom mercy's works delight.<br>"In Aegypt thus with peace and plenty bless'd,<br>I lived (and happy still have lived) a guest. |
| ρ:424–449 | |
| ἀλλὰ Ζεὺς ἀλάπαξε Κρονίων--ἤθελε γάρ που—<br>ὅς μ' ἅμα ληϊστῆρσι πολυπλάγκτοισιν ἀνῆκεν<br>Αἴγυπτόνδ' ἰέναι, δολιχὴν ὁδόν, ὄφρ' ἀπολοίμην.<br>στῆσα δ' ἐν Αἰγύπτῳ ποταμῷ νέας ἀμφιελίσσας.<br>ἔνθ' ἦ τοι μὲν ἐγὼ κελόμην ἐρίηρας ἑταίρους<br>αὐτοῦ πὰρ νήεσσι μένειν καὶ νῆας ἔρυσθαι,<br>ὀπτῆρας δὲ κατὰ σκοπιὰς ὤτρυνα νέεσθαι.<br>οἱ δ' ὕβρει εἴξαντες, ἐπισπόμενοι μένεϊ σφῷ,<br>αἶψα μάλ' Αἰγυπτίων ἀνδρῶν περικαλλέας ἀγροὺς<br>πόρθεον, ἐκ δὲ γυναῖκας ἄγον καὶ νήπια τέκνα,<br>αὐτούς τ' ἔκτεινον: τάχα δ' ἐς πόλιν ἵκετ' ἀϋτή.<br>οἱ δὲ βοῆς ἀΐοντες ἅμ' ἠοῖ φαινομένηφιν<br>ἦλθον: πλῆτο δὲ πᾶν πεδίον πεζῶν τε καὶ ἵππων<br>χαλκοῦ τε στεροπῆς: ἐν δὲ Ζεὺς τερπικέραυνος<br>φύζαν ἐμοῖς ἑτάροισι κακὴν βάλεν, οὐδέ τις ἔτλη<br>στῆναι ἐναντίβιον: περὶ γὰρ κακὰ πάντοθεν ἔστη.<br>ἔνθ' ἡμέων πολλοὺς μὲν ἀπέκτανον ὀξέϊ χαλκῷ,<br>τοὺς δ' ἄναγον ζωούς, σφίσιν ἐργάζεσθαι ἀνάγκῃ. | But Jove, all-governing, whose only will<br>Determines fate, and mingles good with ill,<br>Sent me (to punish my pursuit of gain)<br>With roving pirates o'er the Egyptian main<br>By Egypt's silver flood our ships we moor;<br>Our spies commission'd straight the coast explore;<br>But impotent of mind, the lawless will<br>The country ravage, and the natives kill.<br>The spreading clamour to their city flies,<br>And horse and foot in mingled tumults rise:<br>The reddening dawn reveals the hostile fields,<br>Horrid with bristly spears, and gleaming shields:<br>Jove thunder'd on their side: our guilty head<br>We turn'd to flight; the gathering vengeance spread<br>On all parts round, and heaps on heaps lay dead.<br>Some few the foe in servitude detain;<br>Death ill exchanged for bondage and for pain!<br>Unhappy me a Cyprian took aboard, |

**Table 1** (Continued)

| | |
|---|---|
| αὐτὰρ ἔμ' ἐς Κύπρον ξείνῳ δόσαν ἀντιάσαντι,<br>Δμήτορι Ἰασίδῃ, ὃς Κύπρου ἶφι ἄνασσεν:<br>ἔνθεν δὴ νῦν δεῦρο τόδ' ἵκω πήματα πάσχων."<br>τὸν δ' αὖτ' Ἀντίνοος ἀπαμείβετο φώνησέν τε:<br>"τίς δαίμων τόδε πῆμα προσήγαγε, δαιτὸς ἀνίην;<br>στῆθ' οὕτως ἐς μέσσον, ἐμῆς ἀπάνευθε τραπέζης,<br>μὴ τάχα πικρὴν Αἴγυπτον καὶ Κύπρον ἵκηαι:<br>ὥς τις θαρσαλέος καὶ ἀναιδής ἐσσι προίκτης. | And gave to Dmetor, Cyprus' haughty lord:<br>Hither, to 'scape his chains, my course I steer,<br>Still cursed by Fortune, and insulted here!"<br>To whom Antinous thus his rage express'd:<br>"What god has plagued us with this gourmand guest?<br>Unless at distance, wretch! thou keep behind,<br>Another isle, than Cyprus more unkind,<br>Another Egypt shalt thou quickly find.<br>From all thou begg'st, a bold audacious slave;<br>Nor all can give so much as thou canst crave.<br>Nor wonder I, at such profusion shown;<br>Shameless they give, who give what's not their own." |
| ΙΛΙΑΣ I:381–384<br>οὐδ᾽ ὅσ᾽ ἐς Ὀρχομενὸν ποτινίσεται, οὐδ᾽ ὅσα Θήβας<br>Αἰγυπτίας, ὅθι πλεῖστα δόμοις ἐν κτήματα κεῖται,<br>αἵ θ᾽ ἑκατόμπυλοί εἰσι, διηκόσιοι δ᾽ ἀν᾽ ἑκάστας<br>ἀνέρες ἐξοιχνεῦσι σὺν ἵπποισιν καὶ ὄχεσφιν· | ILIAD<br>Nor *would* I for as many as arrive at Orchomenos, or Egyptian Thebes, where numerous possessions are laid up in the mansions, and where are one hundred gates, from each of *which* two hundred men rush out with horses and chariots. |

800–850°C to produce a turquoise-blue glaze. These authors studied five samples, a Convex tile of the 3rd dynasty, a vase from the "New Empire", a Wedjat eye from the "Late Period" and two cylindrical beads of unspecified period. They report chemical analysis from the surface of the New Empire vase containing copper, sodium, potassium, calcium and magnesium. Table I in their paper lists the colour and appearance of the body and the surface of the first two objects as follows:

Convex tile: Body – white fine grains.
Surface glazed: clear green-blue, glaze & semi-glaze.

New Empire vase: Body – grey-white, rather large grains.
Exterior glazed surface: Blue-turqoise, very vitrious, high glaze.
Interior semi-glazed surface: Violet from Manganese.
Semi-glazed.

They also report that they used X-ray diffraction (diffraction patterns not shown) on the original five samples and refired to 870, 900, 950 and 1000°C in order to determine the original firing temperature by observing the development of cristobalite (it appears at $T > 900$°C in these measurements). Only the fifth sample (cylindrical bead with fine body and grey-blue surface glaze show presence of cristobalite before refiring. No mention is made of other mineral phases in the diffraction data.

Subsequently, pioneering and comprehensive work by Kaczmarczyk and Hedges [17] (and Vandiver in Appendix A) reports XRF analysis data (table in Appendix C) from 1100 analyses of several hundreds of objects covering all periods of Egyptian faience, from the Amratian–Naqada period 4000–3500 BC, to Ptolemaic and Roman times. Amongst them are data for three blue and green glazed Djoser tiles and the white core as well as two measurements on white core by Atomic Absorption Spectroscopy [17, table XXXI] covering all periods of faience production, including two green Djoser tiles. Concentration values for Cl and the oxides of Si, S, K, Ca,

**Table 2** References to the daughters of Danaos in Herodotus, Histories, Euterpe Book 2-91. English translation from [4].

| |
|---|
| 91. Hellenic usages they will by no means follow, and to speak generally they follow those of no other men whatever. This rule is observed by most of the Egyptians; but there is a large city named Chemmis in the Theban district near Neapolis, and in this city there is a temple of Perseus the son of Danae which is of a square shape, and round it grow date-palms: the gateway of the temple is built of stone and of very great size, and at the entrance of it stand two great statues of stone. Within this enclosure is a temple-house and in it stands an image of Perseus. These people of Chemmis say that Perseus is wont often to appear in their land and often within the temple, and that a sandal which has been worn by him is found sometimes, being in length two cubits, and whenever this appears all Egypt prospers. This they say, and they do in honour of Perseus after Hellenic fashion thus,--they hold an athletic contest, which includes the whole list of games, and they offer in prizes cattle and cloaks and skins: and when I inquired why to them alone Perseus was wont to appear, and wherefore they were separated from all the other Egyptians in that they held an athletic contest, they said that Perseus had been born of their city, for Danaos and Lynkeus were men of Chemmis and had sailed to Hellas, and from them they traced a descent and came down to Perseus: and they told me that he had come to Egypt for the reason which the Hellenes also say, namely to bring from Libya the Gorgon's head, and had then visited them also and recognised all his kinsfolk, and they said that he had well learnt the name of Chemmis before he came to Egypt, since he had heard it from his mother, and that they celebrated an athletic contest for him by his own command. |
| 170. Also at Saïs there is the burial- place of him whom I account it not pious to name in connexion with such a matter, which is in the temple of Athene behind the house of the goddess, stretching along the whole wall of it; and in the sacred enclosure stand great obelisks of stone, and near them is a lake adorned with an edging of stone and fairly made in a circle, being in size, as it seemed to me, equal to that which is called the "Round Pool" in Delos. |
| 171. On this lake they perform by night the show of his sufferings, and this the Egyptians call Mysteries. Of these things I know more fully in detail how they take place, but I shall leave this unspoken; and of the mystic rites of Demeter, which the Hellenes call /thesmophoria/, of these also, although I know, I shall leave unspoken all except so much as piety permits me to tell. The daughters of Danaos were they who brought this rite out of Egypt and taught it to the women of the Pelasgians; then afterwards when all the inhabitants of Peloponnese were driven out by the Dorians, the rite was lost, and only those who were left behind of the Peloponnesians and not driven out, that is to say the Arcadians, preserved it. |
| 182. Amasis also dedicated offerings in Hellas, first at Kyrene an image of Athene covered over with gold and a figure of himself made like by painting; then in the temple of Athene at Lindson two images of stone and a corslet of linen worthy to be seen; and also at Samos two wooden figures of himself dedicated to Hera, which were standing even to my own time in the great temple, behind the doors. Now at Samos he dedicated offerings because of the guest-friendship between himself and Polycrates the son of Aiakes; at Lindos for no guest- friendship but because the temple of Athene at Lindos is said to have been founded by the daughters of Danaos, who had touched land there at the time when they were fleeing from the sons of Aigyptos. These offerings were dedicated by Amasis; and he was the first of men who conquered Cyprus and subdued it so that it paid him tribute. |

Fe, and Cu only are reproduced in Table 3. Note that the concentration of the light elements Na, Mg, Al could not be measured by XRF.

We note that the white core contains less copper and potassium than the glaze. The same source [17, p. 149] mentions that "As expected, the purest and darkest blue glazes contain the most copper and the least lead, zinc, antimony, and *more sodium than potassium*" (our emphasis). AAS data show that this is true for the cases where the core of the objects is described as pale-blue, blue-green, bluish or

**Table 3** %Weight values extracted from [17] for Cl and the oxides of Si, S, K, Ca, Fe, and Cu. Determined by (a) X-ray fluorescence and (b) AAS.

| **Object description** | | | | | | | |
|---|---|---|---|---|---|---|---|
| **a. Measured by XRF** | Cl | Si | S | K | Ca | Fe | Cu |
| Blue-green tile 1954.669 | 0.36 | 88.4 | 0.00 | 0.00 | 0.49 | 0.13 | 5.11 |
| White body of tile 1954.669 | 2.30 | 92.7 | 0.29 | 0.20 | 0.51 | 0.18 | 0.01 |
| Green tile 1954.670b | 0.85 | 91.7 | 0.00 | 0.00 | 1.24 | 0.11 | 3.28 |
| White body of blue tile 1942.280 | 0.00 | 94.0 | 0.00 | 0.02 | 0.38 | 0.26 | 0.07 |
| Dark blue tile 1942.280 | 0.48 | 87.2 | 0.00 | 1.50 | 1.17 | 0.14 | 5.80 |
| White body of blue* tile 1937.115 | 0.00 | 94.4 | 0.41 | 0.08 | 1.15 | 0.29 | 0.07 |
| Greenish blue of tile 1937.115 | 0.51 | 87.7 | 0.00 | 0.30 | 2.22 | 0.22 | 5.39 |
| **b. Measured by AAS** | Na | Si | | K | Ca | Fe | Cu |
| White body of green tile 1954.670b | 1.05 | 95 | | 0.19 | 0.49 | 0.18 | 0.01 |
| White body of green* tile 1937.115 | 0.27 | 94.4 | | 0.08 | 1.15 | 0.29 | 0.07 |

*Note that tile 1937.115 is described as blue in the XRF measurement, but green in the AAS table.

just blue. However, in several cases where the core is white but the external glaze is described as blue, we note that the potassium concentration is higher than in cases where the glaze is described as green. Table XXVII in [17] gives the colours induced by the transition metal chromophore which depend on its coordination (octahedral favoured by $Na_2O$ and tetrahedral by $K_2O$) which in turn is affected by the presence of other anions or cations. A more recent study of the effect of copper coordination on colour using EXAFS reports that "The turquoise colour is characteristic of tetragonally distorted $Cu(O)_6$ environments such as hydrated copper(II) cations, copper(II) hydroxide and the mineral turquoise itself ($CuAl_6(PO_4)_4(OH)_8{\cdot}4H_2O$)" [18]. The issue of the effect of sodium and potassium on colour is important in the context of what is discussed in the next section. We continue here with remarks from literature sources relevant to the colour and the chemical composition of faience.

Schiegel in 1988 [19] studied the micro-structure of Djoser tiles by SEM/EDS and WDS and reports that

> The average glaze composition is mainly $SiO_2$, CuO and water (75.7% $SiO_2$, 8.3% CuO and 13% water calculated) with a minor degree: alkalis, iron, phosphorus, chlorine (altogether about 3%).[1]

No mention is made how the water %weight was calculated. He continues by stating that

> In a shallow depression, a sequence of six glaze layers were encountered. This indicates a repeated dipping of the quartzite tile in the glaze liquid to produce

[1] Chemical, i.e. elemental, composition obtained by SEM/EDS is normally given in terms of the corresponding oxides of the elements present. This does not necessarily mean that these oxides are actually present in the material studied.

**Fig. 1** The remains of the pyramid of Djoser. Image from http://www.touregypt.net/featurestories/dsteppyramid1.htm

an intense colour and a smooth surface. In the glaze we found small residues of copper chloride and copper sulphide.

There are three techniques for making faience: Application, cementation and efflorescence [17, 20–22]. In the first, a glazing powder or slurry is applied onto the faience body. With cementation, also known as "Qom technique", the unglazed faience is buried in a glazing powder which reacts with the quartz body to form a glaze on the surface. The glazed faience can then be removed from the cementation material. The last process, efflorescence, is rather special in that it combines the glaze and the core. Water-soluble alkali salts, probably in the form of natron (sodium alkali) and/or plant ash (potassium alkali) are mixed with the siliceous core. This is only necessary for the efflorescent technique. By-products of the copper metallurgy, such as copper oxides, are thought to have been used as the colouring agent [17].

The mixture of these ingredients is moistened and formed into a desired shape. Then, in the process of drying, the salts migrate to the surface of the object to form an efflorescent bloom. When fired at high temperature, usually above 850°C, this layer melts and fuses with the fine quartz body, to create a glassy coating. This latter method is also called self-glazing process and was used for the glazing of the

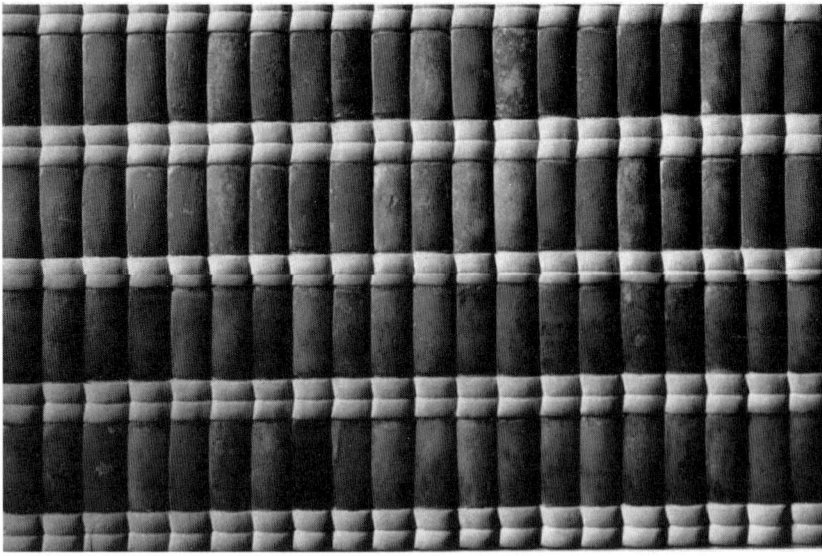

**Fig. 2** Tiles from the pyramid of Djoser, 3rd dynasty, 2750 BC.

Djoser tiles [14, 17, 21]. The pioneering and thorough studies of Kaczmarczyk and Hedges [17] and Vandiver [21] do no report any SEM morphology data on these tiles, although evidence of application of the efflorescence method of self-glazing is evident from other objects of the same chronological period.

Detailed investigations by Tite and coworkers [22] of the microstructures found in these type of ancient faience match well with reproductions. Four ancient Egyptian samples from the British Museum were studied: a ring of New Kingdom date (unspecified Dynasty, c. 1570–1070 BC), a ring from the 18th Dynasty, Armana period (c. 14th century BC), a shabti figure (funerary figurine buried with the deceased), 21st Dynasty (c. 1070–945 BC), and another shabti figure from the Late Period (747–332 BC). The microstructure was characteristic showing the quartz grains from the core, which maybe non-bonded or bonded with some glass depending on the method used, and a surface glaze showing some reaction phases which appear also in the ancient materials. More recent studies by Vandiver [23] and Tite and co-workers [24] have increased our understanding of the morphological differences between the different faience methods. The procedure for the production of Egyptian blue is not all that different from making faience glazes [25]. To our knowledge apart for the work reported in [16, 17, 19] no other experimental work has been reported recently on the Djoser pyramid tiles.

## 3 The Davidovits Proposition [26]

This consists of the following:

> The *glaze of the tiles* is not a ceramic glaze but a geopolymer-type material, (K,Na)-Poly(Sialate-decaSiloxo)(Si-O-Al-O)-(Si-O-)10.
> *Source of Phosphorus*: turquoise, hydrate of aluminium/copper phosphate.
> *Source of CuO*: turquoise + Chrysocolla, hydrated copper silicate.
> *Mafkat*: Turquoise chrysocolla, enormous quantities of which were extracted in the Sinai mines (Figure 3). They could not have been used solely for making jewellery or other treasures.

Throughout Egyptian history, the production of ceramic pottery and glazed tiles is quite prolific. However, it cannot have been all that easy to manufacture 36000 tiles in the 3rd millennium BC in a land where fuel (presumably wood) is so scarce. An alternative explanation has been proposed by Davidovits [26] which involves procedures and material that would have been available to the 3rd dynasty craftsmen. He remarks that "The reported *presence of water* is rather puzzling if such high temperatures were indeed used to produce these tiles, and in such large quantities". It must, of course, be noted that reference to "water" (calculated in [19], not directly determined) can be interpreted as presence of hydroxyl (OH) and not necessarily as adsorbed water that evaporates off at relatively modest temperatures or of chemically bound or complexed $H_2O$ as in a hydrated compound such as turqoise.

According to Davidovits [26], it is possible to replicate the self-glazing process using soluble silicate $SiO_2$, $K_2O$, $Na_2O$ and synthetic turquoise (*mafkat*), made of aluminium phosphate hydrate and copper phosphate hydrate at a firing temperature of 250°C. The resulting vitreous material can be described as a geopolymer-type material [27] where the vitrified matrix is (K,Na)-Poly(Sialate-decaSiloxo), (Si-O-Al-O)-(Si-O-)10. The process can be controlled to produce blue ($K^+$) or green ($Na^+$) at 250°C or black at 350°C in the presence of Cu-phosphate, Cu-silicate and tenorite, or beige in the presence of Fe-salts (phosphate, arsenate). The process results in a vitreous matrix with the visual appearance of faience and with a composition corresponding to the starting materials.

Despite the large-scale exploitation of mafkat (turquoise), which must have run to very high quantities, few turquoise items found by archaeology are displayed in museums [28] when taking into account the extravagant mining expeditions undertaken at great expense, using a lot of personnel and equipment. Davidovits has proposed that a low-tech process requiring low temperatures to produce the vitreous surface on the tiles could have been the following:

- Mix Soluble Potassium Silicate, molar ratio $SiO_2$: $K_2O$ = 12:2, Aluminum Phosphate (5%) and Copper Phosphate (4%).
- Hardening at Room Temperature or 40–60°C for 3 hours in a closed container.
- Demolding and drying at 60°C, then heating in an open oven, at temperatures of bread making, up to 250°C.

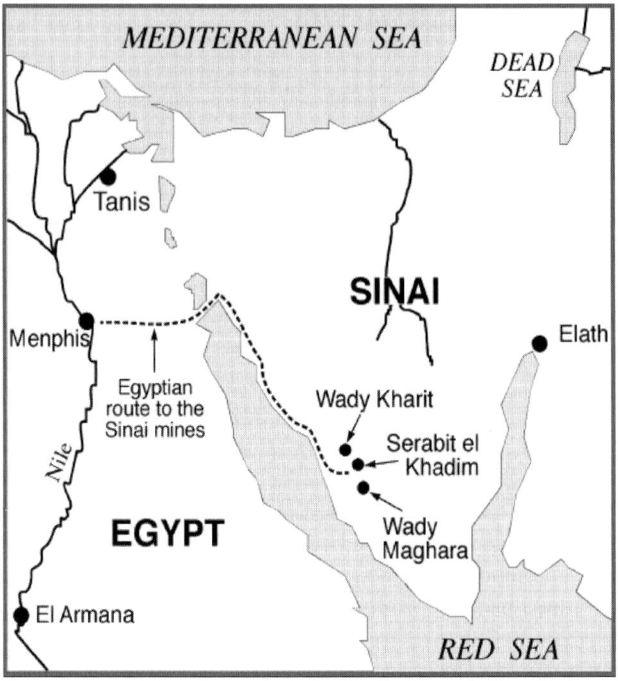


**Fig. 3** Map of northern Egypt and the Sinai showing the location of mines that were worked during the early dynastic times and later.

- Heated in an open fire with flame, the temperature may reach 300 to 400°C and it generates grey-black spots or areas.

This process has been used to produce the material depicted in Figure 4. The study reported here is for a turquoise-blue coloured product which has the visual appearance of the green-blue faience reported by others in the literature quoted above.

It is of course impossible to prove whether such a preparation method was in fact used by the 3rd dynasty craftsmen until and unless the micro-structure and other physical properties of the ancient material are studied and compared with those of the geopolymer product described above, using the same techniques and same equipment protocols and sample handling conditions. We describe below the first results from a variety of materials science techniques employed to characterise a small fragment of a light-blue vitreous geopolymer material selected from Davidovits' product range some of which are shown in Figure 4.

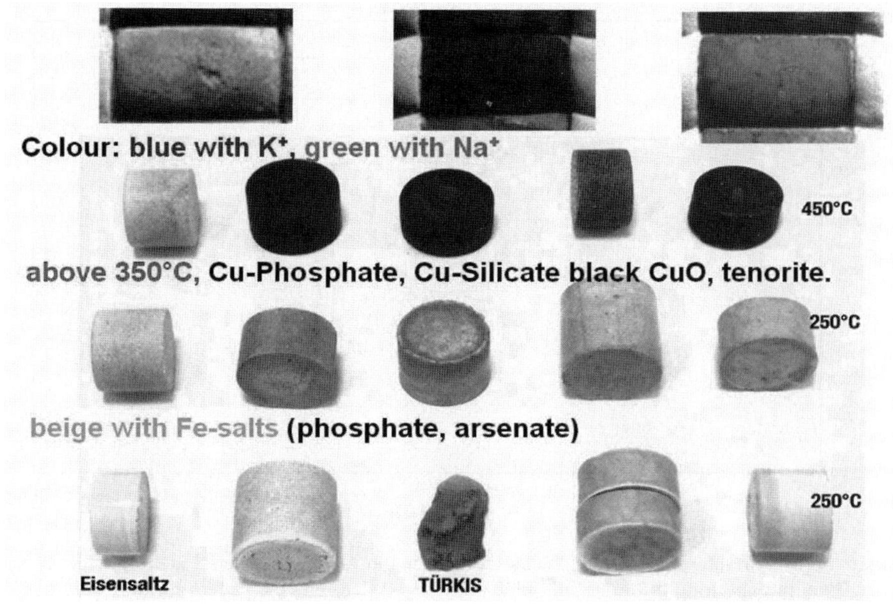


**Fig. 4** Synthetically produced "geopolymer-faience".

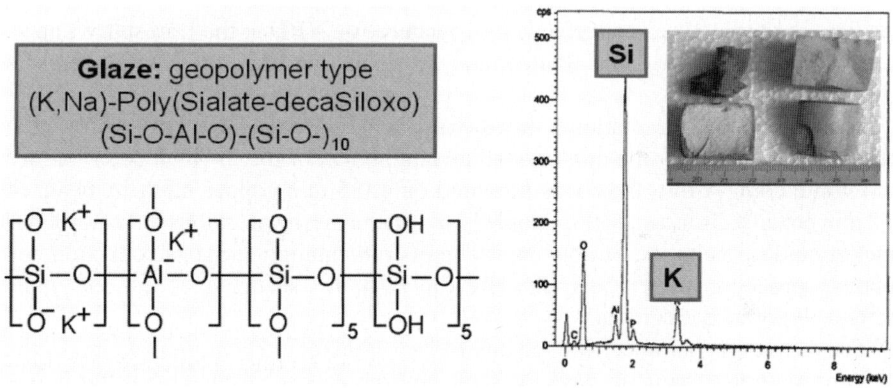


**Fig. 5** The vitreous material produced at low temperatures consists of an alkali alumino-silicate, polymer-like compound with a chemical formula as shown on the left. The XRF spectrum on the right shows that the main elements present for the samples in the inset are O, Si and K. No sodium is present as only potassium is needed to produce the deep blue colour. Sodium would be found in the green-coloured product.

# 4 Characterisation of the Geopolymer Material

Visual examination of the fragment available for experimentation showed that the outer layer, approx. 1 mm in width, was responsible for the light-blue colour while further into the core the colour was a diffuse white. The methods available to us at

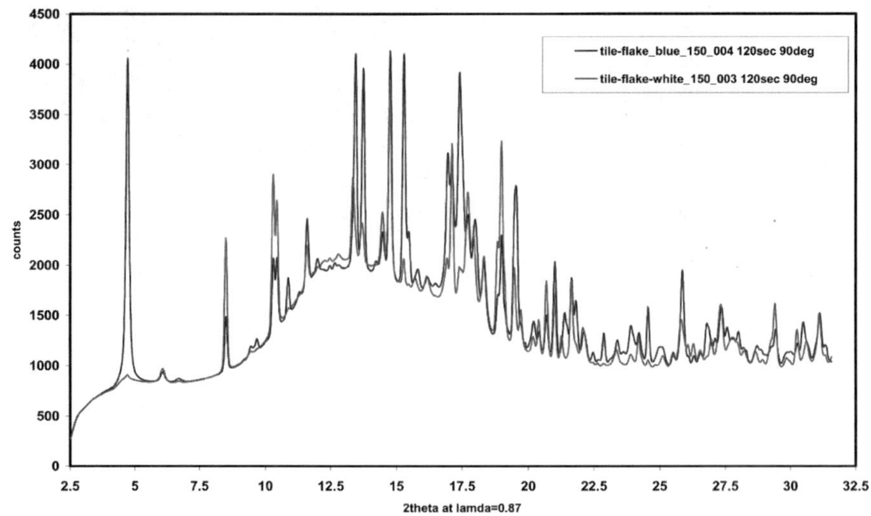


**Fig. 6** X-ray diffraction patterns of the surface of the blue vitreous material and the white inner core. Blue line: the pattern for the light-blue outer layer. Red line: the pattern for the inner, white core.

the time the project was conceived were synchrotron XRD at the Daresbury Laboratory Synchrotron Radiation Source, and SEM/EDS and micro-Raman at Brescia University.

*XRD*: Synchrotron radiation X-ray diffraction was used to identify the diffracting phases and to quantify the non-crystalline component of the the blue outer surface and white core. A thin flake was mounted on a 0.5 mm copper mesh. A beam of 0.2 mm cross-section struck the sample between the mesh wires, thus the amorphous background is due to the sample only. The instrumentation and data collection and analysis procedures have been described in [29]. Air scatter does not contribute noticeably to the background.

The diffraction patterns from the outer layer (blue) and the interior (white) of the vitreous material (Figure 6) have the same amorphous background but differ in one aspect: For the blue surface layer a phase with a strong starting reflection at about 4.4 deg. Further XRD measurements at Brescia University were collected by means of a Panalytical X'Pert Pro$^{TM}$ diffractometer, equipped with the X'Celerator$^{TM}$ detector. The phase identification was carried out using the X'Pert HighScore$^{TM}$ software package (Figure 7). Rietveld analysis as described in [30] was used to determine the %weight of the diffracting phases as well as the amorphous component. The latter was found to constitute approximately 87% of the scattering mass. That is, it is in a vitrified, glassy state. The main phases identified in the blue are libethenite (PDF 72-0572, $Cu_2(PO_4)(OH)$, 2%weight), potassium hydrogen phosphate (PDF 83-0253, K(H2PO4), 2%weight), berlinite (PDF 76-0227, $AlPO_4$, 1%weight), hydroxyapatite (PDF 76-0674, $Ca_{25}(PO_4)3{\cdot}OH$, 1%weight) and Potassium Copper Phosphate

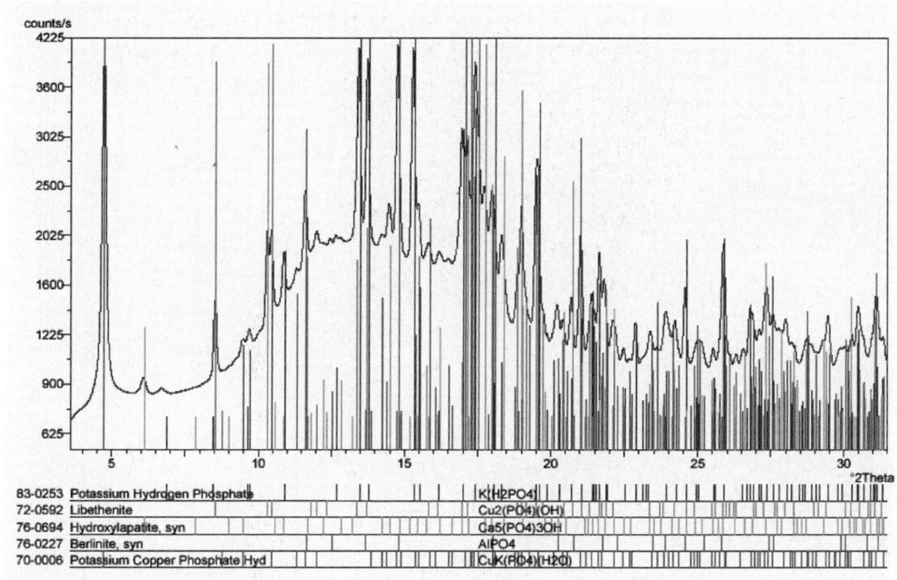


**Fig. 7** Phase identification from the X-ray diffraction pattern of a micro-sample of the blue vitreous coating.

Hydrate, PDF 31-0006, $KCuPO_4{\cdot}H_2O$, 2%weight). This last phase has its strongest reflection in the low angle region. It is much reduced in intensity in the white interior of the vitreous material so it is most likely associated with the blue colour of the surface layer.

Concerning the XRD analysis of the blue and white areas we note (a) the presence of copper phosphates, (b) the difference between the blue and the white areas is the presence of sodium iron oxide, and (c) other phosphates are also present (aluminium, calcium and potassium), which result from the reaction of the aluminium phosphate with the potassium of the potassium silicate solution. The chemical composition is quite different from that reported in the literature for Egyptian faience.

*SEM/EDS*: Scanning Electron microscopy and Energy Dispersive Spectroscopy (SEM/EDS) was employed to study the vitreous matrix morphology as well as the local elemental composition. A fresh-fractured sample was used with no further attempt to polish its surface. Figure 8 shows a representative example of morphology selected from several shots in different areas and at different magnification in the blue and white areas. The data shown in Table 4 confirm the conclusions of the XRD measurements, i.e. the presence of an amorphous matrix. The elemental analysis shows the presence of high percentages of silicon probably as amorphous silicate with Al, Ca, Si and K. The spectrum 1 and 3 carried out on small particles embedded into the amorphous matrix show high copper and phosphorus contents, probably due to the presence of the phase Potassium Copper Phosphate Hydrate, $KCuPO_4{\cdot}H_2O$ detected by X-ray diffraction.

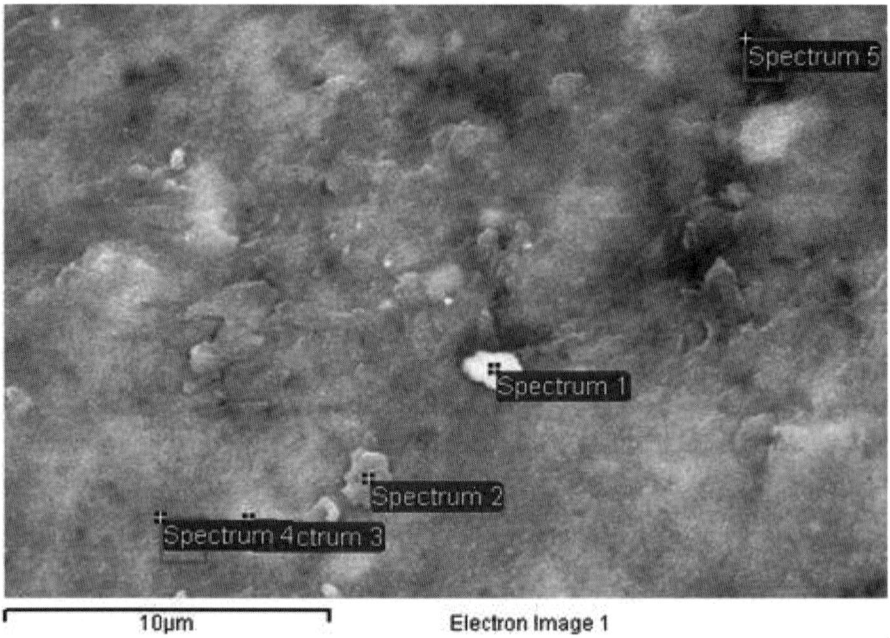


**Fig. 8** Scanning electron microscopy back-scattering image from the blue surface of the vitreous coating.

**Table 4** Chemical composition of different features indicated in Figure 8.

Processing option: All elements analysed (normalised)

| Spectrum | O | Al | Si | P | K | Ca | Cu | Total |
|---|---|---|---|---|---|---|---|---|
| Spectrum 1 | 42.56 | 1.39 | 24.15 | 5.54 | 5.16 | 1.16 | 20.02 | 100.00 |
| Spectrum 2 | 49.20 | 1.99 | 33.34 | 2.11 | 9.39 | 1.74 | 2.23 | 100.00 |
| Spectrum 3 | 39.40 | 1.14 | 24.01 | 6.53 | 10.15 | 0.90 | 17.88 | 100.00 |
| Spectrum 4 | 45.63 | 2.22 | 34.81 | 2.91 | 9.19 | 1.52 | 3.73 | 100.00 |
| Spectrum 5 | 37.76 | 2.14 | 42.78 | 1.81 | 10.42 | 1.97 | 3.13 | 100.00 |
| Mean | 42.91 | 1.78 | 31.82 | 3.78 | 8.86 | 1.46 | 9.40 | 100.00 |
| Std. deviation | 4.64 | 0.48 | 7.92 | 2.13 | 2.13 | 0.43 | 8.77 | |
| Max. | 49.20 | 2.22 | 42.78 | 6.53 | 10.42 | 1.97 | 20.02 | |
| Min. | 37.76 | 1.14 | 24.01 | 1.81 | 5.16 | 0.90 | 2.23 | |

All the data in %weight

It should be noted that the microstructure of the Davidovits products cannot be compared directly with the structure of the blue faience that has been reported in the literature [22, 23] as the SEM image shown in Figure 8 is not of a polished sample. There are some apparent differences between the data for ancient faience reported in the literature and the data from the geopolymer material:

1. The ancient faience contains crystalline quartz. This is determined primarily from SEM observations. No quartz is found in the diffraction data shown in Figure 7.
2. In the ancient glazes, copper is interpreted as being embodied in the glass structure as a cation while in the geopolymer case the copper is still forming phosphates as made evident by the mineral phase analysis of the XRD data (Figure 7).
3. The ancient glaze is a glass with the characteristic compositions of an alkaline glass. In the case of the geopolymer material it has a different composition. All the silicon content shown in Table 4 is due to the silicon contained in the starting material in the form of potassium silicate. No sodium is present since the sample used for this test did not contain any to start with. A green product would have shown sodium but not potassium.

The first two major differences alone should be sufficient to enable one to establish whether the Davidovits proposition has any merit. All that is needed is a very easy to perform measurement on a very small fragment from an archaeologically attested sample. 30 s of counting time on a synchrotron instrument of the type we have used would give a clear and unambiguous yes/no answer.

*Raman Spectroscopy*: Raman spectroscopy was used in an attempt to correlate the above observations to the molecular composition of the areas examined. The rational was that rather than taking archaeological samples to a synchrotron, in-situ measurements using a portable Raman instrument may be just as effective in providing a definitive answer, with no sample extraction and preparation procedures. Portable Raman spectrometers do exist [31, 32] and, in principle, may be used to test ancient material non-destructively and in-situ, in museums where Djoser tiles are kept (Louvre in Paris, Archaeological Museum in Cairo, Metropolitan Museum of Art in New York).

Shots at different positions were taken to study variations along the sample surface. It was possible to execute a map of a region of few hundred of $\mu m^2$. μ-Raman spectra were collected by a high resolution Dilor Labram spectrograph and compared to specialized mineral database for recognition. The exciting source was a HeNe laser (632.8 nm) with a power of less than 10 mW at the sample. The microscope was coupled confocally to the spectrograph. A 50X and 100X objectives were used. Suppression of the exciting line was obtained with a holographic notch filter. The spectra were measured at room temperature positioning the sample on a motorized XY stage so that Raman maps were performed by acquiring a large number of Raman spectra and mapping the intensity of a characteristic Raman band.

Figure 9 shows four representative spectra collected on the surface of the blue vitreous coating on a square of $50 \times 50\ \mu m^2$. The maps in Figure 10 show the area value of a selected Raman band after background subtraction. In this way, by means of μ-Raman spectroscopy it was possible to map the samples surface and analyze the spatial distribution of phases of inhomogeneous samples, even present in small quantity. The Raman maps show the presence of domains of different phases, 5–10 μm in diameter matching the average grain size obtained by SEM observation.

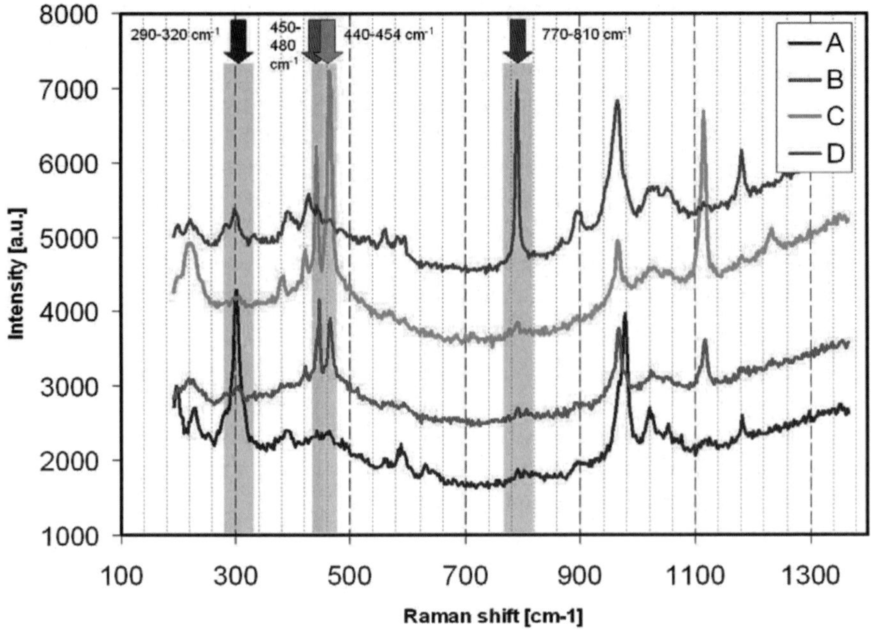


**Fig. 9** Representative μ-Raman spectra collected from four spots on the blue surface from a square grid of 50 × 50 μm.

## 5 The Next Step

The results for the geopolymer product outlined above do not constitute proof that the glaze on the Djoser pyramid tiles was produced by a similar process. To resolve the issue of whether the Djoser dynasty tiles were made by a different procedure than for later faience objects, we propose the following action, which depends on the cooperation of the appropriate museum authorities:

> To perform the SR-XRD measurements as described above on tiny fragments of archaeological material from Egypt, 3rd dynasty and later, and compare properties between them and with those of geopolymer material of similar colour. Micro-XRD and XRF mapping of a thin section of the layer from the surface to the core would be help correlate diffracting mineral phases and amorphous content to elemental composition.

If the results indicate that, for the case of 3rd dynasty Djoser pyramid tiles, major differences from the geopolymer results describe here, i.e., absence of phosphate phases, then the Davidovits proposition remains an "interesting" idea. If, however, it turns out that in the top layer phosphates are present as distinctly as in Figure 7, then a very interesting new insight would have been gained. A further round of measurements would then be justified whereby polished samples are used for SEM/EDX

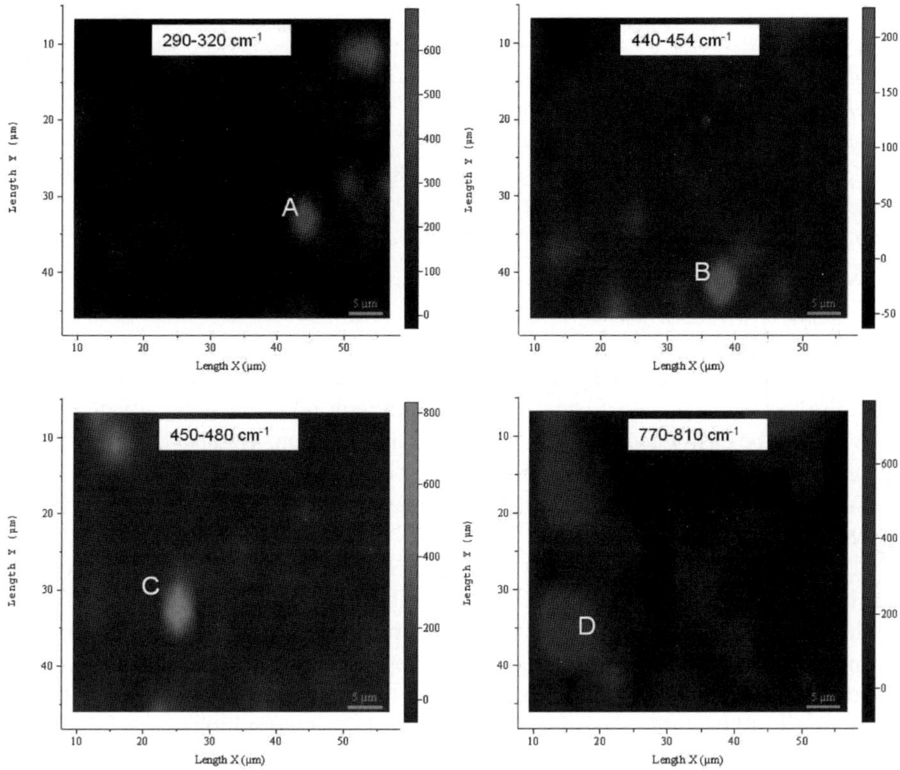


**Fig. 10** μ-Raman maps showing the area value of a selected Raman band after background subtraction.

studies (much preferable to the rough-and-ready method we have employed here for reasons solely to do with time and effort available) to fully characterise the material. This could be followed by a time-resolved X-ray diffraction study of the geopolymer formation as a function of temperature as has been used recently for other studies of the evolution of mineral phases in ceramics [33, 34].

## 6 On Hephaistian Arts and Crafts

To return to Homer, and hopefully still keeping within the scope and objectives of this conference, we now focus our attention to something completely different. The matter of whether Homeric descriptions contain traces of technological information.

One cannot but wonder whether the descriptions in the famous passage in the Iliad of scenes decorating Achilles shield, made by the god-smith himself, Hephaestus (Table 5), is a reference to processes that had been transmitted to Bronze Age craftsmen through contact with Egyptian craftsmen. Of course, Homer was a

**Table 5** Homeric adjectives used for Hephaistos.

| |
|---|
| Ἥφαιστον κλυτοτέχνην (renowned craftsman, *Od.* 8.286), |
| πολύφρονος Ἡφαίστοιο (inventive, *Od.* 8.297, 9.327), |
| Ἥφαιστον κλυτοεργὸν (excellent craftsman, *Od.* 8.345), |
| περικλυτοῦ Ἡφαίστοιο (famous, *Od.* 8.287, 8.349, 8.357, 24.75, *Il.* 1.607, 18.383, 18.393, 18.462, 18.587, 18.590, 18.614), |
| πολυμήτιος Ἡφαίστοιο (knowledgeable, *Il.* 21.357), |
| Ἥφαιστε ἀγακλεές (glorious, *Il.* 21.379) ) |

poet, not a scientist, most certainly not a synchrotron scientist. We believe that imagination is not the preserve of poets alone, although poets have the advantage that they cannot be easily accused of too vivid an imagination. We also believe that science and technology are not strictly prescriptive. At least not always. This is our main defence for delving into matters that may be considered by some as the preserve of scholars – not mere scientists.

Of particular interest to us are the description of melting copper and tin to male the bronze for the shield and, specifically, the inclusion of gold and silver in the crucible (Iliad 18:474–677) and other references further down in the text on decorative scenes that may refer to enamelling. Table 6 reproduces the whole passage from the Iliad where Hephaestos sets to work to make the to make a shield for Achilles, on the request of his goddess mother Thetis. Let us take a close look.

Reading these lines (and between the lines) may lead one to imagine that the country Homer describes where peaceful peasants collect the harvest on fertile brown soil or tend their herds of horned cattle by a big river could be Egypt. Homer hardly ever fails to ignite the imagination of his audience. His description of the scenes in that passage are so vivid, the casual reader can easily forget it is a description of the decorations on the shield, and not an actual account of pastoral life in the countryside. But is there harder information carried in these verses? This wonderful passage potentially contains some real gems of information. Notice for instance the reference to copper and tin as well as gold and silver in the making of the alloy for the shield. Why gold and silver? What kind of special copper alloy was this? What were the ratios of Cu:Sn:Au:Ag? Alas, Homer did not record technical details such as chemical compounds and formulae, percentage compositions or temperatures (how neglectful of him!). Not all that easy or interesting, one might say, to record such prosaic detail in hexameter.

Note that the word *brass*, used in the English translation of the passage quoted in Table 6, instead of *bronze*, is incorrect. Brass (alloy of copper and zinc) was not used until much later times. Although the anonymous translator has given us a beautiful prose rendition of the Iliad in a real treasure of a rare 2-volume edition, this mistranslation of the word χαλκὸς (often encountered in other learned translations) is indicative of other such crucial mistranslations and misinterpretations of the Homeric text that are often encountered in the foreign (as well as modern Greek) literature on the Epics. Another article [35] offers an extended discussion of

**Table 6** The passage from the Iliad 18:474–677. English translation from [12].

| | |
|---|---|
| ὣς εἰπὼν τὴν μὲν λίπεν αὐτοῦ, βῆ δ᾽ ἐπὶ φύσας·<br>τὰς δ᾽ ἐς πῦρ ἔτρεψε κέλευσέ τε ἐργάζεσθαι.<br>φῦσαι δ᾽ ἐν χοάνοισιν ἐείκοσι πᾶσαι ἐφύσων<br>παντοίην εὔπρηστον ἀϋτμὴν ἐξανιεῖσαι,<br>ἄλλοτε μὲν σπεύδοντι παρέμμεναι, ἄλλοτε δ᾽ αὖτε,<br>ὅππως Ἥφαιστός τ᾽ ἐθέλοι καὶ ἔργον ἄνοιτο.<br>χαλκὸν δ᾽ ἐν πυρὶ βάλλεν ἀτειρέα κασσίτερόν τε<br>καὶ χρυσὸν τιμῆντα καὶ ἄργυρον· αὐτὰρ ἔπειτα<br>θῆκεν ἐν ἀκμοθέτῳ μέγαν ἄκμονα, γέντο δὲ χειρὶ<br>ῥαιστῆρα κρατερήν, ἑτέρηφι δὲ γέντο πυράγρην. | So saying, he left her there, and went towards the bellows, which he turned against the fire, and commanded to be worked. Twenty bellows all blew in the furnaces, sending forth a raised, easily inflated blast, sometimes to assist him when in haste, at other times, again, as Vulcan chose, and the work could be completed. He cast into the fire impenetrable brass, and tin, precious gold and silver ; but next he placed the mighty anvil on the stock, in one hand he took his strong hammer, and with the other he seized the forceps. |
| ποίει δὲ πρώτιστα σάκος μέγα τε στιβαρόν τε<br>πάντοσε δαιδάλλων, περὶ δ᾽ ἄντυγα βάλλε φαεινὴν<br>τρίπλακα μαρμαρέην, ἐκ δ᾽ ἀργύρεον τελαμῶνα.<br>πέντε δ᾽ ἄρ᾽ αὐτοῦ ἔσαν σάκεος πτύχες· αὐτὰρ ἐν αὐτῷ<br>ποίει δαίδαλα πολλὰ ἰδυίῃσι πραπίδεσσιν. | First of all he made a shield, large and solid, adorning it on all sides with various workmanship, and threw round it a shining rim, triple and splendid, from which he *hung* a silver thong. Of the shield itself, however, there were five folds ; whilst in it he formed carried workmanship with skilful art. |
| ἐν μὲν γαῖαν ἔτευξ᾽, ἐν δ᾽ οὐρανόν, ἐν δὲ θάλασσαν,<br>ἠέλιόν τ᾽ ἀκάμαντα σελήνην τε πλήθουσαν,<br>ἐν δὲ τὰ τείρεα πάντα, τά τ᾽ οὐρανὸς ἐστεφάνωται,<br>Πληϊάδας θ᾽ Ὑάδας τε τό τε σθένος Ὠρίωνος<br>Ἄρκτόν θ᾽, ἣν καὶ Ἄμαξαν ἐπίκλησιν καλέουσιν,<br>ἥ τ᾽ αὐτοῦ στρέφεται καί τ᾽ Ὠρίωνα δοκεύει,<br>οἴη δ᾽ ἄμμορός ἐστι λοετρῶν Ὠκεανοῖο. | In it he formed the earth, and the sea, the unwearied sun, and the full moon. In it also he *described* all the constellations with which the heaven is crowned, the Pleïades, the Hyades, and the strength of Orion, and the Bear, which likewise they call by the appellation of the Wain, which is there turned round, and watches Orion ; and it alone is deprived of the baths of Oceanus. |
| ἐν δὲ δύω ποίησε πόλεις μερόπων ἀνθρώπων<br>καλάς. ἐν τῇ μέν ῥα γάμοι τ᾽ ἔσαν εἰλαπίναι τε,<br>νύμφας δ᾽ ἐκ θαλάμων δαΐδων ὕπο λαμπομενάων<br>ἠγίνεον ἀνὰ ἄστυ, πολὺς δ᾽ ὑμέναιος ὀρώρει·<br>κοῦροι δ᾽ ὀρχηστῆρες ἐδίνεον, ἐν δ᾽ ἄρα τοῖσιν<br>αὐλοὶ φόρμιγγές τε βοὴν ἔχον· αἳ δὲ γυναῖκες<br>ἱστάμεναι θαύμαζον ἐπὶ προθύροισιν ἑκάστη.<br>λαοὶ δ᾽ εἰν ἀγορῇ ἔσαν ἀθρόοι· ἔνθα δὲ νεῖκος<br>ὠρώρει, δύο δ᾽ ἄνδρες ἐνείκεον εἵνεκα ποινῆς<br>ἀνδρὸς ἀποφθιμένου· ὃ μὲν εὔχετο πάντ᾽ ἀποδοῦναι<br>δήμῳ πιφαύσκων, ὃ δ᾽ ἀναίνετο μηδὲν ἑλέσθαι·<br>ἄμφω δ᾽ ἱέσθην ἐπὶ ἴστορι πεῖραρ ἑλέσθαι.<br>λαοὶ δ᾽ ἀμφοτέροισιν ἐπήπυον ἀμφὶς ἀρωγοί·<br>κήρυκες δ᾽ ἄρα λαὸν ἐρήτυον· οἳ δὲ γέροντες<br>εἵατ᾽ ἐπὶ ξεστοῖσι λίθοις ἱερῷ ἐνὶ κύκλῳ,<br>σκῆπτρα δὲ κηρύκων ἐν χέρσ᾽ ἔχον ἠεροφώνων·<br>τοῖσιν ἔπειτ᾽ ἤϊσσον, ἀμοιβηδὶς δὲ δίκαζον.<br>κεῖτο δ᾽ ἄρ᾽ ἐν μέσσοισι δύω χρυσοῖο τάλαντα,<br>τῷ δόμεν ὃς μετὰ τοῖσι δίκην ἰθύντατα εἴποι.<br>τὴν δ᾽ ἑτέρην πόλιν ἀμφὶ δύω στρατοὶ ἥατο λαῶν<br>τεύχεσι λαμπόμενοι· δίχα δέ σφισιν ἥνδανε βουλή,<br>ἠὲ διαπραθέειν ἢ ἄνδιχα πάντα δάσασθαι<br>κτῆσιν ὅσην πτολίεθρον ἐπήρατον ἐντὸς ἔεργεν·<br>οἳ δ᾽ οὔ πω πείθοντο, λόχῳ δ᾽ ὑπεθωρήσσοντο.<br>τεῖχος μέν ῥ᾽ ἄλοχοί τε φίλαι καὶ νήπια τέκνα<br>ῥύατ᾽ ἐφεσταότες, μετὰ δ᾽ ἀνέρες οὓς ἔχε γῆρας·<br>οἳ δ᾽ ἴσαν· ἦρχε δ᾽ ἄρά σφιν Ἄρης καὶ Παλλὰς Ἀθήνη<br>ἄμφω χρυσείω, χρύσεια δὲ εἵματα ἕσθην,<br>καλὼ καὶ μεγάλω σὺν τεύχεσιν, ὥς τε θεώ περ<br>ἀμφὶς ἀριζήλω· λαοὶ δ᾽ ὑπολίζονες ἦσαν.<br>οἳ δ᾽ ὅτε δή ῥ᾽ ἵκανον ὅθι σφίσιν εἶκε λοχῆσαι<br>ἐν ποταμῷ, ὅθι τ᾽ ἀρδμὸς ἔην πάντεσσι βοτοῖσιν,<br>ἔνθ᾽ ἄρα τοί γ᾽ ἵζοντ᾽ εἰλυμένοι αἴθοπι χαλκῷ.<br>τοῖσι δ᾽ ἔπειτ᾽ ἀπάνευθε δύω σκοποὶ εἵατο λαῶν | In it he likewise framed two beautiful cities of articulate-speaking men. In the one there were marriages and feasts; they were leading the brides from their chambers through the city with lighted torches, and the frequent bridal song arose. The youthful dancers were wheeling around, and amongst them pipes and lyres uttered their sound ; and the women, one and all, standing at the portals, admired *it*. The people were numerous in an assembly, and there a contest had arisen ; for two men disputed about the ransom-money for a man *who was* slain : the one indeed affirmed that he had given all, declaring to the people ; and the other denied that he had received aught : and both desired to end *the dispute* by an umpire. The people *who were* favourers on both sides, shouted in approbation of both ; but the heralds were appeasing the people, whilst the old men sat upon polished stones, in a sacred circle, and held in their hands the sceptres of loud-shouting heralds ; with which they afterwards arose, and gave judgement in turns. Moreover, in the midst lay two talents of gold, to be given to him, who among them should deliver the most correct judgement. But round the other city sat two armies of people, glittering in armour. In two ways was a plan agreeable to them, either to destroy it entirely, or to divide all things into two parts, the wealth, as much as the pleasant city contained within. They, however, did not yet obey, but secretly armed themselves for an ambuscade. In the mean time their beloved wives and young children kept guard, standing upon the wall, and with them the men whom old age possessed. And they marched forth ; and Mars and Pallas Minerva, both golden, led them on, and they were arrayed in golden robes, beautiful and large, along with their armour, distinguished on both sides like gods ; but the people were of humbler size. But when now they had arrived where it appeared convenient for them to place an ambuscade, by a river, where was a watering place for all cattle, there they sat down, wrapt in shining brass. Next, |

**Table 6** (Continued)

<table>
<tr>
<td>δέγμενοι ὁππότε μῆλα ἰδοίατο καὶ ἕλικας βοῦς.<br>οἳ δὲ τάχα προγένοντο, δύω δ᾽ ἅμ᾽ ἕποντο νομῆες<br>τερπόμενοι σύριγξι· δόλον δ᾽ οὔ τι προνόησαν.<br>οἳ μὲν τὰ προϊδόντες ἐπέδραμον, ὦκα δ᾽ ἔπειτα<br>τάμνοντ᾽ ἀμφὶ βοῶν ἀγέλας καὶ πώεα καλὰ<br>ἀργεννέων οἰῶν, κτεῖνον δ᾽ ἐπὶ μηλοβοτῆρας.<br>οἳ δ᾽ ὡς οὖν ἐπύθοντο πολὺν κέλαδον παρὰ βουσὶν<br>εἰράων προπάροιθε καθήμενοι, αὐτίκ᾽ ἐφ᾽ ἵππων<br>βάντες ἀερσιπόδων μετεκίαθον, αἶψα δ᾽ ἵκοντο.<br>στησάμενοι δ᾽ ἐμάχοντο μάχην ποταμοῖο παρ᾽ ὄχθας,<br>βάλλον δ᾽ ἀλλήλους χαλκήρεσιν ἐγχείῃσιν.<br>ἐν δ᾽ Ἔρις ἐν δὲ Κυδοιμὸς ὁμίλεον, ἐν δ᾽ ὀλοὴ Κήρ,<br>ἄλλον ζωὸν ἔχουσα νεούτατον, ἄλλον ἄουτον,<br>ἄλλον τεθνηῶτα κατὰ μόθον ἕλκε ποδοῖιν·<br>εἶμα δ᾽ ἔχ᾽ ἀμφ᾽ ὤμοισι δαφοινεὸν αἵματι φωτῶν.<br>ὡμίλευν δ᾽ ὥς τε ζωοὶ βροτοὶ ἠδ᾽ ἐμάχοντο,<br>νεκρούς τ᾽ ἀλλήλων ἔρυον κατατεθνηῶτας.<br>ἐν δ᾽ ἐτίθει νειὸν μαλακὴν πίειραν ἄρουραν<br>εὐρεῖαν τρίπολον· πολλοὶ δ᾽ ἀροτῆρες ἐν αὐτῇ<br>ζεύγεα δινεύοντες ἐλάστρεον ἔνθα καὶ ἔνθα.<br>οἳ δ᾽ ὁπότε στρέψαντες ἱκοίατο τέλσον ἀρούρης,<br>τοῖσι δ᾽ ἔπειτ᾽ ἐν χερσὶ δέπας μελιηδέος οἴνου<br>δόσκεν ἀνὴρ ἐπιών· τοὶ δὲ στρέψασκον ἀν᾽ ὄγμους,<br>ἱέμενοι νειοῖο βαθείης τέλσον ἱκέσθαι.<br>ἣ δὲ μελαίνετ᾽ ὄπισθεν, ἀρηρομένῃ δὲ ἐῴκει,<br>χρυσείη περ ἐοῦσα· τὸ δὴ περὶ θαῦμα τέτυκτο.<br>ἐν δ᾽ ἐτίθει τέμενος βασιλήϊον· ἔνθα δ᾽ ἔριθοι<br>ἤμων ὀξείας δρεπάνας ἐν χερσὶν ἔχοντες.<br>δράγματα δ᾽ ἄλλα μετ᾽ ὄγμον ἐπήτριμα πῖπτον ἔραζε,<br>ἄλλα δ᾽ ἀμαλλοδετῆρες ἐν ἐλλεδανοῖσι δέοντο.<br>τρεῖς δ᾽ ἄρ᾽ ἀμαλλοδετῆρες ἐφέστασαν· αὐτὰρ ὄπισθε<br>παῖδες δραγμεύοντες ἐν ἀγκαλίδεσσι φέροντες<br>ἀσπερχὲς πάρεχον· βασιλεὺς δ᾽ ἐν τοῖσι σιωπῇ<br>σκῆπτρον ἔχων ἑστήκει ἐπ᾽ ὄγμου γηθόσυνος κῆρ.<br>κήρυκες δ᾽ ἀπάνευθεν ὑπὸ δρυῒ δαῖτα πένοντο,<br>βοῦν δ᾽ ἱερεύσαντες μέγαν ἄμφεπον· αἳ δὲ γυναῖκες<br>δεῖπνον ἐρίθοισιν λεύκ᾽ ἄλφιτα πολλὰ πάλυνον.<br>ἐν δ᾽ ἐτίθει σταφυλῇσι μέγα βρίθουσαν ἀλωὴν<br>καλὴν χρυσείην· μέλανες δ᾽ ἀνὰ βότρυες ἦσαν,<br>ἑστήκει δὲ κάμαξι διαμπερὲς ἀργυρέῃσιν.<br>ἀμφὶ δὲ κυανέην κάπετον, περὶ δ᾽ ἕρκος ἔλασσε<br>κασσιτέρου· μία δ᾽ οἴη ἀταρπιτὸς ἦεν ἐπ᾽ αὐτήν,<br>τῇ νίσοντο φορῆες ὅτε τρυγόῳεν ἀλωήν.<br>παρθενικαὶ δὲ καὶ ἠΐθεοι ἀταλὰ φρονέοντες<br>πλεκτοῖς ἐν ταλάροισι φέρον μελιηδέα καρπόν.<br>τοῖσιν δ᾽ ἐν μέσσοισι πάϊς φόρμιγγι λιγείῃ<br>ἱμερόεν κιθάριζε, λίνον δ᾽ ὑπὸ καλὸν ἄειδε<br>λεπταλέῃ φωνῇ· τοὶ δὲ ῥήσσοντες ἁμαρτῇ<br>μολπῇ τ᾽ ἰυγμῷ τε ποσὶ σκαίροντες ἕποντο.<br>ἐν δ᾽ ἀγέλην ποίησε βοῶν ὀρθοκραιράων·<br>αἳ δὲ βόες χρυσοῖο τετεύχατο κασσιτέρου τε,<br>μυκηθμῷ δ᾽ ἀπὸ κόπρου ἐπεσσεύοντο νομὸν δὲ<br>πὰρ ποταμὸν κελάδοντα, παρὰ ῥοδανὸν δονακῆα.<br>χρύσειοι δὲ νομῆες ἅμ᾽ ἐστιχόωντο βόεσσι<br>τέσσαρες, ἐννέα δέ σφι κύνες πόδας ἀργοὶ ἕποντο.<br>σμερδαλέω δὲ λέοντε δύ᾽ ἐν πρώτῃσι βόεσσι<br>ταῦρον ἐρύγμηλον ἐχέτην· ὃ δὲ μακρὰ μεμυκὼς<br>ἕλκετο· τὸν δὲ κύνες μετεκίαθον ἠδ᾽ αἰζηοί.<br>τὼ μὲν ἀναρρήξαντε βοὸς μεγάλοιο βοείην<br>ἔγκατα καὶ μέλαν αἷμα λαφύσσετον· οἳ δὲ νομῆες<br>αὔτως ἐνδίεσαν ταχέας κύνας ὀτρύνοντες.<br>οἳ δ᾽ ἤτοι δακέειν μὲν ἀπετρωπῶντο λεόντων,<br>ἱστάμενοι δὲ μάλ᾽ ἐγγὺς ὑλάκτεον ἔκ τ᾽ ἀλέοντο.<br>ἐν δὲ νομὸν ποίησε περικλυτὸς ἀμφιγυήεις<br>ἐν καλῇ βήσσῃ μέγαν οἰῶν ἀργεννάων,</td>
<td>apart from the troops, sat down two spies, watching when they might decry the flocks and horned herds. These, however, soon came forward, and with them followed two shepherds, delighting themselves with their pipes, for as yet they foresaw not the treachery. Then they, seeing them before them, rushed upon them, and soon after plundered on all sides the herds of oxen and beautiful flocks of white sheep; and they slew the shepherds besides. The besiegers, however, previously sitting in assembly, as soon as they heard the mighty tumult among the oxen, mounting their nimble-footed steeds, pursued, and speedily overtook them. Then, standing, they fought a battle by the banks of the river, and smote one another with their brazen spears. Amongst them mingled Discord and Tumult, and deadly Fate, holding one yet alive, newly wounded, another unhurt, and by the feet dragged another, slain, through the combat : and had the robe round her shoulders quite empurpled with the blood of men. But the turned about like living mortals, and fought, and dragged away each other's slaughtered carcases.<br><br>On it also he placed a soft fallow field, rich soil, broad, and thrice tilled; and in it many ploughers exercised themselves here and there, turning round their yokes. But when, returning, they came to the boundary of the field, then a man approaching them, gave into their hands a cup of sweetest wine ; upon which they turned to their furrows, eager to reach the boundary of the deep fallow field. But it was black behind like to ploughed land, though being golden ; which was surely very wonderful.<br>On it likewise he placed a private field of high corn, where workmen reaped, holding sharp sickles in their hands. Some handfuls fell thick in ranks upon the ground, and others the binders of sheaves fastened together with cords. Moreover three binders of sheaves followed the reapers, whilst behind them boys gathering the handfuls, and bearing them in their arms, constantly supplied them ; and the master stood among them in silence, holding a staff, delighted in his heart at the rows of corn. But apart, below an oak, servants were preparing feasts, and sacrificing a huge ox, attended to their wants ; while women mixed in much while barley, as a feast for the labourers.<br>On it likewise he placed, of huge size, a vine-yard, heavy with grapes, beautiful and golden ; but the clusters were black ; and they (the vines) stood against silver poles, placed in order. Round it he drew an azure ditch, about which he planted an edge of tin. But one path alone was upon it by which the gatherers went when they collected the vintage. Little maids and youths of tender minds bore the luscious fruit in twisted baskets, in the midst of whom a boy sweetly played upon a shrill lyre ; and, with tender voice, beautifully sung to it a rustic song : whilst they, beating the earth at the same time, with singing and shouts of joy, followed, skipping with their feet.<br><br>Upon it he also formed an herd of oxen, with horns erect. But the oxen were made of gold and of tin, and rushed out with a lowing from the stall to the pasture, beside a river, roaring and flowing, with an abundance of reeds. Four golden herdsmen went along with the oxen, and nine swift dogs followed them at their feet. But two fertile lions seized the bull, groaning sadly among the foremost oxen, and he dragged away, loudly bellowing ; and the dogs and youths followed to his rescue. They indeed, having torn off the skin of the huge ox, licked up his entrails and black blood ; and the shepherds vainly pressed upon them, urging on their fleet dogs. These of a truth refused to bite the lions, but, standing very near, barked, yet shunned them.<br><br>Upon it the illustrious Vulcan likewise formed a pasture in a beautiful glen, full of white sheep, and folds, and tents and covered cottages.</td>
</tr>
</table>

**Table 6** (Continued)

| | |
|---|---|
| σταθμούς τε κλισίας τε κατηρεφέας ἰδὲ σηκούς.<br>ἐν δὲ χορὸν ποίκιλλε περικλυτὸς ἀμφιγυήεις,<br>τῷ ἴκελον οἷόν ποτ᾽ ἐνὶ Κνωσῷ εὐρείῃ<br>Δαίδαλος ἤσκησεν καλλιπλοκάμῳ Ἀριάδνῃ.<br>ἔνθα μὲν ἠΐθεοι καὶ παρθένοι ἀλφεσίβοιαι<br>ὀρχεῦντ᾽ ἀλλήλων ἐπὶ καρπῷ χεῖρας ἔχοντες.<br>τῶν δ᾽ αἳ μὲν λεπτὰς ὀθόνας ἔχον, οἳ δὲ χιτῶνας<br>εἵατ᾽ ἐϋννήτους, ἦκα στίλβοντας ἐλαίῳ·<br>καί ῥ᾽ αἳ μὲν καλὰς στεφάνας ἔχον, οἳ δὲ μαχαίρας<br>εἶχον χρυσείας ἐξ ἀργυρέων τελαμώνων.<br>οἳ δ᾽ ότὲ μὲν θρέξασκον ἐπισταμένοισι πόδεσσι<br>ῥεῖα μάλ᾽, ὡς ὅτε τις τροχὸν ἄρμενον ἐν παλάμῃσιν<br>ἑζόμενος κεραμεὺς πειρήσεται, αἴ κε θέῃσιν·<br>ἄλλοτε δ᾽ αὖ θρέξασκον ἐπὶ στίχας ἀλλήλοισι.<br>πολλὸς δ᾽ ἱμερόεντα χορὸν περιίσταθ᾽ ὅμιλος τερπόμενοι·<br>δοιὼ δὲ κυβιστητῆρε κατ᾽ αὐτοὺς<br>μολπῆς ἐξάρχοντες ἐδίνευον κατὰ μέσσους.<br>ἐν δ᾽ ἐτίθει ποταμοῖο μέγα σθένος Ὠκεανοῖο<br>ἄντυγα πὰρ πυμάτην σάκεος πύκα ποιητοῖο.<br>αὐτὰρ ἐπεὶ δὴ τεῦξε σάκος μέγα τε στιβαρόν τε,<br>τεῦξ᾽ ἄρα οἱ θώρηκα φαεινότερον πυρὸς αὐγῆς,<br>τεῦξε δέ οἱ κόρυθα βριαρὴν κροτάφοις ἀραρυῖαν<br>καλὴν δαιδαλέην, ἐπὶ δὲ χρύσεον λόφον ἧκε,<br>τεῦξε δέ οἱ κνημῖδας ἑανοῦ κασσιτέροιο.<br>αὐτὰρ ἐπεὶ πάνθ᾽ ὅπλα κάμε κλυτὸς ἀμφιγυήεις,<br>μητρὸς Ἀχιλλῆος θῆκε προπάροιθεν ἀείρας.<br>ἣ δ᾽ ἴρηξ ὣς ἆλτο κατ᾽ Οὐλύμπου νιφόεντος<br>τεύχεα μαρμαίροντα παρ᾽ Ἡφαίστοιο φέρουσα. | The illustrious Vulcan likewise adorned it with a dance, like to that which, in the wide Cnossus, Daedalus once composed for the fair-haired Ariadne. There danced the youths and lovely virgins, holding each other's hands near the wrist. Of these, the *maids*, wore fine linen robes, and the *youths* were dressed in well-woven coats, slightly shining with oil : these also had beautiful garlands, and those wore golden swords *suspended* from silver belts.<br><br>Sometimes, with skilful feet, they nimbly ran *the circle* : as when a certain potter, sitting, shall try a wheel fitted in his hands, if it will run ; and at other times again their ranks through one another. But a great crowd stood round the pleasing dance, delighted ; and beside them, two tumblers, beginning their song, wheeled around in the midst.<br><br>Moreover, he placed in it the vast strength of the river Oceanus, near the outer rim of the well-formed shield.<br><br>But when he had completed the shield, huge and solid, he next formed for him a corselet, brighter than the splendour of fire. He also made for him a strong helmet, fitted to his temples, beautiful and variously wrought, and put upon it a golden crest; and formed greaves for him of ductile tin.<br><br>But when the renowned Vulcan had with toil made all the arms, lifting *them* up, he laid *them* before the mother of Achilles ; and she, like a hawk, shot down from the snow-clad Olympus, bearing the glittering armour from Vulcan. |

the topic including on words underlined in the Homeric passage in Table 6. A similar situation is not unlikely to be the case for translations of Egyptian hieroglyphic texts, some of which may contain valuable technical information on metallurgy and chemistry (alchemy) in general [36].

## 7 Conclusions

Davidovits [26] has proposed that blue, green, beige or black glaze-like coatings could have been produced on ceramic tiles at rather moderate temperatures, easily achievable even in a domestic type bread oven in ancient Egypt. We have applied a number of materials science techniques to characterise one of the samples that Davidovits has produced in his laboratory. The results corroborate the claim that an amorphous, *glassy* surface is produced. Whether this can be described as *faience-like* or not, requires close scrutiny and comparison with authentic ancient material from the relevant historical period, 3rd dynasty.

All the same, it is an appealing thought, that ancient craftsmen were as ingenious in matters of solid state chemistry and metallurgy as present-day industrial materials scientists. Could similar skills have been used for decorating metal objects such as mentioned in the Homeric epics? Why not? Are the scenes described in the story of Hephaistos making Achilles' shield set in Egypt? What else could the Greeks have learned from the wise-old Egyptians, a mature and technologically advanced civilisation, well before Homeric times? The great pyramids at Giza had been standing for some 1500 years before *artful Odysseus* [Il. B173] and his *bronze-clad Achaean* warrior friends [Il. A371, Od. α286] *sacked the holy citadel of Troy* [Od. α2].

## Acknowledgements

We are grateful to Dr. Trinitat Pradell of the Technical University of Catalonia, Spain, and Panagiota Manti of the Conservation Department, Cardiff University, U.K., for their many valuable critical comments and for pointing us to literature references on Egyptian faience and ancient glazing techniques that we were not familiar with when this project was conceived.

## References

1. Merrillees, R.S. (1972). Aegean Bronze Age relations with Egypt, *Amer. J. Archaeol.* **76**(3), 281–294.
2. Cline, E.H. (1993). Contact and trade or colonization?: Egypt and the Aegean in the 14th–13th Centuries B.C., *Minos* **25/26** (1990-91 [1993]), 7–36.
3. Robbins, M. (2001). *Collapse of the Bronze Age: The Story of Greece, Troy, Israel, Egypt, and the Peoples of the Sea*, Authors Choice Press.
4. ΗΡΟΔΟΤΟΥ ΙΣΤΟΡΙΑ, *The History of Herodotus*, Loeb Classical Library, Harvard University Press.
5. Manning, S.W., Kromer, B., Kuniholm, P.I. and Newton, M.W. (2001). Anatolian tree rings and a new chronology for the East Mediterranean Bronze-Iron Ages, *Science* **294**(5551), 2532–2535.
6. The Thera Foundation, The Mode of Representation in Egyptian Art in Comparison to Aegean Bronze Age Art, http://www.therafoundation.org/articles/art/ (last accessed 12/11/06).
7. Apollodorus, A Library of Greek Mythology, Oxford World's Classics (1999). Translated with an Introduction and Notes by Robin Hard. Oxford University Press. See also Graves, R., *The Greek Myths – Complete Edition*, Penguin, 1996.
8. Theocaris, P.S., Liritzis, I. and Galloway, R.B. (1994). Dating of two Hellenic pyramids by a novel application of thermoluminescence, *J. Archaeol. Sci.* **24**, 399–405.
9. Hutton, W. (2005). *Describing Greece: Landscape and Literature in the Periegesis of Pausanias*, Cambridge University Press.
10. Stewart, M., People, places & things: Proetus, *Greek Mythology: From the Iliad to the Fall of the Last Tyrant*, http://messagenet.com/myths/ppt/Proetus_1.html (last accessed 13/11/06).
11. Pope, A. (1903). *The Odyssey of Homer*, Grant Richards, London. First published in five volumes in the year 1725. See also http://etext.library.adelaide.edu.au/h/homer/h8op/ (last accessed 14/10/06).
12. *Iliad of Homer*, by "A Graduate of the University of Oxford", 2nd Edn., Whitaker, London, 1825. (First published in 1821.)
13. Lauer, J.-P. (1976). *Saqqara: The Royal Cemetery of Memphis*, Thames and Hudson, London, pp. 86–136 (for the number of tiles and the archaeology of the pyramid and the tiles).
14. Vandiver, P.B. and Kingery, W.D. (1986). Egyptian faience: The first high-tech caremic, in *Ceramics and Civilisation*, Vol. 3, High-Technology Ceramics: Past, Present, and future, The American Ceramic Society, 88th Annual Meeting of the ACS, Chicago, Illinois.

15. Friedman, F.D. (Ed.) (1998). *Gifts of the Nile: Ancient Egyptian Faience*, Thames and Hudson, London.
16. Kiefer, C. and Allibert, A. (1971). Pharaonic blue ceramics, the process of self-glazing, *Archaeololy* **24**, 107–117. See also Kyoko Yamahana, Synchrotron Radiation Analysis on Ancient Egyptian Vitreous Materials, Proceedings of the 25th Linear Accelerator Meeting, Himeji, Japan, July 12–14, 2000, [13C-01].
17. Kaczmarczyk, A. and Hedges, R.E.M. (1983). *Ancient Egyptian Faience: An Analytical Survey of Egyptian Faience from Predynastic to Roman Times*, Aris & Phillips, Warminster, England. Appendix A, pp. 1–146 by P. Vandiver, *The Manufacture of Faience.*
18. Nicholson, D.G. and Nilsen, M.H. (2000). An X-ray absorption spectroscopic study on the local environment of copper in CuAPO-5, *J. Mater. Chem.* **8**, 1965–1971.
19. Schiegel, S. (1988). Investigation of faience tiles from the walls of Djoser's south tomb in Saqqara: An approach to reveal the technique of their manufacture, in *Proceedings 5th International Congress of Egyptology*, Cairo (Abstract).
20. Wulff, H.E., Wulff, H.S. and Koch, L. (1968). Egyptian faience – A possible survival in Iranm, *Archaeology* **21**, 98–107.
21. Vandiver, P. (1982). Egyptian faience technology, in *Archaeological Ceramics*, A.D. Franklin and J.S. Olin (Eds.), Smithsonian Institution Press, Washington, pp. 167–179.
22. Tite, M.S., Freestone, I.C. and Bimson, M. (1983), Egyptian faience: An investigation of the methods of production, *Archaeometry* **25**, 17–27.
23. Vandiver, P.B. (1998). A review and proposal of new criteria for production technologies of Egyptian faience, in *La Couleur dans le Peinture et l'Emaillage d'Egypte Ancienne*, S. Colinart and M. Menu (Eds.), Edipuglia, Bari, pp. 121–139.
24. Tite, M.S., Manti, P. and Shortland, A.J. (2006). A technological study of ancient faience from Egypt, *J. Archaeol. Sci.*, doi:10.1016.
25. Pradell, T., Salvadó, N., Hatton, G.D. and Tite, M.S. (2006). Physical processes involved in the production of the ancient pigment, Egyptian Blue, *J. Amer. Ceram. Soc.* **89**(4), 1426–1431.
26. Davidovits, J. and Davidovits, R. (2005). Why Djoser's blue Egyptian faience tiles are not blue? Manufacturing Djoser's faience tiles at temperatures as low as 250°C, in *Proceedings IXth International Congress of Egyptologists*, Grenoble, France, September 6–11, Session 12.2.
27. (a) Davidovits, J. and Davidovits, M. (1988). Geopolymer: Room-temperature ceramic matrix for composites, *Ceramic Engineering and Science Proceedings* **9**(7/8), 835–841. (b) Davidovits, J. (1989). Geopolymers and geopolymeric materials, *J. Thermal Anal. Calorimetry* **35**, 429–441. (c) Davidovits, J. (1991). Geopolymers: Inorganic polymeric new materials, *J. Thermal Anal. Calorimetry* **37**, 1633–1656. (d) Kriven, W.M., Bell J.L., Gordon M. and Mallicoat, S. (2003). Microstructure and microchemistry of fully-reacted geopolymers and geopolymer matrix composites, *Ceramic Trans.* **153**, 227–252.
28. Aufrère, S. (1991). L'univers Minéral dans la Pensée Égyptienne, *IFAO* **2**, 294.
29. Pradell, T., Molera, J., Roque, J., Vendrell-Saz, M., Smith, A.D., Pantos, E. and Crespo, D., (2005). Ionic-exchange mechanism in the formation of medieval luster decorations, *J. Amer. Ceram. Soc.* **88**(5), 1281–1289.
30. De la Torre, A.G., Bruque, S. and Aranda, M.A.G. (2001). Rietveld quantitative amorphous content analysis, *J. Appl. Cryst.* **34**, 196–202.
31. Van Denabeele, P. and Moens, L.J. (2000). The application of Raman spectroscopy for the non-destructive analysis of art objects, in *Proceedings of the 15th World Conference on Non-Destructive Testing*, Rome, 15–21 October 2000, http://www.ndt.net/article/wcndt00/papers/idn163/idn163.htm (last accessed 13/11/06).
32. Van Denabeele, P., Weis, T.L., Grant, E.R. and Moens, L.J. (2004). A new instrument adapted to in situ Raman analysis of objects of art, *Anal. Bioanal. Chem.* **379**(1), 137–142.
33. Sciau, P., Relaix, S., Goudeau, P., Bell, A.M.T., Jones, R.L. and Pantos, E. (in press). Synchrotron XRD study of phase transforms in illitic clays to extract information on sigillata manufacturing processes, in *Proceedings of the SR2A06 Synchrotron Radiation in Art and Archaeology Conference*, Berlin, 27–30 September 2006, *Appl. Phys. A.*.

34. Pradell, T., Molera, J., Pantos, E., Smith, A.D., Martin, C.M. and Labrador, A., (in press). Temperature resolved reproduction of medieval luster, in *Proceedings of the SR2A06 Synchrotron Radiation in Art and Archaeology Conference*, Berlin, 27–30 September 2006, *Appl. Phys. A*.
35. Pantos E. (2006). Winged words of grand dad Homer, paper presented at *International Symposium on Science and Technology in the Homeric Epics*, Olympia, Greece, 27–30 August 2006, S.A. Paipetis (Ed.).
36. (a) Davidovits, J. and Morris, M. (1988). *The Pyramids, An Enigma Solved*, Hippocrene Books, New York. (b) Davidovits, J. (2002). *Ils ont bâti les pyramides*, Editions Jean-Cyrille Godefroy, Paris. (c) Davidovits, J. (2006). La Nouvelle Histoire des Pyramides, Editions Jean-Cyrille Godefroy, Paris.

# Part 4

# DEFENSIVE WEAPONS

# From Homer to Hoplite: Scientific Investigations of Greek Copper Alloy Helmets

Panagiota Manti and David Watkinson

*Cardiff University, U.K.*

**Abstract.** Homer's Iliad contains the earliest account of Greek armour technology, with heroes such as Hector reported as wearing helmets that are flashing and shiny. Corrosion of helmets during their burial limits understanding of their original appearance in antiquity. Evidence of their original appearance is based mainly on interpretation of ancient literature and numerous artistic representations of helmeted warriors on pottery. Shields decorated with enamel, gold and tin are described in the Iliad and this supports the hypothesis that such surface treatment technology could have been used on helmets. Fragments from two archaic period helmets were analysed using SEM/EDX. One of the helmets was tin-plated. This and a similarly dated helmet in the British Museum are of the earliest recorded examples of tinning in the Mediterranean. This raises questions about the original appearance of Greek helmets, visibility of individuals on the battlefield and their status. A large scale investigation of Greek helmets is underway to address these points and examine the possibility that tinning in armour may go back to Homeric times.

## 1 Introduction

This paper reports the preliminary scientific examination of decoration technologies on two excavated archaic period Greek bronze helmets from Archontiko Cemetery.[1] Results contribute to understanding of the original appearance of copper alloy helmets and the technologies associated with their decoration.[2] Assessment of surface finishes focuses on tin-plating and surface enrichment of tin observed on the helmets.

Low tin bronzes, such as helmets, can be decorated in many ways. Engraving, over-painting, visually contrasted differing metals and surface finishes like tinning, silvering or gilding are all possibilities. The original colour, texture and optical properties of a helmet are vital elements in developing understanding of its visual coher-

[1] Dr. P. Chrisostomou, responsible archaeologist of the excavation and archaeological site, Dr. Maria Akamati and the IZ' Eforia of Antiquities and the Ministry of Culture are sincerely thanked for permitting the sampling and analysis of this material.

[2] This preliminary investigation is part of P. Manti's Ph.D. research at Cardiff University on scientific investigations of Greek helmets' associated technologies, in the aim to provide a better understanding on the production workshops of helmets in the Hellenic world.

*S.A. Paipetis (ed.), Science and Technology in Homeric Epics,* 167–179.

ence and appreciating its importance. Apart from aesthetic appearance, the choice of a particular surface treatment would affect the visibility of a warrior on the battlefield and mark his status or even his affiliation to a particular troop [1]. Tinning would also create a hard surface that affects the mechanical properties of a helmet, which could influence its effectiveness in battle.

# 2 Helmets in Antiquity

## *2.1 From Homer to the Hoplite*

Homer's Iliad is the earliest written source that provides information on the technology of early armour and the appearance of helmets in the Hellenic world. Apart from his famous account describing the making of Achilles' shield by Hephaestus, there are numerous references to the appearance and effectiveness of copper helmets. Borchhardt [2], in his research on the "Homeric helmet" from Bronze Age Mycenaean to orientalising helmet types, provides a detailed study for the ancient words that are used for the word "helmet" and the adjectives that characterise it in the Homeric text.[3] It is evident that the helmet in the Iliad is of copper, which often appears to be flashing and shiny, such as that of Hector (VI:465–474).

Since "there has been some agreement by authorities that the Iliad may have reached its final Homeric form as early as the mid-eighth century" [3], it is likely that the description of helmets in the Iliad were inspired by the technology extant at the time of the writing. It seems that the archaeological record supports this hypothesis, with major material evidence of copper based helmets in Greece starting from the Late Geometric period. One of the earliest examples, is that from a grave at Argos (end of the 8th century BC), which was found with a corslet and sees influences from contemporary Near Eastern helmets [4, 5]. Before the end of the 8th century BC, there is some evidence for the attachment of bronze cheek-pieces on helmets made of organic materials. For example, bronze cheek pieces in association with a boar's tusk helmet were found at the "panoply" tomb at Dentra (15th century BC). Archaeological evidence suggests that there is only a single occurrence of a bronze helmet with cheek-pieces from the Mycenaean period, which was found in a grave near Knossos (dating to around 1400 BC) [4].

The lifespan of a helmet may also need to be considered in relation to technology. It is interesting that the only reference to the most common type of the Mycenaean period helmet, a boar's tusk leather helmet with felt lining (Book X:260–271), is the helmet that Meriones from Crete handed to Odysseus. This helm, which was stolen from Amyntor by Autolykus, had a lifetime of at least three generations and

[3] For example, the words κόρυς, κυνέη, τρυφάλεια refer to the word helmet, the words πήληξ, στεφάνη, ιμάς, κύμβαχος, λόφος, πίλος, φάλος and other refer to helmet parts, whilst common words to characterise the material and appearance of a helmet are amongst other χαλκήρης, χαλκείος, πάγχαλκος, λαμπρός, φωτεινός. For a full list of words and their occurrences in the Iliad, see [2, Beilage C].

was given to Odysseus after changing hands five times, travelling from Boeotia to Kythira, to Crete and thence to Odysseus [6]. When looking at helmets of the Mycenaean period and Dark Ages, Snodgrass notices that a specimen of the boar's tusk helmet is found at Kallithea two centuries after the predominant use of this type of helmet, which leads him to suggest this as possible evidence for the survival of this "old-fashion" type of helmet in the Iliad [4]. The extraordinary circulation and lifetime of Meriones' helmet in the Iliad could offer some support for this argument.

The period starting by the mid 8th century BC is characterised by great development and innovation in the Hellenic world. Within the framework of dynamic change in the political organisation, Greek colonisation and trade growth, there is evidence of obligatory military service being introduced, which would influence Greek warfare for several centuries [7]. The fully-armed Greek hoplite infantryman, the rise of whom is associated with the introduction of the innovative phalanx formation, became the core of the Greek armies and underpinned their reputation in war [4].

## *2.2 Evidence of Helmets and Polychromy*

The hoplite was equipped with a copper alloy helmet, corselet, greaves and a great round shield or hoplon, along with offensive weaponry. Apart from material evidence, such as the great number of helmets excavated from sanctuaries like Olympia and Isthmia, the frequency with which helmets appear in ancient Greek art reveals that their production remained a significant part of ancient Greek metalworking for many centuries. Small copper alloy male figurines wearing helmets appear by the 8th century BC (Olympia Museum), whilst one of the most important pieces of artistic evidence for the making of helmets comes from a late 8th–7th century BC bronze figurine of a bronze-smith hammering out a Corinthian helmet on an anvil (Metropolitan Museum). Pictorial evidence and interpretation of various artistic representations on pottery showing warriors who wear helmets with engraved, painted or plated designs, add to our perspective on the original appearance of helmets. Polychromy was an important aspect of Greek aesthetics [8]. Although its use is predominantly recorded by numerous examples of artistic objects, this tradition could have been utilised in the making of armaments [9]. Helmets may be practical items, but they also have an important aesthetic role as objects of social status. The employment of different metals for achieving a sense of polychromy is supported in the Iliad, with the most vital example being Achilles' shield (XVIII: 468–613). A vineyard decorating part of the shield (XVIII: 561–571) is described by the poet as grapes made of gold with clusters that hung dark purple (or black), silver vine poles, with enamel used for the enclosing ditch and tin for the fence [10]. This may be interpreted as evidence of the attachment of different parts or as evidence of tinning, gilding and artificial patination or painting of the bronze shield. In general, the use of gold for decorative purposes is supported in the Iliad mainly for the decoration of shields, and on one occasion for the crest of a helmet.[4] There are also numerous references to the use of tin for the decoration of shields, and the making of greaves.[5]

Archaeometric evidence of polychrome decoration or plating of helmets is scarce. Scientific examination of a 6th century Illyrian helmet in the British Museum (GR 1914.4–8.1) found at Olympia[6] revealed the use of silver rivets (with 2 wt% Cu) on the low-tin bronze helmet [11]. The bronze would have had a light golden tint if polished, creating a polychromatic effect with the silver decoration. Further analysis of this helmet supports the use of tin solder for the application of a very thin silver foil onto the bronze, whilst silver foil *repoussé* decoration was attached on the cheek-pieces using a calcite based adhesive [11]. In general, there is little known on the use of adhesives for the application of silver or gold foil decoration on bronzes due to the poor preservation of such materials. Corrosion during burial obscures evidence of over-painted decoration, although white, grey and red decoration has been reported on a North Italian bronze helmet (circa 500 BC) [9]. In relation to tinning, the only (known to the authors) reported tin-plated helmet is the 5th century BC Greek helmet in the British Museum (GR 1856.12–26.616), which is of the earliest tinned museum object from the Mediterranean [12].

## 2.3 Methodology

Two Illyrian type helmets from Archaic period burials at Archontiko Cemetery [13] were closely examined during conservation treatment and analysed using metallography and a Camscan Maxim 2040. Scanning Electron Microscope, which was coupled with an Oxford Instruments™ Energy Dispersive X-ray Detector for compositional analysis of the bulk alloy and surface layer features observed in polished cross sections.[7] The two helmets were selected for analysis when, during conservation treatment, a metallic silver-coloured surface was revealed on one helmet (H1) (Figure 1a) and an organic residue, which appeared to have been used as an adhesive to attach gold foil, was seen on another (H2) (Figure 1b). These observations raised questions regarding the occurrence of tinning or silvering (H1) and the nature of the adhesive used for the gold foil decoration (H2). The corrosion layer on H2 was relatively fine, even and uniform (Figure 1b). Small samples were selected from a collection of loose fragments that remained after the conservation treatment of the hel-

[4] Achilles' helmet appears to have a crest of gold (XVIII: 612).

[5] For examples see XI: 25; XI: 34; XVIII: 474, 565, 574, 613; XX: 271; XXI: 592; XXIII: 503, 561.

[6] An identical of which is found on display in Olympia Museum.

[7] We are indebted to Dr. Kilian Anheuser for his advice during the analyses of this material and to Mr Phil Parkes for his kind technical support. Analysis was conducted using facilities at the Conservation Laboratories, School of History and Archaeology, Cardiff University.

mets; most of these were from areas of the crown and the crest-track.[8] Samples were embedded transversely in epoxy resin and then polished using silicon carbide papers (up to No 4000) and diamond pastes (up to 0.25 μm). Metallographic examination and bulk compositional analysis of the core metal and assessment of surface phenomena allowed for interpretation that ruled out surface tin-enrichment due to any particular manufacturing process. A Perkin Elmer$^{TM}$ Fourier Transform Infrared-Attenuated Total Reflectance spectrometer operating in the range 4000–550 $cm^{-1}$ was used for the characterisation of the adhesive sample for the gold decoration (H2). Selected samples were analysed using Synchrotron Radiation X-Ray Diffraction at Daresbury Laboratory,[9] and Neutron Diffraction on the ROTAX instrument at the ISIS Neutron Spallation source, Rutherford Appleton Laboratory.[10]

# 3 Experimental Results and Discussion

## *3.1 Bulk Composition and Manufacturing Process*

Bulk composition analysis using SEM-EDX revealed that both helmets are of low tin content: 11.67 wt% tin for H1 and 11.78 wt% tin for H2, with trace level amounts of impurities. These values are averaged from several analyses of three samples for each helmet.[11] The results are in agreement with published analysis of the bulk metal from Greek helmets of this period, which revealed their composition to be 7.1 to 11.4 wt% tin bronzes [14–18]. A low tin bronze can be easily worked to shape by cold working and annealing.

Metallographic examination[12] revealed heavily worked microstructures for both helmets, with small grain size and severe strain lines, suggestive of several cycles of cold-working and annealing that left the helmets in the cold worked state. This fabrication technology has been reported for other helmets [16, 17]. Evidence for extensive hammering of both helmets also appears as tool-marks across the width of the crown and perpendicular to the cheek-pieces in a concentric pattern. The raised

---

[8] Sampling took place following the guidance given from the Greek Ministry of Culture (Article ΥΠΠΟ/ΓΔΑΠΚ/ΑΡΧ/Α2/Φ30/22268/778/5-3-2004) and ethical standards of Professional Conservation Organisations such as UKIC and AIC.

[9] Dr. E. Pantos at Daresbury Synchrotron is truthfully acknowledged for his support and the opportunity to analyse this material during the COST-G8 "Master Class" on SR in Cultural Heritage week, October 2005. We are also grateful to Dr. M.A. McDonald and Dr. A.T.M. Bell for instruction in the use of stations 14.1 and 2.3.

[10] Dr. W. Kockelmann is sincerely thanked for the analysis of this material.

[11] Composition analysis was at 20 keV beam energy with an electron penetration for copper of 1.1 μm depth and 1.6 μm diameter, using a spot area of 12 × 9.5 μm for analysis of the core metal and a 4.8 × 6 μm spot size for surface layered structures. The system is equipped with ZAF correction.

[12] As this is part of an ongoing investigation, the importance of the surface features and the small number of samples prohibited chemical etching of the cross sections. Observations are based on features revealed by high polishing and corrosion.

(a)

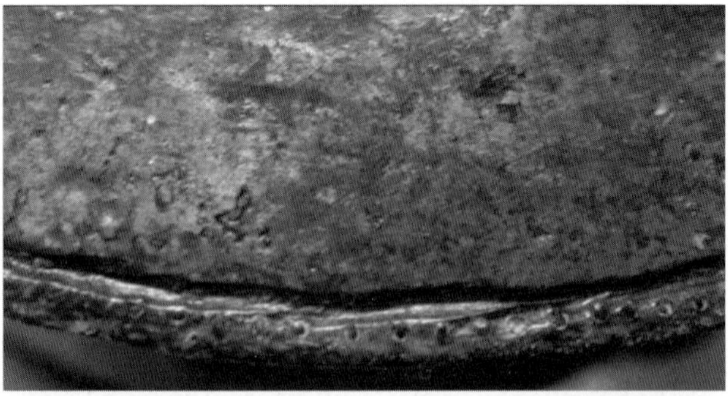

(b)

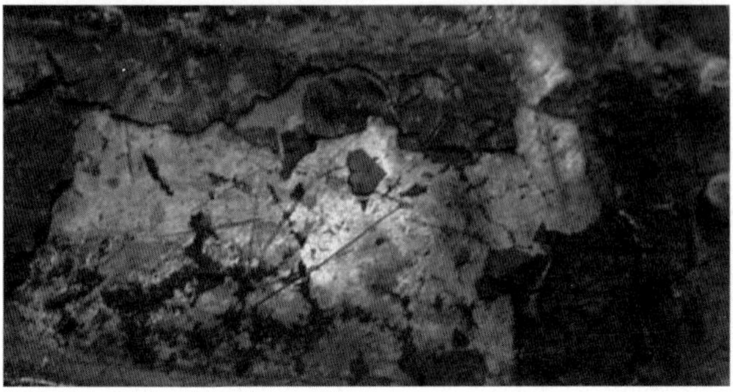

**Fig. 1** (a) Silver coloured surface on the inner side of neck-guard (Helmet 1). The outer side in this area appears the same. (b) Detail of adhesive remains from the application of gold foil decoration on bronze substrate (Helmet 2). Engraved crisscross lines appear on the shiny bronze surface where it was covered with the adhesive. Adhesive remains appear as dark brown/black. A gold foil strip with embossed decoration was applied over this adhesive and is preserved in other areas.

ridge by the row of rivets at the edge of both helmets is hammered out from the interior (see Figure 1a). Polishing marks survive on both helmets but the regularity of the lines on Helmet 2 may even suggest finishing on a lathe.

Shaping of helmets by hammering has been recorded elsewhere [11, 17]. In some cases the use of a roughly formed cast alloy blank, which was then shaped by hammering and annealing, may have been used [15, 16]. The extraordinary thickness of some helmets offers some support for this method of manufacture [18]. The difficulty in establishing this manufacturing process arises from the fact that sufficient annealing of low-tin bronze produces complete homogenisation of the eutectoid microstructure present in the as-cast state. This removes the δ-phase, and any evidence of segregation produced at pre-casting, although it is possible for some remnant coring to appear as a ghost microstructure [15, 19].

Although the use of a cast blank may be possible, the metallic silver-coloured surface on Helmet 1 is neither related to a manufacturing process nor to composition. Any deliberate [20] or unintentional tin sweating or inverse segregation of tin, which occurs in low-tin (8–14 wt%) bronze castings [12] and is more common

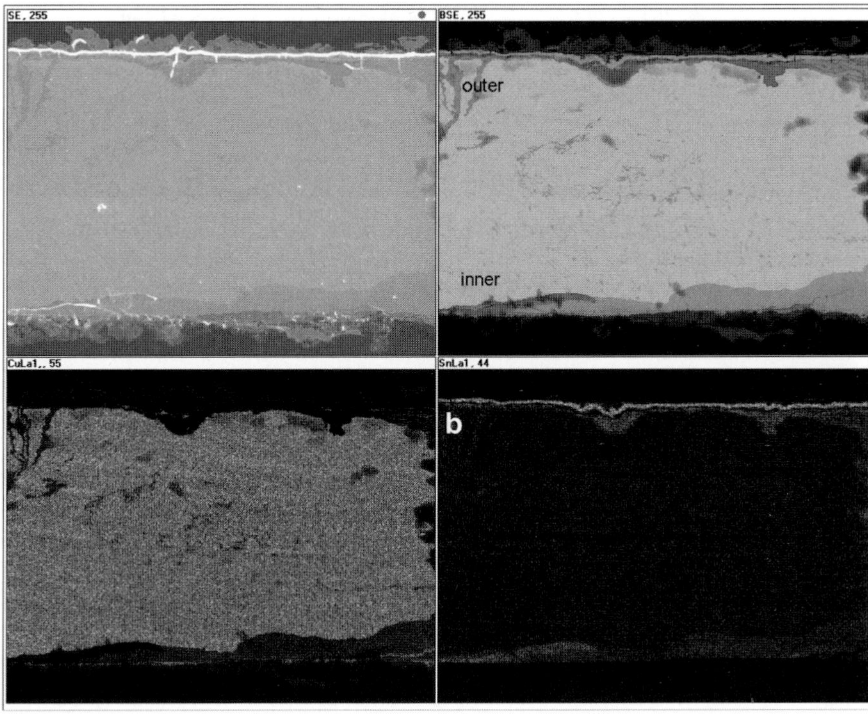


**Fig. 2** Cross section sample of Helmet 1. Top is the outer surface of the helmet and bottom is the inner face. The white bar represents 200 μm. From right to left, from top to bottom: (a) SEM-Secondary Electron Image (SEI). Some charging effects between the voluminous corrosion products and the tin-rich layer may suggest low adherence of this deposit to the high-tin layer. (b) SEM-Backscattered Electron Image (BEI). Compositional contrast shows the presence of a uniform layer on the outer side of the sample. Some cracking of this layer is also visible. (c) Digital X-ray distribution map for CuLa. (d) Digital X-ray distribution map for SnLa.

at 10–14% tin content [21], would be homogenised due to the severe hammering and annealing of the surface of these helmets. Consequently, the tin coating on this helmet arises from other causes.

## *3.2 Surface Corrosion Phenomena and Tinning*

All polished cross sections were investigated using polarised light microscopy, SEM-Back Scattered Electron Imaging and EDX mapping of selected elements. Figure 2 shows a uniform high tin layer of 5 μm thickness which follows surface irregularities across the outer side of a cross-sectioned sample from the crest track of Helmet 1. Comparisons to published research suggest that it is a structure related to tinning [12, 27, 29]. Below this dense high tin line on the outer face, revealed in

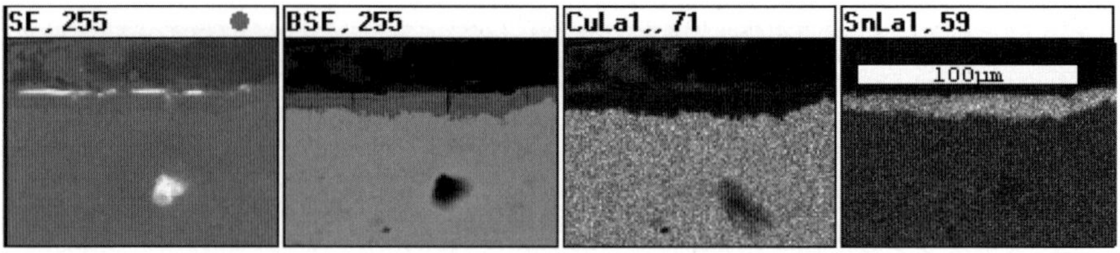


**Fig. 3** Cross section sample of Helmet 2. Top is the outer surface of the helmet. The white bar represents 100 µm. From right to left: (a) SEM-Secondary Electron Image (SEI). (b) SEM-Backscattered Electron Image (BEI). (c) Digital X-ray distribution map for CuLa. (d) Digital X-ray distribution map for SnLa.

the EDX mapping of tin (see Figure 2d), there is a tin-rich band, which follows corrosion patterns in the metal. The same features are visible in other samples from the crown and crest-track areas of this helmet. In contrast, the inside face of this sample from the crest track, visible as the bottom edge in Figure 2 images, lacks a distinct tinning line and the tin enrichment appears to be entirely associated with corrosion phenomena of varying depth (see Figure 2d), which coincides with reduced copper levels (see Figure 2c). The examination of a sample from the neck-guard of Helmet 1 (see Figure 1a), where a metallic silver coloured surface is visible on both the outer and inner sides, was prohibited due to the fine preservation of the specimen at this area. It is expected that this would have revealed distinct evidence of tinning lines on both surfaces on this area of Helmet 1.

Samples from Helmet 2 exhibit tin-rich areas of various thicknesses (5–20 µm) following the corrosion contours on both their inner and outer faces (see Figures 3b, 3c and 3d). There is no evidence of distinct tinning lines and tin enrichment is concurrent with lowered copper levels (see Figure 2c). This tin enrichment appears to result entirely from corrosion.

The tin-rich corrosion layers on both helmets appear water-blue/green when viewed using bright field microscopy. Based on microscopic observations it resembles corrosion patterns of Type I corrosion structures [23], which correspond to even, compact surfaces or passive layers enriched in tin oxides by the oxidation of tin within the bronze to produce a cationic protective layer, and the preferential dissolution of copper followed by migration of $Cu^{+}$ and $Cu^{++}$ towards the external surface [22–25]. The tin oxides have low mobility and remain in-situ; consequently the corroded area is depleted of copper and enriched in tin.

Besides detection of distinct thin metallic tin layers, tinning can be linked to other detectable features. All methods of tinning that involve elevated temperatures lead to the formation of intermetallic compounds according to the binary Cu/Sn equilibrium phase diagram at the interface tin/metal substrate and show typical compositions [26]. The profile of the intermetallic compounds formed depends on the tinning method and the time used to apply it, as well as the application temperature [27]. The $\eta$-, $\varepsilon$- and $\delta$-phases developed are hard and brittle and give a silvery white colour on the surface of the object when they are in an uncorroded state [28, 29]. The $\eta$- and/or $\varepsilon$-phases can be about 2–5 µm thick [12].

Both η- or ε-phases have been identified on tinned archaeological objects by XRD and SEM-EDX analysis [21, 30] and on archaeological objects with no pure tin metal coating in place [20], the presence of η- and/or ε-phases are residual evidence for tinned archaeological copper alloys [12]. Assessment of the corrosion behaviour of bronze in the presence of intermetallic compounds is difficult, however, it is suggested that these intermetallic compounds may be cathodic to copper as well as to surface pure tin [22, 31], which may possibly explain the survival of these phases on archaeological objects.

The tin-rich layer featured on the outer surface of Helmet 1, which it is suggested is related to tinning shows an approximate average compositional value of 51 wt% tin. Its immediate underlying tin-rich layer contained 30–45 wt% tin and all other tin-rich surfaces on both helmets had a 30–50 wt% tin content. All analyses were carried out by SEM-EDX. The composition of the layer thought to be related to tinning has lower tin values than the *eta*-intermetallic phase ($Cu_6Sn_5$ with 61.0 wt% tin) and higher values than the *epsilon*-phase ($Cu_3Sn$ with 38.6 wt% tin). However, any direct comparison of tin content to determine the likely presence or absence of intermetallic compounds associated with tinning is negated, as all recorded tin values lie within the 40–90 wt% tin content that Dorigo et al. [25] encountered within tin enriched patinas formed by copper dissolution from low tin archaeological bronzes. The situation is further complicated by the fact that intermetallic compounds can undergo corrosion that reduces their copper content, which will increase their relative proportion of tin. Consequently other analysis methods were adopted to try and detect the epsilon-phase associated with tinning.

Conventional X-Ray Diffraction (XRD) of a powder sample taken from the surface of Helmet 1 failed to identify the presence of intermetalic compounds or any other tin compounds, due to the small size of the sample and high interferences from other compounds present. Synchrotron Radiation XRD (SR-XRD) [32] at Daresbury Laboratory at station 14.1 (powder samples at micro quantities) and in flat-plate geometry on the high resolution station 2.3 where a small fragment of the sample was examined in situ, similarly failed at a first attempt to produce meaningful results. Neutron Diffraction by Kockelmann [33, 34] on ROTAX at the ISIS facility, Rutherford Appleton Laboratory, did not indicate clear tinning phases, however on the basis of the counting statistics, the data are in agreement with an ε-phase of 0.3 wt% which is at the limit of detection (Kockelmann, pers. commun.). There are no indications of η- and δ-phases. Although it is at the detection limit of the method and considering that neutron diffraction is basically a bulk diagnostic technique, this analysis does not preclude the view that this particular microstructure is remnant of the tinning of Helmet 1.

Visually, this silver coloured surface can be observed on the outer and inner side of the neck-guard of Helmet 1, but analysis of samples from different areas of the helmet (crown and crest-track) shows that tinning is only present on the outer side of the whole helmet. This suggests that tinning may well have been applied only at the visible areas of the helmet if one considers that the neck- guard of this helmet leaned slightly upwards. Although both dip and wipe tinning techniques could have achieved this outcome, Helmet 1 was most probably wipe-tinned as the short

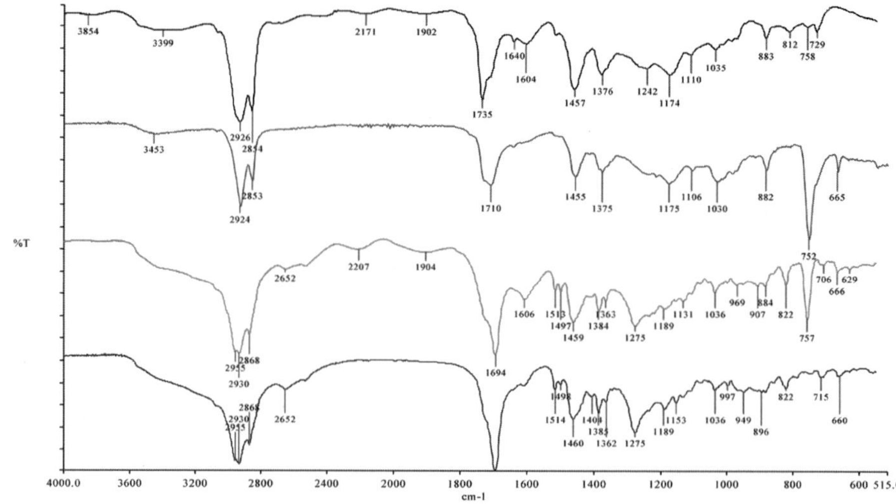


**Fig. 4** Comparison FTIR-ATR spectra in transmission (T%) of organic sample from Helmet 2 (second from top) to reference samples of birch bark pitch (first from top), pine pitch (third from top) and pine tar (fourth from top). All samples were dissolved in chloroform and let dry prior to positioning on the diamond cell of a Perkin Elmer™ FTIR-ATR instrument.

length of application may explain the limited thickness of the intermetallic compound formed.

## *3.3 Adhesive for Gold Foil Decoration*

Analysis with Fourier Transform Infrared-Attenuated Total Reflectance (FTIR-ATR) of a sample from the adhesive used for the application of gold foil decoration on Helmet 2 produces a best match for wood pitch, possibly a pine pitch (distillation residue of pine tar) as compared to analysis of selected reference material (Figure 4). Further analysis with GC/MS is planned in order to assess the FTIR-ATR results [35]. The adhesion of gold foil decoration on helmets is often associated with burial customs of the archaic period in Northern Greece (for example Sindos). For Helmet 2, the adhesive was applied over engraved crossed lines to enhance adherence to the applied bronze substrate [36, 37] (see Figure 1b). Over the adhesive a gold foil stripe with embossed decoration was applied and adhered firmly onto the helmet. It covered the hole at the edge of the cheek-piece, which was used for the strap that secured the helmet onto the head and offers evidence that the gold decoration was applied when the helmet was no longer in use.

# 4 Conclusions

Examination of two archaic period Illyrian type helmets shows that the manufacture of both was related to annealing cycles and hammering, with both helmets left in the cold worked state to increase hardness. It appears that the tin-rich feature seen at the outer side of Helmet 1 is due to tinning that was most probably applied using the wiping technique. Tinning and the development of intermetallic compounds would offer increased wear resistance, hardness and mirror-like reflected properties when polished. The adhesion of gold foil decoration at Helmet 2 using a wood pitch is apparently associated to burial customs. Further evidence for this might be the logic that gold would not have a pronounced decorative effect on a low tin bronze helmet, which is shiny and has a similar colour to gold. It seems that the use of gold foil decoration in this case is likely ceremonial (H2), whilst tinning of Helmet 1 could be for protective (wear resistance), economical (possibly to imitate silvering) or social motives.

Composition analysis for the characterisation of the $\eta$- and $\varepsilon$-phases to identify intermetallic compounds arising from tinning is limited by the problem of differentiating these phases from tin-enrichment due to burial. Neutron diffraction was able to indicate the presence of some $\varepsilon$-phase, although at the limit of its detection (0.3 wt%). More measurements of the same sample at longer counting times to increase the detection limit for the bronze phases is planned.

This is a pilot study. Only by investigation of numerous helmets can the occurrence and frequency of tinned and gilded helmets be examined and placed in a historical context. For that, we have gained sampling permissions from several archaeological sites in Greece and museums in the U.K. Perhaps the use of tinning of armoury may go back to the time of Homer?

# References


1. Van Wees, H. (2004). *Greek Warfare: Myths and Realities*, Duckworth.
2. Borchhardt, J. (1972). *Homerische Helme*, Römisch-Germanisches Zentralmuseum Mainz, Verlag Philipp von Zabern-Mainz am Rhein.
3. Snodgrass, A.M. (1998). *Homer and the Artists: Text and Picture in Early Greek Art*, Cambridge University Press, Cambridge.
4. Snodgrass, A.M. (1967). *Arms and Armour of the Greeks*, Thames & Hudson, London.
5. Dezcö, T. (1998). *Oriental Influence in the Aegean and Eastern Mediterranean helmet Traditions in the 9th–7th Centuries BC: The Patterns of Orientalization*, BAR International Series 691, Archaeopress, Oxford.
6. Homer, *Iliad* (2004). Translation and comments by Th.G. Mavropoulos, Zetros Publications, Athens.
7. Sage, M.M. (1996). *Warfare in Ancient Greece: A Sourcebook*, Routledge, London/New York.
8. Hughes, R. (1993). Artificial patination, in *Metal Plating and Patination*, S. La Niece and P. Craddock (Eds.), Butterworth-Heinemann, pp. 1–18.
9. Born, H. (1990). Patinated and painted bronzes: Exotic technique or ancient tradition?, in *Small Bronze Sculpture from the Ancient World*, Papers delivered at a Symposium organised and held at the J. Paul Getty Museum, Malibu, California, pp. 179–196.

10. Homer, *The Iliad* (1998). Translated by R. Fitzgerald. Oxford University Press, Oxford.
11. Hockey, M., et al. (1992). An Illyrian helmet in the British Museum, in *The Annual of the British School of Athens*, No. 87, BTA, pp. 281-291.
12. Meeks, N.D. (1993). Surface characterization of tinned bronze, high-tin bronze, tinned iron and arsenical bronze, in *Metal Plating and Patination*, S. La Niece and P. Craddock (Eds.), Butterworth-Heinemann, pp. 247–275.
13. Chryssostomou, A. and Chryssostomou, P. (2001). Excavation in the western necropolis of Pella mansion in 2001, in *Archaeological Work in Macedonia and Thrace: 15, 2001*, Ministry of Culture, Archaeological Asset Fund, University of Thessaloniki, Thessaloniki, pp. 477–488 [in Greek].
14. Craddock, P.T. (1976). The composition of the copper alloys used by the Greek, Etruscan and Roman Civilizations: 1. The Greeks before the archaic period, *Journal of Archaeological Science* **3**, 93–113.
15. Blyth, P.H. (1993). Metallurgy of two fragmentary archaic Greek helmets, *Historical Metallurgy* **27**, 25–36.
16. Blyth, P.H. (1988). Cold-working in ancient Greek helmets, in *Aspects of Ancient Mining and Metallurgy: Acta of a British School at Athens Centenary Conference*, J. Ellis Jones (Ed.), University College of North Wales, Bangor, 1986, pp. 151–154.
17. McNamara, M. (2000). Technical studies of four ancient Greek helmets at Harvard University's Arts Museum, Paper given at *The 26th Annual Conference of the Association of North American Graduate Programs in Conservation*, S. Dillon Ripley Center, the Smithsonian Institution, 27–29 April.
18. Swaddling, J. (1987). An unusual Greek bronze helmet, *The Antiquaries Journal* **67**(LXVII), 348–351.
19. Scott, D.A. (1991). *Metallography and Microstructure of Ancient and Historic Metals*, The Getty Conservation Institute, Archetype Books.
20. Hughes, R. and Rowe, M. (1982). *The Colouring, Bronzing and Patination of Metals*, Craftscouncil.
21. Oddy, W.A. and Bimson, M. (1985). Tinned bronze in antiquity, in *Lead and Tin: Studies in Conservation and Technology*, United Kingdom Institute of Conservation, Occasional Paper No. 3, pp. 33–39.
22. Turgoose, S. (1985). The corrosion of lead and tin before and after excavation, in *Lead and Tin: Studies in Conservation and Technology*, United Kingdom Institute of Conservation, Occasional Paper No. 3, pp. 15–26.
23. Robbiola, L. and Hurtel, L.-P. (1997). Standard nature of the passive layers of buried archaeological objects: The example of two Roman half-length portraits, in *Metal 95, Proceedings of the International Conference on Metal Conservation*, I.D. MacLeod et al. (Eds.), James x James (Science Publishers), London, pp. 109–117.
24. Robbiola, L., Blengino, J.-M. and Fiaud, C. (1998). Morphology and mechanisms of formation of natural patinas on archaeological Cu-Sn alloys, *Corrosion Science* **39**(12), 2083–2111.
25. Dorigo, A., Fiaud, C., Labbe, J.P., Brunella, P. and Bocking, H. (1998). Characterisation of the corrosion structures of Roman copper alloys by SEM and EDSX: IMMACO-improvement of means of measurements on archaeological copper alloys for characterisation and conservation, in *Metal 98, Proceedings of the International Conference on Metal Conservation*, France, W. Mourey and L. Robbiola (Eds.), James + James (Science Publishers), London, pp. 145–151.
26. *Guide to Tinplate*, ITRI Publication No. 622, International Tin Research Institute, Greenford.
27. Meeks, N.D. (1986). Tin-rich surfaces on bronze: Some experimental and archaeological considerations, *Archaeometry* **28**(2), 133–162.
28. Oddy, W.A. and Meeks, N.D. (1982). Unusual phenomena in the corrosion of ancient bronzes, in *Science and Technology in the Service of Conservation*, Preprints of the contribution to the Washington Congress 1982.IIC, pp. 119–124.
29. Meeks, N.D. (1993). Patination phenol-mena on Roman and Chinese high-tin bronze mirrors and other artefacts, in *Metal Plating and Patination*, S. La Niece and P. Craddock (Eds.), Butterworth-Heinemann, pp. 63–84.

30. Meeks, N.D. (1988). Backscattered electron imaging of archaeological material, in *Scanning Electron Microscopy in Archaeology*, S.L. Olsen (Ed.), BAR International Series 452, Archaeopress, Oxford, pp. 23–44.
31. Hedges, E.C. (1960). The chemical behaviour of tin, in *Tin and Its Alloys*, E.S. Hedges (Ed.), Edward Arnold Publishers Ltd, London, pp. 78–98.
32. Pantos, E., Kockelmann, W., Chapon, L.C., Lutterotti, L., Bennet, S.L., Tobin, M.J., Mosselmans, J.F.W., Pradell, T., Salvadó, N., Butí, S., Garner, R. and Prag, A.J.N.W. (2005). Neutron and X-ray characterisation of the metallurgical properties of a 7th century BC Corinthian-type bronze helmet, *Nucl. Instrum. Methods B* **239**, 16–26.
33. Kockelmann, W., Pantos, E. and Kirfel, A. (2000). Neutron and synchrotron radiation studies of archaeological objects, in *Radiation in Art and Archaeometry*, D.C. Creagh and D.A. Bradley (Eds.), Elsevier Science.
34. Siano, S., Kockelmann, W., Bafile, U., Celli, M., Iozzo, M., Miccio, M., Moze, O., Pini, R., Salimbeni, R. and Zoppi, M. (2002). Quantitative multiphase analysis of archaeological bronzes by neutron diffraction, *Applied Physics A Materials Science & Processing* **74** (Suppl.), S1139–S1142.
35. Hayek, E., et al. (1990). Identification of archaeological and recent wood tar pitches using gas chromatography/mass spectrometry and pattern recognition, *Anaytical Chemistry* **62**, 2038–2043.
36. Oddy, W.A., et al. (1990). The gilding of bronze sculpture in the Classical world, in *Small Bronze Sculpture from the Ancient World*, Papers delivered at a Symposium organised and held at the J. Paul Getty Museum, Malibu, California, pp. 103–124.
37. Oddy, A. (2000). A history of gilding with particular reference to statuary, in *Gilded Metal: Hstory, Technology and Conservation*, T. Drayman-Weisser (Ed.), Archetype Publications, London, pp. 1–20.

# Defensive Weapons in Homer

S.A. Paipetis and V. Kostopoulos

*University of Patras, Greece*

**Abstract.** Highly interesting structures, i.e. defensive weapons, described in the Homeric Epics and possessing elements of almost modern technology, are investigated. We are mainly referring to the shields of Achilles and Ajax, consisting of successive layers of different metals the former and of metal and leather layers the latter, e.g. they belong to the so-called *laminated structures*, as termed in modern technological terminology. The latter are part of a great class of materials, named *composite materials*. These structures were analyzed on the basis of realistic assumptions, as far their constituent materials are concerned and, by using modern computer codes and on occasions experimentally, their properties were investigated. The results not only confirm with amazing accuracy the Homeric descriptions as regards their battle behaviour, but they also provide indication of very advanced knowledge of science and technology of materials and structures possessed by the Mycenaean Greeks. Comparison of the said structures with the ones of much later times, confirm their technological supremacy. Other defensive weapons, such as helmets and breastplates of the Mycenaeans exhibit equally advanced technical attributes.

## 1 Introduction: Materials in Homer

### *1.1 Metals*

Early civilizations are classified from technology point of view as belonging to copper or bronze era. Copper, a very soft metal, was used for the manufacture of tools and weapons, following the development of hardening techniques and alloy production. Bronze, a yellowish metal made of copper and tin at a 10:1 proportion, was most widely used during the Mycenaean era being in fact the most important material of early civilizations, while its uninhibited supply was the object of many trade and financial arrangement at the time.

The Homeric Epics are believed to belong to the bronze era. The metal is mentioned as copper 128 times in the Iliad against 23 of iron (in fact, on most occasions, to describe qualitative characteristics, such as iron heart or iron courage) while, in the Odyssey, shorter than the Iliad by 3,500 lines, only 28 times against 22 of iron (5 of them qualitative).

*S.A. Paipetis (ed.), Science and Technology in Homeric Epics,* 181–203.

**Table 1** Properties of materials[1] used for the numerical and experimental simulation of the shields of Achilles and Ajax.

| Material and properties | Unit | SAE 40 Bronze (shields) | Tin (Achilles' shield) | Gold (Achilles' shield) | Phosphor Bronze, annealed (projectile) | Calf's leather (Ajax's shield) |
|---|---|---|---|---|---|---|
| Tensile strength | MPa | 255–300 | 27.0 | 10.79 | 276 | – |
| E-modulus | GPa | 93 | 42 | 78 | 110–117 | 1.0 |
| Elongation at break | % | 3.00–4.38 | 40.0 | 50.00 | 3.80% | – |
| Vickers hardness | MPa | 870 | | 216 | | – |
| Poisson's ratio | – | 0.33 | 0.36 | 0.44 | 0.31 | 0.29 |
| Density | $kg/dm^3$ | 8.90 | 7.29 | 19.29 | 8.90 | – |

Detailed references to metals in the Homeric Epics are given by Zeggelis [1]. In particular, metals of interest in the present work are hard bronze, tin and pure gold (shield of Achilles) and also hard bronze and calf's leather (shield of Ajax). The specific properties of these metals are presented in Table 1.

## 1.2 Composite Materials

Among modern structural materials a very important role is played by *composite materials*, i.e. those generated by mechanical composition of two or more simple or monolithic materials and exhibit substantially improved properties. It is a great surprise to meet such materials described in the Iliad, with almost modern texture and mechanical performance.

Composite materials usually consist of a matrix, in which particles, grains, fibres of glass, plastic, metal, graphite etc. are included, in order either to enhance the properties of the matrix or to facilitate the production process or to reduce the production cost. Fibre glass, consisting of a resin matrix and glass fibres and suitable material for the manufacture of recreation boats, vulcanized rubber etc. are typical examples.

Until present, the oldest reference to composite materials was considered the be the one in the Bible (Exodus 5.15–18). It refers to the complaints of the Israelites to Pharaoh, because they were not provided with sufficient quantities of straw for the production of bricks, which was the job assigned to them:

> Then the Israelite foremen came and made this appeal to Pharaoh: "Why do you treat your servants in this manner? No straw is supplied to your servants, and still we are told to make bricks. Look how your servants are beaten! It is you who are at fault." Pharaoh answered, "It is just because you are lazy that

[1] The *kinetic energy* of a spear and of an air-gun projectile, applied for the numerical and experimental simulation of shields' behaviour, was equal to *39.38 Joule*, corresponding to the kinetic energy of a javelin thrown by the world recordman, year 2000.

**Fig. 1** Modern structures consisting by a great percentage of advanced composite materials: A F-18 Hornet fighter and a modern wind generator.

> you keep saying, 'Let us go and offer sacrifice to the Lord.' Off to work, then! Straw shall not be provided for you, but you must still deliver your quota of bricks."

Advanced composite materials respond to very high operational requirements, such as strength, light weight, durability, operability at high temperatures or aggressive environments etc. They are used for the manufacture of load-bearing parts of land, air, sea or space vehicles, etc. (Figure 1).

Advanced composites are mainly laminated materials, i.e. they consist of thin successive layers with widely different properties. Sometimes they are termed as *sandwich structures*. The amazing fact is that the idea of laminated structures appears for the first time in history in the Iliad, in the shields of Achilles and Ajax, with such structural details that their reconstruction and study are possible, both numerically, by means of modern computer codes, and experimentally and eventually the confirmation of their battle behaviour, as described by Homer, most accurately.

## 2 The Homeric Shields

### *2.1 The Shield of Achilles*

The manufacture of Achilles' armour by Hephaestus is described in the Iliad, 18.468–617. Especially, in 18.474–482 a clear technical description of the construction of Achilles shield is given:

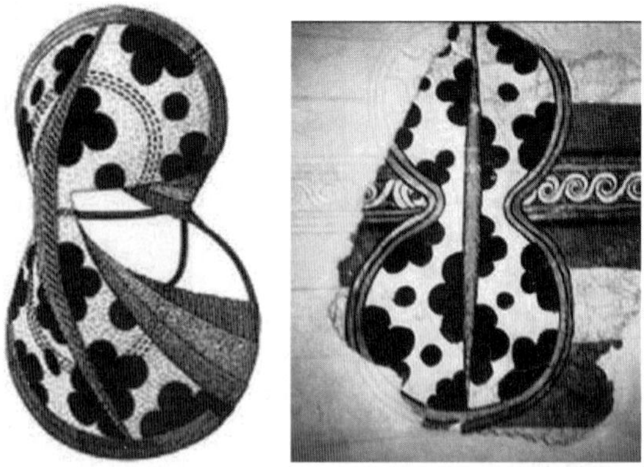

**Fig. 2** Sketch of an eight-shaped shield and its reconstruction on a mosaic.

**Fig. 3** Typical examples of adorned shields of the Homeric period.

[῞Ηφαιστος]
χαλκὸν δ᾽ ἐν πυρὶ βάλλεν ἀτειρέα κασσίτερόν τε
καὶ χρυσὸν τιμῆντα καὶ ἄργυρον· αὐτὰρ ἔπειτα
θῆκεν ἐν ἀκμοθέτῳ μέγαν ἄκμονα, γέντο δὲ χειρὶ
ῥαιστῆρα κρατερήν, ἑτέρηφι δὲ γέντο πυράγρην.
ποίει δὲ πρώτιστα σάκος μέγα τε στιβαρόν τε
πάντοσε δαιδάλλων, περὶ δ᾽ ἄντυγα βάλλε φαεινὴν
τρίπλακα μαρμαρέην, ἐκ δ᾽ ἀργύρεον τελαμῶνα.
πέντε δ᾽ ἄρ᾽ αὐτοῦ ἔσαν σάκεος πτύχες· αὐτὰρ ἐν αὐτῷ
ποίει δαίδαλα πολλὰ ἰδυίῃσι πραπίδεσσιν.

[Hephaestus]
He threw tough copper into the fire, and tin, with silver and gold; he set his great anvil on its block, and with one hand grasped his mighty hammer while he took the tongs in the other. First he shaped the shield so great and strong,

> adorning it all over and binding it round with a gleaming circuit in three layers; and the baldric was made of silver. He made the shield in five thicknesses, and with many a wonder did his cunning hand enrich it.

This account corresponds to a *laminated composite structure*, consisting of five consecutive metal laminates with very different mechanical properties. In fact, the shield consists of two external laminates of hard bronze, two internal ones of tin and a central one of pure (soft) gold. This structure exhibits maximum penetration resistance, as proved by a complete numerical simulation of its elastoplastic behaviour at large deformations, when impacted by the tip of a piercing element, an arrow or a spear. This unique detailed description, which includes the weapon's battle behaviour as well, is also the first known application of laminated structures in human history. By analysis and parameter study based on modern continuum mechanics theory, numerical tools and computer codes, have confirmed the Homeric descriptions concerning the battle behaviour of the shield with amazing accuracy and also revealed important elements of advanced technology, disguised as miraculous power of the gods.

Mycenaean shields appear in various forms, but mainly as eight-shaped (Figure 2), round (Figure 3) and full-body shields.

According to Homer, the shield of Achilles round.

The battle behaviour of Achilles' shield is described in the Iliad on three occasions. The first refers to the duel of Achilles and Aeneias (Il. 20.259–272):

> ἦ ῥα καὶ ἐν δεινῷ σάκει ἤλασεν ὄβριμον ἔγχος
> σμερδαλέῳ· μέγα δ᾽ ἀμφὶ σάκος μύκε δουρὸς ἀκωκῇ.
> Πηλεΐδης δὲ σάκος μὲν ἀπὸ ἕο χειρὶ παχείῃ
> ἔσχετο ταρβήσας· φάτο γὰρ δολιχόσκιον ἔγχος
> ῥέα διελεύσεσθαι μεγαλήτορος Αἰνείαο
> νήπιος, οὐδ᾽ ἐνόησε κατὰ φρένα καὶ κατὰ θυμὸν
> ὡς οὐ ῥηΐδι᾽ ἐστὶ θεῶν ἐρικυδέα δῶρα
> ἀνδράσι γε θνητοῖσι δαμήμεναι οὐδ᾽ ὑποείκειν.
> οὐδὲ τότ᾽ Αἰνείαο δαΐφρονος ὄβριμον ἔγχος
> ῥῆξε σάκος· χρυσὸς γὰρ ἐρύκακε, δῶρα θεοῖο·
> ἀλλὰ δύω μὲν ἔλασσε διὰ πτύχας, αἳ δ᾽ ἄρ᾽ ἔτι τρεῖς
> ἦσαν, ἐπεὶ πέντε πτύχας ἤλασε κυλλοποδίων,
> τὰς δύο χαλκείας, δύο δ᾽ ἔνδοθι κασσιτέροιο,
> τὴν δὲ μίαν χρυσῆν, τῇ ῥ᾽ ἔσχετο μείλινον ἔγχος.

> As he spoke he drove his spear at the great and terrible shield of Achilles, which rang out as the point struck it. The son of Peleus held the shield before him with his strong hand, and he was afraid, for he deemed that Aeneas's spear would go through it quite easily, not reflecting that the god's glorious gifts were little likely to yield before the blows of mortal men; and indeed Aeneas's spear did not pierce the shield, for the layer of gold, gift of the god, stayed the point. It went through two layers, but the god had made the shield in five, two of bronze, the two innermost ones of tin, and one of gold; it was in this that the spear was stayed.

The second case refers to the duel of Achilles and Hector (Il. 22. 290–292):

[Ἕκτωρ]
καὶ βάλε Πηλεΐδαο μέσον σάκος οὐδ᾽ ἀφάμαρτε·
τῆλε δ᾽ ἀπεπλάγχθη σάκεος δόρυ·

[Hector]
He poised his spear as he spoke and hurled it. His aim was true for he hit the middle of Achilles' shield, but the spear rebounded from it, and did not pierce it.

The third case refers to the duel of Achilles and Asteropaeus, the Trojan hero (Il. 21.161–165):

ὣς φάτ᾽ ἀπειλήσας, ὃ δ᾽ ἀνέσχετο δῖος Ἀχιλλεὺς
Πηλιάδα μελίην· ὃ δ᾽ ἁμαρτῇ δούρασιν ἀμφὶς
ἥρως Ἀστεροπαῖος, ἐπεὶ περιδέξιος ἦεν.
καί ῥ᾽ ἑτέρῳ μὲν δουρὶ σάκος βάλεν, οὐδὲ διὰ πρὸ
ῥῆξε σάκος· χρυσὸς γὰρ ἐρύκακε δῶρα θεοῖο·

Thus did he defy him, and Achilles raised his spear of Pelian ash. Asteropaeus failed with both his spears, for he could use both hands alike; with the one spear he struck Achilles' shield, but did not pierce it, for the layer of gold, gift of the god, stayed the point;

These descriptions of Homer on Achilles' shield and its battle behaviour are very enlightening and applicable as a basis for the numerical simulation of the weapon; i.e., a spear with a tip of hard bronze hit the shield. The exact geometry of the shield and the spear are not known, however, based on archaeological findings and pictorial representations, fundamental information can be derived applicable to theoretical analysis. Shield geometry has been discussed already, while typical spear tips appear in Figure 4.

### 2.1.1 Analysis and Results

The problem of contact-impact between solids has been investigated during the last three decades. Themes directly related to this problem are of great technological importance, such as simulation of high-speed impact perforation, high-speed metal forming, nuclear reactor safety, etc., and pressed in the direction of the development of new contact-impact algorithms and of the theoretical formulation of non-linear problems of Mechanics of Continua. The Finite Element Method was used and, in particular, algorithms developed by the *Methods Development Group* of the distinguished research centre *Lawrence Livermore National Laboratory* [2] (LLNL, California, U.S.A.), properly modified. In the present analysis, in the first place, the problem was formulated, emphasizing the theoretical and numerical aspects of impact problems. In the second place, the results of the analysis along with the respective conclusions were evaluated and presented. Although data related to the exact shape and dimensions of spear and shield are not complete, it is possible to de-

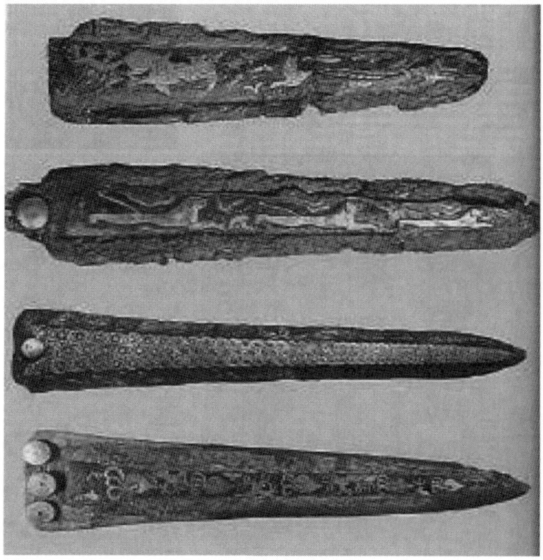

**Fig. 4** Characteristic adorned spearheads used at the Trojan War.

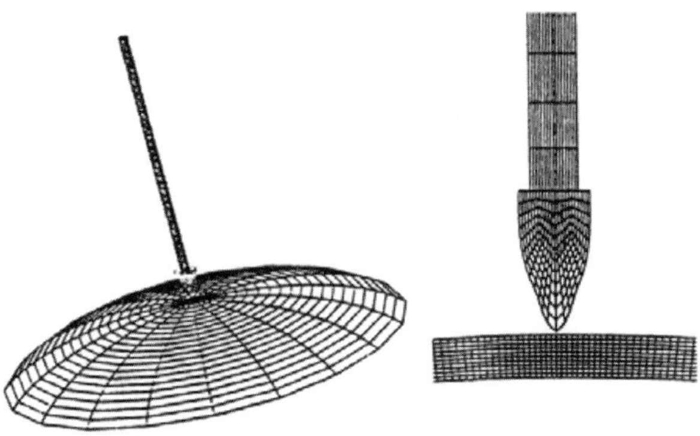

**Fig. 5** Finite-element discretization of the shield-spear system: The complete 3-dimensional model (left) and a detail of spearheads/shield (right).

termine some information by inference, on the basis of certain realistic assumptions and data from the modern sport of javelin throw.

It is assumed that all five shield laminates were 1.5 mm thick, i.e. the (round) shield was 7.5 mm thick. If one considers that the spear hits the shield at the centre, normal to the shield surface, the problem is axisymmetric. In Figure 5 the finite-element 3D discretization of the shield and the spear head is given, along with a detail of a cross section of the shield/spear head system. The shield cross section is of elliptic shape with semi-axes 300 and 120 mm. This leads to a total shield mass of 16.75 kg. The spear is considered as a straight rod 2.2 m long with a circular cross section of 9 mm radius.

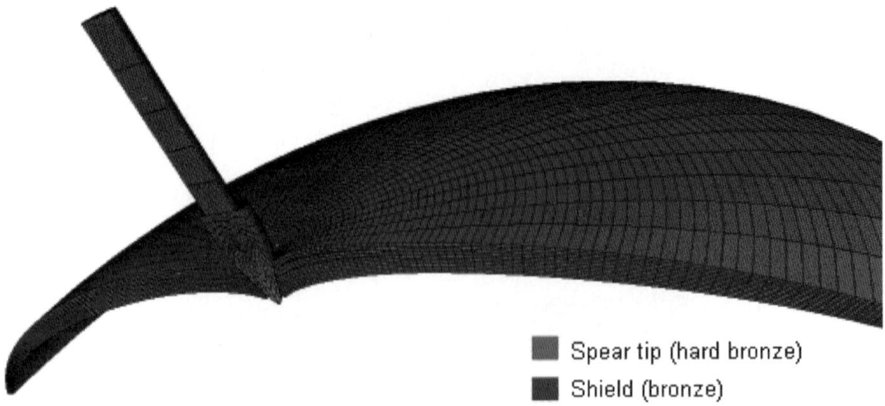


**Fig. 6** Impact behaviour of the spear/shield system, when all shield laminates are made of bronze: in 4 μs time since first contact, the spear penetrates the shield.

As Homer states, at several places in the Iliad, spearheads were made of hard bronze and were fixed on long wooden rods. The dimensions of the spearhead lead to a total mass of the spear equal to 3.25 kg.

A key-point is the way by which the several laminates of the shield are interconnected. It is assumed that the inner laminates of the shield are bent outwards to envelop the bounds of the laminated structure up to the external surface, in fact pocketing the structure. In addition, an array of studs across the thickness was used, to keep the laminates in place and ensure good cooperation. The number and the distribution of the studs may influence impact behaviour and stress concentration in the shield substantially.

For the needs of the present study, it was assumed that studs are arranged circularly at a 120 mm distance from the axis of symmetry of the shield. This arrangement allows the laminates to slip in relation to each other.

The speed of the spear was taken equal to 20 m/s. Two different cases were studied. In the first case, all of the shield laminates were of hard bronze, while, in the second, the real configuration of laminates is considered, i.e. two external bronze laminates, two internal ones of tin and a central one of pure gold. The tip of the spear head is made of hard bronze.

In Figure 6, the response of the shield, when made of identical bronze laminates is presented. In this case, the spear penetrates the shield.

Figures 7 and 8 present respective displacement and velocity of the spear tip knot and of the central shield knot diagrams as functions of time. Penetration is completed at the end of 4 μs time period, while the change of the slope of the velocity diagram of the spear tip corresponds to the ensuing penetration into the bronze laminates. At the end of penetration stage, the spear has reached almost zero velocity, while the deformation of the shield reaches 35 mm.

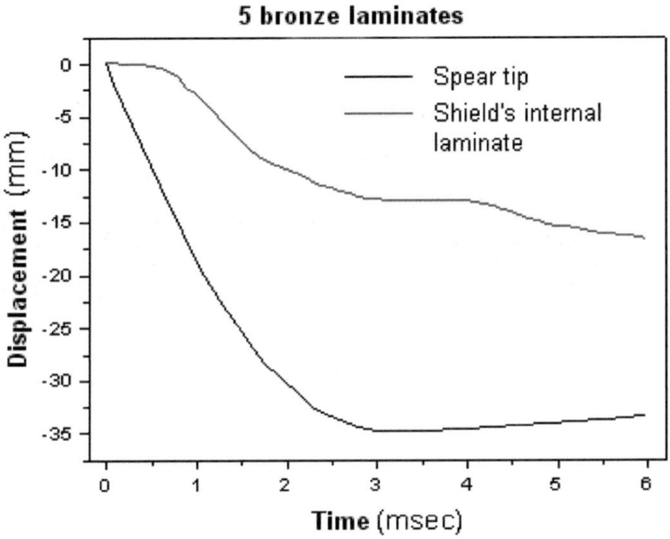


**Fig. 7** Displacements of the spear tip and of the central knot of the shield against time, when the properties of all five laminayes are those of bronze (5 bronze laminates).

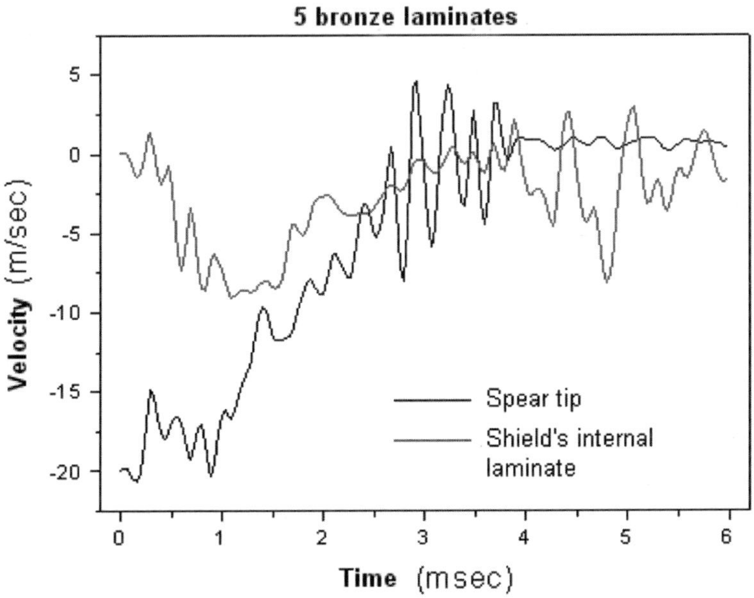


**Fig. 8** Velocities of the spear tip and of the central knot of the shield against time, when the properties of all five laminayes are those of bronze (5 bronze laminates).

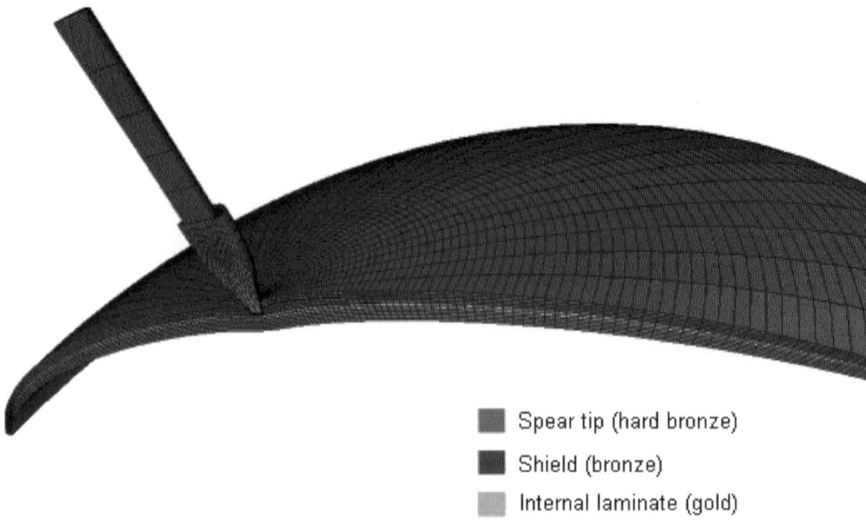


**Fig. 9** Impact behaviour of the spear-shield system based on the real properties of the materials of all layers of the shield (bronze-gold-tin), within 1.5 msec, where the spear pierces the two external layers of bronze and tin and stops at the layer of gold.

Figure 9 presents the response of the simulated shield of Achilles. It is obvious that the spear stopped at the gold layer, having pierced the two external layers of bronze and tin. This occurs within 1.5 msec time, while, in the sequence, the spear is repulsed.

Figure 10 presents spear tip velocity and shield's central knot against time.

Summarizing the results of the present analysis, one may note that (a) the shield consists of a specific laminate sequence of materials with widely different mechanical properties, out of which hard bronze was the strongest, (b) the combination of the said materials in the way stated in the Iliad did not allow the spear to penetrate the shield, since, on one occasion, it was repulsed, while on another, it only managed to penetrate the first two laminates (bronze-tin) and then it stopped at the gold laminate, and (c) if all of the laminates consisted of hard bronze only and under the same impact conditions, the shield would have been penetrated.

This is due to the totally different behaviour of the materials under static and dynamic loading. The problem of a defensive weapon, such as a shield, is to be able to convert the kinetic energy of a fast moving projectile into heat and not just to withstand application of a high static load. This ability is not possessed by hard bronze alone, which undergoes very small deformation, as compared with the rest of the materials and also has very low damping capacity. On the contrary, tin and soft (pure) gold, while undergo plastic deformation, cause the motion to attenuate, by scattering the kinetic energy of the spear. In addition, the laminated structure contributes, to some extent, with further damping due to friction between laminates, which, however, is not the main energy scattering mechanism in the present case.

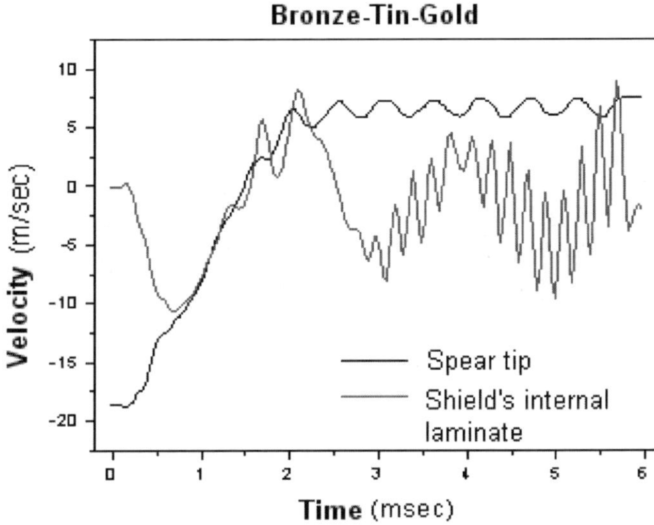


**Fig. 10** Velocities of the spear tip and of the central knot of the shield against time, by using the real properties of the materials (bronze-gold-tin) of Achilles' shield.

Based on these remarks, one may confirm the assumption that the shield manufacturer possessed deep knowledge of the dynamic-mechanical properties of laminated composite structures, i.e. elements widely used in modern technology.

## *2.2 The Shield of Ajax*

Ajax, son of Telamon, King of Salamis, and, according to the Homeric description, a man of enormous stature and a colossal body, was inferior in strength and bravery only to Achilles.

Ajax fought against Hector and, with the help of Athena, he saved Achilles' body from the hands of the Trojans. Ajax lost in a contest with Odysseus for the possession of Achilles armour, which ended up to his death. According to a later version, his disappointment led him to madness. When he came around, he committed suicide, by throwing himself on the sword that he received as a present from Hector.

Ajax was also the patron hero of Salamis island, where a temple with his statue was erected in his honour and *Aianteia*, an annual festivity, was taking place. As a place name, his name survives in *Aianteion* of Salamis.

Ajax Telamonius participated to the siege of Troy, as stated in the Iliad. As with Achilles' shield, his own shield is described with great clarity and sufficient details, so that it can be simulated by the methods so far mentioned.

Since the materials here specified have no high cost, as for example gold in the shield of Achilles, while manufacture is much simpler, it was possible to produce specimens and perform an experimental study of the shield properties, in order to confirm theoretical results and, particularly, to investigate the mechanism of dissipation of the kinetic energy of the impactor, which is completely different than the one of Achilles' shield. The description is the following (Il. 7.219–223):

> Ajax came up bearing his shield in front of him like a wall – a shield of bronze with seven folds of oxhide – the work of Tychius, who lived in Hyle and was by far the best worker in leather. He had made it with the hides of seven full-fed bulls, and over these he had set an eighth layer of bronze.

The battle behaviour of Ajax's shield during his duel with Hector is described in the sequence in the same book (Il. 7. 244–259):

> He poised his spear as he spoke, and hurled it from him. It struck the sevenfold shield in its outermost layer – the eighth, which was of bronze – and went through six of the layers but in the seventh hide it stayed. Then Ajax threw in his turn, and struck the round shield of the son of Priam. The terrible spear went through his gleaming shield, and pressed onward through his cuirass of cunning workmanship; it pierced the shirt against his side, but he swerved and thus saved his life. They then each of them drew out the spear from his shield, and fell on one another like savage lions or wild boars of great strength and endurance: the son of Priam struck the middle of Ajax's shield, but the bronze did not break, and the point of his dart was turned.

Here as well, the description of the shield corresponds to a multi-layered structure, consisting of 8 in total consecutive laminates, namely, of an external laminate of hard bronze and seven layers of calf's leather underneath. As shown in the sequence, by the same advanced numerical methods and, in addition, by a full experimental study, the Homeric descriptions concerning shield's battle behaviour are confirmed with surprising accuracy and, once more, reveal elements of advanced scientific and technological knowledge.

### 2.2.1 Analysis of Results

In this last paragraph the impact of Hector's spear on Ajax's shield is described. Similar assumptions as with Achilles' shield have been adopted, regarding the exact shape and dimensions od spear and shield and data from the modern javelin sport have been utilized.

It is assumed that the seven leather layers of the shield have the same thickness, which varies from 1 to 1.5 mm. For a circular shield and the spear hitting it at a normal direction, the problem is axisymmetric. Details of shield cross section and of the spear head, as well as the 3-dimensional discretization, applied in the present analysis, has already been given in Figure 9. The form of the generatrice of the shield cross section is elliptic with semi-axes 300 and 120 mm.

The model developed allows for the in-depth study of a shield consisting of 1 bronze laminate bronze and a number of layers of calf's leather. The detailed results

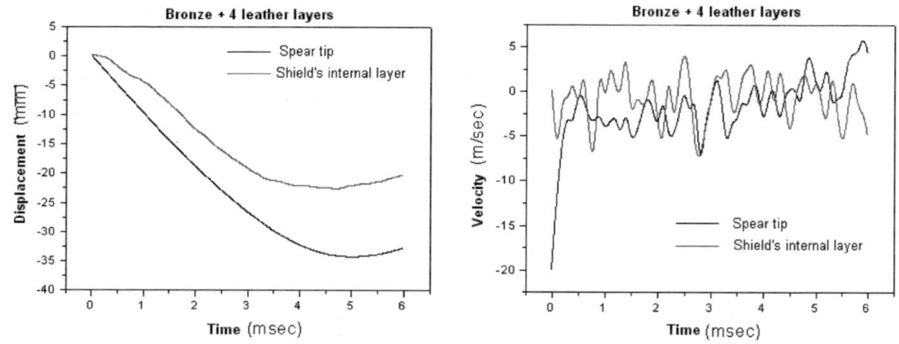


**Fig. 11** Penetration diagram of a simulated shield, consisting of 1 bronze laminate and 4 leather layers.

of the analysis are presented in Figures 5–7 for layers of bronze only (as in the analysis of Achilles' shield, figures are repeated). Also, in Figure 11, as an example, the results for 1 bronze laminate and 4 leather layers are given. It is noted that, in the present analysis, as the main factor of spear motion attenuation, the friction between the leather layers was considered, which in this case is the main conversion mechanism of its kinetic energy into heat.

### 2.2.2 Experimental

To test the penetration resistance, a series of specimens were manufactured, having the same multi-layer structure consisting of one laminate of hard bronze and a number of layers of calf's leather. To investigate the mode of operation of the shield and of spear penetration mechanism, various combinations of number and thickness of layers were examined. In fact, by maintaining the total thickness of leather layers constant, e.g. 7 mm, different forms of specimens, presented in Figure 12, were experimentally tested.

The specimens were subjected to impact with projectiles of hard bronze launched by an air-gun. The properties of the various materials used are given in Table 1.

The experimental setup appears in Figure 13 and consists of the air-gun, a specimen clamping jig and a system to measure projectile velocity. The air-gun has a compressed-air vessel and a barrel.[2] The vessel is equipped with an electric detonation valve and a pressure meter. The air-gun accepts barrels of different diameters and length. In this case, the barrel was q.20 m long and had a diameter of 13.4 mm. Compressed-air was supplied by a central distribution network. The specimen clamping jig with part of the barrel appears in Figure 14.

The air-gun shoots cylindrical projectiles with a conical tip made of bronze harder than shield bronze (Figure 15). To measure the projectile velocity a split

[2] Designed by the late Professor Werner Goldsmith, University of California at Berkeley in 1981 during his sabbatical year in Patras.

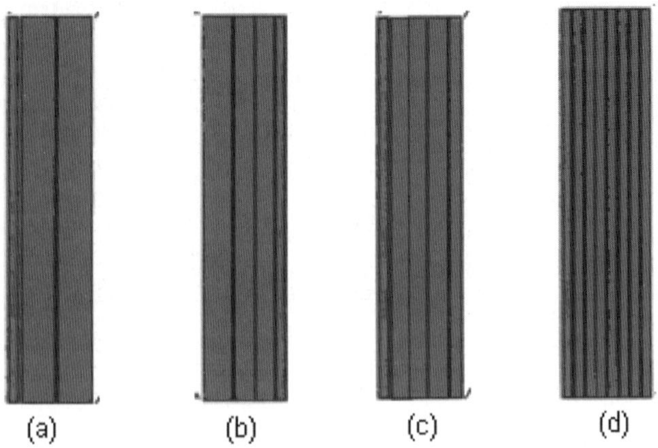


**Fig. 12** Several specimen forms: bronze laminate (red) is 1 mm thick, while total thickness for all leather leyers (green) is 7 mm. There are: in (a) 2 × 3.5 mm leather layers, in (b) 2 × 2 mm + 1 × 3 mm, in (c) 3 × 2 mm + 1 mm and in (d) 7 × 1 mm. Specimen dimensions are 140 × 140 mm. Only specimen (d) resisted penetration.

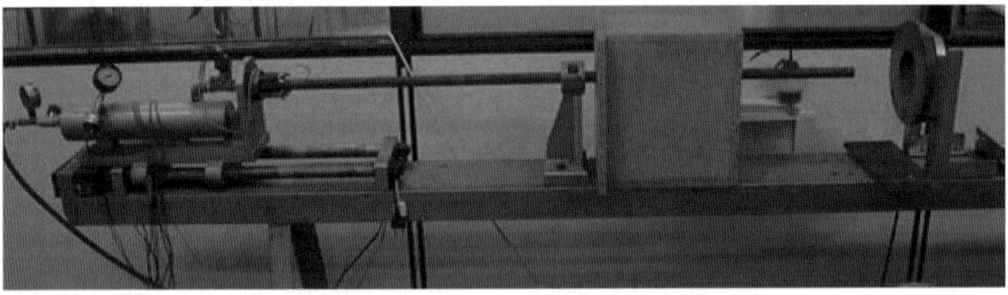

**Fig. 13** The air-gun.

He-Ne laser beam and two photodiodes with their respective outputs feeding a CRT oscilloscope were used. The initial projectile velocity of a given mass projectile, again, corresponded to kinetic energy equal to that of the javelin of year 2000 world champion, equal to 39.38 Joule.

The penetration hole on a specimen bronze surface appears in the photograph in Figure 16.

## 3 Discussion and Conclusions

A measure of the projectile penetration into the respective specimen is the size of the hole created on the surface bronze laminate (Figure 16). The results for the various specimens appear in Figure 17. It is interesting that for the multi-layer shield with seven leather layers, i.e. Ajax's shield, penetration assumes a maximum value.

**Fig. 14** Specimen clamping jig.

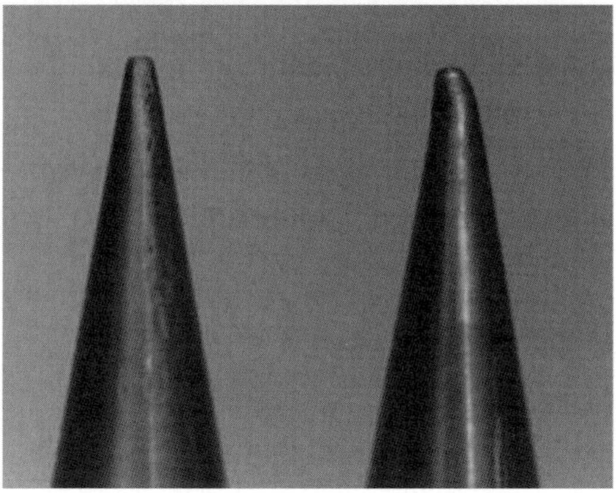

**Fig. 15** The projectile before and after impact.

However, this was the specimen that resisted penetration, and, in reality, was the last leather layer that eventually stopped the projectile. This agrees in admirable way with the Homeric description for its battle behaviour, stating that (Il. 7.244–247):

[Ἕκτωρ]
...ἀμπεπαλὼν προΐει δολιχόσκιον ἔγχος,
καὶ βάλεν Αἴαντος δεινὸν σάκος ἑπταβόειον
ἀκρότατον κατὰ χαλκόν, ὃς ὄγδοος ἦεν ἐπ᾽ αὐτῷ.
ἓξ δὲ διὰ πτύχας ἦλθε δαΐζων χαλκὸς ἀτειρής,
ἐν τῇ δ᾽ ἑβδομάτῃ ῥινῷ σχέτο· ...

[Hector]
poised his spear as he spoke, and hurled it from him. It struck the sevenfold shield in its outermost layer – the eighth, which was of bronze – and went through six of the layers but in the seventh hide it stayed.

The fact is also stated, which appears in Figure 15, e.g. bending of the projectile tip after a failed impact (Il. 7.258–259):

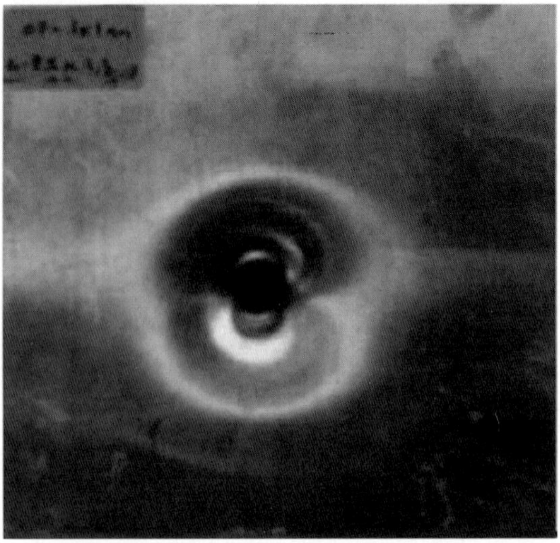

**Fig. 16** Penetration hole on bronze surface.

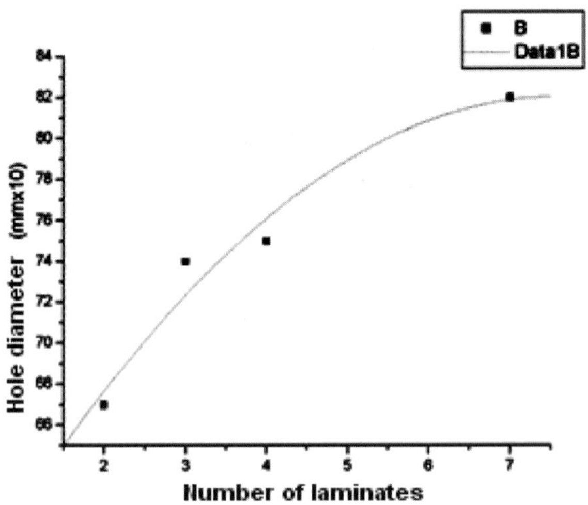


**Fig. 17** Penetration hole diameter on the surface of a bronze laminate against number of leather layers.

Πριαμίδης μὲν ἔπειτα μέσον σάκος οὔτασε δουρί,
οὐδ᾽ ἔρρηξεν χαλκός, ἀνεγνάμφθη δέ οἱ αἰχμή.

> . . . the son of Priam struck the middle of Ajax's shield, but the bronze did not break, and the point of his dart was turned.

In this case, the projectile kinetic energy is absorbed by the friction between layers, acting efficiently in the presence of sufficient deformation which develops with a

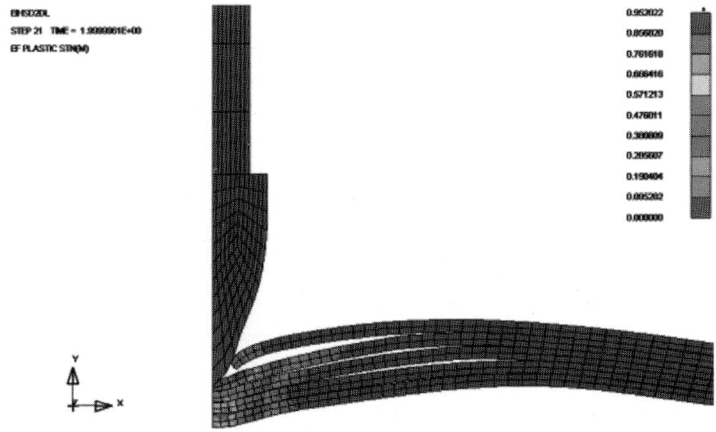


**Fig. 18** Animated representation of penetration of a model shield consisting of a bronze laminate and four leather layers.

number of layers equal to or greater than seven. On the other hand, a polynomial interpolation in the experimental results of Figure 17, shows that, for a greater number of leather layers, probably tends to assume a constant value. This may prove that the arrangement described in the Iliad provides not only maximum resistance towards penetration, but also an optimum solution, provide that, for a greater number of layers, the friction would be too strong and the structure too rigid to deform sufficiently, in order to consume kinetic energy through friction. It is also noted that a small part of the projectile kinetic energy is absorbed by the deformation of the projectile tip, mentioned in 7.259 and appears in Figure 15 of the present experimental results.

Again, the results confirm accurately the assumption that the shield manufacturer, not god Hephaestus anymore, but Tychius, a simple animal hide technician, possessed aldo a deep knowledge of the dynamic mechanical properties of multilayered composite structures, which are valuable structural elements of modern technology.

## 4 The Shield of Heracles

An earlier reference to a shield along with certain structural details, however, not sufficient to make a reconstruction possible, is found in Hesiod, concerning the shield of Heracles and his duel with Cygnus, son of Ares:

> In his hands he took his shield,[3] all glittering: no one ever broke it with a blow or crushed it. And a wonder it was to see; for its whole orb was a-shimmer with

[3] The shield was made by Hephaestus, ordered by Jove, and it was a work to admire, even for the King of Gods.

enamel and white ivory and electrum, and it glowed with shining gold; and there were zones of cyanus(2) drawn upon it. In the centre was Fear worked in adamant, unspeakable, staring backwards with eyes that glowed with fire. His mouth was full of teeth in a white row, fearful and daunting, and upon his grim brow hovered frightful Strife who arrays the throng of men: pitiless she, for she took away the mind and senses of poor wretches who made war against the son of Zeus. Their souls passed beneath the earth and went down into the house of Hades; but their bones, when the skin is rotted about them, crumble away on the dark earth under parching Sirius. (Hesiod, Shield of Heracles, 139–153)

## 5 The Roman Shield

In Polybius'[4] *Historia*, vi. 23.2, a full description of a Roman shield of his time is given. For comparison with the respective Greek technology of a much earlier period, this description is presented in full [3] (Figure 28):

> The next in age, who are called the hastati, are ordered to furnish themselves with a complete suit of armour. This among the Romans consists in the first place of a shield of a convex surface; tile breadth of which is two feet and shelf; and the length four feet, or four feet and a palm of those of the largest size. It is composed of two planks, glued together with bull glue,[5] and covered first with linen, and afterwards with calves' skin. The extreme edges of it, both above and below, are guarded with plates of iron, as well to secure it against the strokes of swords, as that it may be rested also upon the ground without receiving any injury to the surface is fitted likewise a shell of iron; which serves to turn aside the more violent strokes of stones, or spears, or any other ponderous weapon.

## 6 The Shield of Atreid's

An account on the shield of Agamemnon is given in Il. 11.15–46:

> Ἀτρεΐδης δ᾽ ἐβόησεν ἰδὲ ζώννυσθαι ἄνωγεν
> Ἀργείους· ἐν δ᾽ αὐτὸς ἐδύσετο νώροπα χαλκόν.
> κνημῖδας μὲν πρῶτα περὶ κνήμῃσιν ἔθηκε

[4] Polybius (c. 200–118 BC) Greek statesman and historian, who wrote on the development of Rome into a world power.

[5] A glue produced from bull hide.

καλὰς ἀργυρέοισιν ἐπισφυρίοις ἀραρυίας·
δεύτερον αὖ θώρηκα περὶ στήθεσσιν ἔδυνε,
τόν ποτέ οἱ Κινύρης δῶκε ξεινήϊον εἶναι.
πεύθετο γὰρ Κύπρον δὲ μέγα κλέος οὕνεκ᾽ Ἀχαιοὶ
ἐς Τροίην νήεσσιν ἀναπλεύσεσθαι ἔμελλον·
τοὔνεκά οἱ τὸν δῶκε χαριζόμενος βασιλῆϊ.
τοῦ δ᾽ ἤτοι δέκα οἶμοι ἔσαν μέλανος κυάνοιο,
δώδεκα δὲ χρυσοῖο καὶ εἴκοσι κασσιτέροιο·
κυάνεοι δὲ δράκοντες ὀρωρέχατο προτὶ δειρὴν
τρεῖς ἑκάτερθ᾽ ἴρισσιν ἐοικότες, ἅς τε Κρονίων
ἐν νέφεϊ στήριξε, τέρας μερόπων ἀνθρώπων.
χρύσειοι πάμφαινον, ἀτὰρ περὶ κουλεὸν ἦεν
ἀργύρεον χρυσέοισιν ἀορτήρεσσιν ἀρηρός.
ἂν δ᾽ ἕλετ᾽ ἀμφιβρότην πολυδαίδαλον ἀσπίδα θοῦριν
καλήν, ἣν πέρι μὲν κύκλοι δέκα χάλκεοι ἦσαν,
ἐν δέ οἱ ὀμφαλοὶ ἦσαν ἐείκοσι κασσιτέροιο
λευκοί, ἐν δὲ μέσοισιν ἔην μέλανος κυάνοιο.
τῇ δ᾽ ἐπὶ μὲν Γοργὼ βλοσυρῶπις ἐστεφάνωτο
δεινὸν δερκομένη, περὶ δὲ Δεῖμός τε Φόβος τε.
τῆς δ᾽ ἐξ ἀργύρεος τελαμὼν ἦν· αὐτὰρ ἐπ᾽ αὐτοῦ
κυάνεος ἐλέλικτο δράκων, κεφαλαὶ δέ οἱ ἦσαν
τρεῖς ἀμφιστρεφέες ἑνὸς αὐχένος ἐκπεφυυῖαι.
κρατὶ δ᾽ ἐπ᾽ ἀμφίφαλον κυνέην θέτο τετραφάληρον
ἵππουριν· δεινὸν δὲ λόφος καθύπερθεν ἔνευεν.
εἵλετο δ᾽ ἄλκιμα δοῦρε δύω κεκορυθμένα χαλκῷ
ὀξέα· τῆλε δὲ χαλκὸς ἀπ᾽ αὐτόφιν οὐρανὸν εἴσω
λάμπ᾽· ἐπὶ δ᾽ ἐγδούπησαν Ἀθηναίη τε καὶ Ἥρη
τιμῶσαι βασιλῆα πολυχρύσοιο Μυκήνης.

The son of Atreus shouted aloud and bade the Argives gird themselves for battle while he put on his armour. First he girded his goodly greaves about his legs, making them fast with ankle clasps of silver; and about his chest he set the breastplate which Cinyras[6] had once given him as a guest-gift. It had been noised abroad as far as Cyprus that the Achaeans were about to sail for Troy, and therefore he gave it to the king. It had[7] ten courses of dark cyanus, twelve of gold, and ten of tin. There were serpents of cyanus that reared themselves up towards the neck, three upon either side, like the rainbows which the son of Cronus[8] has set in heaven as a sign to mortal men. About his shoulders he threw his sword, studded with bosses of gold; and the scabbard was of silver with a chain of gold wherewith to hang it. He took moreover the richly-dight shield that covered his body when he was in battle – fair to see, with ten circles of bronze running all round see, wit it. On the body of the shield there were twenty bosses of white tin, with another of dark cyanus in the middle: this last was made to show a Gorgon's head, fierce and grim, with Rout and Panic on either side. The band for the arm to go through was of silver, on which there was a writhing snake of cyanus with three heads that sprang from a

---

[6] V. 20. Cinyras, a hero of Cyprus, brought the worship of Aphrodite from Syria to Paphos. Many people, among them Tyrteus, compare him to Midas because of his innumerable wealth. This is the only reference to Cyprus in the Iliad, while in the Odyssey the island is more often mentioned.

[7] V. 24. This is the only detailed description of a breastplate in Homer (compare Ψ, 560–562).

[8] Cronidēs, e.g. Jove.

single neck, and went in and out among one another. On his head Agamemnon set a helmet, with a peak before and behind, and four plumes of horse-hair that nodded menacingly above it; then he grasped two redoubtable bronze-shod spears, and the gleam of his armour shot from him as a flame into the firmament, while Juno and Minerva thundered in honour of the king of rich Mycene.

### Cyanus

Cyanus was a glass-paste of deep blue colour: the 'zones' were concentric bands in which were the scenes described by the poet. The figure of Fear (l. 44) occupied the centre of the shield, and Oceanus (l. 314) enclosed the whole. Here, the offensive character of the shield, i.e. to intimidate the enemy, reappears. Cyanus as a material is repeatedly mentioned in the Iliad (as a constituent of the breastplate of Atreides, 11.23–28. In the same Book, the reception of Patroclus and Machaon to Nestor's tent is described (Il. 11.624–630):

τοῖσι δὲ τεῦχε κυκειῶ ἐϋπλόκαμος Ἑκαμήδη,
τὴν ἄρετ᾽ ἐκ Τενέδοιο γέρων, ὅτε πέρσεν Ἀχιλλεύς,
θυγατέρ᾽ Ἀρσινόου μεγαλήτορος, ἥν οἱ Ἀχαιοὶ
ἔξελον οὕνεκα βουλῇ ἀριστεύεσκεν ἁπάντων.
ἥ σφωϊν πρῶτον μὲν ἐπιπροΐηλε τράπεζαν
καλὴν κυανόπεζαν ἐΰξοον, αὐτὰρ ἐπ᾽ αὐτῆς
χάλκειον κάνεον, ἐπὶ δὲ κρόμυον ποτῷ ὄψον,

Fair Hecamede, whom Nestor had had awarded to him from Tenedos when Achilles took it, mixed them a mess; she was daughter of wise Arsinous, and the Achaeans had given her to Nestor because he excelled all of them in counsel. First she set for them a fair and well-made table that had feet of cyanus; on it there was a vessel of bronze and an onion to give relish to the drink, with honey and cakes of barley-meal.

## 7 The Mycenaean Helmets

Mycenaean helmets, at least in the early period, were manufactured, as mentioned, from boar tusks (Figure 29). The following description in the Iliad concerns the helmet of Odysseus (Il. 10.260-265):

Μηριόνης δ᾽ Ὀδυσῆϊ δίδου βιὸν ἠδὲ φαρέτρην
καὶ ξίφος, ἀμφὶ δέ οἱ κυνέην κεφαλῆφιν ἔθηκε
ῥινοῦ ποιητήν· πολέσιν δ᾽ ἔντοσθεν ἱμᾶσιν
ἐντέτατο στερεῶς· ἔκτοσθε δὲ λευκοὶ ὀδόντες
ἀργιόδοντος ὑὸς θαμέες ἔχον ἔνθα καὶ ἔνθα
εὖ καὶ ἐπισταμένως· μέσσῃ δ᾽ ἐνὶ πῖλος ἀρήρει.

Meriones found a bow and quiver for Ulysses, and on his head he set a leathern

**Fig. 19** Mycenaean armour and helmet (left) and a warrior equipped with eight-shaped shield and a helmet from boar tusks, ivory ανάγλυφο c. 1400–1200 BC from Delos (right), Archaeological Museum of Delos.

> helmet that was lined with a strong plaiting of leathern thongs, while on the outside it was thickly studded with boar's teeth, well and skilfully set into it; next the head there was an inner lining of felt.

This description corresponds to a very balanced design: The natural product surrounding the helmet, the boar tusks, has excellent impact strength, while the internal filling of felt absorbs impact and prevents it from reaching the skull. Fully analogous is the construction of modern military helmets, manufactured from advanced composite materials. However, it appears that, at some later stages, advanced composite laminates were used for the construction of helmets, just like the shields. Thus, their similarity with modern military helmets is even more pronounced (Figure 20).

Now, referring to boar husks, they consist by the major part of *dentine*, a yellowish calcic substance, much denser and harder than bones. Dentine: (a) forms the main bulk of the tooth and is considered a vital tissue, just as the tooth pulp,[9] which provides nutrition, feeling and resistance to fracture, (b) provides the base for the much harder enamel and forms the root (or the roots) of the tooth, (c) is pierced tubules, which extend continuously from the pulp to the external surface, i.e. it is porous, and (d) is a tough material without preferable fracture surfaces. The chemical composition of dentine b.w. is 70% inorganic, 20% organic and 10% water. The

[9] Pulp is the live tissue within a tooth, placed in the special chamber and in the root channels. It contains, in high density, connecting tissue, nerves, lymph and blood vessels.

**Fig. 20** Mycenaean helmet manufactured from boar husks (left) and a modern military helmet of Danish origin[10] (right).

inorganic phase is oxyapatite and the organic collagen. Its structure within the tooth is fairly complex.

The major part of the teeth of mammals is covered by another substance, *enamel*, which is the hardest substance of the body. It consists mainly from apatite crystals, containing calcium and phosphoric salts. Enamel is harder at the points that the tooth bites.

## 8 Review and Conclusions

In the present work, a whole series of defensive weapons, which, according to the Homeric descriptions, were used in the Trojan War, mainly from the side of the Achaeans. In particular, for the shields of Achilles and Ajax, sufficient information was provided, to have them reconstructed on the basis of realistic assumptions, both as computer models and as physical objects (test pieces), allowing for a complete investigation of their properties. The results not only confirmed the Homeric descriptions of their battle behaviour accurately, but also support the hypothesis that in Mycenaean Greece scientific knowledge and advanced technology, in the modern sense of the words, were available. Structures of much later times, e.g. Roman, appear to be products of much inferior, if not primitive, technology. Other defensive weaponry, armours, helmets, breastplates, etc., although they are not the object of a similar analysis, appear to confirm, at least qualitatively, the above findings.

[10] Danish M/96 Combat helmet (CGF Gallet Combat Helmet TC "D", from http://en.wikipedia.org/wiki/Image:M96_helmet_Denmark_001.jpg

## References

1. Zeggelis, C.D., *The Science of Nature in Homer*, Athens, 1891 [in Greek]. Republished in 1977 by University of Patras Editions.
2. Steinberg, D.J. and Guinan, M.W., A high strain rate constitutive model for metals, University of California, Lawrence Livermore National Laboratory, Report UCRL-80465, 1978.
3. Polybius, The military institutions of the Romans, in *The Library of Original Sources*, O.J. Thatcher (Ed.), University Extension Co., Milwaukee, WI, 1901, pp. 172–186.

# How the Greeks Got Ahead: Technological Aspects of Manufacture of a Corinthian Type Hoplite Bronze Helmet from Olympia

A.J.N.W. Prag[1], R. Garner[1], E. Pantos[2], S.L. Bennett[2], J.F.W. Mosselmans[2,*], M.J. Tobin[2,**], W. Kockelmann[3], L.C. Chapon[3], N. Salvado[4] and T. Pradell[5]

[1] *The Manchester Museum, The University of Manchester, Manchester, U.K.*
[2] *CCLRC, Daresbury Laboratory, Keckwick Lane, Warrington, U.K.*
[3] *CCLRC, Rutherford-Appleton Laboratory, ISIS Neutron Spallation Source, U.K.*
[4] *Departamento de Física i Enginyeria Nuclear, Univ. Politècnica de Catalunya, Barcelona, Spain*
[5] *Departamento d'Enginyeria Quimica, EPSEVG Univ. Politècnica de Catalunya, Vilanova, Spain*

First Aias son of Telamon, bulwark of the Achaians, brake a battalion of the Trojans and brought his comrades salvation, smiting a warrior that was chiefest among the Thracians, Eussoros' son Akamas the goodly and great. Him first he smote upon his thick-crested helmet ridge and drave into his forehead, so that the point of bronze pierced into the bone; and darkness shrouded his eyes.

(Homer, *Iliad* VI 5-11. Transl. by A. Lang, W. Leaf and E. Myers, Macmillan 1912. Used as the introduction of two research proposals submitted to CCLRC for Synchrotron and Neutron work at the SRS and ISIS facilities).

**Abstract.** The object of this study is a battle helmet of Corinthian type, now in the collections of The Manchester Museum. The Corinthian helmet has been called "one of the great independent achievements of early Greek technology". It was manufactured out of a single piece of bronze, probably on a rod-anvil, and like all body-armour it was made to measure. This required exceptional skill on the part of the smith, but once discovered the design was so efficient that it was still being used in fifteenth-century Italy, more than 2000 years after its invention around 700 BC. However, by the seventeenth century the art had been lost and had to be re-invented for modern replicas.

Victorious Greek cities often set up trophies of armour from the defeated as thank-offerings at temples. Like animals being sacrificed, weapons were "killed" to be offered to the other world: so helmets had their cheek-pieces and nose-guards bent back. When the sanctuary became too

* Present address: Diamond Light Source Ltd, Chilton, Didcot, U.K.
** Present address: Australian Synchrotron, Level 17, 80 Collins St, Melbourne VIC 3000, Australia.

*S.A. Paipetis (ed.), Science and Technology in Homeric Epics,* 205–220.

crowded the trophies were cleared away and buried. The Manchester helmet was most likely found at such a site, probably Olympia itself. In order to sell it, the finder – probably in the 19th century – straightened out the cheek-pieces, which cracked at the edges, but the nose-guard was either missing or snapped off altogether: in its present form it is too short and too close to the face to be practical.

X-ray fluorescence has shown that the main body of the helmet is a copper-tin alloy of varying composition at different places and containing iron and lead while the nose-guard contains zinc in high abundance. This key compositional difference indicates that the nose-guard is not the genuine broken-off piece retrieved from the finding place but is a modern substitute fabricated for restoration purposes. Synchrotron XRD at glancing angle and variable wavelength on several spots on the head and the nose-guard itself shows a shift of the copper Bragg reflections which can be related to Cu-Sn or Cu-Zn percentage composition. In addition, several corrosion products are identified. Small samples of corrosion extracted from the inside of the helmet have been used to obtain powder XRD patterns. The same samples have been studied with synchrotron micro-FTIR. Neutron diffraction sampling the bronze volume at different areas has also been used to quantify the composition. Crystallographic texture data obtained on a neutron diffractometer with large angular detector coverage was used to draw conclusions about the processes used to manufacture this precious example of archaic military technology. Some other interesting questions remain to be investigated in the near future.

# 1 Introduction

The object of this study is an ancient Greek helmet of Corinthian type, the only one for which we know the ancient Greek name. The Corinthian helmet has been called "one of the great independent achievements of early Greek technology". It was manufactured out of a single piece of bronze, probably on a rod-anvil, and like all body-armour it was originally made to measure. This required exceptional skills on the part of the smith, but once discovered the design was so efficient that it was still being used in fifteenth-century Italy, more than 2000 years after its invention around 700 BC. However, by the seventeenth century the art had been lost and had to be re-invented for modern replicas. This type of helmet was part of the Classical Greek infantryman's armour. Such men formed the core of the citizen armies of Greek city-states, and had to be rich enough to provide their own equipment. Victorious Greek cities often set up trophies of armour from the defeated as thank-offerings at temples, and when the sanctuary became crowded these were cleared away and buried. Like animals being sacrificed, weapons were "killed" to be offered to the other world and the helmets had their cheek-pieces and nose-guards bent back. The Manchester helmet was most likely found at such a site, probably Olympia itself. In order to sell it, the finder – probably in the nineteenth century – straightened out the cheek-pieces, which cracked at the edges, but the nose-guard snapped off altogether: in its present form it overlaps on the inside and is too short and too close to the face to be practical. The bronze helmet was acquired in November 2002 by the Manchester Museum as a teaching, display and research object [1], and is illustrated in Figure 1b. It is flanked by Miltiades' helmet dedicated at the temple of Zeus in Olympia, of a later design, the so-called "Attic type", and a modern (attempt at) reproduction.

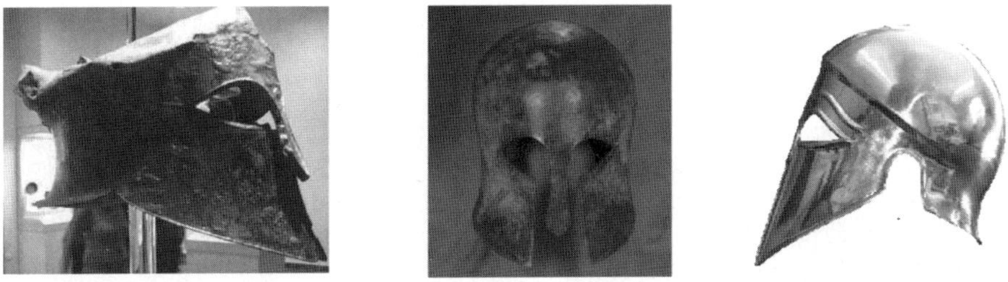

a b c

**Fig. 1** (a) The helmet of Miltiades, the Athenian general who defeated the Persians at Marathon, dedicated to the temple of Zeus in Olympia, now at the Olympia Museum (photo: E. Pantos). (b) The Manchester Museum helmet (photo courtesy of The Manchester Museum). (c) Modern, "fancy ball" reproduction. Armed warriors are usually depicted in Classical Greek sculpture and pottery wearing helmets of this type.

**Fig. 2** Different types of bronze helmets (images from images of helmets from http://www.hellenic-art.com/armour/helmets.htm and http://www.ncl.ac.uk/shefton-museum/).

## 2 Techniques Applied

The starting question posed by the museum partners in this project was to investigate the authenticity of the restored part of the nose-guard using non-destructive methods. X-ray fluorescence can easily characterise the composition of the alloy

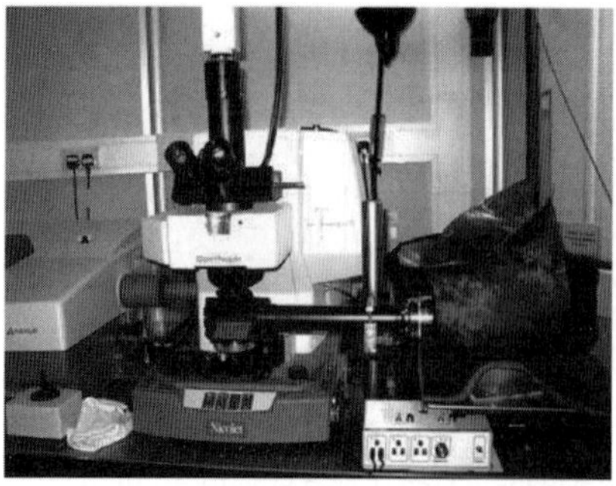

a

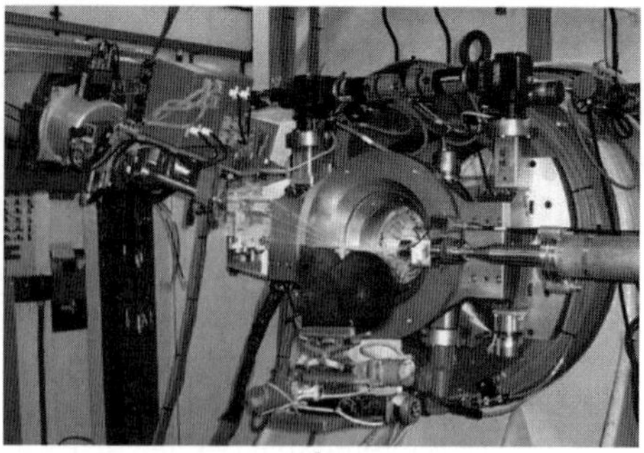

b

**Fig. 3** (a) The helmet at the SRS 11.1 FTIR station, set for measurement in reflection mode. Reproduced from figure 2 of reference [3]. (b) The helmet set for XRD in reflection mode with variable wavelength for depth profile analysis at the SRS station 9.4.

and establish whether the repaired nose-guard was made of the same material as the rest of the object. The original bronze was expected to be a copper/tin alloy and the objective was to compare the relative abundance in tin composition of the main part of the helmet with the restored nose-guard. XRF measurements were carried out on station 9.2 at the SRS. The tin content could be evaluated from diffraction measurements. The shift in the position of the copper diffraction lines depends on the percentage composition of the alloying element. Neutron diffraction was chosen as the appropriate technique as neutrons penetrate through coatings and corrosion phases deep into the thickness of the helmet wall and illuminate a considerable volume portion, thus delivering representative microstructural information and avoiding problems associated with single-spot analyses. The corrosion products lining the surface could be characterised with X-ray diffraction. Two modes of measurement were possible: (a) off-the-surface XRD (on station 9.4 of the SRS) from areas easy to access within the geometrical constraints; in this mode, changing the wavelength of the incident X-rays alters the penetration depth and thus allows a measure of depth-profiling to be carried out; (b) micro-sampling of corroded areas on the outside and inside surfaces could be examined by powder diffraction (on station 9.6 of the SRS), the sampled area being of the order of 0.2 mm. Further investigation of corrosion products could be studied with FTIR (Figure 3a).

Of greater interest was whether information could be obtained on the bulk structural properties of the alloy as well as of microstructural features arising from processes involved in producing the helmet. Neutron diffraction is a suitable non-destructive diagnostic tool for obtaining average structural information from the interior of large, undisturbed archaeological objects. A particularly promising application is texture analysis, which is used for determining the grain orientations in artefacts. The orientations of crystallites change in a characteristic way if a material undergoes plastic deformations or thermal treatments during manufacturing. Mapping of the grain orientation distributions by neutron texture analysis in terms of the so-called pole figures could therefore provide important clues to the deforma-

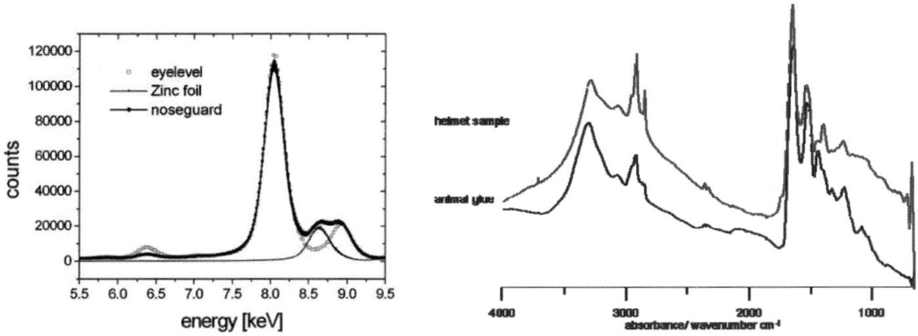


**Fig. 4** (a) X-ray fluorescence from nose-guard (thick solid line) and helmet's eye level (symbols) compared to a zinc foil reference material (thin solid line). The peaks at 8.6 and 8.9 keV are the $K_\alpha$ and $K_\beta$ lines of copper. The peak at 6.4 keV is that for the iron $K_\alpha$ line. Reproduced from figure 2 of reference [3]. (b) FTIR spectra showing that the helmet had been coated (by a previous owner?) with an animal glue. In glue-free locations, the composition of some of the surface deposits could be identified.

tion history. The pole figures are the maps of the grain orientation distribution and can be regarded as fingerprints of the working processes involved the production of the object. Neutron texture analysis was carried out on the GEM diffractometer at the ISIS facility equipped with banks of detectors surrounding the object, where pole figures are collected without any sample movements or rotations. Additional diffraction patterns were collected on ROTAX at ISIS in order to obtain information on the alloy composition of the nose-guard.

# 3 Results

The experimental details of application of all these techniques and the analysis steps performed are described in two research papers [2, 3]. We concentrate here on the summary outcome of these studies but have included some additional information and relevant quotations from the Homeric epics.

## *3.1 Corrosion Products*

XRF spectra were obtained at the SRS at four places on the nose-piece and four places around the helmet at eye level. The data showed clearly that zinc was detected only on the nose-guard. All other measurement points determined as alloying elements only tin, traces of iron (Figures 4a and 8) and in some cases traces of lead.

Several corrosion products were identified by SR-FTIR in reflection mode (Figure 3a). In the nose-guard piece the presence of hydrocerussite in addition to

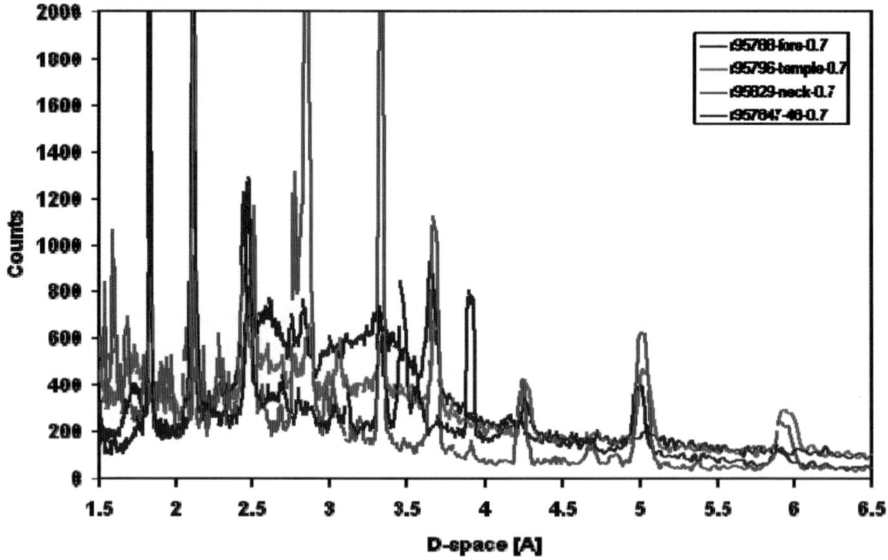


**Fig. 5** X-ray diffraction patterns from different locations on the surface of the helmet (forehead, temple and neck) at the same wavelength, 0.7A. The surface diffractometer on SRS station 9.4 (Figure 3b) can collect high resolution data with the incident X-rays at a small angle (1–3 deg) and variable wavelength. This allows depth profiling of the corrosion layers. Differences in surface alloy composition can also be measured by comparing the shifts of the Cu reflections with respect to those in pure Cu.

malachite has been detected. The spectra obtained from a sample showed the presence of a protein compound, associated with animal glue (Figure 4b), very likely as a means of preventing further corrosion before the helmet was acquired by the museum. SR-XRD in reflection mode (Figures 3b and 5) detected the presence of malachite, cuprite and antlerite, the last two in a low proportion. The relative amount is a function of wavelength, indicating layering of the corrosion products, malachite being more dominant on the outside.

The powder X-ray diffraction patterns from the surface samples extracted from the helmet yielded a richer mixture of corrosion phases (Figure 6). The corrosion products are malachite, brochantite, antlerite, romarchite, chalcocite and digenite. On the nose-guard, we determined malachite, anglesite and cuprite as well as hydroxided nitrate of copper and zinc. In addition, secondary minerals such as quartz, calcite, hematite, clay minerals and feldspars deposited from the soil during burial were identified. It is important to emphasize the presence of nitrates, which are not products to be expected amongst those created by natural corrosion phenomena. The presence of these nitrates can be related to the use of materials employed for artificial ageing of the piece.

The neutron diffraction patterns are dominated by the bronze peaks belonging to a copper-type face-centred cubic structure (Figure 7b). The peak positions and hence the lattice parameter of the bronze are increased compared to pure copper

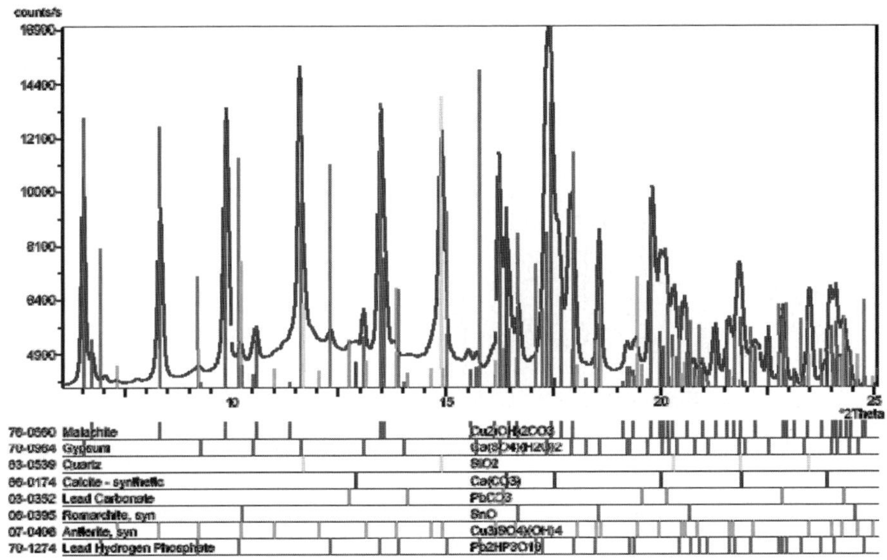


**Fig. 6** Example of phase id from one of the corrosion flakes. Tiny flakes (ca. 0.1 mm) extracted from the corrosion layers inside the helmet were examined with SR-XRD using a CCD detector for fast acquisition (30 s). Corrosion phases identified: Malachite $Cu_3(CO_3)_2\ (OH)_2$, Cerusite $PbCO_3$, Romarchite SnO, Anglesite $PbSO_4$, Cuprite $Cu_2O$, Antlerite $Cu_3(SO_4)(OH)_4$, Brochantite $Cu_4SO_4{\cdot}(OH)$, Chalcocite $Cu_2S$, Digenite $CuS_{1.8}$, Azurite $2CuCO_3{\cdot}Cu(OH)_2$, Hydroxided Nitrate of copper and zinc $Zn_3(OH)_4(NO_3)_2$ and $Cu_2(OH)_3NO_3$. Soil minerals: Quartz, Calcite, Gypsum, Hematite, Illite, Feldspars.

because of the replacement of Cu by bigger atoms. Measurement of the lattice parameter shift allows one to estimate the amount of the alloying element, e.g. Sn, in the bronze. The refined lattice parameters are translated into Cu and Sn fractions using a Vegard-type calibration curve. It should be emphasised that whilst diffraction methods are good at determining changes of lattice spacings, the "chemical analysis" through Vegard's law is based on an assumption about which alloying elements are present. Bragg reflections of much smaller intensities can be attributed to the surface minerals malachite, copper oxide, quartz and calcite (the last two presumably from burial soil trapped under the varnish layer), which were included in the refinement procedure. The Sn contents of the bronze vary slightly between 11–12 wt% for the analysis spots. The average lattice parameters of the alloy of 3.6822 Å corresponds to a Sn content of 11.7 wt% on the side of the helmet. The nose-guard, however, exhibits a distinctly different lattice parameter of 3.6378(2) Å indicating that it is made of a different type of alloy. The expansion of the copper lattice for the nose-guard can be interpreted in terms of different alloying elements. Assuming a binary Cu/Sn alloy, we obtain a Sn content of 4.2 wt%. Using a Vegard-type relation for a Cu/Zn alloy, the neutron patterns yield a zinc content of about 11 wt%. Considering that the XRF data clearly show a large presence of zinc and negligible amount of tin on the nose-guard, we conclude that the nose-guard was made of a Cu/Zn bronze

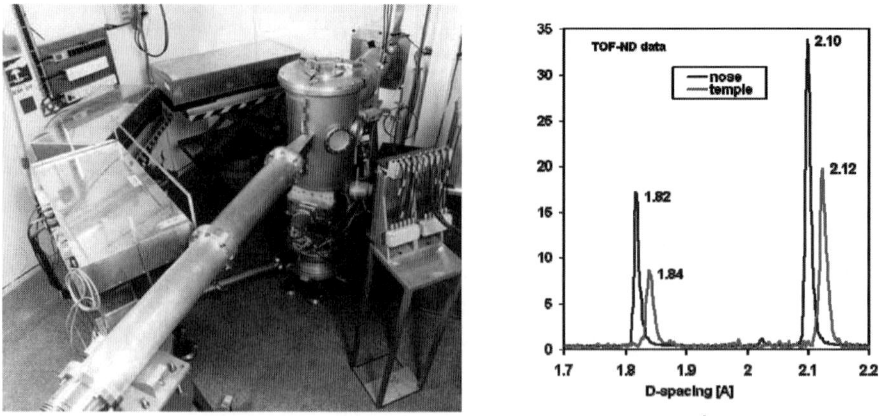


a b

**Fig. 7** (a) The ROTAX Time-of-Flight neutron diffractometer at ISIS. (b) Neutron diffraction data from the nose-guard and the temple, showing the shift of the copper reflections indicating different alloy composition.

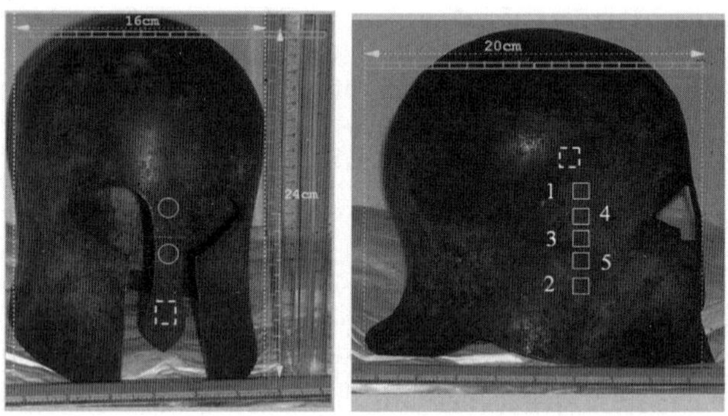


**Fig. 8** Bronze helmet in front view (left) and side view (right). Analysis spots are indicated by solid squares (neutron diffraction, GEM, scan-1=top, scan-2 bottom), dashed squares (neutron diffraction, ROTAX), and solid circles (XRF, station 7.1 and 9.2). Reproduced from figure 1 of reference [3].

with 11 wt% Zn. This implies that the nose-guard is not part of the original helmet but is a later replacement.

## 3.2 Microstrain Broadening

Peak broadening is observed for all bronze peaks of all 5 neutron measurements taken on GEM on the side of the helmet. Generally, the peak broadening may have different causes such as (i) very small particle/grain sizes, (ii) microstrain broadening due to cold-working or thermal treatment, or (iii) distributions of lattice parameters of the alloy due to a strong variation of the Sn content. Microstrains may be induced by working processes such as cold-working, or thermal treatments such as quenching, introducing lattice defects and distributions of lattice plane distances around an average value, thus becoming visible as broadening of Bragg peaks.

The analysis detailed in [3] clearly concluded that the diffraction peak broadening indicates the presence of residual microstrains. It is to be noted that the alloy peaks display smooth distributions of lattice planes, i.e. smooth Bragg profiles in contrast to typical structured Bragg peak shapes of as-cast materials. It is therefore reasonable to assume that the helmet alloy was subjected to both annealing and working processes such as hammering for hardening. Hammering may have been used to shape the helmet and to harden the alloy. Since thorough annealing would wipe out microstrains the helmet was probably produced by repetitive annealing-hammering cycles. The presence and the magnitude of the microstrains indicates that the final working step involved hammering of the alloy.

### *3.3 Texture Analysis*

The neutron diffraction data contain information on the volume texture of the alloy, i.e. the orientations of grains. Well-defined textures are produced by specific conditions during primary crystallization from a melt, and by thermal and mechanical treatments of the cast such as annealing, drawing, rolling or hammering. Any preferred orientation manifests itself in Bragg intensity changes when the sample is rotated, or equivalently, when the scattered neutrons are collected at different detector angles. In terms of the GEM multi-detector arrangement, this means that intensity ratios do not vary from bank to bank for a texture-free sample. The texture evaluation of GEM data from 160 detector groups, representing 160 different sample orientations, was performed on 5 analysis spots. The texture is displayed in terms of pole figures of representative lattice planes (111), (200), and (220). Figure 9 shows the pole figures reconstructed for the highest (scan 1) and the lowest (scan 2) analysis points on the helmet. Maximum pole densities are between 1.6 and 1.8, indicating a rather weak texture for the bronze. There is no systematic variation of the texture strengths detectable on the five spots. The (220) pole figures are characterised by elevated pole density in the centres, demonstrating preferential alignment of (220) lattice planes parallel to the wall of the helmet. Asymmetries of pole densities along the vertical for some of the analysis spots may indicate a working direction from top to bottom of the helmet, or vice versa. The texture is likely to be the result of the hardening processes which, through plastic deformation, preferentially aligned the copper/tin (220) planes perpendicular to the working direction. This finding is in agreement with the detection of microstrains discussed in the previous paragraph.

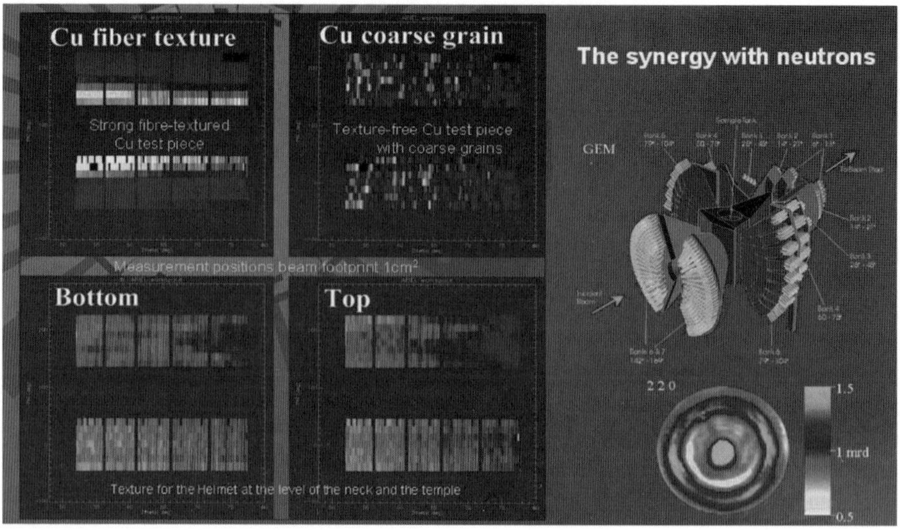


**Fig. 9** Left: Intensity distribution of the Cu 1.8A Bragg peak from the detector bank centred at 2theta 60 deg on the GEM instrument at ISIS. Right: Schematic drawing of the GEM detector bank and a pole figure of the Cu 220 reflection. The Bragg peak shifts correspond to an average Sn content of 11.5wt%, i.e. the bronze is relatively hard. The lack of any great variation of diffraction intensities indicate a fairly statistical distribution of the Cu/Sn grains with some characteristics of (110) alignment parallel to the helmet wall, typical of weak texture, resulting from a hammering process perpendicular to the surface.

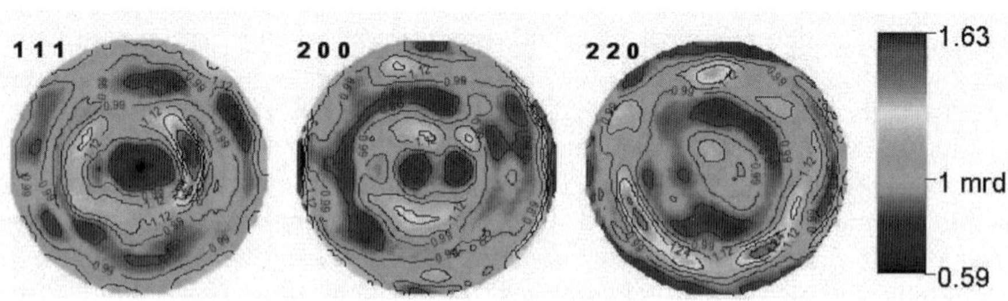


**Fig. 10** (111), (200), and (220) pole figures from neutron diffraction of scan 1 (top) and scan 2 (bottom). The pole figures indicate a non-statistical distribution of Cu/Sn grains due to plastic deformation of the alloy with characteristic pole density maxima in the centre of (220). mrd stands for 'multiples of a random distribution'. mrd=1 marks the average pole density for a random distribution of grains. Reproduced from figure 5 of reference [3].

## 4 What Does Homer Say about Helmets?

Homer is quite informative on weaponry, as one might expect, particularly in the Iliad. In many respects, he is the only source of knowledge about technology in an era for which historical records and archaeological evidence are rather incomplete.

**Table 1** Verses from the Odyssey referring to helmets and their attributes. English translation translation from [8].

| ΟΔΥΣΣΕΙΑ | ODYSSEY |
|---|---|
| τ32 ἐσφόρεον κόρυθάς τε καὶ ἀσπίδας ὀμφαλοέσσας<br>τ33 ἔγχεά τ' ὀξυόεντα | The plumy-crested helms and pointed spears,<br>With shields indented deep in glorious wars. |
| φ432 Τηλέμαχος, φίλος υἱὸς Ὀδυσσῆος θείοιο,<br>φ433 ἀμφὶ δὲ χεῖρα φίλην βάλεν ἔγχεϊ, ἄγχι δ' ἄρ' αὐτοῦ<br>φ434 πὰρ θρόνον ἑστήκει κεκορυθμένος αἴθοπι χαλκῷ. | He said, then gave a nod; and at the word<br>Telemachus girds on his shining sword.<br>Fast by his father's side he takes his stand:<br>The beamy javelin lightens in his hand. |
| χ122 αὐτὸς δ' ἀμφ' ὤμοισι σάκος θέτο τετραθέλυμνον,<br>χ123 κρατὶ δ' ἐπ' ἰφθίμῳ κυνέην εὔτυκτον ἔθηκεν,<br>χ124 ἵππουριν, δεινὸν δὲ λόφος καθύπερθεν ἔνευεν<br>χ125 εἵλετο δ' ἄλκιμα δοῦρε δύω κεκορυθμένα χαλκῷ. | And now his shoulders bear the massy shield,<br>And now his hands two beamy javelins wield:<br>He frowns beneath his nodding plume, that play'd<br>O'er the high crest, and cast a dreadful shade. |
| ω522 αἶψα μάλ' ἀμπεπαλὼν προΐει δολιχόσκιον ἔγχος,<br>ω523 καὶ βάλεν Εὐπείθεα κόρυθος διὰ χαλκοπαρῄου.<br>ω524 ἡ δ' οὐκ ἔγχος ἔρυτο, διαπρὸ δὲ εἴσατο χαλκός,<br>ω525 δούπησεν δὲ πεσών, ἀράβησε δὲ τεύχε' ἐπ' αὐτῷ. | Full at Eupithes drove the deathful spear:<br>The brass-cheek'd helmet opens to the wound;<br>He falls, earth thunders, and his arms resound. |

Apart from the Homeric epics our best source of information is the artistic record, as expressed in literature, sculpture and scenes on pottery [4, 5].

In several translations one soon notices the lack of consistency in the translation of the word κόρυς (helmet) and of the adjective that often accompanies the word χαλκός (copper or bronze). The reasons are very understandable: Translators, wonderful and knowledgeable scholars as they may be, concentrate their attention to poetic expression and linguistic style rather than scholastic details to technological semantics. This happens in all the translations with few exceptions. It is however possible that significant information may be lost or misunderstood. For instance, in several English translations consulted the word χαλκός is sometimes translated as brass, an alloy of copper and zinc, instead of the correct name for that historical period, bronze, which is an alloy of copper and tin. In the Homeric language the word χαλκός is used either for just pure copper or for bronze. This ambiguity remains in the modern Greek language where the word ορείχαλκος (mountain copper) is often used erroneously for either bronze or bras.

A more general problem of the faithful and accurate rendering of Homeric words is that often words that carry significant information, technological, historical or geographical, are either paraphrased or missed out altogether and new ones inserted to suit the translator's poetic licence. Homer remains very consistent, particularly in the use of adjectives.

Κόρυς is the usual word for helmet in the Homeric epics [6]. Κυνέη is the Homeric word for a soldier's cap or helmet made of leather or weasel's skin, also of metal stiffened or adorned with metal, χαλκήρης also fitted with metal plates to protect the cheeks. Sometimes entirely of bronze. Figure 11 shows an example of a Κυνέη very similar to what is described in Il. K260–265.

**Table 2** Verses from the Iliad referring to helmets and their attributes. English translation translation from [7]. There are 109 references to the word κορυς in different forms. The most common phrase is κορυθαίολος (39), followed by κεκορυθμένος αίθοπι χαλκώι (8). We list here only a few of the verses. Notice that this translation (but often other, more recent translations) often miss the direct translation of the word κόρυς or mistranslate the word for bronze, χαλκός, as brass.

| ΙΛΙΑΣ | **ILIAD** |
|---|---|
| B816 Τρωσὶ μὲν ἡγεμόνευε μέγας **κορυθαίολος** Ἕκτωρ<br>B816 Πριαμίδης· ἅμα τῷ γε πολὺ πλεῖστοι καὶ ἄριστοι<br>B816 λαοὶ θωρήσσοντο μεμαότες ἐγχείῃσι. | The **mighty** Hector, son of Priam, fierce in fight, commander of the Trojans, with whom marched the most numerous and bravest troops, skilful with the spear. |
| Γ18 καὶ ξίφος· αὐτὰρ δοῦρε δύω **κεκορυθμένα χαλκῷ**<br>Γ19 πάλλων Ἀργείων προκαλίζετο πάντας ἀρίστους<br>Γ20 ἀντίβιον μαχέσασθαι ἐν αἰνῇ δηϊοτῆτι. | And, shaking two **brazen-pointed** spears, he challenged all the bravest of the Greeks to fight him in horrid combat. |
| Γ361 Ἀτρεΐδης δὲ ἐρυσσάμενος ξίφος ἀργυρόηλον<br>Γ361 πλῆξεν ἀνασχόμενος **κόρυθος φάλον·** | Then the son of Atreus, drawing his silver-studded sword, raised it, and smote the **cone of his helmet**. |
| Γ369 Ἠ καὶ ἐπαΐξας **κόρυθος** λάβεν ἱπποδασείης | seized him by the **helmet** crested with horse-hair |
| Δ495 βῆ δὲ διὰ προμάχων **κεκορυθμένος αἴθοπι χαλκῷ** | and advanced among the foremost of the combatants, **armed in glittering brass** |
| E4 δαῖέ οἱ ἐκ **κόρυθός** τε καὶ ἀσπίδος ἀκάματον πῦρ<br>E5 ἀστέρ' ὀπωρινῷ ἐναλίγκιον, ὅς τε μάλιστα<br>E6 λαμπρὸν παμφαίνησι λελουμένος ὠκεανοῖο | She made unwearied fire blaze from his **helmet** and his shield, like to an autumnal star, which shines the greatest sp[lendour when washed by the ocean. |
| H206 ὣς ἄρ᾽ ἔφαν, Αἴας δὲ **κορύσσετο νώροπι χαλκῷ.** | but Ajax **armed himself in splendid brass.** |
| Λ43 εἵλετο δ᾽ ἄλκιμα δοῦρε δύω **κεκορυθμένα χαλκῷ** | He likewise took two tough spears, **headed with brass**, and sharp |
| Λ351 ἄκρην κὰκ κόρυθα· πλάγχθη δ᾽ ἀπὸ χαλκόφι χαλκός | But the spear was shaken off by the brass |
| M182 ἔνθ᾽ αὖ Πειριθόου υἱὸς κρατερὸς Πολυποίτης<br>M183 δουρὶ βάλεν Δάμασον **κυνέης διὰ χαλκοπαρῄου·**<br>M184 οὐδ᾽ ἄρα **χαλκείη κόρυς** ἔσχεθεν, ἀλλὰ διὰ<br>M185 αἰχμὴ χαλκείη ῥῆξ᾽ ὀστέον | Then the son of Pirithous, the brave Polypoetes, smote Damasus with his spear, through the **braze-cheeked helmet**; nor did the **brazen casque** resist, but the brazen point burst quite through the bone |
| N130 φράξαντες δόρυ δουρί, σάκος σάκεϊ προθελύμνῳ·<br>N131 ἀσπὶς ἄρ᾽ ἀσπίδ᾽ ἔρειδε, **κόρυς κόρυν**, ἀνέρα δ᾽ ἀνήρ·<br>N132 ψαῦον δ᾽ ἱππόκομοι **κόρυθες** λαμπροῖσι φάλοισι<br>N133 νευόντων, ὡς πυκνοὶ ἐφέστασαν ἀλλήλοισιν·<br>N134 ἔγχεα δ᾽ ἐπτύσσοντο θρασειάων ἀπὸ χειρῶν<br>N135 σειόμεν᾽· οἳ δ᾽ ἰθὺς φρόνεον, μέμασαν δὲ μάχεσθαι. | joining spear to spear, shield to shield, inclosest order, so that shield supportd shield, **helmet helmet**, and man man. Their **helmets**, crested with horse-hair. touched each other with their splendid cones as they nodded, so close they stood to one another; |
| N304 τοῖοι Μηριόνης τε καὶ Ἰδομενεὺς ἀγοὶ ἀνδρῶν<br>N305 ἤϊσαν ἐς πόλεμον **κεκορυθμένοι αἴθοπι χαλκῷ** | So advanced to battle Meriones and Idomeneus, leaders of heroes, armed in **glittering brass**. |
| Π130 ὣς φάτο, Πάτροκλος δὲ **κορύσσετο νώροπι χαλκῷ** | This he spake ; and Patroclus **armed himself in glittering brass** |
| P3 βῆ δὲ διὰ προμάχων **κεκορυθμένος αἴθοπι χαλκῷ** | but he advanced through the foremost combatants, **armed in shining brass** |
| Y117 Αἰνείας ὅδ᾽ ἔβη **κεκορυθμένος αἴθοπι χαλκῷ** | This Aeneas hath advanced against Pelides, **armed in glittering bras** |

Il. K260-265
Μηριόνης δ᾽ Ὀδυσῆϊ δίδου βιὸν ἠδὲ φαρέτρην
καὶ ξίφος, ἀμφὶ δέ οἱ κυνέην κεφαλῆφιν
ἔθηκε
ῥινοῦ ποιητήν· πολέσιν δ᾽ ἔντοσθεν ἱμᾶσιν
ἐντέτατο στερεῶς· ἔκτοσθε δὲ λευκοὶ ὀδόντες
ἀργιόδοντος ὑὸς θαμέες ἔχον ἔνθα καὶ ἔνθα
εὖ καὶ ἐπισταμένως· μέσσῃ δ᾽ ἐνὶ πῖλος
ἀρήρει.

And Meriones gave a bow, quiver, and sword, to Ulysses, and put upon his head a helmet made of skin; and within, it was firmly fastened with many thongs; whilst without, the white teeth of an *ivory-tusked boar* in rows on either side covered it well, and skilfully; and in the middle wool was sewed. (English translation translation from [7])

**Fig. 11** Boar's tusk helmet and armour.

## 5 Threads to Follow

To our knowledge this work, the first of its kind, has brought to light several other issues which need to be followed up.

*Other helmets in the U.K., Greece and elsewhere*: A much larger project is in progress at Cardiff University Conservation Department following preliminary results from ISIS and the SRS [9] to study the corrosion, surface finish and technology of over 100 ancient Greek Helmets. Initial results have been obtained on ancient Greek helmets from both U.K. museums and micro-samples from ancient Olympia itself, involving amongst other techniques the use of neutron and synchrotron methods to study helmets from British museum collections and micro-samples from Greece.

**Fig. 12** Miniature bronze group of an armourer working on a helmet (8th century BC). From *Ancient Greece, A Concise History*, Peter Green, Thames and Hudson, 1973.

*Closer look at black spots on helmets*: Are they simply due to corrosion during burial or was there an intentional black patination? Homer repeatedly uses the stock phrase κεκορυθμένος αίθοπι χαλκώι which could be translated as *wearing a dark-red bronze* (or *copper*?) *helmet* (or does αίθοψ mean *black*?), just as he calls the wine given to the Cyclops, αίθωπα οίνον – *dark red* (or *black*?) wine. According to mythology, Αίθωψ was a son of Hephaistos who became king of the Ethiopians. What kind of dark-red or black bronze/copper is Homer talking about?

*Bronze patination issues*: Polychromy on bronze is hinted at by Homer and portrayed in pottery with Homeric scenes. Whether this is the poet's or the potter's poetic licence is an open question. Of particular interest, scientifically, is the possibility, just a possibility, that special helmets for special leaders (Achilles, Hector, Agamemnon, etc.) were patinated in kyanos.

*Bronze provenance issues*: This is a really thorny issue that has been around for quite some time. Bronze recasting was fairly wide-spread in antiquity, as it still is now in some societies. This makes it a thankless task trying to provenance bronze objects from trace elements or lead isotopes. But the approach may work for objects that were used once. Where are such objects? Can modern science provide evidence for bronze objects having been recast? The answer is, probably yes. Some experimentation is needed.

*Making a helmet the Corinthian way*: This really would be an experiment that is worth doing. Not quite the way depicted in Figure 12, safety regulations, not to mention laws of decency, may limit somewhat the fidelity of the reconstruction, but it is worth doing it.

There's work to be done.

## 6 Conclusions

Neutron and synchrotron X-ray analytical techniques were used to characterize a Corinthian-type bronze helmet in the Manchester Museum. The alloy of the helmet consists predominantly of Cu with a varying Sn content between 11 and 12 wt%. The neutron data contain clear indications of the working processes involved in the production of the helmet. The observed degree of microstrain broadening hints to repetitive annealing-hammering working cycles in order to harden the alloy. The last step was most likely a hardening step. The preferred orientation of grains, as displayed in the texture maps, agrees with hammering in one direction. The object was more than likely cast as a 'skull-cap', then beaten and heated in an iterative cycle and dressed down to its final thickness and shape to fit the customer's head, allowing for padding, probably in the form of a felt or leather cap sewn or more likely riveted to the inside of the helmet [10]. Considerable effort was undertaken by the makers of the helmet to harden the alloy. Hence, the object was surely produced for battle rather than just cast for ceremonial purposes.

The nose-guard of the helmet is made of a different alloy, namely Cu/Zn. One can assume that the nose-guard, being made of a different material, is most likely not part of the original helmet but is a later replacement, maybe by the 19th-century finder of the object. This confirms earlier suggestions [1] that the shape of the nose-guard is unusual and that the present angle at which it is set is not functional and therefore not authentic, and that the edges of the nose-guard itself and of the holes for fixing the lining are much sharper than on the rest of the helmet.

Our results are in agreement with the existing knowledge in the field of ancient bronze working. With the exception of the XRF data that could, in principle, have been obtained with conventional equipment, we have determined material properties of the alloy that cannot be simply obtained by other methods or with a portable system. This is certainly the case for the crystallographic texture of the alloy and for the microstrain broadening. Quantitative information on material properties such as the texture type, the texture strength and the degree of microstrain broadening is important, even more so if results on other helmets and objects become available in the near future.

## Acknowledgements

We wish to thank CCLRC for beamtime grants to use ISIS and the SRS. Particular thanks are due to our colleagues at Daresbury, Glenys McBain and Christine Ramsdale for the software and Alan Porter and Joe Salvini for the xyz-stage on 9.4. At RAL, John Dreyer and Chris Goodway for their support on GEM at ISIS. Many others, including management, also contributed in so many ways to the successful completion of this project and the dissemination of the outcomes.

## References


1. Jackson, A.H. (2004). An early Corinthian helmet, *Annual of the British School at Athens* **99**, 273–282.
2. Salvadó, N., Butí, S., Tobin, M.J., Pantos, E., Prag, A.J.N.W. and Pradell, T. (2005). Advantages of the use of SR-FTIR microspectroscopy: Applications to cultural heritage, *Anal. Chem.* **77**(11), 3444–3451.
3. Pantos, E., Kockelmann, W., Chapon, L.C., Lutterotti, L., Bennet, S.L., Tobin, M.J., Mosselmans, J.F.W., Pradell, T., Salvadó, N., Butí, S., Garner, R. and Prag, A.J.N.W. (2005). Neutron and X-ray characterisation of the metallurgical properties of a 7th-century BC Corinthian-type bronze helmet, *Nuclear Instruments and Methods B* **239**, 16–26.
4. Snodgrass, A.M. (1998). *Homer and the Artists: Text and Picture in Early Greek Art*, Cambridge University Press, Cambridge.
5. Snodgrass, A.M. (1967). *Arms and Armours of the Greeks*, Thames & Hudson, London.
6. Autenrieth, G. (2004). *Homeric Dictionary*, Translated by Robert Keep, Duckworth, London.
7. *Iliad of Homer*, by "A Graduate of the University of Oxford", 2nd Edn., Whitaker, London, 1825. (First published in 1821.)
8. Pope, A. (1903). *The Odyssey of Homer*, Grant Richards, London. First published in five volumes in the year 1725. See also http://etext.library.adelaide.edu.au/h/homer/h8op/ (last accessed 14/10/06).
9. Manti, P. and Watkinson, D. (2008). From Homer to Hoplite: Scientific investigations of Greek copper alloy helmets, in *Proceedings International Symposium on Science and Technology in the Homeric Epics*, Olympia, Greece, August 27–30, 2006, S.A. Paipetis (Ed.), Springer, Dordrecht (this volume).
10. Swaddling, J. (1987). An unusual Greek bronze helmet, *Antiquaries Journal* **LXVII**, Part II, 348–351.

# Part 5

# TELECOMMUNICATIONS

# Theoretical Analysis of Telecommunication through "Friktories"

Nikolaos Uzunoglu

*National Technical University of Athens, Greece*

**Abstract.** The use of long distance communication links during the siege of Troy has been reported in historical times and in particular by Aeschylus in his tragedy *Agamemnon*. The analysis of the possibility of such long distance communication links using fire towers and relaying of information is examined using electromagnetic theory principles in this paper. Furthermore possible communication protocols employed in this type of communication network is reviewed taking into account similar systems used even today incorporating very low rate data transmission links.

## 1 Introduction

In the *Iliad*, the use of optical communication means by the Trojans is indicated (Il. 19.209–213), while asking for support from neighboring tribes during the siege of Troy. More concrete information of the use of a long distance optical communication link between the Hellenic expedition army and Mycenae is mentioned by Aeschylus in his tragedy *Agamemnon* (written 300 years after Homer). According to this account, such a link has been used to announce the fall of Troy to Mycenae within one hour only. That link applied the principle of cascaded "Friktories", i.e. Fire Towers, starting from Ida mountain on Asia Minor coast, opposite to Lesvos island, using consecutive transmissions through Lemnos island-Mount Athos and several mountain-tops along the eastern coast of the Greek mainland, to reach Mycenae. The use of similar "Fire Tower" technology in latter times, during Ancient Hellenic and Byzantium Empire, confirms the use of this technology. In the present work:

(a) The electromagnetic signal propagation at visible spectrum on the Aegean Sea and the above signal path are analyzed. The analysis is based on human eye signal detection sensitivity and also on signal power available as generated by a fire source. The variability of propagation because of troposphere conditions is also considered, as well as propagation of visible spectrum during normal and ducting troposphere conditions.

*S.A. Paipetis (ed.), Science and Technology in Homeric Epics,* 223–229.

(b) The possible coding techniques used by this Optical Communication System is examined, considering that was not only used to inform the Mycenaeans of the fall of Troy, but also at latter times, as, for example, the code of Cleoxenes-Demokletus (4th century BC). The transmission of the news from Troy to Mycenae within approximately one hour is considered, in order to evaluate the coding technique used. The use of prehistoric Linear B alphabet in coding the information transmitted is taken into account.

## 2 Electromagnetic Analysis of a Troy-Mycenae Communication Link

Optics is a very ancient science, evolved in parallel with Astronomy. Geometry the "miracle" of Ancient Greek science, making use of thousands of years old observations of Mesopotamia and Egypt, was highly developed in association with observation of optical phenomena. The science of Optics itself is one of the most fascinating stories of scientific endeavors of mankind. In 1874 J.C. Maxwell, one of the most brilliant scientists of 19th century, provided a complete explanation of electromagnetic wave phenomena by proving the electromagnetic nature of optical phenomena. However, in 1900, a scientific scandal erupted, while he was trying to explain how a heated body emitted radiation. The problem was named "black body radiation". This lead into the discovery of Quantum Nature of energy exchange between matter and electromagnetic radiation fields by Max Planck in the year 1900. The basics of Quantum Theory of light is based on the principle that exchange of energy between material media and radiation occurs in finite amounts of energy known as *quanta* or *photons*, equal to $E = h\nu$, where $h = 6.628 \times 10^{-34}$ Joule/sec, and $\nu$ is radiation frequency.

In our case the signal source is fire made by burning wood. This means that a black body used was with a spectral peak at the red light of average wavelength of 700 nm. The receiver is a naked human eye, with very high sensitivity, comparable to that of very sensitive electromagnetic receivers we use today, provided that no environmental optical noise is present or it is very small, occurring when no other optical sources near the observer exist.

As with modern communication systems, the analysis must start from the receiver. The important parameter is the sensitivity of human eye. Recent physiological studies have shown [1–3] that the human eye, although its cells are capable of detecting even a single photon, because of brain integrative processing, in order to sense incident radiation, the human eye needs least 80–120 photons within 100 ms time. Of course, a human eye capable of detecting single photons, would be a great nuisance, as it would have to observe a continuous noise signal. Since, in our case, the human observer would have to look continuously at a distant fire, the observation would take several seconds if not minutes. Therefore, the number of photons needed to detect a distant fire source, is approximately 1000 photons/sec. Accordingly, one must examine, if and at what distance a human observer can sense a fire source.

Let it be assumed that the fire source corresponds to a 10 kW thermal source, which is a realistic assumption. Then, by considering the isotropic radiation of energy at a distance $R$, the power flux, taking into account the atmospheric absorption because of air gaseous particles, one yields the basic equation giving the energy flux incident on the human eye:

$$P_{\text{received}} = P_{\text{emitted}} \times \exp(-aR) \times A_p/(4 \times \pi \times R^2), \quad (1)$$

where $R$ is the distance, $\alpha$ the atmospheric absorption coefficient and $A_p$ the human eye pupil area. It should be emphasized that $\exp(-\alpha R)$ plays a decisive role, as shown latter, and depends on the visibility conditions of the atmosphere. Then the number of photons incident on the human eye per second is computed by

$$N = P_{\text{received}}/(h\nu). \quad (2)$$

However the above equation is valid as long as fire source and eye "see" each other or otherwise both are within the "common geometrical horizon". Assuming that the levels of fire source and observer above sea surface are $h_1$ and $h_2$ respectively and by taking into account the spherical shape of the earth, the "common geometrical horizon" is found to correspond to a distance on earth surface equal to:

$$R_{\text{comm. horizon}}\ (\text{km}) = 3.1\left(\sqrt{h_i\ (\text{m})} + \sqrt{h_2\ (\text{m})}\right). \quad (3)$$

Therefore, a successful information transfer from a fire source transmitter to a human observer receiver requires that both conditions $N > 1000$ photons/sec and $R_{\text{comm. horizon}} > R$ be simultaneously satisfied.

Let us now turn our attention to the historical account given primarily by Aeschylus in *Agamemnon*, which was written 300 years after Homer and 700 years after the fall of Troy. There, the message of the fall of Troy is reported to be sent to Mycenae following a path through Fire Towers as shown in Figure 1. In Table 1, the positions of Fire Towers are given showing the level of corresponding stations and the distance between adjacent positions.

It is evident that the longest path is between Mt. Athos and Mt. Kandilion on the island of Euboea, equal to 177 km. As far as the common geometrical horizon of the two Fire Towers is concerned, one finds that $R_{\text{comm. horizon}} = 247$ km, i.e. both stations can see each other. The validity of $R_{\text{comm. horizon}} > R$ has been checked in all of the cases of Table 1.

Now, the number of photons incident to the observer's eye, using Equations (1) and (2), must be computed. To this end, the following assumptions are necessary:

$P_{\text{emitted}} = 1000$ Watt (emitted power from fire source)
$\lambda = 0.70\ \mu\text{m}$, average wavelength of fire source
pupil diameter of human eye $= 4$ mm.

**Fig. 1** Map showing the communication link between Troy and Mycenae [9].

**Table 1** Fire Tower positions of the Troy-Mycenae communication link.

| Location | Modern name | Modern name altitude (m) | Distance (km) |
|---|---|---|---|
| Troy | Troy | 100 | 0 |
| Mt. Ida | Kaz Dagi | 1774 | 55 |
| Lemnos | Skopia at Limnos | 430 | 154 |
| Mt. Athos | Athos | 2033 | 70 |
| Macistus | Kandilion at Euboea | 1209 | 177 |
| Messapius | Ktipas | 1020 | 30 |
| Cithaeron | Elatias | 1410 | 25 |
| Mt. Aegiplanetus | Mt. Jeraneia | 1370 | 30 |
| Arachnaean | Hgt. Arna | 1199 | 50 |
| Mycenae | Mycenae | 150 | 20 |

Introducing these into Equations (1) and (2), the number of photons received by the observer at a distance $R$ from the fire source is found by

$$N = 3.5 \times 10^9 \exp(-\alpha R)/R^2, \tag{4}$$

where the distance $R$ is expressed in kilometers.

Coefficient $\alpha$ is determined by the Kruse algorithm [4]:

$$a = \frac{3.912}{V}\left(\frac{\lambda_{nm}}{550}\right)^{-q},$$

$$q = 1.6 \text{ for } V > 50 \text{ km},$$

$$q = 1.3 \text{ for } 6\text{ km} < V < 50\text{ km},$$

where coefficient $V$ expresses the meteorological visibility conditions of the atmosphere. With $V > 50$ km excellent visibility conditions prevail, very common in the Aegean Sea, while 15 km $< V <$ 40 km corresponds to medium visibility and $V < 6$ km corresponds to low visibility.

With $R = 50$ km (i.e. the link between Arna-Gerania mountain) the following results are obtained:

- High visibility conditions, $\alpha = 0.0532$ Np/km, $R = 50$ km, $N = 98.000$ photons/sec.
- Medium visibility conditions, $\alpha = 0.0572$ Np/km, $R = 50$ km, $N = 80.100$ photons/sec.
- Low visibility conditions, $\alpha = 1.06$ Np/km, $R = 50$ km, $N = 0$ photons/sec.
- Clearly with good visibility conditions, communication for 50 km is more than feasible.

In case of the longest path, between the mountains Athos-Kandylion, the results are as follows:

- High visibility conditions, $\alpha = 0.0532$ Np/km, $R = 177$ km, $N = 9$ photons/sec.
- Medium visibility conditions, $\alpha = 0.0572$ Np/km, $R = 177$ km, $N = 1.4$ photons/sec.
- Low visibility conditions, $\alpha = 1.06$ Np/km, $R = 177$ km, $N = 0$ photons/sec.

It seems that the Athos-Kandylion link communication is questionable. To make it possible, the fire source at Athos should be increased to 10 kW (100 kW), leading to a number of photons/sec received $N = 90$ (900).

The above analysis clearly shows that the communication link between Troy and Mycenae was feasible, in terms of signal transmission under good visibility conditions. This result is important, since the possibility of such a link has been argued by several researchers in the past, such as Henning [5], while Darmstaeder [6] suggested that a link even up to 240 km is possible under the very high visibility conditions in Greece. Furthermore, Stamatis [7], one of the pioneers in ancient Greek Technology research, claimed that an in-between Fire Tower station on one of the Sporades Islands could have been used, which Aeschylus, in his poetic approach, fully neglected. However, the present analysis shows the clear possibility of achieving an extraordinarily long distance of approximately 180 km between Mt. Athos and Mt. Kandylion in Euboea.

## 3 Communication Analysis of Troy-Mycenae Link

Reviewing possible communication protocols and based on historical accounts, one can state the following.

It is known that torches were used in signaling at short distances in the siege of Troy. A sort of IFF (interrogation of Friend or Foe) based on waving/non-waving of torches was used by the defenders of Troy. Such a technique is not applicable with the Troy-Mycenae link.

The Achaean Army in Tenedos Island received the information to return by a torch signal of Odysseus (10 km). This was a predefined signal.

In the case of Troy-Mycenae Communication link, at each regeneration node, the observer(s) should wait to receive the message and then light up the fire to relay the signal. This is necessary since eye sensitivity works only in absolute darkness at the vicinity of the receiver. Moon or star light does not affect sensitivity. Therefore, the 1 hour transmission time mentioned by Aeschylus is reasonable.

In communication systems of this type, signal levels are very low and one has to resort to predefined signal protocols. An example is the Extremely Low Frequency (ELF) [8] submarine communication link of USA/USSR during Cold War. The fact that Clytemnestra was expecting the signal, makes this argument possible.

In the case of Troy-Mycenae, possible signaling mechanisms could be argued, such as:

- Fire duration (short/medium/long) could be used to code signals.
- A sort of modulation (obstructing) the fire radiation towards the observer at the fire point could be used with the help of a copper shield or even a wooden table. This type of "modulation technique" could provide an efficient transmission system.

One has to remember that during the Troy expedition, an alphabet was known to the civilizations of Greece and Anatolia. In Mycenae, Linear B script was used, while the Hittites also had an alphabet similar to the Mesopotamian one. Both alphabets have been decoded. Therefore the basic concept of coding with a finite set of symbols was very well known and this means that it was not difficult to develop a linear coding method using 2 or 3 symbols (short, medium and long fire duration).

## 4 Conclusions

The present study leads to the following main conclusions:

- Analysis of Electromagnetic Radiation Propagation (Fire Source Emitted) in in the atmosphere shows that the possibility of a Troy-Mycenae Communication Link is feasible. Historical Accounts are verified by using Electromagnetic Theory.
- Predefined final set message technique is the most probable communication protocol used in this multi-hop, very-long-distance communication link.

## References

1. Schnapf, J., How photoreceptors respond to light, *Scientific American*, April, 1987.
2. Hecht, S., Schlaer, S. and Pirenne, M.H., Energy, quanta and vision, *Journal of the Optical Society of America* **38**, 1942, 196–208.
3. Baylor, D.A., Lamb, T.D. and Yau, K.W., Response of retinal rods to single photons, *Journal of Physiology, Lond.* **288**, 1979, 613–634.
4. Kruse, P.W., et al., *Elements of Infrared Technology: Generation, Transmission and Detection*, John Wiley and Sons, New York, 1962.
5. Henning, R., *Die älteste Entwickhung der Telegraphie und Telephonie*, Leipzig, 1908, p. 87.
6. Darmstaeder, E., Feur-Telegraphie im Alterum, *Die Umschau* **28**, 1924, 505–507.
7. Stamatis, E., *The Secret Telecommunications of Ancient Greece*, Athens, Greece, 1969 [in Greek].
8. Bansal, R., ELF communications: An obituary, *IEEE Antennas Propagation Magazine* **46**(6), 124, December 2004.
9. Lazos, C., *Telecommunications of Ancient Greeks*, Aiolos Publ., Athens, 1997 [in Greek].

# Part 6

# GEOLOGY – GEOMECHANICS

# Elements of Engineering Geology and Geotechnical Engineering in the Homeric Poems

Dimitrios Zekkos[1], George Athanasopoulos[2], Adda Athanasopoulos Zekkos[3] and Ioannis Manousakis[4]

[1]*GeoSyntec Consultants, U.S.A.*
[2]*University of Patras, Greece*
[3]*University of California at Berkeley, U.S.A.*
[4]*Dromos Consulting, Greece*

**Abstract.** This paper examines the information provided by Homer related to the fields of engineering geology and geotechnical engineering. The information is subsequently compared against archaeological findings. The Homeric descriptions of battle scenes, of ancient cities and the geologic environment suggest that an understanding of some of the geologic processes already existed at the time of Homer, probably as a result of observations of the geologic environment. A study of the geologic descriptors reveals that Homer was particularly careful in the selection of the words he used in his accounts. The poems also provide some insight into issues related to geotechnical engineering. Homer describes a deep excavation constructed by the Achaeans for defensive purposes that failed during the Trojan war. Based on the Homeric characterizations, the defensive excavation appears to have been an interesting and impressive technical project. In addition, a variety of engineering projects are mentioned, such as roads, embankments, and harbors. Of particular interest are also Homer's comments on the causes of various failures. The information provided on topics related to the field of engineering geology and geotechnical engineering is remarkable given the scope of the poems.

## 1 Introduction

The Homeric poems have been a source of inspiration since the ancient years: Virgil was inspired by Homer to write the Aeneid; Alexander the Great used to sleep with a copy of the Iliad next to him; and the historian Diodorus of Sicily refers with great respect to Homer and considers him the oldest and most celebrated poet. Diodorus also considered Homer and Hesiod the most distinguished of all men. Themes from Homer's poems have inspired numerous works of art from the ancient years until present. In addition to their literary value, the Homeric poems are also important because they refer to a period of time for which very limited written information is preserved, the Mycenaean Civilization. The Homeric poems are probably the most important source of information regarding the values, traditions and habits of the Mycenaeans. However, it is estimated that the Trojan war took place sometime early in the 12th century BC while Homer lived about 400 years later (8th century BC), so a recurrent problem when collecting information from these poems is whether Homer refers to facts of the Mycenaean Civilization or is influenced by his own

*S.A. Paipetis (ed.), Science and Technology in Homeric Epics,* 233–242.

time. However, in general, much of the information provided by Homer on the Mycenaean civilization has been consistent with archaeological evidence. Schliemann used Homer's poems to identify the location of the city of Troy and unearth his findings in Mycenae and Tiryns [1, 2]. Of particular interest is the information provided by Homer on topics related to science and technology, such as physics, hydraulics, psychology, naval engineering, astronomy, civil engineering, and military science. These have been the topic of study by many researchers. In particular, this paper examines information provided by Homer in the fields of engineering geology and geotechnical engineering.

## 2 Geological Descriptions

When Homer refers to various cities or regions, he commonly uses geologic descriptors. For example, in the Iliad, the poet refers to sandy Pylos (IX, 153), the canyons of mountains Olympus (XI, 76–77) and Parnassus (XIX, 431–432), the many springs of Ida (XX, 218), and the wide and deep river Axios (XXI, 141–142). Similarly, in Odyssey, the poet refers to sandy Sparta and Pylos (I, 93), to the many canyons of Sparta (IV, 1), "εριβόλαξ" Scheria (V, 34), rocky Ithaca (IX, 27, X, 416–417), rugged Ithaca (X, 462–463, XIII, 242–247, O510). The word "εριβόλαξ" (i.e. soil with clodds, very fertile) is repeatedly used to characterize rich, flourishing cities or places. It is used for Askania (Iliad XIII, 793), Lycia (Iliad XVII, 172), Scheria (Odyssey V, 34), Tarni (Iliad V, 44), twice for Paionia (Iliad XXI, 154 & XVII, 350), twice for Thrace (Iliad XI, 222, XX, 485), twice for the city of Larisa (Ilias II, 841, XVII, 301) three times for the city of Achilles, Phthia (Iliad I, 155, IX, 362, IX, 478) and seven times for the city of Troy (Iliad III, 74, III, 257, VI, 314, IX, 329, XVI, 460, XVIII 67, XXIV, 86).

On the other hand, the word "κραναός" which is translated to "rugged" or "rocky" [3–5] is used a total of 5 times in the Iliad (III, 201) and Odyssey (I, 247, XV, 510, XVI, 124, XXI, 346). In all occurrences, it is used to describe Ithaca. It is not used to describe any other city or place. This epithet is also one of the common descriptors used for Ithaca. The word "παιπαλόεις" (i.e. craggy, rugged) is used also for Ithaca (Odyssey XI, 480) but also for the islands of Sami (Odyssey, XV, 29, IV, 670, IV, 845), Chios (Odyssey III, 170), for a mountain in Samothrace (Iliad XIII, 17), and Imvros (Iliad XIII, 33, XIV, 280, XXIV, 78). In fact, the island of Imvros is only mentioned four times in the two poems, of which only two times a descriptor is used and in both cases the same epithet is used. From the above observations, it becomes apparent that Homer is consistent in his geologic descriptions in both poems and is also very careful in the selection of the attributed epithets, which are also accurate descriptions of the geologic environments of the places mentioned.

## 3 Geologic Processes

The Homeric descriptions also suggest the existence of the first elements of knowledge of various geologic processes by Homer's time. This knowledge was likely the result of observations of the geologic environment. For pre-industrial societies, observing the environment was an integral part of the survival and continued growth of the society as ancient economies depended primarily on agriculture and sea commerce. Some of the geologic information provided by Homer is reviewed in the following sections.

### *3.1 Riverine Environment and Deposits*

Homer's texts refer on several occasions to the riverine geologic environment. Simple observations such as that rivers form in the mountains and empty in the seas were certainly made early on in the history of mankind and are mentioned in the Iliad. Of particular interest is the description in the Iliad where Skamander river becomes "infuriated" by the number of dead bodies that Achilles throws in the river's waters during the battle. The river threatens the hero that if he continues his action, he will be covered first with a layer of sand, then with a layer of gravel and finally with a layer of mud. The poet refers to three different soil layers, all of which are typical riverine deposits.

In another example (Iliad, XII, 10–33) the poet refers to the destruction of the defensive wall of the Achaeans after the end of the war by the combined forces of Poseidon, Apollo and Zeus. Homer describes in great detail that the courses of eight rivers were changed so that their combined flows could destroy the walls, while at the same time Zeus arranged for rain to support the effort. Homer describes that after the flood and the destruction of the walls, the entire area was leveled and covered with sand (VII, 459–463). River flooding and intense rainfall were already two correlated events at Homer's time. In another instance (Iliad, XI, 492–495) Homer mentions that rainfall results in flooding of the rivers which transfer wood, soil and bushes up to the coast, while in Iliad XIX, 205–207 Homer describes that as the snow melts, the rivers swell.

Again, it is not surprising for agriculture-dependent societies that these observations were already made. In fact, flooding protection, river course modification, and drainage of lakes to create more land for irrigation were among the first civil engineering infrastructure built. There is ample Mycenaean archaeological evidence of large-scale hydraulic systems that consisted of embankments, tunnels and excavations that allowed the drainage of lakes and the creation of land for agriculture [6–8].

## 3.2 Soil Types

In the numerous occasions that the poet refers to earth materials there is a clear discrimination between sand, soil, gravel and rock. Sand is mentioned repeatedly to describe the coastlines near Troy, as well as the coasts of other areas. The poet refers to the anchoring of the ships on the sand beaches at the shallow waters (Odyssey, IV, 426, IX, 546–547) and differentiates between the sandy coasts (Iliad, IX, 182, Odyssey, III, 39, IV, 432, IX, 74, IX, 552) and the coasts with gravel or pebbles (Odyssey, Z94–95). In the Iliad, the poet describes the fall of Mydon's coachman from the chariot. The warrior falls head first and the poet describes that he was stuck there with his head in the sand for some time because the sand was "deep". Sands are also mentioned as part of the riverine deposits, as discussed previously. In a distinctly different type of soil refers Homer when he talks about "εριβόλαξ" (i.e. soil with clods, very fertile) soils. This epithet is used to describe the wealth of a city as discussed earlier. Achilles' shield also illustrates a freshly cultivated, thick, wide field (Iliad, XVIII, 541–543), while in the Odyssey (IX, 134–135) the productivity of the soils is attributed to their thickness.

## 3.3 Landslides and Erosion

The characteristic steep rock coastlines of many of the Ionian and Aegean islands are used by the poet to characterize strength and endurance against the eroding force of waves (Iliad XV, 619–621, XVI, 34–35, Odyssey III, 293) (Figure 1). In the Odyssey, the poet describes the death of hero Ajax. The steep rock where the hero was standing was broken by Poseidon, ruler of the sea, and the hero was killed (Odyssey IV, 501–510). The Phaeacians are also scared of Poseidon, because, according to an old prophecy, their city would be destroyed by being buried by the large mountain near the city.

Of great interest is another description in the Iliad (XIII, 137–143). The attack of Hector is compared to a rounded rock that detaches from the end of a precipice where it was standing and rolls downhill. The poet explains that the rock detaches due to the intense rainfall and because "the soil that holds the rock breaks due to the plentiful water". The epithet "rounded" is accurately used to describe a rock eroded by water. More importantly, the explanation of the mechanism of rock fall is provided. The cause of a natural phenomenon is identified, without being attributed to a Godly power, as would be expected 300 years before natural philosophers first tried to explain the causes of the various natural phenomena in the history of mankind.

**Fig. 1** View of the steep rock coastline of Corfu, considered to be the island of Phaeacians.

## *3.4 Earthquakes*

Greece has the highest seismic risk in Europe. Earthquakes have shaped human societies in Greece even from the beginning. Over the centuries, Greeks have experienced many earthquakes, some of which resulted in the destruction of civilizations, cities, and the initiation, or interruption of wars [8]. In the Iliad (XV, 177–183) Poseidon mentions that in the early times the sky was awarded to Zeus, the Kingdom of the Dead to Hades and the sea to Poseidon. It was decided that the three God-brothers would share the earth. However, Homer refers to Poseidon as the earth's master (Iliad, XV, 174 & 222, Odyssey VIII, 350). The most common characterization of Poseidon is "world shaker" (Iliad XIII, 10, Odyssey V, 340, VI, 326, VII, 34–36, IX, 518). This characterization is also found in Hesiod [9] and Greek Mythology mentions that Poseidon caused earthquakes. During one of the battles in Troy (Iliad XX, 57–67) Poseidon causes an earthquake. The shaking is so intense that, according to Homer, Hades feared that the earth would split in two and the Kingdom of the Dead (also known as the Under World) would be revealed to the mortals. To the authors' knowledge this is the first time in literature that there is a direct relation of cause and effect between earthquakes and faults. However, it appears that the relation is reversed, with the cause of the earthquake being Poseidon and the fault being the effect of intense shaking not its cause. It is possible that in a highly seismic region such as Greece, the relation between earth ruptures and earthquakes was established already by Homer's time based on post-earthquake observations. Greece's surface is scarred with numerous faults at short distances from populated areas. Of interest also is the fact that the poet, in an effort to provide a sense of the magnitude of the event, mentions that the shaking was strongly felt even to the high mountains of Ida, which are located about 60 km from Troy. Thus, as early as the 9th century BC, distance was used as a measure of the size of an earthquake.

# 4 Geotechnical Engineering

Information on issues related to the field of geotechnical engineering is also provided in the Epics. While the amount of information is not large, it is still significant, considering the epic scope of the poems.

## *4.1 The Trench of the Achaeans*

The Achaeans, in an effort to reinforce their defense and protect their ships, stationed along the coast, constructed a defensive fortification wall and a trench. Achilles, who did not participate in the project because of his disagreement with King Agamemnon, refers with bitterness to this trench specifically, because the Achaeans built "a wide and great trench" (Iliad IX, 348) without his support. Homer describes the trench as deep, wide and large, while he also informs us that the Achaeans drove into the sides of the trench large wooden piles (Iliad VII, 435–441).

During the battle, the horses of the Trojans not only could not jump over the trench, but could not even cross it because the trench sides were vertical. This information is of particular interest because we know that vertical deep trenches in sandy soils, such as the soils that would be expected near the coast, would normally not stand without the necessary support, while vertical trenches of limited depth in more clayey soils would probably stand for only limited time. The poet explains that the Achaeans had placed very large pointed piles, one next to the other in a dense configuration. An interpretation of the Achaean trench based on Homer's description is shown in Figure 1. The trench configuration was such that, as the poet explains, not only the cavalry but also the infantry was wondering how they were going to overcome this obstacle (Iliad XII, 50–59). The piles were certainly a second line of defense, but it is also possible that they intended to support the deep trench and its vertical sides. If the piles were to serve only for defense then there would be no need to place piles on both sides of the trench as Homer suggests they did. During the battle, the trench failed, allowing the Trojans to attack. Homer provides also details of the failure reporting that Apollo kicked the sides of the trench, which fell into its bottom and created a bridge for the enemy to cross. Homer reports that the length of the failed trench was greater than the distance a human can throw

**Fig. 2** Possible interpretation of the Achaean trench.

the spear. Thus, the failure is not an isolated local failure, but a global failure of significant size. Is the intervention of a God a way for Homer to explain an apparently inexplicable failure of a project that has been constructed before or was it the first time that a trench of that scale was constructed? We do not know. The second explanation is however, supported by the epithets attributed to the trench and the expressions of admiration not only of the enemy, but also of Achilles.

## *4.2 Pavement Construction and Road Infrastructure*

There are numerous references of road infrastructure that connected the Mycenaean cities, but also of the road network within each city. Telemachus travels with a chariot from Sparta to Pylos to collect information about the whereabouts of his father, Odysseus. Characteristic is the repeated expression that at the end of the day "the sun sets and all the roads darkened" (Odyssey XV, 182–185). Such references are used for the roads outside the cities of Chalkis and Pylos.

Homer also describes the numerous trails of Ithaca, but again he carefully does not use the same epithets for all the roads. The road that connected Eumaeus' barn to the palace is systematically attributed characteristics such as "τραχείαν" (i.e. rugged) (Odyssey XIV, 2), "αρισφαλές" (i.e. slippery) (Odyssey XVII, 196) and "παιπαλόεις" (i.e. craggy, rugged) (Odyssey XVII, 204–205) while other roads are characterized as crowded with people (Iliad XV, 682). This difference in characterization is not again random.

However, the most interesting remark regarding road infrastructure can be found in the Iliad XXIII, 419–422, where Menelaus and Antilochus are racing with chariots as part of athletic games to honor the death of Patroclos. At some point the road is partly destroyed and becomes narrower and Menelaus warns Antilochus not to attempt to overtake him by exiting the narrow road because he will destroy his chariot. This interesting piece of information suggests that different types of roads with different specifications existed and chariots could only travel on roads that met certain specifications. The poet also explains that the destruction of the road was caused by heavy rainfall that induced soil settlement and damage to the road. The information provided by Homer is consistent with the archaeological findings of the Mycenaean highway infrastructure as presented by Jansen [10]. Jansen performed a survey of the currently preserved Mycenaean highways with particular focus on the road system in the vicinity of the city of Mycenae. He identifies at least four main roads, that he calls highways. These main roads consisted of high terracing walls and bridges in an effort to avoid steep grades and facilitated the transportation of wheeled traffic. Wheel ruts are still preserved in several locations along these roads. However, Jansen did not identify main roads connecting the various Mycenaean states, but it is likely that such roads were continuously used, improved and subsequently overlayed by roads during the classic, the Roman, the Byzantine period and eventually were covered by the present highway infrastructure.

### *4.3 Port Facilities*

Ports are mentioned numerous times in the Odyssey. However, it is not entirely clear what the poet means when he uses the word "port". On some occasions the term is used to describe a natural gulf protected from the waves and winds, while other times it is used to describe an engineered harbor, possibly by fill placement with organized facilities. Homer mentions that Ithaca had wide ports while the island of the Phaeacians does not have ports or other mooring areas except steep rocky coastlines (Figure 1). When Odysseus enters the city of the Phaeacians, he is impressed by the double port and the supporting infrastructure that is used to protect their ships. Next to the port, stones were brought and were embedded in the soil so that the ships' masts and sails can be repaired (Odyssey VI, 262–271). This is a well-organized port of a strong naval power. The Phaeacians mention that their power is based on their fast ships and not in their land army. In another occasion, Eumaeus refers to his city of origin and reports that it has a well-known port and many roads (Odyssey XV, 471–474). Finally, Odysseus reports that the island of Pharos has a protected port and a spring from which ships can collect water prior to embarking (Odyssey IV, 358–360).

### *4.4 Embankments*

In the Iliad (V, 87–92) the attacking fury of the hero Diomedes is compared to the fury of a flooding river that cannot be controlled even by reinforced embankments. Homer refers to the type of projects that Mycenaeans are well known for: The construction of embankments and hydraulic works. These types of projects have been studied by various researchers [6, 11, 12]. Knauss [6] presented different types of Mycenaean retaining walls and embankments based on archaeological findings. More recently, Aravantinos et al. [13] presented archaeological findings of a Mycenaean embankment near the city of Orchomenos that could explain the embankment reinforcement that Homer is referring to. The embankment had a width of 8 m and consisted of two parallel stonewall faces 2 m wide and a core of yellowish mud in between. The element of interest is the existence of four internal transverse walls that connected the two sides of the embankment and operated as struts.

## 5 Conclusions

The information provided by Homer on topics related to engineering geology and geotechnical engineering is remarkable, given the scope of the poems. Homer is apparently careful in the selection of words used to describe various geologic environments and phenomena. The words are selected not only based on literary criteria, but are also accurate. Homer's remarks also suggest that some basic understanding of the geologic processes existed at the time, probably as a result of observations of the geologic environment. This is not surprising, as preservation and growth of agricultural societies were greatly dependent on environment. Of interest is also the information collected on various types of geotechnical engineering projects mentioned in the poems, which include deep excavations, road infrastructure, embankments and port facilities. In the poems, there are indications of the existence of construction practices and projects built to certain specifications. The information provided by Homer is largely consistent with the archaeological findings.

## Notes

The Perseus Digital Library of Tufts University (http://www.perseus.tufts.edu/) was used for the study of the frequency of occurrence of various epithets in the Homeric poems. The authors would also like to thank Dr. Edmund Medley of GeoSyntec Consultants for his valuable review and comments. Additional studies related to the history of geotechnical engineering and engineering geology are available at the Geoengineer website at http://www.geoengineer.org.

## References

1. Schliemann, H., *Troy and Its Remains*, Reprint of the 1875 edition, Arno Press, New York, 1976.
2. Schliemann, H., *Mycenae: A Narrative of Researches and Discoveries at Mycenae and Tiryns*, Reprint of the 1880 edition. Arno Press, New York, 1976.
3. Homer, *The Iliad*, translated by Robert Fagles, Penguin Classics, 1998.
4. Homer, *The Odyssey*, translated by Robert Fagles, Penguin Classics, 1997.
5. *The Perseus Digital Library*, Ancient Greek versions of the Iliad and Odyssey http://www.perseus.tufts.edu/.
6. Knauss, J., *Post-Helladic Hydraulic Works: Studies for the Infrastructure of Hydraulic Works for Water Management during the Mycenaean Times*, Society for the Study and Promotion of the Hellenic History, 2002 [in Greek].
7. Zekkos, D., Manousakis, J. and Athanasopoulos, A.G., Geotechnical engineering practice in the Mycenaean civilization (1600–1100 BC), in *Proceedings 2nd International Conference on Ancient Greek Technology*, Athens, 17–21 October 2005. Also available at http://www.geoengineer.org.
8. Spyropoulos, P.I., *Chronicle of Earthquakes in Greece from Antiquity to Present*, Dodoni Publications, 1997 [in Greek].
9. Hesiod, *Complete Works* (ancient text and translation), Publisher Kaktos, 1993.
10. Jansen, A.G., *A Study of the Remains of Mycenaean Roads and Stations of Bronze-Age Greece*, The Edwin Mellen Press, 2002.
11. Iakovidis, S.E., *Gla and the Kopais in the 13th Century BC*, Library of the Archaeological Society at Athens, No. 221, 2001.
12. Balcer, J.M., The Mycenaean dam at Tiryns, *American Journal of Archaeology* **78**(2), April 1974, 141–149.
13. Aravantinos, B., Kountouri, E. and Fappas, I., The Mycenaean drainage system of Kopaids: New facts and first appraises, in *Proceedings 2nd International Conference on Ancient Greek Technology*, Athens, 17–21 October 2005.

# Geological Knowledge of Greeks in the Era of Trojan War

Ilias D. Mariolakos

*National & Kapodistrian University of Athens, Greece*

**Abstract.** Among the many important historical, cultural and geographical elements found in the two Homeric epics, Iliad and Odyssey, there are many that allow the present-day geoscientist to draw indirect conclusions about the geological knowledge of the inhabitants of the Aegean and Circum-Aegean region.

By the end of 19th century, K. Zeggelis, published a monograph, entitled *The Science of Nature in Homer* (1891), where, among others, he mentioned and commented on the poet's references on minerals (metals and non-metals), their origin and the metallurgical knowledge of the people of that era. The opinion of Zeggelis that the metallurgical processes used, although known at the time of Homer, were not performed in Greece, but in other (probably Oriental) countries has been rejected by the newest archaeological and archaeometric studies, showing that metallurgy and smelting had begun in Greece long before the Trojan war, even before the Mycenaean times.

In this paper, we shall refer to the indirect conclusions to be drawn by the modern geoscientist, regarding the technical knowledge of the prehistoric Greeks, by studying drainage – anti-flooding works and dams constructed in Arcadia, during the Mycenaean times.

Arcadia was chosen because, as mentioned, the Arcadian king Agapeinor, son of Lycurgus, who lived in the town of Tegea, lead 6,000 Arcadians against Troy. The army was carried on 60 ships, offered by Agamemnon.

In the greater area of the Arcadian Plateau, a series of basins constitute a geologically "composite" polje. These basins are: the Takka basin, the Mantineia basin, the Argon Field (Nestani plain), the Levidion – Ancient Orchomenus basin and, finally, the Kandela basin. In three out of these five basins, the prehistoric people of Minyans had constructed a series of earth dams and other drainage works, as mentioned by Pausanias. These works were studied in great detail by J. Knauss, Professor of Hydraulic Engineering in Munich Polytechnic School.

These works aimed at:

(i) protecting great parts of the basins against flood waters coming from the surrounding mountains and the many karstic springs of the areas, thus increasing the land suitable for cultivation (land reclamation);
(ii) securing irrigation water; and
(iii) draining the many small swamps formed in the various plains, thus reducing the risk of malaria.

The detailed study of these works by Knauss, by a hydraulic engineer's point of view, show that Minyans were not only skilled engineers, but that they also had excellent capacity and knowledge on construction-site management, project management (very similar to the knowledge of modern-day engineers ), and that they were also capable of "diplomatic" interventions between cities, etc.

The scientific and technological knowledge of the Minyans are comparable to those of modern scientists in matters related to the study of:

*S.A. Paipetis (ed.), Science and Technology in Homeric Epics,* 243–255.

- the meteorological and climatic conditions of an area,
- the river-yields and their sediment load,
- the topography,
- the physical and mechanical characteristics of soils,
- the geology of the flood basin and of the greater area, as well as the hydrogeological characteristics of the alpine and post-alpine geological formations,
- the karstification and the hydraulic behavior of the karstic forms (caves, sinkholes, karstic springs, etc.),
- and many more.

## 1 Introduction

It is known that many ancient civilizations, probably the oldest civilizations of the world, developed in the greater area of the East Mediterranean Sea, as well as in Mesopotamia.

The reason for this is that favourable climate prevailed in the East Mediterranean until 10,000 years ago, whereas a big part of the rest of Europe was covered by glaciers and the non-glaciated ground was frozen for most of the year. As regards Mesopotamia, the reason was not the warm climate that prevailed, but the abundance of water carried by the two great rivers, the Tigris and the Euphrates.

It is remarkable that the inhabitants of the countries of these two areas have developed culturally since the Late Palaeolithic era up to the present day. However, the actual intervention in the geoenvironment began much later, especially after the stabilization of climate, which coincides with the beginning of the Climatic Optimum of the Holocene Epoch (6th millennium B.P.)

Of course, Mycenaean times, which have been marked, *inter alia*, by the Trojan War described by the two unsurpassed Homeric epics, are much later. As a result, the Homeric epics and archaeological discoveries allow us to draw many conclusions regarding the technological knowledge of the ancient Greeks of this era.

Among the many important historical, cultural and geographical elements found in the two Homeric epics, Iliad and Odyssey, there are many that allow the present-day geoscientist, to draw indirect conclusions about the geological knowledge of the inhabitants of the Aegean and Circum-Aegean region.

By the end of the 19th century (1891), K. Zeggelis, published a monograph entitled *The Natural Science in Homer*, where, among others, he mentioned and commented on the poet's references on minerals (metals and non-metals), their origin and the metallurgical knowledge of the people of that era. The opinion of Zeggelis that the metallurgical processes used, although known at the time of Homer, were not performed in Greece, but in other (probably Oriental) countries has been rejected by the newest archaeological and archaeometrical studies, which show that metallurgy and smelting had begun in Greece long before the Trojan war, even before the Mycenaean times.

In this paper, we shall refer to the indirect conclusions to be drawn by the modern geoscientist, regarding the technical knowledge of the prehistoric Greeks, by studying the drainage – anti-flooding works and dams constructed in Arcadia, during the Mycenaean times.

Arcadia was chosen because the Arcadian king Agapeinor, son of Lycurgus, who lived in the town of Tegea, led 6,000 Arcadians against Troy. The army was carried on 60 ships, offered by Agamemnon.

Another reason for this choice is associated with the role played by Poseidon in the Trojan War and Arcadia.

It is known that Poseidon is not only the God of the Sea, but also master of the geological phenomena and processes occurring both in the interior of the lithosphere and on Earth's surface. Therefore, vertical movements, earthquakes, fluvial processes, all the physicogeological processes in general, which are known in Geology as "endogenous" and "exogenous" forces, are controlled by Poseidon. It is said that Poseidon, along with Apollo, having the intention of destroying the walls of the Greeks, "engineered" the change of the position of many river channels, as well as *river flooding*, the Scamandros River included. It is also said that after the flooding the beach *became level* and the floodwater withdrew to their normal river channels. Homer refers to another case in which the Scamandros overflowed its banks. This flood was also caused by Poseidon, in his unsuccessful effort to sidetrack Achilles.

Oddly enough, Poseidon was born in Mantinia, at a slope of Alission Mountain, near Nestani situated near *Tegea*.

After this parenthesis, which at first glance may appear unnecessary, although it is interesting, let us return to the greater area of Arcadia and discuss the geological, engineering geological, hydrological etc. knowledge of the Minyans in Mycenaean times, long before the Trojan War.

## 2 The Minyans

The native land of the Minyans is considered to be either Thessaly or the greater area of Orchomenus, mainly Lake Copais (Boeotia). The Minyans are considered to belong to the Pelasgians, whereas Ploutarch associates them with the Leleges. They are also regarded to be Thracians or Aeolians. However, little is known about this Pelasgian tribe.

The Argonauts were also called Minyans, because either most of them were sons of the daughters of Minyas or Jason, the ruler of Argos and the leader of the Argonauts and the expedition, belonged to the Minyans. It is also argued that all the inhabitants of Iolcus were called Minyans.

Minyas was king of Orchomenus city, where is located his Tholos tomb, known as the *Treasury of Minyas*. He is considered to be son of Poseidon or Ocean. It is also maintained that he is grandson of Poseidon, etc. Minyas was very rich and it is said that he used to store his riches in underground galleries-stores. His wealth derived mainly from the cultivation of Lake Copais, which had already been drained in the middle of the 2nd millennium BC, or, as it is also argued, much earlier, in the later 3rd millennium BC

The galleries in which Minyas stored his riches might have been natural or artificial. The natural ones were a multitude of caves lying in the margins and the mountains surrounding Copais. These mountains consist of intensely karstified carbonate rocks. As regards the artificial galleries, the Minyans have shown that they knew how to construct excellently, from an engineering point of view, underground galleries, like the one located north of the present-day settlement of Neo Kokkino. Part of the floodwater was conducted to the Euboean Gulf through this underground gallery, whereas the rest of it was conducted through the cave – sinkhole lying at the present-day settlement of Neo Kokkino, known as "the cave of Hercules". Remains of the great engineering structures constructed by the Minyans are found in nearly all of the basins of Greece.

## 2.1 Works Constructed by the Minyans in the Greek Region

Since Pausanias' time, it has been known that in prehistoric times many and great geotechnical works were constructed in various lowlands of the Greek mainland, such as in Thessaly, Boeotia, Argolis, etc.

These works aimed at:

(i) protecting great parts of the basins against flood waters coming from the surrounding mountains and the numerous karstic springs of the area, increasing thus the land suitable for cultivation (land reclamation).
(ii) securing irrigation water, and
(iii) draining the numerous small swamps formed in the various plains, thus reducing the risk of malaria.

All of these works have been studied in detail by Knauss, who has devoted all of his research activities to these studies for the last 30 years.

The detailed study of the works by J. Knauss,[1] from a hydraulic engineer's point of view, show that the Minyans were not only skilled engineers, but that they also had excellent experience and knowledge on construction-site management, project management (very similar to the knowledge of modern engineers), and also that they were capable of "diplomatic" interventions between cities, etc.

The site and the remains of these works are described by Pausanias, and many of them have been located by the contemporary archaeological research. Of course, some works are not mentioned by the traveler, probably because they have been destroyed. The greatest and best preserved anti-flooding work is the great drainage ditch in Copais, which was used to drain Lake Copais (Boeotia).

The Minyans constructed many geotechnical works, such as earth dams, draining ditches, river diversion works, taping of sinkholes etc. The most known of these land reclamation works are a dam at *Taka*, the drainage works close to the basin of *ancient Orchomenus – Kafies*, the anti-flooding works in the *Kandela plain*, in

[1] Jogst Knauss, Professor of Hydraulic Engineering and Water Management in Munich Polytechnic School, Germany.

the *Feneos plain* and in *Stymphalia*. There are probably other works, not known to the public yet, such as Mantineia dam, which was unknown until quite recently. Similar works were constructed in *Argolis*, in particular at *Mycenae*, *Tiryns* and in the greater area of the *Lerni springs*.

From all these, to a large extent, unknown works, which, as already mentioned, have been studied in great detail by Knauss, who has made them known to other scientific circles apart from archaeologists, reference shall be made only to Mantineia dam and the drainage trench at ancient Orchomenus.

### 2.1.1 Mantineia Dam

In the area lying south of ancient Mantineia, the Minyans constructed a dam that retained the water coming from the south. In this way, a *reservoir* was formed, which, as Knauss [7] estimated, could retain a volume of water as large as 15 million cubic meters.

This dam (Figure 1), like all Mycenaean dams, was an embankment 3 m high, with a totally impervious clay core surrounded by additional clay material. It was covered by processed boulders, placed in such a way that it would have been protected from erosion. This earth dam was constructed at a narrow site in a long and narrow valley lying NE of city of Tripolis, in the flat lowland lying in front of the foot of the hill, where the present day settlement of Skopi is situated. This natural narrow site, having a width of about 300 m, is located between *Mytikas* mountain and *Ag. Nikolaos* hillock. It should be noted that a Mycenaean settlement has been found in the next hill lying south of Skopi.

As regards the purpose of this dam, it should be said that the dam served two purposes, since it protected Mantineia lowland from flooding and secured irrigation water. However, it should be mentioned that in the reservoir basin is the great *Kanatas sinkhole* (Figure 2). It is doubtful, therefore, whether this dam could hold all of that quantity of water and form the reservoir, as is the case with modern dams. Besides, the Minyans, on the basis of our present knowledge of them, constructed dams not only to collect water.

Moreover, the Minyans knew the behaviour of the karstic formations very well. This is inferred from the great works that they constructed to drain Copais Lake, where there are many karstic formations. Therefore, given that the conditions in the reservoir basin in the era of the Minyans had been the same as today, the dam would have served only the purpose of controlling floodwater, since it would have been impossible to hold the water leaking through the sinkhole. No matter whether or not the Minyans managed to form the reservoir, the dam prevented floodwater of Sanovista stream, known as *Lahas* in ancient times, from flooding Mantineia lowland.

It should be noted that before the construction of the dam, the whole lowland and the surrounding area were periodically flooded. Besides, as it follows from the topographical map of the H.M.G.S., between hill Gortsuli and ancient Mantineia there is an area called *Limni Lepidi*, whereas to the west another area lies called

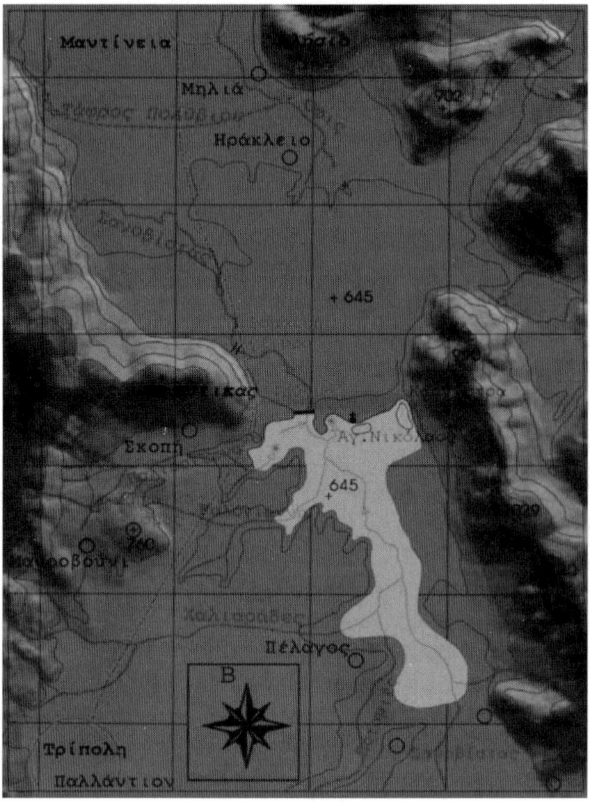


**Fig. 1** The Mycenaean dam and the reservoir in the Mantineia valley (based on Knauss [7]).

Amoudera. Moreover, an area lying between Mantineia and Kapsa is called *Megali Limni*. Furthermore, Phillipson [18], on his visit to Arcadia in the late 19th century, mentioned that he found Mantineia in a flooding state. Thus, from the fact that the area is flooded even today (Figure 3), when the water table has been lowered due to excessive pumping of the wells, and, in addition, from the fact that the area was flooded in the previous centuries, when the climatic conditions were not favourable, it is inferred that the area was flooded easily in early times, since the climatic conditions were more favourable, i.e. as they were in Mycenaean times, when Hercules performed his labours, many of which took place in Peloponnesus and relate to water, such as the Lernean Hydra, the Stymphalian Birds, the Augeian stables, the destruction of the anti-flooding – drainage works of the Minyans at Copais, his fight with the Achelous river, etc. Probably this is why it is argued that *Ptolis* is a settlement or city that come into existence before Mantineia and was erected on the nearby hill called *Gortsuli*. This hill, located north of ancient Mantineia, is 400 m wide and rises 100 m above the plain. Remains of urban development of the Protohelladic Era (2800–2000 BC) and the Mesohelladic Era (2000–1500 BC) have been found in Ptolis.

**Fig. 2** Kanatas sinkhole and modern engineering structures constructed to conduct sewage of Tripolis.

**Fig. 3** Ancient Mantineia area flooded, due to heavy rainfall, winter 2004–2005.

### 2.1.2 The Drainage Trench at Orchomenus

Between the Upper Orchomenus Plain, known as the Levidion Plain, and the Lower Plain, which stretches between settlements Kandyla, Kafies and Hotousa sinkhole, the Minyans, in early Mycenaean era, maybe earlier than 3rd millennium BC, excavated an artificial ditch characterized by Pausanias as "trench", through which they

**Fig. 4** The drainage trench that Minyans dug to conduct floodwater from Upper Field (Levidion plain) to Lower Field.

managed to drain the Upper Field, conducting the floodwater to the Lower Field (Figure 4).

The eastern edges of this artificial "gully", whose width varies between 5.5 and 6.5 m, have been excavated in formations of the Olonos-Pindos geotectonic unit. Some portions of it coincide with a fault surface. At its western edges, a wall, about 2 m wide and a remaining part 1–1.5 m high, has been constructed. According to Knauss, the construction remaining in the central part of the valley may actually constitute the remains of a watermill, either of the Mycenaean period or later (Figure 5).

In the Lower Plain, also known as the Kafies Plain, the Minyans constructed a reservoir. In this reservoir the water coming from the Orchomenus Plain, the surrounding mountains and the Kandela Springs was gathered. The reservoir was formed behind of a dam constructed at the NW narrow site of the valley. Pausanias characterized this narrow site as "earth's soil". Also, there was a smaller embankment at the eastern part, and the overflow of water was conducted to the great Hotousa sinkhole. In this way they achieved many goals, such as:

(i) Drainage of the Orchomenus basin (Upper Field).
(ii) Partial drainage of a great part of Kafies basin (Lower Field).

**Fig. 5** The prehistoric hydraulic works of Minyans, on the slopes of the artificial trench, may constitute watermill installations. In the background is the big Kandela polje – Pausanias' Lower Field – where a dam and a reservoir have been constructed as well.

(iii) Reclamation of the drained areas for agricultural purposes.
(iv) Adequacy of water for irrigation purposes in summer.
(v) Cleansing of the swamps, thus reduced malaria risk.

## *2.2 Technological Knowledge of Minyans*

The systematic and in-depth study of the works, not only from a geotechnical point of view, but also from the aspect of the purpose of construction, raises a series of questions about the Minyans, such as: Who are those Minyans? Why did they come to Peloponnesus? Were they invited? They cannot have come on their own. Their arrival must have been approved by local authorities. One cannot admit that they themselves made the decision to construct a dam at Skopi or at Taka. The construction of a dam has consequences. Firstly, huge cultivated areas and probably settlements are inundated. Moreover, the water stored in the reservoir is actually taken from the settlements lying downstream of the dam.

For instance, in the case of Skopi dam, water storage caused the area of Mantineia to be drained, thus areas can be cultivated. On the other hand, a whole area is inundated and transformed into a lake. This area cannot be cultivated. All of these interventions did not cause reclamation of land in general, since the area of Mantineian Field was favoured, but, at the same time, the conditions of the area extending from Skopi to north of the present-day village Pelagos got worse, since this area was inundated. Was this area inhabited until then? Had it been cultivated? Who decided to construct such a work with so many consequences?

It should be born in mind that, at that time, for reasons already mentioned, Mantineia and any other city in the reservoir basin could not have been founded. This is an indirect conclusion, since it is highly unlikely that the authorities of a city made the decision, their own houses to be inundated and their city obliterated, just to drain marshes of another area, i.e. of Mantineia, in this case. This would be completely irrational.

Anyway, someone must have taken this grand and important decision. However, to take such a decision, one must be aware of the consequences in advance, no matter who decides. This means that a detailed study of the greater area of the Arcadian Plateau was performed, before the final decision. This study, e.g. of a dam, should consist of three basic parts: (i) a detailed study of the topography, (ii) a hydrological study, and (iii) a geological-hydrogeological study.

The *study of topography* is absolutely essential, since engineers/planners must know the dimensions of the dam, i.e. width, height and mainly length. Furthermore, it is essential, since engineers must know the extent of the reservoir basin, as well as height and position of spillway, so that, if the quantity of floodwater exceeds storage capacity of the reservoir, conduct floodwater out of the reservoir, preventing overflowing the top of the dam, which may destroy the dam and cause disastrous flood downstream.

A *detailed hydrological study* is essential, since planners must know the quantity of water flowing into the reservoir basin and the elevation, to which water will rise, particularly with earth dams. The quantity of water to be stored in the reservoir can be calculated in detail, provided that the following are known:

(i) The *surface area of the drainage basin*, i.e. of the river basin collecting and funnelling water to the reservoir.
(ii) The *distribution of precipitation in space and time*, i.e. distribution of rain and snow with elevation during the year, flood discharge, and many more. To appreciate the importance of this, it should be mentioned that, at present days, although there is a network of meteorological stations in Arcadia, a detailed hydrological study for the construction of a dam cannot be performed, as the number of the stations is not enough.
(iii) The *geological structure and the rocks*, constituting the foundation of the dam, as well permeability of rocks of the reservoir must also be known. This is essential, since, if the rocks of the floor were permeable, the water stored would leak and the dam itself would encounter many problems, possibly disastrous.
(iv) Special soil-rock materials are required for the construction of an earth dam. Not all soils are suitable. Locating these suitable soils is not always feasible and special knowledge and experience is required. Nowadays, this job is carried out by *experts in soil mechanics and engineering geology*, and the samples collected are tested in special soil mechanics laboratories.
(v) Dam construction is such a difficult and elaborate job, requiring special theoretical knowledge and lots of experience. Of course, the dams of the Minyans were not higher than 3 to 4 m. However, even those dams required special study and the same level of knowledge, since the elevation of the water level in the reservoir was much higher than the ground surface downstream. Thus, the dam had to be waterproof. This is accomplished by applying a special technique, e.g. *soil compaction*. Nowadays, compaction is carried out by placing the soil

in layers of a specific thickness, next it is wetted with quantities of water according to its grain size distribution, and, finally, it is tamped.

All of the above have been mentioned to make clear that the Minyans, in those ancient times, had a wealth of knowledge, on which these unprecedented engineering structures, preserved for at least 3,500 years, were based. The fact that the many kilometres-long earth dam or embankment, on which an important road of province road network was founded, i.e. the road connecting Kastro village at Copais with Orchomenus, has not exhibited any problems due to consolidations, means that this work is technically flawless. As far as I know, there are no public works having operated and been preserved for 3,500 years, as the embankment on Melanas River, on the northern margin of the Copais basin (Boeotia).

## *2.3 Economic and Managerial Capacities of the Minyans*

Apart from an official decision, dam construction requires not only scientific and engineering knowledge, but also *funds and manpower*: The cost was enormous, even by present-day standards. Who financed these works? What did the Minyans get in exchange? Who worked for their construction? Were they slaves or free people? In addition, they had to manage a construction site.

And the great question is, where and when did Minyans learn all these things? Is mere experience sufficient? Would the implementation of a project of such a scale have been possible, without even a rudimentary plan?

Such questions have neither been raised nor studied, i.e. they have not yet been considered by the scientific community, and, as a result, are still unknown. I believe that we would get conscious of the way that ancient Greek civilization reached this level, only if we study the works of the Minyans systematically. Why did the Minyans come to Arcadia and began constructing these enormous, for this era, works? Were many people living there, needing these works to increase their yield? Certainly, there is no information about the population of the area; also, there are no ruins of cities or settlements allowing us to draw any conclusion indirectly. Of course, in pre-Mycenaean times, there must have been many settlements that, as mentioned in the Greek mythology, were founded by the sons of Lycaon.

Obviously, these enormous works benefited not all of the settlements of Arcadia. It is almost certain that the inhabitants of the settlements developed on the boundaries of basins, such as of Nestani, Kapsa, Tegea, Orchomenus etc. were favoured.

There are, however, many great works, if the population is taken into consideration. Furthermore, such works have not been constructed even in the greater Megalopolis basin. The works constructed even in Argolic plane, regarded as the heart of Mycenaean world, are comparatively small; they are mainly anti-flooding works, not dams and reservoirs, which, apart from anti-flooding protection, are used for irrigation purposes. It thus appears that there is a contradiction, since dams for irrigation purposes have not been constructed in the Argolic plane, which is much more fertile than Mantineia plane, whereas these dams have been constructed in the

Arcadian planes, undoubtedly having lower yield capacity. Why did the Minyans not construct such dams in Argolis?

In my opinion, the reason is that, from a hydraulic and engineering geological point of view, the area of Argos is not suitable for the construction of dams, since all the river systems have torrential flow and high sediment discharge. Of course, a dam could have been constructed on Erasinos River, fed by the water of Kefalari springs; consequently, it does not have torrential flow and its sediment discharge is zero. However, along the channel of Erasinos no sites suitable for the construction of a dam and the filling of a reservoir exist, except for the plane entrance area. According to J. Knauss, a great embankment has been constructed there, mainly an anti-flooding work, least used for irrigation purposes, since the elevation is low. In addition, since sea level in the Argolic Plane in Mycenaean times was around 1–1.5 higher, it is clear that the reservoir could not be used for irrigation purposes.

Thus, although in the Argolic Plain, climatic conditions, mainly temperature, are favourable for agriculture, no systematically cultivation can take place due to water shortage, e.g. in summer. However, this is not a big problem for the cultivation of cereals: Basic species, such as cereals (wheat, oats, barley), olive and vineyard, not requiring irrigation, could be cultivated.

In the Arcadic plateau, all of the above species, except for olive, can be cultivated. However, the cultivation of them requires manpower and stretches of flat lowlands. In the Arcadic Plateau, in Mycenaean or Pre-Mycenaean times, there are flat and fertile lowlands, but they are inundated and transformed into lakes or marshes for most of the year. These lakes or marshes are not suitable for agricultural purposes and favour malaria. Therefore, it is possible that the construction of the works had taken place after agreements between the powerful and rich Minyans and the Arcadians, who "financed" the works and assigned the technologically advanced Minyans to erect them.

## References

1. Curtius, E. (1892). Die Deichbauten der Minyer, *Sitzungsbericht der Berliner Akademie der Wissensch., Philosophisch-Historische Klasse* **55**, 1181–1193.
2. Kakridis, J. (1986). *Greek Mythology*, Ekdotiki Athinon, Vol. 4 [in Greek].
3. Kenney, I. (1935). The ancient drainage of the Kopais, *Annals of Archaeology and Anthropology, University of Liverpool* **22**, 189–206.
4. Knauss, J. (1984). Die Wasserbauten der Minyer in der Kopais – die älteste Flussregulierung Europas (Kopais 1), *Wasserbau und Wasserwirtschaft*, No. 50, Technische Universität München.
5. Knauss, J., Heinrich, B. and Kalcyk, H. (1986). Der Damm bei Kaphyai und Orchomenos in Arkadien, *Archeologischer Anzeiger*, 538–611.
6. Knauss, J. (1987). Die Melioration des Kopaisbeckens durch die Minyer im 2 Jt. v. Chr. *Wasserbau und Siedlungsbedingungen im Altertum (Kopais 2) – Wasserbau und Wasserwirtschaft*, No. 57, Technische Universität München.
7. Knauss, J. (1989). Mykenische Wasser-banten in Arkadien, Böotien und Thessalien – mutmassliche Zielsetzung und rekonstruierbare Wirkungsweise, in *Akten Kongress Wasser*, Berlin, pp. 31–70.

8. Knauss, J. (1996). Argolische Studien: Alte Strassen – Alte Wasserbauten, *Wasserbau und Wasserwirtschaft*, No. 77, Technische Universität München.
9. Leontiadis, I. and Dimitroulas. C. (1972). The use of radioisotopes in tracing karst groundwater in Greece, III, Investigation on the possible interconnection between Nestani sinkhole and submarine springs of Argos area, *Democritos Nucl. Res. Cent., Athens*, Demo 72/3E, Athens.
10. Leontiadis, I. and Dimitroulas. C. (1973). The use of radioisotopes in tracing karst groundwater in Greece, IV, Investigation on the possible interconnection between sinkholes of Milia and Taka lake with various springs of the near area, *Democritos Nucl. Res. Cent., Athens*, Demo 73/4E, Athens.
11. Lüttig, G. (1966). Die nichtmarinen Neogen – Becken im Mittelmeerraum und ihre Bedeutung für die Stratigraphie, in *Comm. Neogen Strat., Proceedings, 3rd Session*, Bern, 1964, Leiden.
12. Mariolakos, I. (1998). The geomythological geotope of Lerni Springs (Argolis, Greece), *Geologica Balcanica* **28**(3/4), 101–108.
13. Mariolakos, I. (2002). The geoenvironmental dimension of Greek mythology, in *Proceedings of the 9th International Congress of the Geological Society of Greece*, Athens, September 2001, *Bull. Geological Soc. Greece* **XXXXIV**(6), 2065–2086.
14. Mariolakos, I.D. and Mariolakos, D.I. (2004). The argon field in Arcadia, the sinkhole of Nestani Village, the God Poseidon and the submarine Dini Springs in Argolic Gulf (Peloponnesus, Greece). A geomythological approach of the Poseidon birth, in *Proceedings of the 10th International Congress*, Thessaloniki, April 2004, *Bull. Geological Soc. Greece* **XXXVI**, 2004, 1146–1153 (http://www.geo.auth.gr/ege2004/articles/GA1_160.pdf)
15. Milankovitch, M. (1941). Kanon der Erdbestrahlung und seine Anwendung auf dem Eiszeitenproblem, *Royal Serbian Sciences*, Spec. Publ. 132, Section of Mathematical and Natural Sciences, Vol. 33, Belgrade, 633 pp.
16. Paepe, R. (1986). Landscape changes in Greece as a result of changing climate during the quaternary, in *Desertification in Europe*, R. Fantechi and N. Margaris (Eds.), D. Reidel Publ. Co., Dordrecht.
17. Pausanias, *Arcadica*.
18. Philippson, A. (1892). *Der Pelopones*. Berlin 1891–92, 642 pp.
19. Zeggelis, K.D. (1891). *The Science of Nature in Homer*, University of Patras Publications (Preface by S.A. Paipetis), Patras, 1997.

# Static and Dynamic Analysis of the Atreus Vaulted Tomb in Mycenae

P.K. Askouni, H.A. Agelopoulou, M.G. Sfakianakis and D.E. Beskos

*University of Patras, Greece*

**Abstract.** A finite element methodology for the nonlinear static and dynamic analysis of historical masonry structures is described and applied to the case of the Atreus vaulted tomb, known as "Treasury of Atreus", in Mycenae, Greece. This monument belongs to the most famous masterpieces of prehistoric architecture and it has often been the subject of scientific discussions in the past. The stone-masonry structure is statically and seismically analyzed by the Finite Element Method taking into account its nonlinear material behaviour. According to this methodology, the inelastic material behaviour is simulated with the aid of the theory of continuum damage mechanics. More specifically, the developed damage model is a combination of the theories of Mazars and Faria-Oliver and is characterized by its successful modelling of the mechanical behavior of quasi-brittle materials such as concrete or masonry. In addition, the model permits an easy calculation of damage indices at various locations of a structure and a global damage index for the whole structure as well. Furthermore, in order to achieve an estimation of the contact phenomenon between the stone surfaces, due to the absence of any kind of mortar, dynamic analyses were repeated assuming half rigidity and strength of the material. The analyses were carried out using the general purpose computer program MINOS for three-dimensional static and dynamic nonlinear analysis, which has been developed at the University of Patras. Stress contours and displacement histories were produced for various cases of loading and type of analyses. The results of this work show that the Atreus tomb exhibits such a high structural integrity and strength that its behaviour remains always elastic, thereby demonstrating the ability of the Mycenaean Greeks to construct very safe structures.

## 1 Introduction

Referring to the past research on the subject of using finite and boundary elements for the nonlinear analysis of monuments and special structures, the existing linear and nonlinear models up to 1993 are reviewed in the extensive article of Beskos [1]. Also, interesting articles on the topic of the analysis of historical structures by analytical or experimental methods can be found in the books of Brebbia and Lefteris [2], Sanchez-Beitia and Brebbia [3], Rocca et al. [4], Crocci [5], Brebbia and Jager [6] and Bull [7]. The existing models for the mechanical behaviour of masonry materials can be classified into two basic categories, the discrete and the continuum ones. Discrete models are used for the analysis of monumental structures

*S.A. Paipetis (ed.), Science and Technology in Homeric Epics,* 257–265.

composed of large discrete parts, such as stone arches or ancient temples with vertebrate columns. In this case, the discrete large elements of the structure are assumed to behave fully elastically, while the behaviour of the contact interface between them is assumed to be described by a unilateral friction law. Continuum models are used for masonry structures composed of a combination of bricks or stones and mortar at their interfaces of contact. The mechanical behaviour of the continuum models can be described by a stress-strain law, which is derived from an one-phase or a two-phase model of the masonry material. The one-phase models consider the masonry as consisting of a single material, while the two-phase models take into account the different inelastic behaviour between the components of the masonry (brick or stone and mortar). It is obvious that the one-phase models are much simpler than the two-phase ones and for this reason they can be very successfully used for the analysis of three-dimensional structures of large size and great geometrical complexity.

In the present work a one-phase damage model is adopted to investigate the static and dynamic nonlinear response of masonry structures. It is an anisotropic damage model, which can be thought of as a combination of the elastic-damage part of the elastoplastic-damage model of Faria and Oliver [8] with two damage indices (one for compression and one for tension) and the damage theory of Mazars [9], which unifies appropriately the two indices into one common damage index. It is called the FOM (Faria–Oliver–Mazars) model. The strain rate effect is also taken into account in a simple manner. A detailed description of the model can be found in the work of Hatzigeorgiou et al. [10].

The above damage model is used herein to analyze dynamically (seismically) the masonry vaulted tomb of Atreus in Mycenae under three-dimensional (3D) stress conditions. The axonometric view of the tomb from two different antidiametrical optical views is shown in Figure 1. Its diameter is 14.20 m and its height is 13.20 m. The main doorway and the doorway leading into a small rectangular side chamber are roofed with two linted-blocks, above which are the usual relieving triangles. In this figure one can also observe its FEM discretization. The tomb is analyzed for its own weight and seismic loading, applying it statically and dynamically and assuming inelastic material behavior.

## 2 The FOM Damage Model

The theory of continuum damage mechanics attempts to model macroscopically the progressive mechanical degradation of any quasi-brittle material under different stages of loading. It is assumed that the process is governed by a damage index $d$, which physically represents the damaged area of a unit surface of the body, cut by a given plane, i.e. the area with existing cracks and cavities. By assuming homogeneous distribution of microvoids and invoking the hypothesis of strain equivalence, the total stress tensor, $\sigma$, is represented as

$$\sigma = (1 - d) \cdot \bar{\sigma}, \tag{1}$$

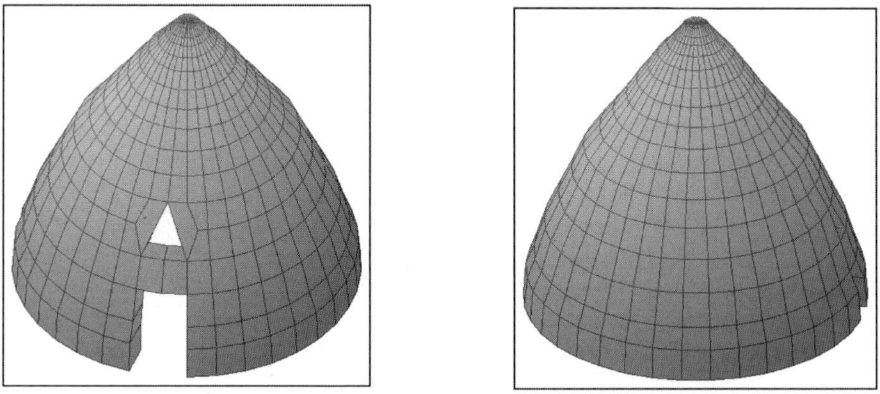

**Fig. 1** Axonometric view of the tomb from two different optical views.

where $\bar{\sigma}$ is the effective stress tensor for the undamaged material, while the damage index $d$ takes values between 0 (undamaged state) and 1 (full loss of material coherence or failure). The proposed damage model works on the basis of the following procedure in the framework of a 3D finite element analysis.

At every load step and at each Gauss integration point of an element, the strain and effective stress vectors $\{\varepsilon\}$, $\{\bar{\sigma}\}$, are computed by the known relations

$$\{\varepsilon\} = [B] \cdot \{u\}, \quad \{\bar{\sigma}\} = [D] \cdot \{\varepsilon\}, \tag{2}$$

where $\{u\}$ is the displacement vector and $[D]$ the elasticity matrix. Then the computation of principal strains $\varepsilon_i$ and principal effective stresses $\bar{\sigma}_i$ $(i = 1, 3)$ follows. From those principal magnitudes, the corresponding effective strains $\langle \varepsilon_i \rangle$, $\rangle \varepsilon_i \langle$, and effective stresses $\langle \bar{\sigma}_i \rangle$, $\rangle \bar{\sigma}_i \langle$, are computed by the formulas

$$\langle x_i \rangle = (x_i + |x_i|)/2, \quad \rangle x_i \langle = (x_i - |x_i|)/2, \quad i = 1, 3 \tag{3}$$

by setting $x_i = \varepsilon_i$ or $\bar{\sigma}_i$, respectively. Afterwards, the total equivalent strains $\tilde{\varepsilon}^+$, $\tilde{\varepsilon}^-$, $\hat{\varepsilon}^+$ and $\hat{\varepsilon}^-$, defined as

$$\tilde{\varepsilon}^+ = \sqrt{\sum_{i=1}^{3} \langle \varepsilon_i \rangle^2}, \quad \tilde{\varepsilon}^- = \sqrt{\sum_{i=1}^{3} \rangle \varepsilon_i \langle^2}, \quad \hat{\varepsilon}^+ = \sqrt{\frac{\max(\langle \varepsilon_i \rangle)}{\sum_{i=1}^{3} \langle \varepsilon_i \rangle}}, \quad \hat{\varepsilon}^- = \sqrt{\frac{\min(\rangle \varepsilon_i \langle)}{\sum_{i=1}^{3} \rangle \varepsilon_i \langle}} \tag{4}$$

are computed, where the superscripts "+" and "−" refer to tension and compression, respectively. According to Mazars [9], the equivalent strains $\varepsilon_i^+$ and $\varepsilon_i^-$ of the strain tensor are defined as

$$\varepsilon_i^+ = \frac{\left[(1+\nu)\langle \sigma_i \rangle - \nu \sum \langle \sigma_i \rangle\right]}{E}, \quad \varepsilon_i^- = \frac{\left[(1+\nu)\rangle \sigma_i \langle - \nu \sum \rangle \sigma_i \langle\right]}{E}, \quad i = 1, 3 \tag{5}$$

and from these, the parameters $H_i^+$ and $H_i^-$ are computed as

$$H_i^+ = 1 \text{ if } \varepsilon_i = \varepsilon_i^+ + \varepsilon_i^- \geq 0 \text{ and } H_i^+ = 0 \text{ if } \varepsilon_i = \varepsilon_i^+ + \varepsilon_i^- < 0, \quad i = 1, 3$$
$$H_i^- = 1 \text{ if } \varepsilon_i = \varepsilon_i^+ + \varepsilon_i^- \leq 0 \text{ and } H_i^- = 0 \text{ if } \varepsilon_i = \varepsilon_i^+ + \varepsilon_i^- > 0, \quad i = 1, 3. \tag{6}$$

Parameters $k^+$, $k^-$ are defined as

$$k^+ = \frac{\left[\sum_{i=1}^3 H_i^+ \varepsilon_i^+ (\varepsilon_i^+ + \varepsilon_i^-)\right]}{(\tilde{\varepsilon}^+)^2} \text{ if } \tilde{\varepsilon}^+ \neq 0 \text{ or } k^+ = 0 \text{ if } \tilde{\varepsilon}^+ = 0,$$
$$k^- = \frac{\left[\sum_{i=1}^3 H_i^- \varepsilon_i^+ (\varepsilon_i^+ + \varepsilon_i^-)\right]}{(\tilde{\varepsilon}^-)^2} \text{ if } \tilde{\varepsilon}^- \neq 0 \text{ or } k^- = 0 \text{ if } \tilde{\varepsilon}^- = 0, \tag{7}$$

and from these the strain parameters $\alpha^+$, $\alpha^-$ are computed from

$$\alpha^+ = \min\left(\frac{2k^+ + k^-}{2k^+}, 1\right), \quad \alpha^- = \min\left(\frac{k^+ + k^-}{k^-}, 1\right). \tag{8}$$

Computation of the tensile $\bar{\tau}^+$ and compressive $\bar{\tau}^-$ stress norms is then accomplished by

$$\bar{\tau}^+ = \sqrt{\{\langle\bar{\sigma}\rangle\}^T [D]^{-1} \{\langle\bar{\sigma}\rangle\}}, \quad \bar{\tau}^- = \sqrt{\sqrt{3}(K\bar{\sigma}_{\text{oct}}^- + \bar{\tau}_{\text{oct}}^-)},$$
$$K = \sqrt{2}(1 - R_0)/(1 - 2R_0), \quad R_0 = f_{c(2D)}^- / f_{c(1D)}^-, \tag{9}$$

where $f_{c(1D)}^-$ and $f_{c(2D)}^-$ are the uniaxial and biaxial compressive strengths. Then, the damage threshold values $r_0^+$ and $r_0^-$ are computed from

$$r_0^+ = \frac{f_t}{\sqrt{E}}, \quad r_0^- = \sqrt{\sqrt{\frac{2}{3}} \frac{R_0}{1 - 2R_0} f_o}, \tag{10}$$

where $f_o$ is the elastic limit of the compressive strength. The damage evolution laws for the two damage indices $d^+$ and $d^-$ are given by [8, 10],

$$d^+ = 1 - \frac{r_0^+}{\bar{\tau}^+} \exp\left[A^+ \left(\frac{\bar{\tau}^+}{r_0^+}\right)\right],$$
$$d^- = 1 - \frac{r_0^-}{\bar{\tau}^-}(1 - A^-) - A^- \exp\left[B^- \left(1 - \frac{\bar{\tau}^-}{r_0^-}\right)\right], \tag{11}$$

where terms $A^+$, $A^-$ and $B^-$ are parameters depending on material properties, such as the tensile and compressive strengths, $f_t$, $f_c$, the fracture energy, $G_f$, the Young modulus of elasticity, $E$, and the characteristic element length $l^*$. The evaluation of those parameters is an easy process which can be found in the work of Hatzigeorgiou et al. [10]. Finally, the unification of the two damage indices into one is made by

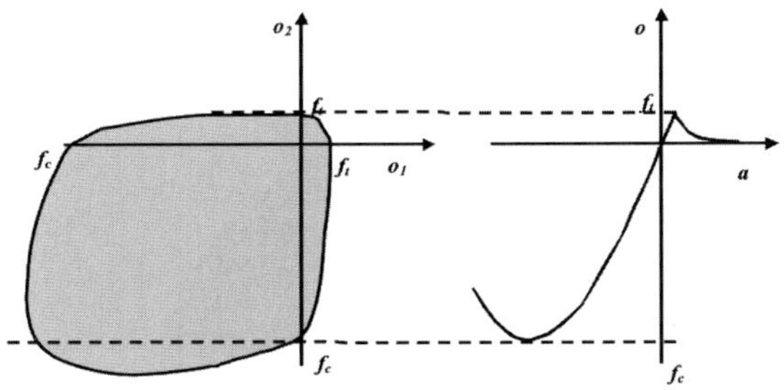

**Fig. 2** Limit damage surface and uniaxial stress-strain curve for the FOM model.

$$d = \alpha^+ d^+ + \alpha^- d^- . \tag{12}$$

Using this final damage index, the total stress tensor $\sigma$ is computed from Equation (1).

Figure 2 shows the limit damage surface and the uniaxial stress-strain curve as produced by the model for two-dimensional (2D) conditions.

## 3 Introductory Remarks about the Structure

The above damage model, developed by Hatzigeorgiou et al. [10] has been applied to the analysis of the Atreus tomb of Figure 1. The structure is divided into 580 3D 8-noded solid finite elements with 1258 nodes in total (Figure 3). For the analyses, the MINOS FEM code, developed by Sfakianakis [11], was used.

The estimated material parameters are shown in Table 1, where $G_f$ is the fracture energy, $f_{o,t} = f_{u,t}$ is the tensile strength, $\sigma_1$ the compression value at point 1 of the diagram $\sigma$–$\varepsilon$ (Figure 4), $\sigma_2$ the corresponding value at point 2 of the same diagram, $\varepsilon_1$ and $\varepsilon_2$ the strain values at points 1 and 2 respectively, $f_{o,c1} \cong -f_c/3$ the uniaxial compressive strength for the onset of damage, $f_{o,c2} \cong -(1.16/3) f_c$ the corresponding biaxial compressive strength, $R_o$ the strength ratio $f_{o,c2}/f_{o,c1}$, $\nu$ the Poisson ratio, $\rho$ the weight density and $E$ Young's modulus of elasticity.

## 4 Static and Dynamic Analysis of the Tomb

The tomb is initially loaded by its own weight and the corresponding seismic force is considered to act statically along the $x$-axis. Two static analyses were performed, considering or not the presence of the covering ground. The reason is that there is some doubt about the exact quantity of soil that covers the tomb. According to the literature, the tomb was initially constructed from masonry and afterwards was

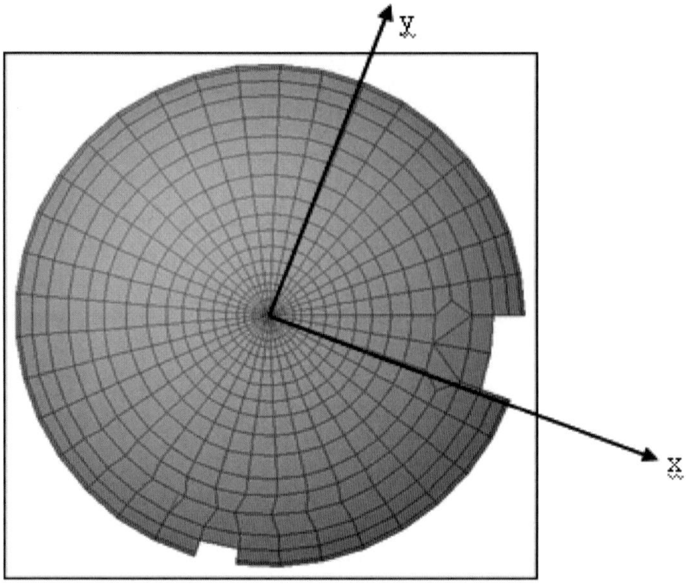


**Fig. 3** Plan of the tomb and its discretization.

**Table 1** Material parameters of the masonry tomb.

| | | |
|---|---|---|
| $G_f$ | kN/m | 0.105 |
| $f_{o,t} = f_{u,t}$ | kPa | 2000 |
| $\sigma_1$ | kPa | –20000 |
| $\varepsilon_1$ | | –0.002 |
| $\sigma_2$ | kPa | –10000 |
| $\varepsilon_2$ | | –0.0035 |
| $f_{o,c1}$ | kPa | –6666.7 |
| $f_{o,c2}$ | kPa | –7733.3 |
| $R_0$ | | 1.16 |
| $\nu$ | | 0.20 |
| $\rho$ | kN/m$^3$ | 22 |
| $E$ | GPa | 25 |

covered with soil. The response of the tomb to this static loading is purely elastic with zero damage index.

As for the dynamic analysis, the structure is assumed to be subjected to a seismic excitation described by the first 5 secs of the N-S component of the 1940 El Centro earthquake. According to the Greek Seismic Code, the peak ground acceleration (pga) in the area of Mycenae is equal to 24% of the gravity acceleration, while the El Centro accelerogram gives a pga of 34%. Hence, the El Centro accelerogram is multiplied by a reduction factor 0.24/0.34 for the horizontal and by $(0.24/0.34) \times (2/3)$ for the vertical component of the imposed excitation. The two components are imposed together in conjunction with simultaneous support excita-

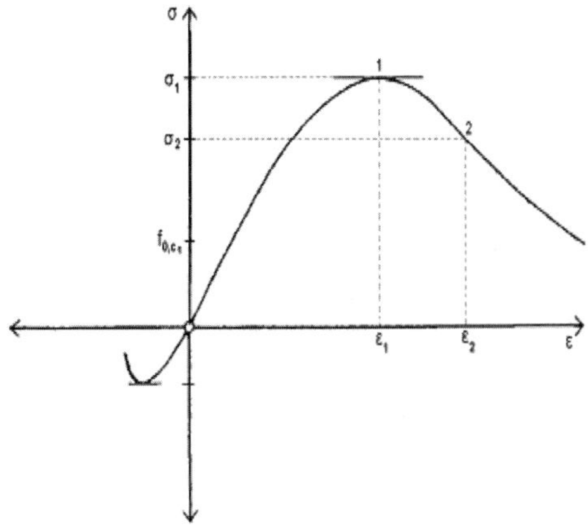

**Fig. 4** Graph of tension-deformation.

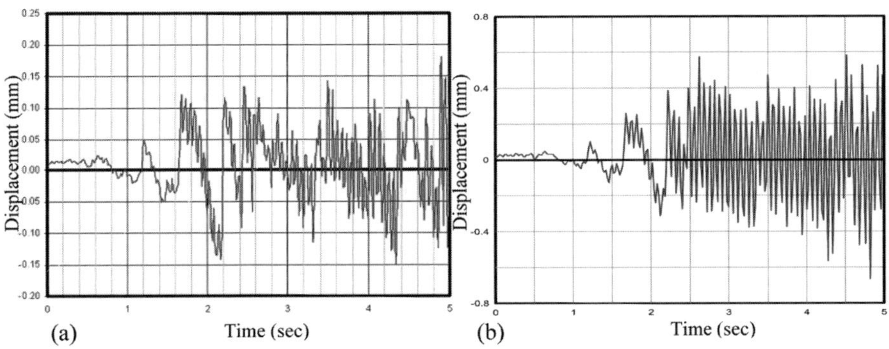


**Fig. 5** Time-displacement history of the top point of the tomb in the case of (a) $E = 25$ Gpa (b) $E = 12.5$ GPa combined with half strength.

tion. To be more conservative, an additional case was examined, where the El Centro accelerogram is multiplied by unity. These analyses took place, considering or not the contribution of damping ratio which can be calculated from the mass damping constant $a_m$, associated with the Rayleigh type of damping.

It is worth to be mentioned that due to the absence of any kind of mortar between the surfaces of the masonry elements, "contact" phenomena occur. In order to take approximately into account these phenomena, one more dynamic analysis was carried out, assuming half the values of Young's modulus of elasticity ($E = 12.5$ GPa) and material strength ($f_c = 10000$ kPa and $f_t = 1000$ kPa). This way, a qualitative evaluation of the behavior of the elements through the reduction of the structural rigidity and strength was achieved.

The analyses results show that the structure behaves full elastically under all assumptions of rigidity and strength. Diagrams in Figure 5 show the displacement history of the top point of the tomb (covered with soil) along the $x$-axis for the

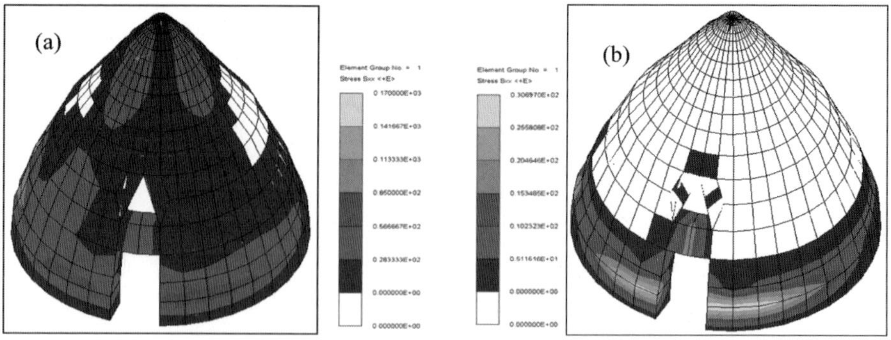


**Fig. 6** Contours of tensile envelope values of stress $\sigma_x$ for the cases of (a) $E = 25$ GPa and (b) $E = 12.5$ GPa.

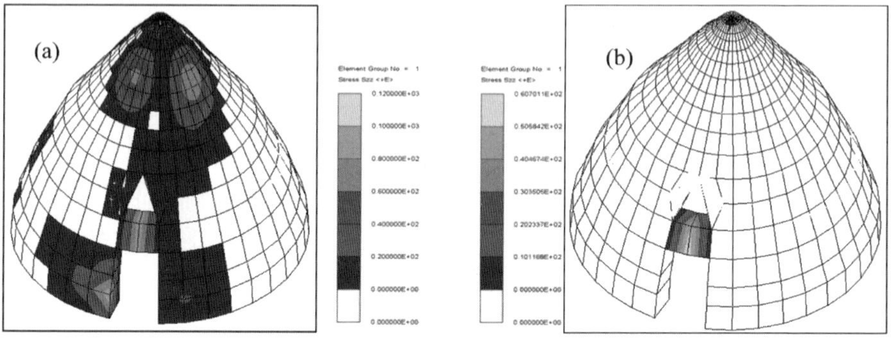


**Fig. 7** Contours of tensile envelope values of stress $\sigma_z$ for the cases of (a) $E = 25$ GPa and (b) $E = 12.5$ GPa.

case of the El Centro accelerogram with pga = 34% g and zero damping. Figure 6 shows the tensile envelope values of stress $\sigma_x$ (along the seismic loading direction) for the two cases of the different modulus of elasticity, while Figure 7 shows the corresponding values of the stress $\sigma_z$ (along the vector of gravity acceleration). It is observed that in all cases the behavior of the monument remains elastic and no damage is developed.

## 5 Conclusions

In the present work a finite element methodology for analyzing historical masonry structures exhibiting linear elastic or inelastic material behavior under static or dynamic loading was presented. The inelastic behavior is successfully simulated in an efficient manner by a continuum damage theory for brittle materials. This method was used to analyze the response of the Atreus tomb in Mycenae to seismic loads under 3D conditions. Dynamic analyses show that the structure behaves purely elastically with no presence of damage noticed. The compressive strength that is connected with the beginning of the damage is 6666.67 kN/m$^2$, while the maximum tensile strength is 2000 kN/m$^2$, as it has already been mentioned. According to the results, the maximum compressive and tensile stresses are $\sigma_x = -260$ kN/m$^2$, $\sigma_z = -740$ kN/m$^2$ and $\sigma_x = 170$ kN/m$^2$, $\sigma_z = 120$ kN/m$^2$, respectively. These results clearly show that no damage occurs and the behavior is linear elastic. In conclusion, the present results demonstrate that the ancient Greek designers and builders possessing excellent technical skills and experience, were able to construct such a great structure that is being kept in perfect condition until our days.

## References

1. Beskos, D.E., Use of finite and boundary elements for the analysis of monuments and special buildings, *Bull. Greek Soc. Civil Engrs., Athens* **216**, 1993, 31–43, 1993 & **217**, 1994, 15–32.
2. Brebbia, C.A. and Lefteris, B. (Eds.), *Structural Studies of Historical Buildings IV*, Computational Mechanics Publications, Southampton, 2 Vols., 1995.
3. Sanchez-Beitia, S. and Brebbia, C.A. (Eds.), *Structural Studies, Repairs and Maintenance of Historical Buildings V*, Computational Mechanics, Southampton, 1997.
4. Roca, P., Conzalez, J.L., Mari, A.R. and Oñate, E. (Eds.), *Structural Analysis of Historical Constructions: Possibilities of Numerical and Experimental Techniques*, CIMNE, Barcelona, Spain, 1997.
5. Crocci, G., *The Conservation and Structural Restoration of Architectural Heritage*, Computational Mechanics Publications, Southampton, 1998.
6. Brebbia, C.A. and Jager, W. (Eds.), *Structural Studies, Repairs and Maintenance of Historical Buildings VI*, Computational Mechanics, Southampton, 1999.
7. Bull, J.W., *Computational Modelling of Masonry, Brickwork and Blockwork Structures*, SaxeCoburg Publications, Edinburgh, 2001.
8. Faria, R. and Oliver, X., *A Rate Dependent Plastic – Damage Constitutive Model for Large-Scale Computations in Concrete Structures*, Monografia No 17, CIMNI, Barcelona, Spain, 1993.
9. Mazars, J., A description of micro-and macroscale damage on concrete structures, *Journal of Engineering Fracture Mechanics* **25**, 1986, 729–737.
10. Hatzigeorgiou, G.D., Beskos, D.E., Theodorakopoulos, D.D. and Sfakianakis, M.G., A simple concrete damage model for dynamic FEM applications, *International Journal of Computational Engineering Science* **2**, 2001, 267–286.
11. Sfakianakis, M.G., *MINOS – A FEM General Purpose Computer Program for Nonlinear Analysis of Structures*, Theory Manual & User's Guide, University of Patras, Greece, 2001–2006.

# Part 7

# MEDICINE

# Homeric Injury Scenes on Ancient Greek Pottery Reveal Medical Knowledge

S. Geroulanos, A. Tasouli, E. Lymberopoulou and K. Papadopoulos

*Onassis Cardiac Surgery Center, Athens, Greece*

**Abstract.** Most of the 114 injury descriptions in the Iliad and the Odyssey are illustrated on numerous ancient and Hellenistic Greek vases. The present work, by collecting and analyzing as many such injury scenes on vases as possible, investigates whether they correspond to advanced knowledge as far as anatomy of vital organs is concerned.

## 1 Introduction

The Homeric poems, mainly the Iliad, have bequeathed to the ensuing generations many sources of inspiration, and this influence has been preserved on ancient pottery. The numerous descriptions from the Trojan War prove that, in that era, the war was part of every day reality. Mutinies, slavery, social agitation and, above all, Persian and Peloponnesian wars had a profound effect on society and consequently on art.

This work investigates whether the artists of the Classic Hellenic Period, undoubtedly influenced by Homer, and absolute experts of the "external" anatomy, had also knowledge of internal anatomy of human body, a knowledge probably existing even in the Homeric era.

## 2 Illustrations Proving Medical Knowledge

Homeric poems, greatly popular works, along with accurate depictions of wounds on ancient vases were used as reference. A conflict is usually illustrated in a way reflecting instruction to the inexperienced warrior as to how he should act. More precisely, the Homeric texts indicate where and how the unskilled warrior should hit his adversary, in order to inflict lethal wounds more efficiently. This is how we can interpret the large number of produced artworks with such subjects as duel, killing

*S.A. Paipetis (ed.), Science and Technology in Homeric Epics,* 269–274.

and war in general. It is important to notice that ancient painters avoid depictions of wounds and of the frightening and unpleasant view of bloodshed.

Despite the fact that preparation for war, war itself and its consequences play an important role in the life of ancient Greeks, the creations of ancient painters are seldom inspired by real wars. The majority of painted representations draw their subjects from mythological scenes. The stories of Homeric heroes were particularly popular. The knowledge of morphology and anatomy of human body allow the exact reproduction of reality in their works. In those depictions, the gestures are expressive, the body at the moment of death creates emotion, blood seems to spout. Most of the 114 injury descriptions in the Iliad and the Odyssey are recognizable still today by means of the many illustrations on ancient and Hellenistic Greek vases.

The aim of this investigation was to gather as many injury scenes depicted on vases as possible and try to investigate whether those injuries relate to contemporary knowledge of the anatomy of vital organs of the human body.

In general, an ancient painter depicts moments just before the lethal wound. Most probably, he tries to avoid the unpleasant and frightening scenes of bleeding. It is striking, however, that in more than 20 painted vases the injury itself is depicted in a way that today is recognized as anatomically correct, as to the hitting a vital organ.

Injuries of the supraclavicular region, the heart, the thoracic cavity, the liver, the aorta, the iliac or femoral artery, prove that the painter should have exact anatomical knowledge of the underlying vital organs (or at least would have consulted experts). He depicts injuries in points of the human body that would result lethal. This conclusion can be drawn also from other mythological scenes depicted on ancient vases, e.g. of wars between Gods and Titans, of Hercules's labors or of the achievements of Theseus.

The results are obvious and impressive, as practically all lethal injuries are depicted at the correct anatomical point, proving that even anonymous painters, who might be illiterate, were not mistaken in their depictions.

The conclusions drawn are based on the artworks of Antiquity till the classical period. However, the numerous specific words used in the Homeric texts with medical terminology recognizable and in use still today, lead us also to the conclusion that the knowledge of human body anatomy should preexist and date back to Homeric times. In contrast to the above, this knowledge was completely lost in Medieval times. The artists of Christian times have practically no knowledge of human anatomy – a kind of knowledge which had to be rediscovered at the times of Michelangelo and Leonardo da Vinci and is constantly broadened ever since.

Even the expression of feelings is depicted in an admirable way. A representative artwork of this expressive ability is the painting of the killing of Penthesilia, the queen of Amazons, by Achilles in front of the walls of Troy. The artwork is attributed to the so-called "painter of Penthesilia". In this painted scene, Achilles plunges his sword deep into the supra-clavicular region, wounding lethally the underlying major vessels. The queen, a tender figure, drops in her knees and raises her right arm offering a desperate resistance, but her face is expressing supplication. She stares her killer into the eyes and, according to the Iliad, she falls in love with him. The eyes of Achilles look astonished, admiring the beauty of his victim and

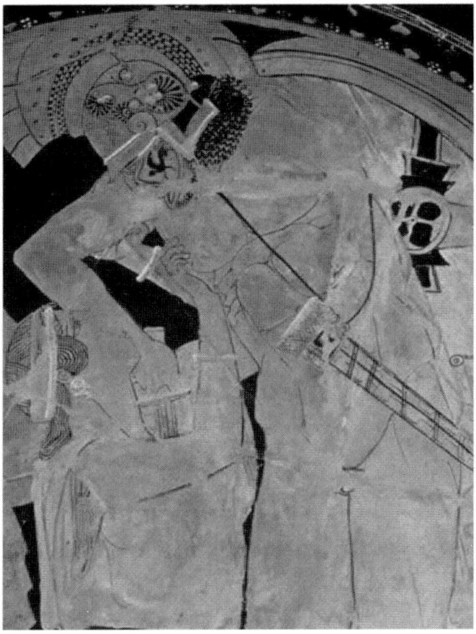

**Fig. 1** Achilles stabs Penthesilia, Queen of Amazons, with his sword. Attica vase, approx. 460 BC. Archaeological Museum, Munich.

**Fig. 2** Battle over the dead body of Achilles, who has two arrow-injuries: one in the heel and one in the back of the thorax. Vase from Chalkis, approx. 520–550 BC.

repenting for this deed. However, love arrives too late. The Amazon succumbs to her severe pulmonary and vessel wound (Figure 1). The depiction of the same scene by Execias, the painter of black-painted vases, one of the greatest artists of his time (late 6th century BC), is analogous. Achilles, however, does not remain unpunished. The great hero of the Greeks, the invulnerable fighter will be wounded by the Paris' arrow, guided accurately by Apollo himself. The latter represents an act of personal revenge (a "vendetta"), since Achilles had killed Hector, a protégée of the God, and had being carrying his lifeless body around the battlefield. Obviously, the artist is fully aware that an arrow-wound at the heel could not be fatal and for this reason he depicts Achilles hit by a second arrow in the thorax (Figure 2).

**Fig. 3** The dead body of Sarpedon is lifted in the air by the two brothers Hypnos (Sleep) and Death. He is fatally wounded in the heart, the iliac and the femoral artery. Attica vase by Euphronius, approx. 510 BC, Metropolitan Museum, New York.

Euphronius, the painter, presents an elegant image of the lethal wounding of Sarpedon, King of Lycia, who is depicted bleeding from the heart and the iliac artery. Concerning the injury of the femoral artery, he is not certain where the artery exactly is, consequently he draws three circles around the points from which blood gushes up (Figure 3).

On the contrary, Douris, a painter of red-painted vases, pinpoints the exact positions of the two femoral arteries, while describing the injuries of Memnon (Figure 4).

Many other illustrations on vases bring us mentally into the fighting scenes of the Trojan War, described most realistically not only by Homer or the physicians of the time, but also by the artists of the Classical Hellenic Period. The acquired knowledge of internal anatomy of the human body is consistent with the work of all artists, namely, the unknown painters of famous artworks, such as the ones designated as of Epidrome, Penthesilia, Berlin, Eretria, Kleophrades, Dokimasia, Brygos, Polyphemos and the well-known ones, such as Execias, Douris, Euphronius and others.

## 3 Medical Terminology

Of course, the said depictions of Homeric stories were painted during the classical period. Did however, this knowledge exist in Homer's time already? Anatomical knowledge, and hence terminology used in the Homeric times, was so accurate that there was no need for those terms to be changed or redefined and were used identically by Hippocrates (460–370 BC) and later on by Galen (130–199 AD).

**Fig. 4** Godess Io transfers the dead body of her son Memnon, wounded in the heart and the two femoral arteries. Attica vase of approx. 480 BC, Louvre Museum, Paris.

Even today, the Western medical terminology includes words rooted in the Iliad and the Odyssey. Professor G. Rigatos in his dictionary *Homeric Words in Contemporary Medicine* [4] reports at least 680 ancient Greek medical words, constituting the basis for thousands of other contemporary anatomical and medical derivatives used in Greek and International Medical Terminology. For example, the word *haema* (blood) with numerous derivatives, such as *haematology*, *haemorrhagy* (bleeding), *haemostasis*, *haemorrhoids*, *haemangioma*, *haematocrit*, *haemolysis*, *haemophilia*, etc., are identically transferred and enrich the international medical terminology. Another example is the word *chole* with its derivatives *cholidochos*, *choliphora*, *cholesterole*, as well as other widely used words, such as *melancholy* etc.

Words like *eye*, *nose*, *throat*, *haema* are easy to describe, since they are visible on the outside. But words like *cardia*, *chole*, *hepar*, *myelos*, *cyste*, etc. need a better anatomical approach. Finally, the names of nerves, tendons, arteries or veins, show that the anatomical knowledge in Homeric times was far more advanced than expected in that remote era.

Cypriot students of the University of Ioannina have traced at least 2450 medical terms in use in modern English, stemming from Homeric words. In French, the medical terms relying on ancient Greek terminology exceed 5000. The anatomical knowledge was broadened and developed further during the Hellenistic Period, especially in Alexandria, Egypt. This knowledge was preserved in the early Byzantine era (330–647 AD) and survived till present through translations in Latin and Arabic, constituting the basis of modern Medical Terminology.

During Middle Ages, the knowledge of anatomy of the human body remained obscured and lots of things had to be rediscovered, almost from the beginning, during Renaissance. However, even the "re-discovered" parts of the human body were named after the same terms that were used in Homeric times and in Antiquity in general.

## 4 Epilogue

In conclusion, judging by the great number of illustrations on ancient pottery depicting body injuries at the correct position to cause a lethal wound, the artists of the Classic Hellenic Period had complete knowledge of the internal anatomy of the human body. Additionally, it is also proven that the Homeric terms for anatomy are still used in contemporary Greek and International Medical Terminology, indicating that essential knowledge of human body anatomy existed already in Homeric times. Finally, the newly created terms in medical vocabulary frequently originate from words of the Homeric vocabulary.

## References

1. Geroulanos, S. and Bidler, R., *Trauma; Wundentstehung und Wundpflege im antiken Griechenland*, Philipp von Zabern, Münster, 1994.
2. Koutrouvelis, K., The Homeric medical terminology, *Davlos* **175**, 1996, 10623–10630.
3. Papayiannopoulos, I., *Subjects of History of Medicine*, Ioannina, 1992, pp. 70–99.
4. Rigatos, G., *Homeric Words in Contemporary Medicine*, Kaktos, Athens, 1996.

# The Healing Art in the Iliad

S.G. Marketos and G.J. Androutsos

*Athens University Medical School, Greece*

**Abstract.** The oldest sources of information about Hellenic medicine are the two Homeric epics, the *Iliad* and the *Odyssey* (7th–8th century BC). *Iliad* provides an unforgettable picture of army surgery and anatomy at the time of Trojan War in Asia Minor. It contains realistic descriptions of wounds and injuries of widely differing types. According to a careful statistical analysis [1] the overall mortality rate due to traumatic lesions was 77.6%. The most dangerous wounds were sword and spear thrusts and the less dangerous ones those inflicted by arrows. It is obvious that in Greek expeditionary force, apart from the amateur surgeons, were also professional healers, skilled in the extraction of embedded weapons, the arrest of hemorrhage and the relief of suffering. The most eminent professional healers were the two sons of the god of healing art Asclepius, Machaon and Podaleirius, famous for their skill as healers, straddling the fine line between professionalism and amateurism. Through their medical knowledge they occupy a special place in healing art and are called "ïetroi" (physicians). In the Homeric world, a physician "ïatros" was a respected figure performing no heroic deeds other than medical caring and healing art.

## 1 Introduction

The most ancient source of knowledge on Greek medicine are the Homeric epics, the *Iliad* and *Odyssey*, composed around 700–750 BC, but based on events occurred five or six centuries earlier.

## 2 The Origin of Traumatology

According to the *Iliad*, mortality from battle wounds was very high. A calculation of all the wounds named in Homer's epic has shown a total of 147 cases [1]: 106 spear-thrust, 17 sword-chops, 12 arrow-shots, and 12 sling-shots. Out of those wounded by spear, four in five died; of sling victims, about two in three; of sword victims, every one; and of arrow victims, nearly half. The most dangerous wounds were sword and spear thrusts while less dangerous were those inflicted by arrows.

*S.A. Paipetis (ed.), Science and Technology in Homeric Epics,* 275–281.

According to a statistical study [2], in the *Iliad*, the various types of wounds are localized mainly on the uncovered parts of the body (head, neck, arms and legs), because only these areas are accessible targets. The location of injuries in the urogenital system was found in 12 cases only, with 92% mortality rate (11 out of 12 cases).

In the *Iliad*, Homer refers to the removal of arrowheads and javelins, and to bandaging, compresses, methods of stopping bleeding and of curing wounds with balm, and to medicines made of herbal extracts [3]. Wine and other liquids are also used to revive the injured. Undoubtedly, medical information given by Homer must reflect contemporary practice in the Minoan and Mycenaean civilizations and in the late Bronze Age. Like most ancient peoples, the Greeks recognized the importance of blood, though not its true functions. The practice of bleeding was used at the time and went on for centuries and for various complaints, either by cutting veins or by cupping.

Despite the frequent references of *Iliad* to gods and prayers of the dying, it is clear that medicine in the time of Homer was not based on magic, but was an independent discipline practised by experts who earned a living from it.

It is evident that a Homeric surgeon could best treat arrow injuries [3]. For example, according to the *Iliad*, Machaon removed the arrow that hit Menelaos. And, later, when Machaon's shoulder is pierced by an arrow, wise Nestor recommended that great measures must be taken quickly, to save the famous professional healer, saying that "a physician is of the worth of many other men for the cutting out of arrows from wounds and the spreading of soothing medicaments".

It is necessary to looking in two directions for the basic features of Homeric surgery. Greek military forces were skilled in the extraction of embedded weapons, the arrest of hemorrhage, and the alleviation of pain. Wounds were washed and picked clean of detritus. Conversely, there was a psychological need for mystic-religious rites, which were important adjuncts to therapies.

## 3 Two Case Reports of Battle-Wounds

The *Iliad* presents one of the world's earliest literary accounts of a battle wound [4]. Two illustrative cases demonstrate the capabilities of a Greek military surgeon.

(a) In the first episode [5], King Menelaus (Helen's forsaken husband) was hit by an arrow, evidently in the waist. Agamemnon, the Mycenaean Commander-in-chief of the Greeks, orders Machaon, son of Asclepius, to help the injured warrior. Machaon approached Menelaus, who lay on the ground. The skillful physician removed the arrow "and when he drew it out, its sharp barbs broke. He loosend the shining belt and the mail-coat below, and last the girdle, bronzed by wondrous smiths. But seeing the wound where the painful arrow entered, he sucked out the blood, then skilfully smoothed on salve, which Cheiron once had kindly given his father". Besides the fact that it would have been better to undress the patient before

**Fig. 1** Achilles (right) bandaging a wound to the arm of his friend Patroclos during the Trojan war.

removing the arrow, this treatment must have worked, since we read that Menelaus went away alive.

Machaon "sucked out" the wound, i.e. presumably he cleaned it somehow – and then laid on a healing salve, which was an heirloom in the physician's own family. Cheiron was a Centaur, or man-horse, who had brought up god Asclepius and taught him the art of healing. We shall never know what the salve contained, nor are there any other clues about wound-healing salves from the days of Trojan War. Menelaus was lucky, because his bronze-clad girdle weakened the force of the shot, or because he was treated by the army's best professional physician.

(b) In another episode [3], Eurypulus, a warrior, is injured by an arrow at the thigh. His friend Patroclos attends to him: "The attendant ... spread the oxhide couch: then as he lay reclined, Patroclos with his dagger, from the thigh cut out the biting shaft; and from the wound with tepid water cleansed the clotted blood; then pounded in his hand, a root applied, astringent, anodyne, which all his pain allayed; the wound was dried, and stanched the blood".

**Fig. 2** Sthenelus (left) bandages Diomedes' hand.

## 4 Discussion

The *Iliad* and the *Odyssey* present the death of more than 200 well-known individuals. In the majority of cases, death results from combat wounds. In accord with the differences in their subject matter, martial violence plays a more important role in the *Iliad* than in the *Odyssey*. In the *Iliad*, a great number of highly varied wounds inflicted on the warriors fighting before the walls of Troy is presented with remarkable anatomical precision, and not without an elengance of style (Figures 1–3). In some ways, these descriptions constitute the oldest surgical report of losses in a military campaign. Homeric warriors know where to strike, to finish off the enemy quickly.

The severity of a wound depends more on the region and the organs hit than the weapon used [4–6]. The Homeric descriptions of wounds, attest to an excellent knowledge of vulnerable points in the human body, of the anatomical arrangement of principal organs, and of the most likely consequences of lesions for each of them. There are thirteen instances, when mortally wounded warriors fall forward, and fourteen when they fall backward. Each time there is a good reason, either physiological or physical, to justify what the poet says. Men die in the *Iliad* falling backward or forward, stiff or sagging, gasping or crying out, but always in a way compatible with what a modern physician would predict, given the exact position of the wound in question. Head wounds are rightly considered especially dangerous. With a sufficiently severe lesion at that part of the body, death is inevitable. Wounds at the face from a spear or sword are particularly shocking. Wounds at the

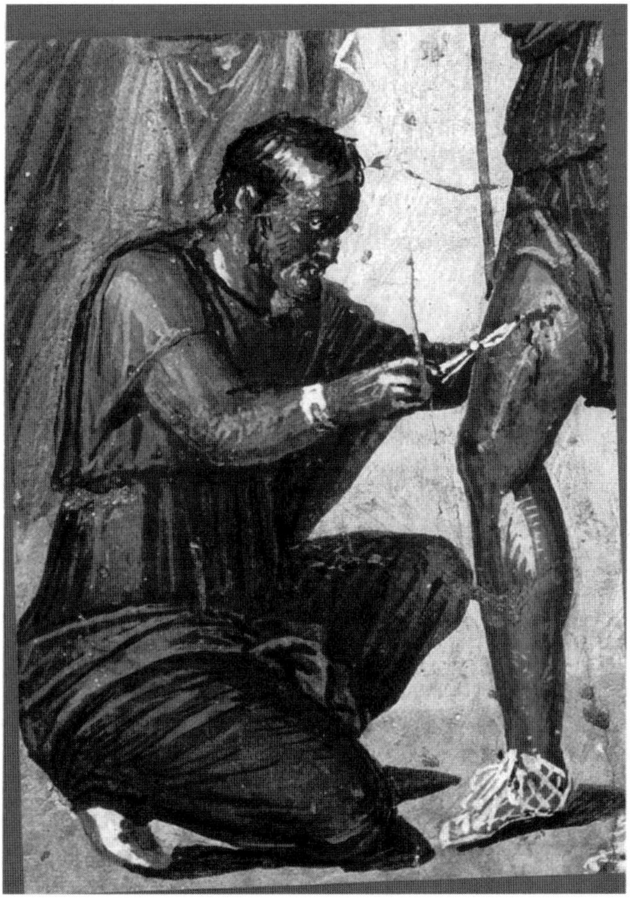

**Fig. 3** Aeneas receives surgery for a leg trauma from a professional healer before leaving Troy with his family.

neck are relatively frequent, considering the small area of this region of the body, which, however is very delicate and poorly protected by the warrior's armor. In several other instances, a warrior plunges his spear into his enemy's throat, cutting the carotid arteries or the whole neck from one side to the other. Sometimes decapitation is the *coup de grâce* administered to a powerless, fallen enemy. The number of wounds to the trunk or abdomen is particularly high in the epic. It is explicable, in view of the central position and relative size of the trunk as a target for blows and arrows. The chest itself was heavily protected, both by the breastplate and by the shield, which was held by the heft hand and covered the cardiac region. The belly was also a region where wounds were fatal. It is a surprising and completely abnormal fact that Homeric heroes never get sick as a result of their wounds. They either die or return to normal activity very quickly. The rule in Homer is that the heros either die or reemerge in good health; wounded men can be in a state of traumatic shock, but never have any fever; cuts bleed but are never inflamed; tissues are torn and smashed but never suppurate.

It is obvious that the Greek expeditionary force, apart from amateur surgeons, included professional healers also, skilled in the extraction of embedded weapons, arrest of hemorrhage and relief of suffering. The most eminent of them were Machaon and Podaleirius, the sons of "the blamed physician" Asclepius [6]. They serve, like their father, as physicians and they straddle the fine line between professionalism and amateurism [5]. They did not work for a reward, through their medical knowledge occupy a special position in the healing art and are called "ïetroi".

## 5 Concluding Remarks

1. The *Iliad* provides a realistic picture of surgery and an illuminating account of anatomy at the time of the Trojan War.
2. Many of the names of the various human organs and bones – still in use in anatomical terminology – are found in the *Iliad*.
3. Anatomical knowledge of Homer's time was attributed to critical observation of human injuries, rersulting from wounds received in battle and not through dissections.
4. Homer seems to have an intimate knowledge of human anatomy and describes it diligently, so that some medical historians and writers [1, 7, 8] maintained that the author of the greatest epics "was an army surgeon himself!"
5. It is evident that wounds were not the only occasions to observe human body and skeleton, and it is certain that Greek expeditionary force, apart from amateur surgeons, included very experienced professional healers ("ïetroi") as well.
6. According to the epic, healing of trauma has a special usefulness. It was frequently quoted as the glorification of the whole of healing art.
7. Homeric Asclepius [6] was depicted as a physician-hero in the first place and then was transformed to the god of healing art.
8. The two eminent sons of Homeric Asclepius, Machaon and Podaleirius, cross the fine line between professionalism and amateurism [3, 5].
9. The reprentatives of surgery (Machaon) and of internal medicine (Podaleirius) were skilled craftsmen [5], performing no heroic deeds other than medical care and the healing art.
10. Homer's assertion that "a physician is worth many of laymen" reveals that in the Homeric world a physician was an important figure, and medicine was the most respected and useful of all human arts.

## References

1. Frölich, H., *Die Militärmedizin Homers*, Stuttgart, 1879.

2. Poulakou-Rebelakou, E., Rebelakos, A.G. and Marketos, S.G., Urological references in the Homeric epics, in *De Historia Urologiae Europaeae*, J.J. Mattelaer (Ed.), 1998, Kortrijk, Belgium, pp. 249–257.
3. Bishop, W.J., *The Early History of Surgery*, Robert Hale Ltd., London, 1960.
4. Edelstein, E.J. and Edelstein, L., *Asclepius: A Collection and Interpretation of the Testimonies*, Vol. 2, John Hopkins Press, Baltimore, 1945.
5. Horstmanshoff, H.F.J., The ancient physician: Craftsman or scientist?, *J. Hist. Med.* **45**, 1990, 176–197.
6. Bailey, J.E., Asklepius: Ancient hero of medical care, *Ann. Intern. Med.* **124**, 1996, 257–263.
7. Iliopoulos, S., The traumatology of the Trojan war, *Hellenic Surgical Orthopedics and Traumatology, Gr.* **43**(2), 1992, 60–73.
8. Godquin, B., Homère était-il chirurgien?, *Chirurgie* **116**, 1990, 136–147.

# Medicinal Herbs and Plants in Homer

G. Klimis

*Athens, Greece*

**Abstract.** This paper is a brief survey of references found both in the *Iliad* and the *Odyssey* on herbs and medicinal plants, as well as to their use for healing wounds and the treatment of many diseases in Homeric Epics era. Textual references are used to describe several herbs and plants, easily identifiable today. The analysis of Homeric verses though, proves that modern translations are often less than accurate. The description of herbs and medicinal plants indicates that their use was common during the previous two or even three thousands years. It points out a natural adaptation of the healing process to the particular environment of Greece, although several modern researchers tend to question the use of the plants. However, despite specialists' disagreement, it is quite certain that the extensive use of plants and herbs in a way that seems particularly attached to the Greek civilisation contributed to further development of herbology, pharmacology and medicine.

## 1 Introduction

Contrary to common belief, recent studies have proved that the works of Homer record substantial knowledge for several disciplines that Greeks possessed. References to medical matters with emphasis to healing processes, denote a whole range of beliefs, knowledge and practices, which, despite their religious origins, constitute the foundation of Greek medical science.

People of the Aegean Sea use their everyday observation and experience to establish, rather quickly as it is, some fundamental scientific notions. Based on descriptions of scenes with wounded or dying heroes, one can assume that Homeric people had definitely acquired some elemental notions of anatomy. It is also important to point out that they followed a distinct treating protocol for wounds inflicted during battle.

The use of herbs and medicinal plants, quite extensive during the Homeric era, was not only limited to the treatment of wounds. It was also a means to soothe physical and, possibly, even psychological pain.

They can thus be divided to those which treated and soothed and those destined to poison. Also, the Epics bear traces of preventive medicine, just as in the conservation

*S.A. Paipetis (ed.), Science and Technology in Homeric Epics,* 283–291.

of dead bodies and the burning of the dead, to avoid contamination. In this latter case, burning the bodies in a timely manner prevents epidemics from spreading.

To be able to survive through the numerous battles they fought, Homeric heroes had to master some basic notions of wound treatment. The battles themselves provided everyone with this notion. However, there were specialised doctors for this job. We meet them in the epics, where they are called "ιητῆρες" (healers) and belonged to the same respectable social category as "δημιοεργοί (skilled workmen) – along with craftsmen and bards – all of them most welcome in every royal court.

The Centaurs were particularly familiar with the properties of plants and knew how to use them for healing, although only one of them, Cheiron, became the very image of the healer through time. His family origins suffice to justify this development, as he was thought to be son of Cronus (Saturn) and Phillyrea, the nymph after whom the lime tree was named, and was widely used in ancient Greek medicine. The name "Cheiron" is associated with "χειρ", i.e. "hand", and a magical one at that. Cheiron, contrary to other Centaurs, is portrayed as just, calm, even sweet, and beneficial, attributes he has in common with other healers, Asclepius and Apollo.

## 2 Homeric References to Plants and Magical Herms

Pharmaceutical substances in the epics can be categorised to those which heal or soothe and those which poison. In Od. X, 234–236, Circe is said to use "Λυγρά Φάρμακα" slipping them in some kind of soup, the Κυκεών" (a mixture of Πράμνιος Οίνος" (Pramnian wine), flour and goat cheese), which she flavoured with honey, to mask the bitter taste of the plants, in order to offer it to Ulysses' crew (Od. X, 264–266, X.329):

> ἐν δέ σφιν τυρόν τε καὶ ἄλφιτα καὶ μέλι χλωρὸν
> οἴνῳ Πραμνείῳ ἐκύκα· ἀνέμισγε δὲ σίτῳ
> φάρμακα λύγρ᾽, ἵνα πάγχυ λαθοίατο πατρίδος αἴης.

(... and mixed them a mess with cheese, honey, meal, and Pramnian but she drugged it with wicked poisons to make them forget their homes.)

Another plant mentioned is (Od. IV, 219–221), the *Nepenthes*, combined with wine, acted as a powerful medicine, while it was also soothing and killing pain. It is described as a substance, suspending conscious response to external stimuli as well as recalling events through memories, although not causing any confusion or limited perception of stimuli.

This plant, certain researchers believe, was a narcotic substance, specifically opium, whereas others maintain that it was a collective noun for a whole class of cheering medicines. According to Homer, *Nepenthes* originates from Egypt, and it was *Polydamne*, *Thon*'s wife who taught its uses to Menelaus' wife, Helen of Sparta. *Nepenthes* is reminding the cult of Osiris (Od. IV, 219–221):

ἔνθ᾽ αὖτ᾽ ἄλλ᾽ ἐνόησ᾽ Ἑλένη Διὸς ἐκγεγαυῖα·
αὐτίκ᾽ ἄρ᾽ εἰς οἶνον βάλε φάρμακον, ἔνθεν ἔπινον,
νηπενθές τ᾽ ἄχολόν τε, κακῶν ἐπίληθον ἁπάντων.

(Then Jove's daughter Helen bethought her of another matter. She drugged the wine with an herb that banishes all care, sorrow, and ill humour.)

"Μυρίκη" or "Αρμυρίκη" (Myrike-Armyrike) was associated with beauty and youth and was devoted to Aprodite (Venus). Tradition says that Myrike, daughter of Cinyras, king of Cyprus, and sister of Adonis, was transfigured to this tree. Homer describes (Il. X, 466) how Ulysses hang Dolon's armour on such a tree, to signify respect for the memory of the young man, killed by Diomedes.

However, the number of adjectives describing these plants indicates that Homer was aware of the dangers following their use. When Menelaus was wounded during the war, healer Machaon treated him (Il. IV, 212–219):

πάσσε, τά οἵ ποτε πατρὶ φίλα φρονέων πόρε Χείρων.
κυκλόσ᾽, ὃ δ᾽ ἐν μέσσοισι παρίστατο ἰσόθεος φώς,
αὐτίκα δ᾽ ἐκ ζωστῆρος ἀρηρότος ἕλκεν ὀϊστόν·
τοῦ δ᾽ ἐξελκομένοιο πάλιν ἄγεν ὀξέες ὄγκοι
Λῦσε δέ οἱ ζωστῆρα παναίολον ἠδ᾽ ὑπένερθε
ζῶμά τε καὶ μίτρην, τὴν χαλκῆες κάμον ἄνδρες.
Αὐτὰρ ἐπεὶ ἴδεν ἕλκος ὅθ᾽ ἔμπεσε πικρὸς ὀϊστός,
αἷμ᾽ ἐκμυζήσας ἐπ᾽ ἄρ᾽ ἤπια φάρμακα εἰδὼς

(He undid the burnished belt, and beneath this the cuirass and the belt of mail which the bronze-smiths had made; then, when he had seen the wound, he wiped away the blood and applied some soothing drugs which Chiron had given to Aesculapius out of the good will he bore him.)

Another of the older gods-healers, according to Homer, was Paeon, who healed Ares (Mars) in Olympus mountain, using "Φάρμακα" (painkillers) (Il. V, 401–402) [1]:

Τῷ δ᾽ ἐπὶ Παιήων ὀδυνήφατα φάρμακα πάσσων
ἠκέσατ᾽· οὐ μὲν γάρ τι καταθνητός γε τέτυκτο.

(... whereon Paeeon spread pain-killing herbs upon his wound and cured him, for he was not of mortal mould.)

Ancient Greeks named the plant "Παιωνία" (peony) after him, having the power to stop bleeding. Dioskourides (III, 140) was calling "Παιωνία" also "Γλυκυσίδη, Πεντόροβον", adding that its root is given to women having just given birth and in need of purging. When drunk with wine, it helps those with pains in the abdomen or urethra or suffering of jaundice or kidney diseases; children could eat the seeds, in order to be cured from the disease of the stone.

Given our actual knowledge of medicine, a careful reader of *Odyssey* (Od. X, 340–342) reaches the conclusion that there is a reference to an "anti"-aphrodisiac plant. "Μώλυ" (Moly) [2] is mentioned only once in the *Odyssey* and never in the *Iliad* (Od. X, 302–306):

ὣς ἄρα φωνήσας πόρε φάρμακον ἀργεϊφόντης
ἐκ γαίης ἐρύσας, καί μοι φύσιν αὐτοῦ ἔδειξε.
ῥίζῃ μὲν μέλαν ἔσκε, γάλακτι δὲ εἴκελον ἄνθος·
μῶλυ δέ μιν καλέουσι θεοί· χαλεπὸν δέ τ᾽ ὀρύσσειν
ἀνδράσι γε θνητοῖσι, θεοὶ δέ τε πάντα δύνανται.

(As he spoke he pulled the herb out of the ground and showed me what it was like. The root was black, while the flower was as white as milk; the gods call it Moly, and mortal men cannot uproot it, but the gods can do whatever they like.)

According to the Homeric text, Ulysses did not know how to counteract both magical plants and the beauty of Circe, who had already turned his men into pigs (this is a possible metaphor as to how inebriation can make men to lose their humanity). Unsuspecting, he was heading to her palace, when he met Hermes (Mercury) who effortlessly, unlike humans, takes "Μώλυ" (Moly) with its black roots and white flowers off the ground and offers it to Ulysses. Thus, as Homer masterly portrays the scene, the hero remained immune to Circe, who in utter surprise, watched her magical plants failing her for the first time. Not only had her beauty and knowledge of plants proven ineffective, but Ulysses, enraged, tried to kill her by his sword. The witch-goddess begs of him to have mercy on her. Popular practices along with the description of the plant (black roots and white flowers) help to identify Μώλυ" (Moly) as the black hellebore.

"Στραμώνι ον" (Stramonion) is another of Circe's magical plants, rich in a mixture of alkaloids, such as atropine and hyoscyamine, both powerful poisons. In modern pharmacology *galanthamine* is another alkaloid extracted from the eyes of "γάλανθος ο χιονώδης" (Snowdrop).

In the underworld, Hades welcomes the ghostly figures of the dead in a field of asphodels, whose pale flowers match the sadness and loneliness of the place. It is in such a field (since they are not rare to see, even nowadays) that the souls of the heroes slain during the Trojan War met (Od. XI, 538–539):

"ὣς ἐφάμην, ψυχὴ δὲ ποδώκεος Αἰακίδαο
φοίτα μακρὰ βιβᾶσα κατ᾽ ἀσφοδελὸν λειμῶνα,

(… whereon they reached the meadow of asphodel where dwell the souls and shadows of them that can labour no more,)

White asphodel, a member of Liliaceae family, was also the symbol of god Dionysus. To Ancient Greeks, it signified mourning, and they used to plant it in graveyards for the dead to feed on it. They believed the dead to reside in asphodel fields in the underworld. Asphodel flowers were believed to make forget (herdsmen say goats eating them become dizzy, engorged and can even die), while asphodel leaves were used to stuff cheap mattresses.

"Ασφόδελος ο Κοίλος" (Asphodel the Hollow), that is to say asphodels which are bending, yielding, must stand for another type of the plant, decorating with its white flowers and fine leaves the Elysian Fields, the land in the west side of the Earth, where the sons of gods and heroes killed during battles resided after death.

Asphodel roots were used to cure diarrhoeas and as a means to treat high blood pressure and circulation problems.

Timely burning of bodies during the plague sent by Apollo, was a basic form of protection and prevents from further spread of the epidemic (Il. I, 52):

βάλλ᾽· αἰεὶ δὲ πυραὶ νεκύων καίοντο θαμειαί.

(... and all day long the pyres of the dead were burning.)

The epics also bear traces of preventive medicine: there is a process to conserve dead bodies and to prevent contamination by burning the dead. Under divine supervision a special salve is spread on Partoclus' dead body to counteract decay (Il. XIX, 38–39):

Πατρόκλῳ δ᾽ αὖτ᾽ ἀμβροσίην καὶ νέκταρ ἐρυθρὸν
στάξε κατὰ ῥινῶν, ἵνα οἱ χρὼς ἔμπεδος εἴη.

(... and she then dropped ambrosia and red nectar into the wounds of Partoclus, that his body might suffer no change.)

The warriors themselves seem to be familiar with this knowledge. In both of the epics we find the description of a medicine art with no mythical or god inspired elements, based on experience and functioning as an independent art. An art practiced by skilled practitioners, who get paid for their services. Homer portrays them when using their skills and enjoy public appraisal for their job (Il. XI, 514–515):

ἰητρὸς γὰρ ἀνὴρ πολλῶν ἀντάξιος ἄλλων ἰούς τ᾽ ἐκτάμνειν ἐπί τ᾽ ἤπια φάρμακα πάσσειν.

(A physician is worth more than several other men put together, for he can cut out arrows and spread healing herbs.)

Descriptions of wounded and dead warriors lead to the conclusion that people of Homeric age knew basic anatomy and plant-based treatments (Il. IV, 518–521):

χερμαδίῳ γὰρ βλῆτο παρὰ σφυρὸν ὀκριόεντι
κνήμην δεξιτερήν· βάλε δὲ Θρῃκῶν ἀγὸς ἀνδρῶν
Πείρως Ἰμβρασίδης ὃς ἄρ᾽ Αἰνόθεν εἰληλούθει.
Ἀμφοτέρω δὲ τένοντε καὶ ὀστέα λᾶας ἀναιδὴς

(... for he was struck by a jagged stone near the ancle of his right leg. He that hurled it was Peirous, son of Imbrasus, captain of the Thracians, who had come from Aenus; the bones and both the tendons were crushed by the pitiless stone ...)

Also (Il. V, 290–294):

Ὣς φάμενος προέηκε· βέλος δ᾽ ἴθυνεν Ἀθήνη
ῥῖνα παρ᾽ ὀφθαλμόν, λευκοὺς δ᾽ ἐπέρησεν ὀδόντας.
Τοῦ δ᾽ ἀπὸ μὲν γλῶσσαν πρυμνὴν τάμε χαλκὸς ἀτειρής,
αἰχμὴ δ᾽ ἐξελύθη παρὰ νείατον ἀνθερεῶνα·
ἤριπε δ᾽ ἐξ ὀχέων, ἀράβησε δὲ τεύχε᾽ ἐπ᾽ αὐτῷ

(With this he hurled his spear, and Minerva guided it on to Pandarus's nose near the eye. It went crashing in among his white teeth; the bronze point cut through the root of his to tongue, coming out under his chin, and his glistening armour rang rattling round him as he fell heavily to the ground. The horses started aside for fear, and he was reft of life and strength.)

Again, in Il. V, 73–74:

βεβλήκει κεφαλῆς κατὰ ἰνίον ὀξέϊ δουρί·
ἀντικρὺ δ᾽ ἀν᾽ ὀδόντας ὑπὸ γλῶσσαν τάμε χαλκός·

(... drove a spear into the nape of his neck: it went under his tongue all among his teeth, so he bit the cold bronze, and fell dead in the dust.)

As to the protocol of treating wounds in the middle of the battle (Il. XI, 828–830):

ἀλλ᾽ ἐμὲ μὲν σὺ σάωσον ἄγων ἐπὶ νῆα μέλαιναν,
μηροῦ δ᾽ ἔκταμ᾽ ὀϊστόν, ἀπ᾽ αὐτοῦ δ᾽ αἷμα κελαινὸν
νίζ᾽ ὕδατι λιαρῷ, ἐπὶ δ᾽ ἤπια φάρμακα πάσσε

(But save me and take me to your ship; cut out the arrow from my thigh; wash the black blood from off it with warm water, and lay upon it those gracious herbs which ...)

And in Il. V, 112–113:

πὰρ δὲ στὰς βέλος ὠκὺ διαμπερὲς ἐξέρυσ᾽ ὤμου·
αἷμα δ᾽ ἀνηκόντιζε διὰ στρεπτοῖο χιτῶνος.

(... the arrow went right through the metal and pierced the flesh, so that the cuirass was covered with blood.)

According to how deep they are, wounds are qualified as more or less dangerous. Mortality rates are high though. Treatment starts cleansing of the wound and washing with lukewarm water. Special care is taken to stop the bleeding by means of leaves, powder or other effective concoctions. Then the wound had to be bandaged.

Life and healthiness were cherished above everything by the Homeric heroes, whereas nothing was more despised than an unworthy death out of the battlefield. Disease was thus considered as a major misfortune. To prevent it they tried to stay healthy by means of a frugal diet, physical exercise and personal hygiene.

The ancient city of *Ephyra*, near the river *Selleis*, having its springs near *Pholoy*, needs to be mentioned as related to these questions. It was the kingdom of mythical king *Augeias*, whose stables were cleaned by Hercules. The city name was given to honour a city in Epirus, which stood as its metropolis. Lots of magical plants were believed to be found in the outskirts and people from all over Greece would come to collect them. Even Telemachus, on its way to Sparta, came here first, to find the ingredients for the potion he needed (Il. XV.514–519):

ὣς εἰπὼν ὄτρυνε μένος καὶ θυμὸν ἑκάστου.
ἔνθ᾽ Ἕκτωρ μὲν ἕλε Σχεδίον Περιμήδεος υἱὸν
ἀρχὸν Φωκήων, Αἴας δ᾽ ἕλε Λαοδάμαντα
ἡγεμόνα πρυλέων Ἀντήνορος ἀγλαὸν υἱόν·
Πουλυδάμας δ᾽ Ὦτον Κυλλήνιον ἐξενάριξε
Φυλεΐδεω ἕταρον, μεγαθύμων ἀρχὸν Ἐπειῶν.

(With these words he put life and soul into them all. Hector then killed Schedius son of Perimedes, leader of the Phoceans, and Ajax killed Laodamas captain of foot soldiers and son to Antenor. Polydamas killed Otus of Cyllene a comrade of the son of Phyleus and chief of the proud Epeans.)

*Augeias*' daughter, *Agamede*, a renowned witch, lived in the same city and was married to Moulius, a distinguished warrior. Augeias' son, Phyleas had made the armour of a friend of Otos killed by Polydamas in the Trojan War.

In the epics one can find the description of no less than 140 types of wounds inflicted with weapons of that age; wounds of the scull, neck and throat, thorax, traumas of the inner organs considered as particularly dangerous, fractures of upper and lower ends and of skull bones etc. Most of the anatomical terms are still used in modern Greek and are in the basis of international medical terminology.

In *Iliad* X, Paeon is named as the healer of the gods and is seen to treat first Hades and then Ares (Mars).

Machaon and Podaleirios, sons of Asclepius, heads of the army and of a thirty boat fleet of the cities Trikke, Ithome and Oichalia, fought also in the Trojan War. To Homer they are first of all "Αγαθοί Ιητήρες", kind healers, who learned the precious art from their father. Machaon's services prove to be most valuable, as he knows how to extract arrows from the body and heal the wound with the appropriate herbs and plants (Il. IV, 210–219):

Ἀλλ᾽ ὅτε δή ῥ᾽ ἵκανον ὅθι ξανθὸς Μενέλαος
βλήμενος ἦν, περὶ δ᾽ αὐτὸν ἀγηγέραθ᾽ ὅσσοι ἄριστοι
κυκλόσ᾽, ὁ δ᾽ ἐν μέσσοισι παρίστατο ἰσόθεος φώς,
αὐτίκα δ᾽ ἐκ ζωστῆρος ἀρηρότος ἕλκεν ὀϊστόν·
τοῦ δ᾽ ἐξελκομένοιο πάλιν ἄγεν ὀξέες ὄγκοι.
Λῦσε δέ οἱ ζωστῆρα παναίολον ἠδ᾽ ὑπένερθε
ζῶμά τε καὶ μίτρην, τὴν χαλκῆες κάμον ἄνδρες.
Αὐτὰρ ἐπεὶ ἴδεν ἕλκος ὅθ᾽ ἔμπεσε πικρὸς ὀϊστός,
αἷμ᾽ ἐκμυζήσας ἐπ᾽ ἄρ᾽ ἤπια φάρμακα εἰδὼς
πάσσε, τά οἵ ποτε πατρὶ φίλα φρονέων πόρε Χείρων

(... and went on till they came to the place where Menelaus had been wounded and was lying with the chieftains gathered in a circle round him. Machaon passed into the middle of the ring and at once drew the arrow from the belt, bending its barbs back through the force with which he pulled it out. He undid the burnished belt, and beneath this the cuirass and the belt of mail which the bronze- smiths had made; then, when he had seen the wound, he wiped away the blood and applied some soothing drugs which Chiron had given to Aesculapius out of the good will he bore him.)

In the *Iliad* there is no mention of Podaleirios' services, but in a fragment of the lost epic of Arctinus *The Sacking of Ilion*, Machaon appears to be keen in surgery, while Podaleirios in treating diseases.

In Il. XX, 478–485, one reads:

Δευκαλίωνα δ᾽ ἔπειθ᾽, ἵνα τε ξυνέχουσι τένοντες
ἀγκῶνος, τῇ τόν γε φίλης διὰ χειρὸς ἔπειρεν
αἰχμῇ χαλκείῃ· ὃ δέ μιν μένε χεῖρα βαρυνθεὶς
πρόσθ᾽ ὁρόων θάνατον· ὃ δὲ φασγάνῳ αὐχένα θείνας
τῆλ᾽ αὐτῇ πήληκι κάρη βάλε· μυελὸς αὖτε
σφονδυλίων ἔκπαλθ᾽, ὃ δ᾽ ἐπὶ χθονὶ κεῖτο τανυσθείς.
αὐτὰρ ὃ βῆ ῥ᾽ ἰέναι μετ᾽ ἀμύμονα Πείρεω υἱὸν
Ῥίγμον, ὃς ἐκ Θρῄκης ἐριβώλακος εἰληλούθει·

(... wounded Deucalion in the fore-arm where the sinews of the elbow are united, whereon he waited Achilles' onset with his arm hanging down and death staring him in the face. Achilles cut his head off with a blow from his sword and flung it helmet and all away from him, and the marrow came oozing out of his backbone as he lay. He then went in pursuit of Rhigmus, noble son of Peires, who had come from fertile Thrace, ...)

In Il. XVI, 345–350, one reads:

Ἰδομενεὺς δ᾽ Ἐρύμαντα κατὰ στόμα νηλέϊ χαλκῷ
νύξε· τὸ δ᾽ ἀντικρὺ δόρυ χάλκεον ἐξεπέρησε
νέρθεν ὑπ᾽ ἐγκεφάλοιο, κέασσε δ᾽ ἄρ᾽ ὀστέα λευκά·
ἐκ δ᾽ ἐτίναχθεν ὀδόντες, ἐνέπλησθεν δέ οἱ ἄμφω
αἵματος ὀφθαλμοί· τὸ δ᾽ ἀνὰ στόμα καὶ κατὰ ῥῖνας
πρῆσε χανών· θανάτου δὲ μέλαν νέφος ἀμφεκάλυψεν.

(Idomeneus speared Erymas in the mouth; the bronze point of the spear went clean through it beneath the brain, crashing in among the white bones and smashing them up. His teeth were all of them knocked out and the blood came gushing in a stream from both his eyes; it also came gurgling up from his mouth and nostrils, and the darkness of death enfolded him round about.)

So, both healers and warriors in the epics knew a lot about painkillers and ways to cleanse and sterilise. They knew how to stop bleeding, how to perform operations and disposed of an impressive variety of bandages. They also had stretchers, while there is a mention, in the first book of the *Iliad* of using *paeanes* (choral songs) therapeutically: all day long performance of hymns and paeanes prevented from total damage induced by the disease, spread by the "talking arrows" of Apollo.

## 3 Conclusions

Substancial medical knowledge is presented and catalogued in the epics of Homer. Particular emphasis is given to medicinal plants and substances which are either healing or poisoning. Though more empirical than based in thorough scientific background, this knowledge constitutes however the foundation of scientific medicine and healing methods which will later develop in the Greece.

## References

1. Klimis, K.G., *Herbs and Medicines in Ancient Greece*, Publ. Georgiades, Athens, 2006.
2. Vlachakis, N., Moly, The antimagic herb of ancients Gods of Greece, *Bulletin I.K.M.* **20**, 1996.
3. Fougias, P., The magic Moly, *Bulletin I.K.M.* **28**, July–December, 2000.
4. Chaviara-Karachaliou, S., Is the Homeric Moly the black mandragora, poster presented at *History of Medicine in Athens*, 2 December 1995.
5. Chaviara-Karachaliou, S., The ophthalmology knowledges of Homer, *Bulletin Ophthalmology* **7**(2), Thessaloniki, 1995. Also, *Medical Science, Historical Society, Bulletin* **13**, Summer, 1995, 3–9.
6. Koliopoulos, J., Symbology proportion for a magic herb, *Bulletin I.K.M.* **22**, July–December, 1997, 60.

# Part 8

# FLORA AND FAUNA

# Agricultural Development in the Homeric Era

C.C. Thanassoulopoulos

*Aristotelian University of Thessaloniki, Greece*

**Abstract.** In this work, the Homeric references to plants and their way of cultivation in the Iliad and the Odyssey are briefly presented. The plant species are described, as appear in a table, and are similar to those cultivated nowadays. However the analysis of references concerning the plant species zia (ζεια) and olyra (όλυρα) indicate that translations given to day do not correspond to the real meaning. The two paragraphs referred to these plants indicate undoubtedly that are not species of barley, as usually translated by the scholars. The probability that they were species of the genus Sorgum is the most significant, as this genus is indigenous in Greece since antiquity and still exists under several names. From the above mentioned we could draw the conclusion that the agricultural practicing in the Homeric period is similar to that existing during two or three millennia before. It is described as typically specialized and intensive Mediterranean multi-cultivation, due to the natural adaptation of cultivated plants to the particular Greek peninsula environment. Other scholars however, consider that little evidence survived concerning specificity of cultivations in the areas described in Homeric poems which by no means shows cultivation intensity. In spite of academic disagreements the model of this agricultural development which was applied for almost five millennia – from at least the Late Neolithic to our days – is the one that contributed to the establishment of the Greek civilization and its dramatic evolution.

## 1 Agricultural Development

The Homeric period covers few years of the late Bronze Age at the end of the Mycenaean period. According to the existing data, events described by Homer took place some time between the end of the second millennium or, according to some authors, possibly even earlier. Therefore, everything referring to the agriculture of that time should be examined within the frame of the wider Bronze Age period, in order a more objective and exact picture to be obtained. Considering that farmers are regarded as the most conservative part of any population in the sense of preserving traditions, it is not surprising that in a number of cases the agricultural practices traced during the middle Neolithic Age, i.e. about 6000–8000 years ago, are similar to those applied in mountainous and isolated Greek villages nowadays. Regarding the above mentioned, an effort was undertaken in order, everything reported in the Homeric poems concerning agriculture, to be collected. The data were classified ac-

*S.A. Paipetis (ed.), Science and Technology in Homeric Epics,* 295–301.

cording to the most recent concepts of agricultural practice and compared to those of the Bronze Age [1–5, 7, 10–16, 18, 19]. In this project only the Homeric references reporting plants and the way of their cultivation are presented.

The decoding of the Linear B script, found in the ruins of the Mycenaean and Minoan palaces, revealed the farming practices during the Bronze Age. The accounting system by the king's administration service, the way of practicing the farming as well as the cultivated plant species was more or less evident in these scripts. For that reason the agricultural events described in the Homeric poems are compared to those of the afore-mentioned data provided by the Linear B script. The cultivated as well as the other wild plants shown in the wall painting and mainly the seed remnants found in excavations, or mentioned both in the Linear B script and in the Homeric texts are presented in Table 1. It is easy for any reader to find the verse of the Homeric poems to which we refer since it is reported in a parenthesis in Table 1.

Thus it is evident that 39 genus of plants are reported from the excavations, whereas 28 from the Linear B script and 52 from the Homeric texts. The difference in the reported figures is justified due to the fact that most of the times in the scripts only the plants or plant products that could be stored [17] are reported. The pertaining of agricultural and forest species indicated rather clearly that there are no differences among these three sources, which furthermore supports that the Homeric period is enrolled into the Bronze Age and not in a later period as it is hypothesized by some authors.

For instance the analysis of references of the words for the plants zia (ζεια) and olyra (όλυρα) indicate that translations given to day do not correspond to the real meaning. In the Odyssey (given as Od. in the next) δ 602–604 it is reported the following:

> ... σύ γάρ πεδίοιο ανάσσεις ευρέος, ω ένι μέν λωτός πολύς, εν δέ κύπειρον πυροί τε και ζειάι τε ιδ' ευρυφυές κρι λευκόν, ...
>
> (for you have much flat ground in your kingdom where lotus thrives, as also meadowsweet and wheat and barley, and oats with their white and spreading ears).

In the Iliad (given as Il. in the next) E 196 it is also reported:

> ... ιπποι εστασι κρι λευκόν ερεπτόμενοι καί ολύρας, ...
>
> (a pair of horses, champing barley and rye).

These two paragraphs indicate undoubtedly that these two species are not the species of barley, as are usually translated by the scholars. The probability that these two plants were species of the genus Sorgum is the most significant, as this genus is indigenous in Greece, since antiquity-still existing under several names, such as skoupohorto (grass used for brooms), dari (using as bird food), white mays and also mays. Concerning the plant "moly", no student was able to identify it with certainty up to recent times.

**Table 1** Cultivated and wild plants referred to in the Chalcolithic period.

| |
|---|
| **Cereals** |
| Wheat, (*Triticum* dicoccum[1] *T.* monococcum[1], *T. aestivum*[1], *T.* durum[1]) ( η 104, ι 9, 110, Λ 67, μ 8, K 569, δ 603)[3], Barley (*Hordeum* vulgare[1], *H.*distichum[1], *H. tetrastichum*[1]), (δ 604, E 196, δ 41, μ 358, γ 443), Zia (oats) (sorgum?) (δ 41, δ 602 ), olyra (sorgum?) ( E 195, Θ 564), rye [1], millet[1]. |
| **Legumes and other edible and industrial plants** |
| Peas [1], vetch [1], vetch-pea [1], large beans, *Ervum ervilia*[1], lens[1] (*Lens esculentum*[1]), clover[1](lotus) (B 775, δ 602), chick-pea[1] ( N 589), anise[1], coriandrum[1,2], parsnip[1], linen [1,2], linen-cloth (B 529, 831, I 661, ν 73), watercress[2], celery seeds[2], herbage of celery (ε 72, B 776), cumin[2], fennel[2], rush[2] meadows weed (δ 603, Φ 531), ginger[2], mint, spearmint[2], Calamintha pastinaca[2], carthamus[2], safflower[2], sage[2], sesame[2], moly?, undetermined (κ 287, κ 304), vegetables (η 127), onion (Λ 630, τ 232), Vicia faba[1], faba bean (N 589). |
| **Grapevine and other fruit-bearing plants** |
| Grapes (*Vitis vinifera*)[1], grapevine[2] (ι 106, ι 131, η 122, ω 235), grapes (ε 69, Σ 561), raisin (η 122), souer grape (η 125), wine [2] (ι 111, 358). Olive tree[1], olive fruit[1] (*Olea europea*)[1], olive tree, olive oil[2] (ν 372, λ 588, κ 364, ζ 79, ε 234, N 611, ι 319), wild olive tree (ε 476). Wild pear fruit[1], pear (η 114, ω 340, λ 589), pear, probably other variety (ω 234), wild pear tree (ξ 10), almond[1], figs[2], fig tree[1] (η 114, η 120, ω 340), caprifig tree (Φ 37, μ 103), pistachio nut tree[1], plume tree[1], pomegranate fruit [1], pomegranate tree (η 115, λ 589, ω 340), apple tree, apple fruit (η 114, ω 340, I 514). |
| **Grasses, Ornamentals, Dyeing and other wild plants** |
| Sweet juicy herbage (ζ 90), thistle (ε 329), thistles ( for fence)(ω 224, σ 357), cane[1] (Ω 448, Σ 576), rush marshy herbages(Φ 351, ε 462) daffodil (λ 538, 572, ω 13), rose[1,2] (Ψ 186, β 1), hyacinth (ζ 230, Ξ 347), violet (ε 72), poppy[1] (Θ 306, Ξ 499, δ 219), saffron, crocus[1,2] (Ξ 347), , meadows[1] (ι 134), rainbow[1], ivy[1], kissyvion, wooden cup from ivy wood (ι 346, ξ 78), lily[1]. |
| **Forest trees and bushes** |
| Poplar tree (aspen) (ι 140, ε 63, ρ 208, ζ 290, Δ 485, 482, η 103), oak fruit[1], oak tree[2] (Ψ 117, ι 84, 34, φ 42, ξ 11, N 389, M 131), oak tree (figos), probably other species (E 837, Z 237, H 58, E 692, Π 765), laurel ( ι 182-4), fir[1], taxus [2], fir (Ξ286, ε 237, β 424, Ω448, E 560), , salix[2] (willows) (κ 510), wicker ( ε 256), cedar (Ω 191), alder (ε 63), cornel-tree ( κ 241), cypress[2] (ε 63-64, ρ 339), linden tree (;)[2], ash (Π 143, 719, ρ 339), tamarisk (Z 38, Φ18), pine[1] (ι 186, N 389, Π 482), plane tree (B 305), elm[1,2] (Z 419), date-palm[1,2] (ζ 162, Δ 141, ζ 219), byblus (φ 390), box tree (Ω 269), osier (ε 257), rush (ε 462, Δ 393) beech tree[1], sycamore[1](;). |

1. It is referred in findings in excavations and paintings.2. It is referred in Linear B. 3. The number into parenthesis are referred in Homeric fragments, capital letters Iliad small letter Odyssey.

Genadius, a Greek botanist, considered that this plant was that known under the modern name of mandrake [1, 8, 9]. From excavation data, large beans and black beans are reported by Vermeule. This is another questionable point of translation, as the origin of this species is the American continent and consequently the real bean species did not exist in Europe at the time. According to Genadius [1], probably, the one reported is another legume, either a kind of peas or of lupine, used by the Spartans.

Moreover, in Od. 7.123–124 regarding grapes, Homer refers to the following

... της ετερον μέν θ' ειλόπεδον λευρω ενί χώρω τέρσεται ηελίω, ...

(the grapes are being made into raisin)

which clearly indicates that at those times drying of grapes was applied.

Farming practice during the Homeric era appears to be quite similar to the practice applied in previous prehistoric times, e.g. 1500–2500 years before, which are still in use in underdeveloped mountainous areas. The main Mediterranean farming practice was almost the same as the one used in the Homeric times and the only differentiation was the significant presence of fruit and forest trees. According to Halstead [6], from the archaeological and archaeobotanical evidence of the early Bronze Age, it is not possible to trace whether fruit tree cultivation was extending all over the Greek mainland or was limited to some small domestic areas during that period.

In the description of Alkinoos' farm, Od. 7.113–132, according to what Ulysses said to his father Laertis, Od. 24.222–250, and the description of the decorations of

Achilles' shield, Il. 18.542–573, it is concluded that, besides the cereal cultivation in extensive field areas, orchards and vineyards also existed in several places. More specifically, farming is described in detail in several lines of the two epic poems.

## *1.1 Cereal Farming*

The basic cultivation was that of the cereals. The several works, from the preparation of the field up to the harvest, as described in the poems are the following.

*Plowing*: This work is important for the good development of any cereal plant. It is referred to four times in the poems, as follows: In the Il. 18.542, information is provided of the proper times of ploughing a field. It is also reported that plowing a field three times is necessary. The number of plowmen and the care taken by the farmers to keep them in good health are also described. Triple plowing is also indicated in Od. 5.127: "νειω ενι τριπόλω" ("in a new field thrice ploughed"). In the Od. 18.371 plowing with a pair of cows is clearly described, while in the Il. 10.351 there is evidence that mules instead of cows were used.

*Reaping* (*Harvesting*): The reaping practice is presented in the Iliad with two excellent descriptions, in which the reader could recognize a modern scene. In the first, Il. 18.550, the description in translation is the following:

> He wrought also a field of king, where the reapers were reaping with sharp sickles in their hands. Swathe after swathe fell to the ground in a straight line behind them, and the binders bound them in bands of twisted straw. There were three binders, and behind them there were boys who gathered the cut corn in armfuls and kept on bringing them to be bound: among them all the king owner of the land stood by in silence and was glad.

In the second, Il. 11.68, another good description is given:

> And now as a band of reapers mow swathes of wheat or barley upon a rich man's land, and the sheaves fall thick before them, even so did the Trojans and Achaeans fall upon one another; they were in no mood for yielding but fought like wolves.

*Threshing*: Homer refers only once in Il. 20.495 describing Achilles' impulse in fighting and the way he killed his enemies, similar to threshing the wheat.

*Winnowing*: A very important practicing of wheat cultivation is winnowing. It is used for the cleaning of wheat and barley as well as legumes from their chaffs. This practicing is referred to twice in the Iliad in 5.499 and 13.589.

*Grinding*: The final work, grinding, is referred to once in Il. 20.106 and clearly indicates the way the cereal seeds were grinded. The same work was practiced in several ancient periods, as it is well known by archaeological evidence. Furthermore, the word "alfita", which means barley flower, is mentioned in many other cases. Consequently the existence of flour means that a method of grinding was available.

Finally, a very interesting detail is the comparison between the end of the work of reaping or threshing with cows in the end of the day to the end of battle between the two opposite armies, as it is described in Il. 16.778, "but when he went down towards the time when men loose their oxen, ... ".

## 1.2 Grapevine Cultivation

Another interesting cultivation was that of grapevine and consequently wine production. The most characteristic description of grapevine cultivation is presented on Achilles' shield, Il. 18.542–573, depicting private farms similar to those still existing in Greece, mainly in some islands. It is significant that the vines are propped by pillars, a very innovative practice during those times. The description also of vintage is not at all different of that of recent times, except probably of the lyrics of the songs and the music. In the description of the Alkinoos' farm, Od. 7.122–126, besides the existence of grape vine cultivation, drying grapes in the sun is described accurately, as performed nowadays for Black Corinth and Sultana varieties. Furthermore, information is given concerning pressing of grapes by foot, evidently to produce wine. It is significant that grapes for harvesting, grapes in the beginning of colour changing and grapes in the time of flowering existed simultaneously. That leads to the assumption that at this period several varieties were cultivated, e.g. for wine production, edible varieties, etc.

*Olive tree cultivation*: In spite the fact that this tree was very significant for the agricultural economy of the Homeric period there are only three references, Od. 7.115, Il. 17.53 and also Od. 5.476, in which it is stated that Ulysses

> υπήλυθε θάμνους εξ ομόθεν πεφυωτας ο μέν φμλίης, ο δ' ελαίης
>
> (there he crept beneath two shoots of olive that grew from a single stock- the one an ungrafted sucker, while the other had been grafted),

from which there is no doubt that it was a wild olive tree a branch of which was grafted with a variety of cultivated olive tree.

## 1.3 Field Irrigation

The irrigation of cultivated plants is referred to as a fundamental work of agricultural practicing. In Od. 7.128 it is stated that: "... Two streams go through it, the one turned in ducts throughout the whole garden, while the other is carried under the ground of the outer court to the house itself, and the town's people draw water from it" and in Od. 13.243 that: "... all kinds of timber grow here, and there are watering places where the water never runs dry". In the Il. 21.257 an excellent description of tree irrigation is given:

> ... As one who would water his garden leads a stream from some fountain over his plants, and all his ground-spade in hand he clears away the dams to free the channels, and the little stones run rolling round and round with the water as it goes merrily down the bank faster than the man can follow.

This description is so vividly presented, that anyone can clearly realize the way it is practiced.

## 2 Conclusion

Agricultural practicing in the Homeric era is similar to the one existing during two or three millennia before. It is described as typically specialized and intensive Mediterranean multi-cultivation, due to the natural adaptation of cultivated plants to the particular Greek peninsula environment [14]. However, other scholars consider that little evidence survived concerning specificity of cultivations in the areas described in the Homeric poems, which by no means shows cultivation intensity [6.] In spite of academic disagreement, the model of this agricultural development, which has been applied for almost five millennia, e.g. from at least the Late Neolithic to our days, is the one that contributed to the establishment of the Greek civilization and its breaking evolution.

## References


1. Genadius, P.G., 1914. *Phytological Lexikon*, Vol. 2, M. Giourdas Publ., Athens, 1959 [in Greek].
2. Chadwick, J., 1992. *Linear B and Related Scripts*, transl. N. Konomis, Papadimas Publ., p. 78 [in Greek].
3. Dickinson, O.T.P.K., 1999. *The Origin of Mycenaean Civilization*, M. Kardamitsas Publ., p. 283 [in Greek].
4. Doucas, K., 1993. *Homer's Great Secret*, Vol. 2, Dodoni Publ. [in Greek].
5. Doucas, K., 1996. *Ancients Reports of Homer and the Language*, Free Thinking Publ., p. 181 [in Greek].
6. Halstead, P., 1990. Agricultural in the Bronze Age Aegean, in *Agricultural in Ancient Greece, Proc. 7th Internat. Symp.*, B. Wells (Ed.), Swedish Instistute of Athens, pp. 105–117.
7. Thanassoulopoulos, K., 2005. *Agricultural Development in the Homeric Era*, Davlos, p. 47 [in Greek]
8. Kavvadas, D.S., 1938. *The Flora of Greece*, Vol. I, Theodoridis Publ., Thessaloniki, Greece [in Greek].
9. Kofiniotis, E.K., 1886. *Homeric Lexicon*, Creation Publ., Athens, 1992, p. 415 [in Greek].
10. Homer, *Iliad*, English translation by Samuel Butler.
11. Homer, *Odyssey*, English translation by Samuel Butler.
12. Renfrew, C., 1972. The emergence of civilization: The Cyclades and the Aegean in the third Millenium B.C., in *Studies in Prehistory*, J.M. Coles (Ed.), Methuen and Co. Publ., p. 595.
13. Renfrew, C., 1973. Trade and craft specialasisation, in *Neolithic Greece*, D. Theocharis (Ed.), Cultural Foundation of National Bank of Greece, pp. 179–191.

14. Renfrew, J.M., 1973. Agriculture, in *Neolithic Greece*, D. Theocharis (Ed.), Cultural Foundation of National Bank of Greece, pp. 147–164.
15. Ruiperez, M.S. and Melena, J.L., 1996. *Mycenaean Greeks*, M. Kardamitsas Publ., p. 285.
16. Treuil, R., Darcque, P., Poyrsat, J.Cl. and Touchais, G., 1996. *Aegean Civilizations*, M. Kardamitsas Publ., p. 657 [in Greek].
17. Sarpaki, A., 1990. Paleothnobotanical approach. Aegean, in *Agricultural Development in Ancient Greece. Proc. 7th Internat. Symp.*, B. Wells (Ed.), Swedish Institute of Athens, pp. 61–76.
18. Vermeule, E., 1972. *Greece, Bronze Age*, transl. Xenos Th., M. Kardamitsa Publ. p. 437 [in Greek].
19. Hourmouziades, G.H., 1993. *Neolithic Dimeni*, Vanias Publ., Thessaloniki, p. 192 [in Greek].

# The Fauna of Greece and Adjacent Areas in the Age of Homer*

Eleni Voultsiadou and Apostolos Tatolas

*Aristotle University of Thessaloniki, Thessaloniki, Greece*

**Abstract.** Aim of the present work is to study the composition of the fauna in Greece and adjacent areas around 3000 years ago through the knowledge of the Homeric man about the animal kingdom in Greece and adjacent areas. The method consists of the analysis of information derived from a thorough study of the first written documents of the Greek literature, the epics, attributed to Homer and Hesiod. Records of 2442 animals were found, corresponding to 71 different animal names. All animal names were attributed to recent taxa at different category levels; the majority (65%) were assigned to taxa of the species level and the rest to supraspecific taxa. Most of the animal names recorded in the epics have been retained, as integral words or roots in Modern Greek and they have been used in the formation of Latin scientific taxa names. Five animal phyla appear in the texts: (1) Chordata (mostly birds and mammals), (2) Arthropoda, (3) Mollusca, (4) Porifera, and (5) Annelida. Information in the epics also includes morphology, biology, ecology (habitat and prey-predator relationships) and behavior. The presence of several species in the area in that period is documented on the basis of archaeological and/or palaeontological findings from various Greek localities.

The knowledge of Homeric man about animals, as reflected in the epics, seems to concentrate mainly, but not exclusively, on animals involved in human activities. The populations of some common animal species of the Homeric age in the Greek populated areas have become extinct or reduced at the present time. On the other hand, some common animals of the present time do not appear in the epics, since they were introduced later. Useful zoological information can be derived from the study of classical texts, which may help historical biogeographers, as a supplement to archaeology and art, in the reconstruction of faunas of older periods.

## 1 Introduction

Scientists have searched for paleo-faunistic and paleo-environmental evidence in their attempt to gain an understanding of animal life and the environment in the past and for evidence of human intervention and subsequent impact on regional faunas. Palaeontologists and archaeologists have studied osteological material found in Greek archaeological sites during excavation. Based on these findings, they have

---

* The present article has appeared in English in the *Journal of Biogeography* (2005) 32, 1875–1882, and is reproduced in the Proceedings of the International Symposium Science and Technology in Homer's Epics with the kind permission of Blackwell Publishing Ltd.

*S.A. Paipetis (ed.), Science and Technology in Homeric Epics,* 303–315.

drawn interesting conclusions concerning the relationships between man and animals in previous epochs, such as the Pleistocene (e.g. Jarman, 1996) or the early historical periods such as the geometric period (e.g. Tsoukala and Hatzi-Vallianou, 1996; Wilkens, 1996). Another angle of approach has been through the study of art. Iconographical sources have commonly given zoologists useful evidence. For instance, various authors have tried to recognize and record the animals depicted in Minoan wall paintings, vases or figurines (Vanschoonwinkel, 1996).

In addition to archaeology and art, useful information can be derived from written documents from early civilizations. The Age of Homer is a critical moment in Greek history. It is the late geometric period, covering the eighth and seventh centuries BC. During this period the Greek alphabet was developed and the first written documents of Greek literature, the epics, appeared in two main forms: the heroic epic represented by Homer, and the didactic epic represented by Hesiod. The epics, in spite of being questionable as historical documents, constitute a useful source for historians and archaeologists in their attempt to understand early historic community culture. The "Homeric community" reflects three different time periods (Latacz, 1997; Mazarakis Aenian, 2000): (1) prehistoric (mostly Mycenaean), intermediate (corresponding to the dark Ages, covering the eleventh to ninth centuries BC), and the eighth century BC (the period in which Homer lived). The latter two form the so-called "geometric period". A fourth layer, which is mythical, appears to be an invention of the poet's imagination.

Langdon (1993) considered that the dynamic relationship between art and poetry signaled both the climax and the end of the Age of Homer. The presence of animals is conspicuous in both the art and the epics of this late geometric period. These first written documents of Greek language contain valuable information on the relationship of Homeric man with animals. Furthermore, the results of their study should help zoologists and historical zoogeographers reconstruct the fauna of this age. This paper presents an annotated list of all animals appearing in the epics, makes an attempt to assign the classical animal names to recent taxa and gives a comparison of fauna illustrated in the texts of that age to present fauna.

## 2 Classical Names and Recent Taxa

As mentioned above, an attempt was initially made to record all animals appearing in the texts of the Age of Homer and to correlate them to recent taxa. To achieve this goal, according to the historical documentation of Greek classical literature of that period (Lesky, 1971; Easterling and Knox, 1985), the following texts were studied: *Iliad* (I) and *Odyssey* (O) by Homer, *Theogony* (T), *Works and Days* (D), *Aspis* (A) and *Catalogue of Women* (C) by Hesiod and the *Homeric Hymns* (H) written after Homer's death. The standard editions of Oxford classical texts were used together with valid translations in Modern Greek, such as those by Doukas (2000), Lecatsas (1941, Zacharopoulos Press, Athens) and Papaditsas and Ladia (1977, Estia Bookstore Press, Athens). First, all lines were checked for animal records. The correlation of animal names to recent taxa was made possible by combining the information embodied in animal names (most of which have been retained in Modern Greek) with morphological descriptions and/or information on the ecology, biology and behaviour given in the texts. Wherever possible, this was supported by available archaeological or palaeontological evidence. All the above data were evaluated by employing a variety of zoological books and papers (e.g. Nowak, 1991; Handrinos and Akriotis, 1997), as well as encyclopedias and general or specialized lexicons (Encyclopedia Papyros Larousse Britannica; Liddell and Scott: Great Dictionary of the Greece Language, Kofiniotis: Homeric Lexicon; etc.). The works of Aristotle (*History of Animals*, etc.) were also consulted. In addition to this material, various specialists in different taxonomic groups were consulted.

Care was taken to be as accurate as possible when attributing classical names to modern taxa. As a result, in various cases for which no clear evidence existed in the texts, we preferred to suggest a higher taxon instead of a specific species or genus. The fact that a classical name has been retained in Modern Greek was not in itself sufficient to identify a species, as shown through the examples given below. We should however, stress the continuity of Greek language from the Homeric poems up to the present, pointed out by various authors (Browning, 1983; Babiniotis, 2000; Doukas, 2000). It is remarkable that the majority of the Homeric words are used in the same way and many of them with the same meaning in Modern Greek. Additionally, etymology proves in several cases to be critical in recognizing animals. Some examples: κυνοραϊστής (κύων = dog + ραίω = destroy), ἡμίονος (ἡμί = half + όνος = donkey), πάρδαλις (παρδαλός = spotted), πολύπους (πολύς = many + πούς = leg), etc.

Detailed justification was not considered necessary for the names of well known domestic animals such as the horse, cattle, goat, sheep, donkey, pig, as well as lion, wolf, honeybee, brown bear or the monk seal. However, in some cases the recent scientific name of an animal was not so clearly evident, or the authors had to decide among several closely related species. Several examples are given below showing how the identification of some animals was traced through various clues given in the texts.

The name ἡμίονος ἀγρότερος (*Iliad*, II.852) was attributed to the species *Equus onager* (wild mule) due to the adjective ἀγρότερος (wild) and the information about its area of origin, somewhere in northern Asia Minor. The known distribution of this species covers the area around the Caucasus, the Black Sea, Anatolia and Iran, although its populations are very restricted nowadays due to hunting and habitat destruction. This animal was distinguished from ἡμίονος (= mule) which was frequently reported in the epics and was recognized as the hybrid of *Equus asinus* and *E. caballus*. The animal called ἐρωδιὸς (= heron) (*Iliad*, XI.274) was identified as the night heron, *Nycticorax nycticorax* since it is reported to appear during the night. Μυῖα (= fly) was recognized as a species of the family Calliphoridae according to the fact that it lays its eggs on wounds or decaying bodies. Αἴξ ἀγρίη was attributed to the wild goat *Capra aegagrus* as a result of the description of its habitat (*Odyssey*, IX.118–124) and the length of its horns (*Iliad*, IV.105). The name δελφὶς (= dolphin) most possibly corresponds to the most common dolphin species *Tursiops truncatus*, for its silver color and the fact that it appears in schools very near the coast (*Aspis*, 207–212; *Iliad* XX.I22). Ἶψ (= woodworm), although as a name has been used in the construction of the genus name *Ips* of the family Scolytidae (the members of which live in wood) was attributed to a member of the family Dermestidae, since it is reported to eat into objects made of animal body parts, such as Odysseus' bow (*Odyssey*, XXI.393). The species *Falco peregrinus* was assigned to the name κίρκος (= a kind of harrier) because it was described as being very fast, nesting on rocky mountain slopes and praying on woodpigeons. *Cuculus canorus* was recognized under the name κόκκυξ, from its characteristic call mentioned by Hesiod (*Works and Days*, 486). *Cygnus cygnus* is the species behind the name κύκνος (= swan), due to its piercing cry (*Homeric Hymns*, XXI.1). Πολύπους (= having many legs) most probably refers to the species *Octopus vulgaris*; its distribution in shallow waters makes the observation of its thalamus easier (*Homeric Hymns*, III.77). The name τήθεον, although in some lexicons is described as "a kind of bivalve", was recognized as a sea squirt of the class Ascidiacea, since a detailed description of these animals is given by Aristotle under the name τήθυα (*History of Animals*, 531a9-31). The common commercial sponge species *Hippospongia communis*, which bears the most numerous and large canals, possibly hides behind the name σπόγγος (= sponge) (*Odyssey*, I.111, XXII.439), while χέλυς is the marginated tortoise, an endemic Greek species (*Testudo marginata*) having a large shell which was used by Hermes for the construction of his lyre (*Homeric Hymns*, IV.24–48). Finally, χελιδὼν (= swallow) should be attributed to the swallow *Hirundo rustica* which is observed even inside buildings (*Odyssey*, XXII.239) and has been depicted on wall paintings in Thera (Masseti, 1997).

Overall 2442 records of animals, corresponding to 71 different animal names were revealed after a thorough study of the Homeric and Hesiodic epics. All animal names were correlated to current animal taxa (Table 1) at different category levels; the majority (65%) was assigned to species and the remaining to supraspecific taxa.

Out of the total of 2442 mentioned records of animals, 1283 were found in *Iliad*, 783 in *Odyssey*, 178 in *Homeric Hymns* and 195 in Hesiodic works. Only three of the animals recorded, the ant, the cuckoo and the carrion crow, were found exclusively in Hesiod, while a lot of animals found in the Homeric epics do not appear in the works of Hesiod. The tortoise appears exclusively in the 4th *Homeric Hymn*.

Additionally, c. 100 records that were not included in the above calculations were either common animal group names such as ιχθὺς (fish) and ὄρνις (bird), or zoological terms describing animal body parts, e.g. κέρας (horn), λοφίη (mane), οὖθαρ (udder), μηρίον (thigh) and γαμφηλὴ (jawbone).

As seen in Table 1, 56 of the 71 classical Greek animal names (79%) found in the epics appear in Modern Greek, according to the *Lexicon of Modern Greek Language* (Babiniotis, 2000). Although in some cases they do not have exactly the same form, they still retain the same main theme. Furthermore, 61 of them (86%) have been used in the formation of Latin scientific names. These estimations were based on species catalogues of certain publications, such as Honacki et al. (1982), Howard and Moore (1991), etc. The wide use of Greek words in the zoological nomenclature is reflected by the set of rules found in *The International Code of Zoological Nomenclature*

**Table 1** List of taxa names attributed to the classical Greek animal names recorded in the epics. Common English names, total records found and one selected record for each item are also given.

| Scientific name | Classical Greek name (modern Greek name)*† | Common English name | Number of records | Selected record‡ | Example of scientific name formation |
|---|---|---|---|---|---|
| Accipitridae sp. | αἰετός, πέρκνος*† (αετός) | Eagle | 19 | I(XVII.674) | ***Gypaetus*** |
| Acrididae sp. | ἀκρὶς*† (ακρίδα) | Locust, grasshopper | 1 | I(XXI.12) | **Acrididae** |
| *Anguilla anguilla* (Linnaeus, 1758) | ἔγχελυς*† (έγχελυς, χέλι) | European eel | 2 | I(XXI.203) | *Scolec**enchelys*** |
| *Anser anser* (Linnaeus, 1758) | χὴν*† (χήνα) | Graylag goose | 9 | O(XV.160) | ***Chenoptera*** |
| *Apis mellifera* Linnaeus, 1758 | μέλισσα*† (μέλισσα) | Honeybee | 107 | O(XIII.105) | *L.* ***melissa*** |
| Araneidae sp. | ἀράχνη*† (αράχνη) | Spider | 4 | D(777) | **Arachnida** |
| Ascidiacea sp. | τῆθος† | A kind of tunicate | 1 | I(XVI.747) | ***Tethya*** |
| *Athene noctua* (Scopoli, 1769) | γλαῦξ*† (γλαύκα) | Little owl | 112 | I(X.553) | *T.* ***glaucopis*** |
| Aves sp. | αἴθυια† | Unknown bird | 2 | O(V.352) | ***Aethia*** |
| Aves sp. | κὴξ | Unknown bird | 1 | O(XV.479) | |
| Aves sp. | κύμινδις, χαλκὶς | Unknown bird | 1 | I(XIV.290) | |
| *Bos taurus* Linnaeus, 1758 | βοῦς*† (βους, βόδι) | Domestic cattle | 340 | I(XX.466) | ***Bos*** |
| Calliphoridae sp. | μυῖα*† (μύγα) | Blow fly | 9 | I(XIX.25) | *C.* ***myia*** |
| *Canis aureus* Linnaeus, 1758 | θὼς*† (θως=τσακάλι) | Jackal | 4 | I(XI.473) | *O.* ***thous*** |
| *Canis familiaris* Linnaeus, 1758 | κύων*† (κύων=σκύλος) | Dog | 183 | D(604) | *Chryso**cyon*** |
| *Canis lupus* (Linnaeus, 1758) | λύκος*† (λύκος) | Wolf | 14 | I(X.334) | ***Lyc**alopex* |
| *Capra aegagrus* Linnaeus, 1758 | αἶξ ἀγρίη*† (αίγα άγρια) | Wild goat | 8 | O(XX.118) | *C.* ***aegagrus*** |
| *Capra hircus* Linnaeus, 1758 | αἶξ*† (αίγα) | Domestic goat | 170 | O(XX.173) | *C.* ***aegagrus*** |
| *Capreolus capreolus* Linnaeus, 1758 | πρόξ | Roe deer | 2 | O(XVII.294) | |
| *Cervus elaphus* Linnaeus, 1758 | ἔλαφος*†, ἐλλὸς (ελάφι) | Red deer | 33 | O(X.158) | *C.* ***elaphus*** |
| *Chlidonias niger or Sterna* sp. | κορώνη εἰναλίη | (Black) tern | 3 | O(XII.418) | |
| Cicadidae sp. | τέττιξ*† (τέττιξ=τζίτζικας) | Cicada | 3 | A(393) | ***Tettig**onidae* |
| *Columba livia* Gmelin, 1789 | πέλεια† | Rock dove | 12 | I(V.778) | *Scoto**pelia*** |
| *Columba palumbus* Linnaeus, 1758 | φάσσα* (φάσσα=αγριοπερίστερο) | Woodpigeon | 1 | I(XV.237) | |
| Corvidae sp. | κόραξ*† (κόρακας) | Crow | 1 | C(54-1) | *C.* ***corax*** |
| *Corvus corone* Linnaeus, 1758 | κορώνη*† (κουρούνα) | Carrion crow | 2 | D(746) | *C.* ***coronae*** |
| *Corvus monendula* Linnaeus, 1758 | κολοιὸς*† (κολοιός=καλιακούδα) | Jackdaw | 2 | I(XVII.755) | ***Colius*** |
| *Cuculus canorus* (Linnaeus, 1758) | κόκκυξ† | Cuckoo | 1 | D(486) | ***Coccyzus*** |
| *Cygnus cygnus* (Gmelin, 1789) | κύκνος*† (κύκνος) | Whooper swan | 5 | H(XXI.1) | ***Cygnus*** |
| Dermestidae sp. | ἴψ† | Dermestid beetle | 1 | O(XXI.393) | **Ip**idae |
| Elephantidae sp. | ἐλέφας*† (ελέφαντας) | Elephant | 11 | O(XXIII.199) | ***Elephas*** |
| *Equus asinus* Linnaeus, 1758 | ὄνος*† (όνος) | Domestic donkey | 1 | I(XI.558) | *E. hemi**onus*** |
| *Equus caballus* Linnaeus, 1758 | ἵππος*† (ίππος=άλογο) | Domestic horse | 748 | I(X.436) | ***Hippo**potamus* |
| *Equus caballus* × *E. asinus* hybrid | ἡμίονος, οὐρεὺς*† (ημίονος) | Mule or hinny | 55 | I(XXIII.265) | *E.* ***hemionus*** |
| *Equus onager* Boddaert, 1785 | ἡμίονος ἀγρότερος* (ημίονος άγριος) | Onager | 1 | I(II.852) | |
| *Falco peregrinus* Gmelin, 1788 | κίρκος*†, ἶριξ (κιρκινέζι) | Peregrine | 13 | I(XXII.138) | ***Circus*** |
| Falconiformes sp. | ἅρπη† | Falcon | 1 | I(XIX.350) | ***Harpia*** |
| Formicidae sp. | μύρμηξ*† (μυρμήγκι) | Ant | 2 | C(7-5) | ***Myrmex*** |
| *Grus grus* (Linnaeus, 1758) | γέρανος*† (γερανός) | Common crane | 4 | I(II.459) | *G. leuco**geranus*** |
| *Gypaetus barbatus* (Linnaeus, 1758) | αἰγυπιὸς† | Lammergeyer | 7 | A(405) | ***Aegypius*** |
| *Gyps fulvus* (Hablizl, 1783) | γὺψ*† (γύπας) | Griffon vulture | 7 | I(XVIII.270) | ***Gyps*** |
| *Haliaeetus albicilla* (Linnaeus, 1758) | φήνη | White-tailed eagle | 2 | O(III.372) | |
| *Hippospongia communis* (Lamarck, 1814) | σπόγγος*† (σπόγγος) | Common sponge | 5 | I(XXII.438) | ***Spongia*** |
| *Hirundo rustica* Linnaeus, 1758 | χελιδὼν*† (χελιδόνι) | Swallow | 3 | O(XXII.239) | ***Chelidoptera*** |
| *Ixodes ricinus* Linnaeus, 1758 | κυνοραϊστὴς | Tick | 1 | O(XVII.300) | |
| *Larus* sp. | λάρος*† (λάρος, γλάρος) | Gull | 1 | O(V.51) | ***Larus*** |
| *Lepus capensis* Linnaeus, 1758 | λαγωὸς*†, πτὼξ (λαγός) | Brown hare | 6 | I(XVII.676) | **Lago**morpha |
| *Lumbricus* sp. | σκώληξ*† (σκουλίκι) | Earthworm | 1 | I(XIII.654) | ***Scolex*** |
| *Luscinia megarhynchos* Brehm, 1831 | ἀηδὼν*† (αηδόνι) | Nightingale | 2 | O(XIX.518) | *T.* ***aedon*** |
| *Lynx lynx* (Linnaeus, 1758) | λὺγξ*† (λύγξ, λύγκας) | Lynx | 1 | H(XIX.24) | ***Lynx*** |
| *Martes* sp. | ἰκτὶς*† (ικτίδα) | Marten | 2 | I(X.334) | ***Ictid**omys* |
| *Monachus monachus* (Hermann, 1779) | φώκη*† (φώκια) | Monk seal | 8 | O(IV.405) | ***Phoca*** |
| *Nycticorax nycticorax* (Linnaeus,1758) | ἐρωδιὸς*† (ερωδιός) | Night heron | 1 | I(X.274) | *A.* ***herodias*** |

**Table 1** Continued

| Scientific name | Classical Greek name (modern Greek name)*† | Common English name | Number of records | Selected record‡ | Example of scientific name formation |
|---|---|---|---|---|---|
| *Octopus vulgaris* Cuvier, 1797 | πολύπους*† (πολύποδας) | Common octopus | 2 | O(V.432) | ***C. polypus*** |
| Ophidia sp. | ὄφις*†, δράκων (όφις, φίδι) | Snake | 18 | A(161) | **Ophidia** |
| Ophidia sp. | ὕδρος | Water snake | 1 | I(II.722) | |
| *Otus scops* (Linnaeus, 1758) | σκὼψ*† (σκωψ=γκιώνης) | Scops owl | 1 | O(V.65) | ***O. scops*** |
| *Ovis aries* Linnaeus, 1758 | ὄις†, ἀρνειὸς*, ἀρήν (αρνί) | Domestic sheep | 162 | O(XX.1) | ***Ovis*** |
| *Panthera leo* (Linnaeus, 1758) | λέων*†, λῖς (λέων, λιοντάρι) | Lion | 86 | I(V.161) | ***P. leo*** |
| *Panthera pardus* (Linnaeus, 1758) | πάρδαλις*† (λέο-πάρδαλη) | Leopard | 7 | I(X.29) | ***F. pardalis*** |
| *Passer domesticus* (Linnaeus, 1758) | στρου0ὸς*† (στρου0ί) | House sparrow | 3 | I(II.311) | ***Struthio*** |
| *Rhinolophus* sp. or *Myotis* sp. | νυκτερὶς*† (νυκτερίδα) | Bat | 2 | D(6) | ***Nycteris*** |
| *Scyliorhynus* sp. | κύων | Cat shark | 1 | O(XII.96) | |
| *Sturnus vulgaris* (Linnaeus, 1758) | ψὰρ, ψὴρ*† (ψαρόνι) | Starling | 2 | I(XVII.775) | ***Leucopsar*** |
| *Sus scrofa* Linnaeus, 1758 | σῦς†, χοῖρος*, κάπρος* (χοίρος, κάπρος) | Wild boar or pig | 192 | O(XIV.5) | ***Sus*** |
| *Tabanus* sp. | οἶστρος*†, κυνάμυια (οίστρος) | Gadfly | 3 | O(XXII.229) | ***Oestrus*** |
| *Testudo marginata* Schoeptt, 1792 | χέλυς*† (χέλυς, χελώνα) | Marginated tortoise | 5 | H(IV.25) | ***Chelys*** |
| *Turdus* sp. | κίχλη† | Thrush | 1 | O(XXII.468) | ***Cichlornis*** |
| *Tursiops truncatus* (Montagu, 1821) | δελφὶς*† (δελφίνι) | Bottle-nosed dolphin | 8 | I(XXI.22) | ***Delphinus*** |
| *Ursus arctos* (Linnaeus, 1758) | ἄρκτος*† (άρκτος) | Brown bear | 10 | H(V.157) | ***U. arctos*** |
| Vespidae sp. | σφὴξ*† (σφήκα) | Wasp | 2 | I(XII.167) | ***Sphigidae*** |

*Words appearing either as roots or as integral words in modern Greek names are given in parentheses.
†Words used as integral words, prefixes, bases or suffixes, in the formation of Latin scientific taxa names (examples are given separately).
‡I, *Iliad*; O, *Odyssey*; T, *Theogony*; D, *Works and Days*; A, *Aspis*; C, *Catalogue of Women* and H, *Homeric Hymns*.

(Ride et al., 1985) governing their transliteration and latinization. The contribution of the Greek element to the nomenclature and terminology of some animal groups reaches 80% (Voultsiadou and Gkelis, 2005).

## 3 The Homeric Man and His Knowledge of the Animal Kingdom

The data given in Table 1 show that man's knowledge of animals in the Age of Homer based on the epics, concentrates around 5 phyla; of the animal taxa reported in the examined texts, c. 81% were Chordata, 14% Arthropoda, 1.5% Mollusca, 1.5% Porifera and 1.5% Annelida (Figure 1). Within Chordata, birds and mammals prevailed, including 48% and 41% respectively, of the total chordates recorded. Tunicates, fish and reptiles participated with low percentages (2%, 4% and 5% respectively).

Mammals and birds, besides being easily recognizable, are more familiar to man at that age due to their involvement in human activities: agricultural works and transportation (cattle, horses and mules), hunting (wild boar, deer, brown hare, lion, etc.), food and clothing (goat, sheep, pig, etc.), construction of household and war objects (cattle, dog, goat, etc.), offerings to gods (cattle, sheep, etc.) and symbolism (e.g. lion, owl, common crane). This is further supported by the number of records of each animal in the studied texts. The most frequently appearing animals are: the horse (contributing with 30% of the total records), domestic cattle (14%), the pig (8%), the dog (7%), the goat (7%) and the sheep (7%). These six animals constitute 73% of the total records. Some animals appear frequently in metaphors and similes

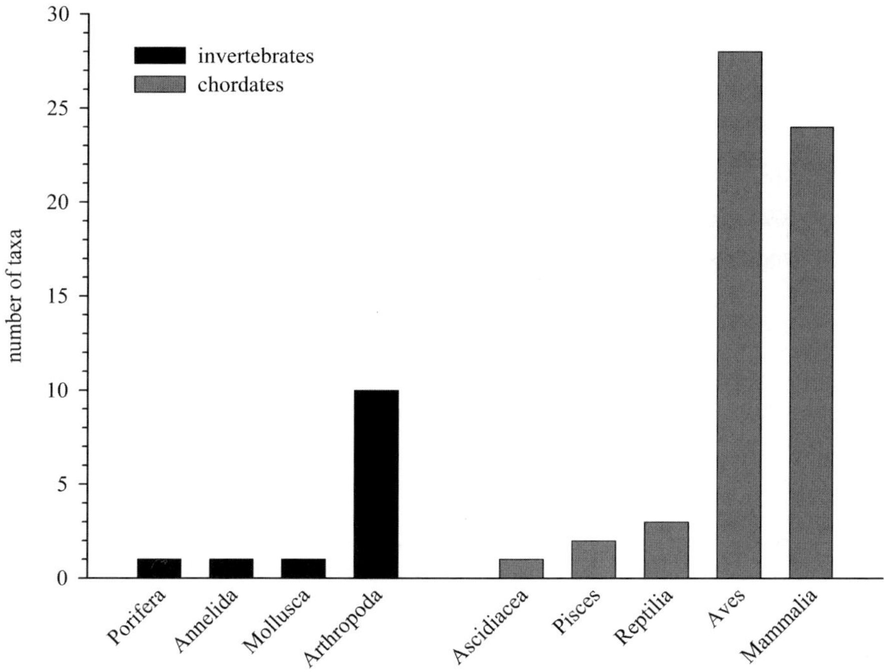


**Fig. 1** Number of animal taxa per higher taxonomic group found in the texts of the Age of Homer.

used to describe human behaviour or appearance. The owl, for example, appears in the texts only through the adjective γλαυκῶπις meaning "having shining eyes" or "grey-eyed". Obviously, the prominence of these animals in the texts does not necessarily mean high densities of populations at that time.

Invertebrates and small sized vertebrates do not seem to be of interest to man, with the exception of some species of economic importance (either useful or harmful to people or their domestic animals). Examples are the honeybee, the common sponge, the sea squirt, the gadfly, and the tick. The knowledge of marine fauna also seems to be limited. The only marine animals mentioned are the monk seal, cat shark, European eel, bottle-nosed dolphin, common sponge, common octopus, and an edible ascidian species. Fish are reported in general as ἰχθύες and the sea is often called ἰχθυόεις πόντος meaning "sea full of fish".

Knowledge about the ecology and behaviour of animals is illustrated in the texts. Most of the time, information is given on the habitat and prey-predator relationships (Figure 2). Behaviour, diet and migration are also discussed in several cases. Thirteen taxa are reported more than three times for their ecology or behaviour (Figure 3): the dog is the most frequently reported of all (17%) due to its involvement in hunting and protection of domestic animals. Some animals (wolf, jackal, peregrine, lion, leopard) are distinguished as active predators, others as prey (red deer, roe deer, wild goat, brown hare, domestic cattle, sheep and pig), hunted by both

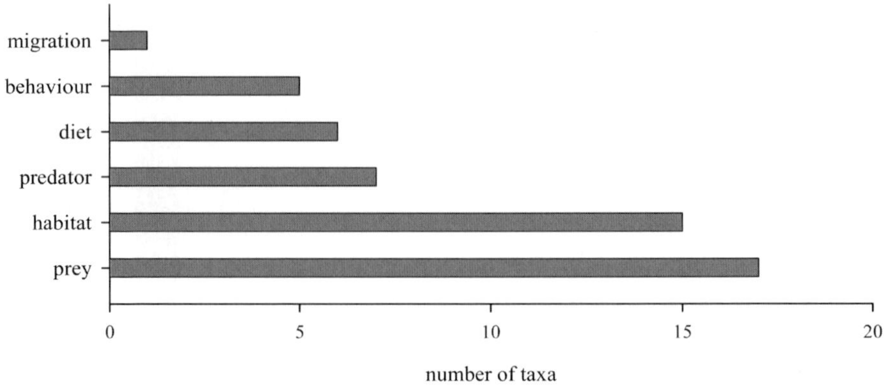


**Fig. 2** Histogram showing the number of records on various aspects of ecology and behaviour for animals given in the texts of the Age of Homer.

animals and humans. In this latter category we find several bird species, such as the jackdaw, rock dove, house sparrow, woodpigeon, graylag and starling. Parasitic relationships are also mentioned, such as those encountered in dermestid beetles and ticks. Interesting information is given on the habitat of various animals, sometimes contributing to their identification. Graylags, whooper swans and common cranes are reported to be living in groups by the rivers, monk seal populations in coastal marine caves, octopuses in their thalami, peregrines on steep mountain slopes, scops owls in areas covered with poplars and cypresses. In some cases, the behaviour of animals is illustrated in metaphors and similes, as mentioned above: people are chatting like cicadas, persistently protecting their land like honey bees or wasps, being cowardly like doves or brave as lions or wild boars, trapped like thrushes, being tough and persistent like mules, and so on. Man at that age was interested in the lives and habits of animals that signal agricultural activities. The arrival of swallows was a sign for pruning the vine-yards (*Works and Days*, 568), while the clang of cranes signified the beginning of ploughing (*Works and Days*, 448). Overall, the information given in the epics on the ecology and behaviour of animals does not seem to contradict current knowledge.

For some of the animals, both the domesticated and the wild forms are reported under the same name such as in the case of the graylag (χὴν). In others, the two different forms are found under different names, as in the case of the wild boar (κάπρος) and the domestic pig (χοῖρος); obviously people were aware of the close relationship between the two forms since they used a third name (σῦς) describing both.

Although in the present work a thorough study of the animals reported in the first written documents of the western civilization was attempted, it should not be considered that a comprehensive picture of the fauna of ancient Greece at that age has been given. Actually, in this work we collect and decode the information regarding animal life recorded in the texts from a biological point of view, in an attempt to

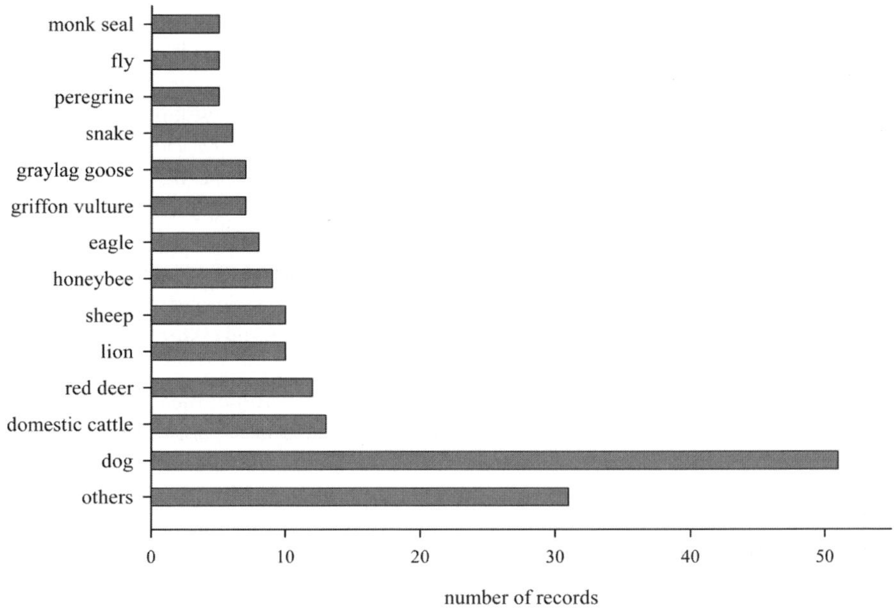


**Fig. 3** Histogram showing the number of records of the most frequently recorded animals for their ecology and behaviour in the texts of the Age of Homer.

show human knowledge as is reflected therein. Although the Homeric heroic poems are far from being historical documents (since they reflect different historical periods), they still remain a rich source of knowledge on human life and interests of a time around 3000 years ago. The Hesiodic *Works and Days* have long been considered as a valuable document on the agricultural life of the same age (Mireau, 1954). It is highly possible however, that Homeric man was familiar with many more animal species than those reported in the epics. This is supported by the fact that four centuries later, Aristotle gives a catalogue of c. 500 animal species, more than 160 of which are fish and marine invertebrates that hardly appear in the texts of the 8th century BC (Voultsiadou E. and S. Kiousis, unpublished data).

## 4 Past and Present: Comments on Faunal Composition

Some of the species appearing in the studied texts and constituting basic faunal elements of the late geometric period have now become extinct in Greece and adjacent regions. Lions were widely distributed in Northern Greece, as reported in various classical texts (Xenophon, *Hunting*, IX.1; Herodotus, *Histories*, VII.126; Aristotle, *History of Animals*, 6.579b). Nowak (1991) claimed that lions disappeared from the Balkan Peninsula around 2000 years ago. Lions roamed Europe in early antiquity but had disappeared by the first century B.C., possibly due to the forestation (Hughes, 2003). Leopards and other wild animals, such as lions, lynxes and bears are mentioned by Xenophon (*Hunting*, IX.1) to inhabit various areas of Northern Greece (e.g. Mount Pangaion and Pindos). Populations of the endangered subspecies *tulliana* of the genus *Panthera pardus* still exist in western Turkey (Hughes, 2003). Bones of both *P. leo* and *P. pardus* from the Pleistocene have been found in various Greek sites (Tsoukala, 1989; Guest-Papamanoli, 1996). However, while lion bones have been found at archaeological sites in Southern Greece, the presence of the leopard in the Balkan Peninsula in historical times has not been confirmed. The common crane *Grus grus* which appears both in Homeric and Hesiodic texts and is reported by Aristotle in *History of Animals* (e.g. 597a4), stopped resting in Greece during its migrations after 1965 (Handrinos, 1992).

The populations of several species known in the age of Homer have been reduced in the present time and are restricted to certain areas. Examples are the populations of lynx, brown bear, red deer and wild goat. As far as the species *Lynx lynx* is concerned, its populations have been dramatically reduced in Europe because it has been severely hunted for its fur until recently; isolated populations have survived in the southern Balkans among other European areas (Nowak, 1991). It is the biggest cat in Greece today and no information on its population size and distribution exists (Paraschi, 1992). *Ursus arctos* and *Cervus elaphus* have been recently included in the endangered species list of the Greek fauna (Mertzanis, 1992; Poirazidis and Paraschi, 1992). Populations of wild goat, *Capra aegagrus* have been restricted to a few Aegean islands, with the largest population in Crete, where the subspecies *Capra aegagrus cretica* (Schinz, 1838) is currently protected by law (Paragamian, 1992). Bones of this species were found in various excavations at Cretan locations (Jarman, 1996). Hughes (2003) claims that the extinction or decline of biodiversity in some areas in antiquity was the result of the reckless collection and consumption of animals by Greeks and Romans.

It has to be mentioned that the elephant was not present in Greece during the Homeric Age, but was well known for its ivory, which was a much-appreciated decorative material. People had come across elephant bones found as fossil remains that as a result of their imaginative interpretations were thought to be the skeletons of mythical giants and monsters (Mayor, 2000; Hughes, 2003). Fossil bones of a pygmy elephant species of *Elephas, E. falconeri* Busk, 1867 that had lived in the late Pleistocene and the early Recent epochs, have been found on some Aegean islands (Nowak, 1991); radiocarbon dates as late as 4390 BC have been reported for specimens from the Greek island of Telos.

Archaeological excavations in various sites over the Greek territory (such as in Crete and Peloponnesus) have revealed bone material of several species, confirming their presence in the area during the Homeric Age. *Bos taurus*, *Canis familiaris*, *Capra aegagrus*, *C. hircus*, *Equus asinus*, *E. caballus*, *Martes foina* (Erxleben, 1777), *Sus scrofa* and *Ovis aries* were recognized from bones and/or horns as significant elements of the domestic fauna and the contemporary economy (Jarman, 1996; Tsoukala and Hatzi-Valianou, 1996; Wilkens, 1996). Besides these, the fox *Vulpes vulpes* (Linnaeus, 1758), the wild cat *Felis sylvestris* (Schreber, 1777), the rabbit *Oryctolagus cuniculus* (Linnaeus, 1758), and the badger *Meles meles* (Linnaeus, 1758), although not mentioned in the epics, have been found. Furthermore, fossil bone collections from the late Pleistocene and Holocene add information on the presence in the Greek territory (mainland and islands) of species such as *Canis lupus*, *C. aureus*, *Ursus arctos*, *Cervus elaphus*, *Testudo marginata*, *Lynx lynx*, *Cervus dama* (Linnaeus, 1758), *Lepus timidus* (Linnaeus, 1758), *Rhinolophus sp.* and *Myotis sp.* (Trantalidou, 1996; Tsoukala, 2001, 2003). To these should be added some species of rats and mice, e.g. *Apodemus mystacinus* (Danford and Alston, 1877), *Mus musculus* (Linnaeus, 1766) and *Rattus rattus* (Linnaeus, 1758), as well as frogs (e.g. *Bufo viridis*) which seem to be continuously present in Greek territory through time though they were totally absent from the studied texts.

Two animal species that were very common later in Greece and Europe, the domestic cat and hen, are not included in the texts. Their absence strengthens the assumption that they were subsequently introduced to Europe from other areas. It is suggested that the cat came to Greece from Egypt, where it was an object of worship and had been domesticated (Herodotus, *Histories*, 2) by the fifth century BC (e.g. Hughes, 2003). Hens were introduced to Greece from India where the species *Gallus gallus* (Linnaeus, 1758) had been domesticated 4000 years ago. Hughes (2003) suggests that this introduction took place in the 7th century BC. Based on the above, the report of cats and cocks in the *Batrachomyomachy*, a text attributed to Homer by authors in antiquity (Easterling and Knox, 1985), strengthens the modern opinion suggesting that it was written at a later date, around 500 BC. Aesop's fables, in which records of these animals also exist, are also believed to have been written at the later date. It is clear that most of the domestic animals kept by man in the Age of Homer were aliens: horses, donkeys, sheep, goats, cattle and pigs had all been introduced from Asia, during the Neolithic period (McNeil, 2003).

Finally, we should stress the importance of animal iconography as seen in the artwork of the Geometric or earlier Minoan period. Animals such as the dolphin, monk seal, wolf, bull, horse, wild goat, red deer, dog, lion and the swallow are depicted on wall paintings, vases and other objects (Langdon, 1993; Vanschoonwinkel, 1996). The red deer, monk seal and the little owl appear on early coins of the sixteenth, the seventh and the fifth century BC respectively. Sometimes, when animal representations are naturalistic, the image becomes an effective source of paleo-faunistic evidence for the zoologist (Masseti, 2000).

## 5 Conclusions

Summarising, the following conclusions can be drawn: (1) The knowledge of Homeric man about the animal kingdom concentrates mainly on animals involved in human activities: domestic animals are most frequently recorded in the epics. In addition, terrestrial fauna attracts more attention than marine or freshwater fauna. (2) Animals reported in classical texts can be assigned to recent taxa on the basis of diverse information on their morphology, ecology, or behaviour given in the texts.

Animal names can be of great help since most of them are retained in Modern Greek or were used in the formation of Latin scientific names. (3) Some populations, mainly wild animal species, common at that time in the Greek areas have disappeared or are now reduced, while some currently common animals do not appear in the texts since they were introduced at a later time.

Useful zoological information can be derived from the study of classical texts, which may help historical biogeographers, as a supplementary approach in addition to archaeology and art, in the reconstruction of the faunas of older periods.

## Acknowledgements

We thank T. Sofianidou and V. Goutner for their help with animal species recognition, as well as S. Gkelis and D. Vafides for encouragement and improvement of the text and figures. Gratitude is owed to S. Kokkini, N. Krigas and A. Dardioti for inspiring the first author to search for biological information in the classical texts.

## References

Babiniotis, G. (2000) *Lexicon of Modern Greek Language*, Centre of Lexicology Press, Athens [in Greek].

Browning, R. (1983) *Medieval and Modern Greek*, Cambridge University Press. Greek translation by M. Konomi 2002. Papademas Press, Athens.

Doukas, K. (2000). Homeric dialect: The common language of Greeks, in *Homer's Iliad*, Translation and interpretative comments, Ideotheatron & Georgiadis Press, Athens [in Greek].

Easterling, P.E. and Knox B.M.W. (1985) *The Cambridge History of Classical Literture. I: Greek Literature*, Cambridge University Press. Greek translation by N. Konomi, C. Griba and M. Konomi (1999), Papademas Press, Athens.

Guest-Papamanoli, A. (1996) Hunting and trapping in prehistoric Crete: A proposal for ethnoarchaeological research, in *Pleistocene and Holocene Fauna of Crete and Its First Settlers*, Reese (Ed.), Prehistory Press, Madison, Wisconsin, pp. 337–349.

Handrinos, G. (1992) Birds, in *The Red Data Book of Threatened Vertebrates of Greece*, Hellenic Zoological Society, Athens.

Handrinos, G. and Akriotis, T. (1997) *The Birds of Greece*, A. & C. Black, London.

Honacki, J., Kinman K. and Koepple, J. (Eds.) (1982) *Mammal Species of the World. A Taxonomic and Geographic Reference*, Allen Press and Association of Systematic Collections, Kansas.

Howard, R. and Moore, A. (1991) *A Complete Checklist of the Birds of the World*, Academic Press, London.

Hughes, D. (2003) Europe as consumer of exotic biodiversity: Greek and Roman times, *Landscape Research* **28**, 21–31.

Jarman, M. (1996) Human influence in the development of the Cretan mammalian fauna, in *Pleistocene and Holocene Fauna of Crete and Its First Settlers*, Reese (Ed.), Prehistory Press, Madison, Wisconsin, pp. 211–229.

Langdon, S. (Ed.) (1993) *From Pasture to Polis. Art in the Age of Homer*, University of Missouri Press, Columbia, Missouri.

Latacz, J. (1997) *Homer. Der erste Dichter des Abendlandes*, Artemis & Winkler Press. Greek translation by E. Sistakou (2000), Papademas Press, Athens.

Lesky, A. (1971) *Geschichte der griechischen Literature*. Greek translation A. Tsopanakis (2003), Kyriakidis Press, Thessaloniki.

Masseti, M. (1997) Representations of birds in Minoan art, *International Journal of Osteoarchaeology* **7**, 354–363.

Masseti, M. (2000) Did the study of ethology begin in Crete 4000 years ago?, *Ethology Ecology and Evolution* **12**, 89–96.

Mayor, A. (2000) *The First Fossil Hunters*, Princeton University Press, Princeton, New Jersey.

Mazarakis Ainian, A. (2000) *Homer and Archaeology*, Book Institute A. Kardamitsas, Athens [in Greek].

Mireau, E. (1995) *La vie quotididienne au temps d'Homere*. Greek translation K. Panagiotou (1995), Papademas Press, Athens.

McNeill, J.R. (2003) Europe's place in the global history of biological exchange, *Landscape Research* **12**, 33–39.

Merzanis, G. (1992) *Ursus arctos* (Linnaeus, 1758), in *The Red Data Book of Threatened Vertebrates of Greece*. Hellenic Zoological Society, Athens, Greece.

Nowak, M.R. (1991) *Walker's Mammals of the World*, The John Hopkins University Press, Baltimore and London.

Paragamian K. (1992) *Capra aegagrus cretica* (Schinz, 1838), in *The Red Data Book of Threatened Vertebrates of Greece*, Hellenic Zoological Society, Athens, Greece.

Paraschi, L. (1992) *Lynx lynx* (Linnaeus, 1758), in *The Red Data Book of Threatened Vertebrates of Greece*, Hellenic Zoological Society, Athens, Greece.

Poirazidis, K. and Parschi, L. (1992) *Cervus elaphus* (Linnaeus, 1758), in *The Red Data Book of Threatened Vertebrates of Greece*, Hellenic Zoological Society, Athens, Greece.

Ride, W., Sabrosky, C., Bernardi, G. and Melville, R. (Eds.) (1985) *International Code of Zoological Nomenclature*, Great Britain International Trust for Zoological Nomenclature and University of California Press, London.

Trandalidou K. (1996) The animal world, in *The Palaeoloithic Period in Greece*, G. Kourtesi-Philippakis (Ed.). *Archaeology and Arts* **58**, 45–53 [in Greek].

Tsoukala, E. (1989) *Contribution to the study of the Pleistocene fauna of large mammals (Carnivora, Perissodactyla, Artiodactyla) from Petralona cave Chalkidiki (N. Greece)*, Doctorate Degree Thesis, Thessaloniki [in Greek].

Tsoukala, E. (2001) Quaternary faunas of Greek Islands. *Bulletin de la Société des Sciences Historiques et Naturelles de la Corse*, Nos. 686–697.

Tsoukala, E. (2003) *Palaeontological research in Pella. Cave bears and late Pleistocene associated faunal remains from Loutra Arideas (Pella, Macedonia, Greece)*, Prefecture of Pella, pp. 1–44 [in Greek].

Tsoukala, E. and Hatzi-Valianou, D. (1996) Fauna and human diet in the Acropolis of Smari in the geometric and the palaeoanactoric period, in *Proceedings of the 8th Cretan Congress*, Heraklion, Crete (in Greek).

Vanschoonwinkel, J. (1996) Les animaux dans l'art minoen, in *Pleistocene and Holocene Fauna of Crete and Its First Settlers*, Reese (Ed.), Prehistory Press, Madison, Wisconsin, pp. 352–422.

Voultsiadou, E. and Gkelis, S. (2005) Greek and the phylum Porifera: A living language for living organisms, *Journal of Zoology* **267**, 143–157.

Wilkens, B. (1996) Faunal remains from Italian excavations on Crete, in *Pleistocene and Holocene Fauna of Crete and Its First Settlers*, Reese (Ed.), pp. 241-254. Prehistory Press, Madison, Wisconsin.

# Part 9

# ASTRONOMY

# “Eneoros Minos” and the Minoan Calendrical Abacus

P.D. Gregoriades

*Athens, Greece*

**Abstract.** While studying the Homeric text, is noted an enigmatic epithet given to Minos called “Eneoros” (Od. 19.178–179) with the oldest and general meaning of hour as a time period, repeating itself based on the number nine (9). The above epithet is possibly related with the Minoan Calendrical Abacus which is based in the number 9 (9 days a “week”, $4\times 9 = 36$ days a month $\times 10$ months = 360 days a year + 5.25 epagomenal). This Abacus is kept today in the Herakleon Museum in Crete and it is the oldest calendar working even today. All the Ancient Calendars co-align themselves every 9 years as it is proved by the following study.

## 1 The Minoan Calendar

During the excavations at Knossos a Calendrical Abacus was found, which is kept today in the Museum of Iraklio, in Crete. It is mentioned with the explanation “game matrix” by Dimopoulou [A]. Evans [8] and other archaeologists called it “great chess”. It is made of ivory, stone, blue glass as a covering and sheets of gold and silver. It was found along with 4 pawns made of ivory, and is 104 cm long and 61 cm wide.

As mentioned in detail in [11], the item was found at the southern end of the narrow corridor of the palace at Knossos, a corridor named accordingly “corridor of the chess”; the ivory pawns were found at a small distance to the south. The corridor lies on the North-South direction.

All archaeologists subscribed to the view that the finding is a kind of “royal game, analogous with chess or backgammon”. However, after a careful study, it can be shown that it cannot be a game, but instead it constitutes a perfect calendar, upon which one can represent 4 years. To this conclusion, the following hypotheses were made:

1. The Minoans were led to construct “chronometer” water-clocks measuring 20-minute time intervals, which formed their basic unit of time.
2. On the same grounds of practicality, they divided the year into ten parts of 36 days each, without any other change; this reduced the size of the Abacus. Using the term *Dactyl* in this Minoan subdivision of the year, we have 10 Dactyls of 36

*S.A. Paipetis (ed.), Science and Technology in Homeric Epics,* 319–324.

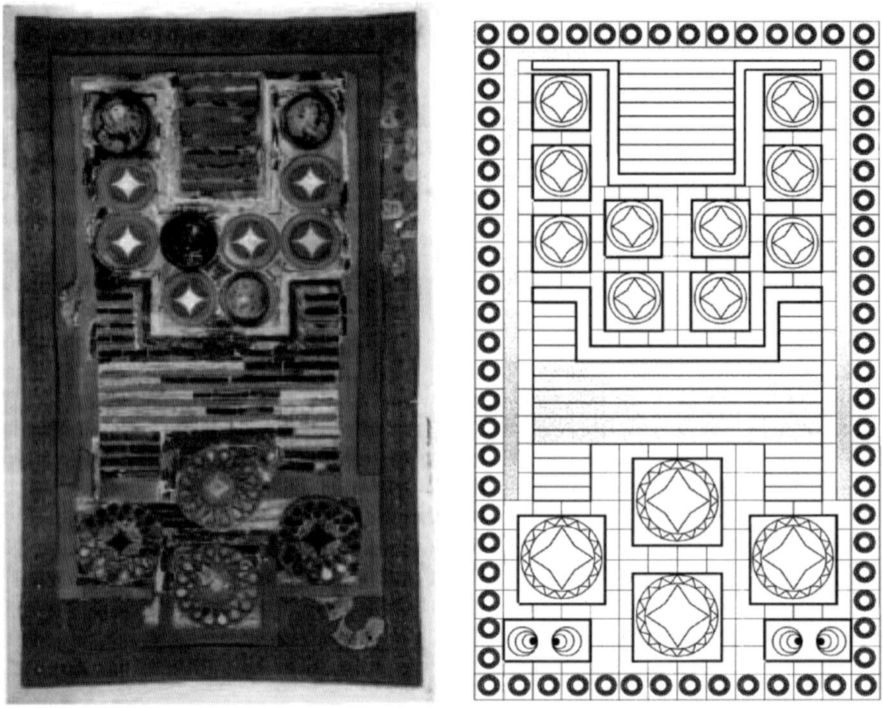

**Fig. 1** (Left) Minoan calendrical abacus (Archaeological Museum of Herakleion, Crete). (Right) Accurate reconstruction by the author. Reprinted by permission from "The Archeological Museum of Herakleion", N. Dimopoulo-Rethimniotaki edit. I. Latsis, 2005, pp. 332–333.

days each. Thus, the year is divided into 10 36-day parts. One dactyl consists of four 9-day intervals (weeks).

3. The error of 5 1/4 days per year caused by the initial adoption of the 360-day year, was being added at the end of every fourth year in the form of a "lame" or leap Dactyl of 21 days. The induced days are 21, therefore they correspond exactly to $5.25 \times 4$ days; however, the Abacus adds linearly 23 days in total. This happens if we consider the last of the six long lines as dashed, with four parts instead of three, as it is presented (Table 1) due to erosion of the original (Table 1A).

This view does not alter the use and operation of the Minoan Calendrical Abacus (MCA) at all. Instead, it increases its accuracy from the fact that in this case is taken into consideration the error of the time measurements with water-clocks. That is, the difference of the two days $(21 + 2)$ that arises is a corrective addition according to the following quantitative description of the function of the MCA, and based on the assumption of the 20-min basic unit of time measurement.

The difference of 2 days every 4 years covers the difference produced by the inaccuracy of the water-clock; then, according to the modern units of time:

**Table 1** Memorandum.

| | |
|---|---|
|  | 20 minutes time laps each rosette x 72  = 1440min = One day |
| 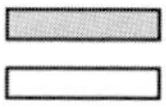 | One Day (24 hours) |
| 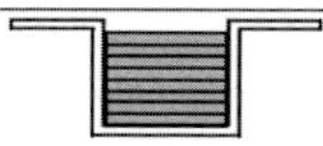 | 8 + 1 = 9 days ONE "WEEK" |
|  | One Month = 4 Weeks x 9 days/week = 36 Days |
| 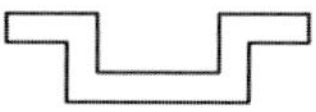 | Symbol of EQUINOXES |
| 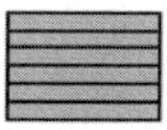 | Six Days Epagomenal |
| 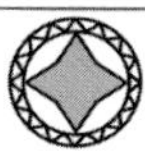 | One Year = 360 Day + 5,25 EPAGOMENAL |
|  | One Week = 9 Days |
|  | Solstices a) Summer 15,33 sun hours b) Winter 15,33 night hours |
| | |
| 72  = or 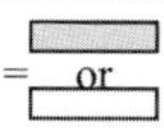 One Day x 36 = 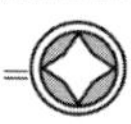 One Month x 10 = 360 Days =  One Year + (plus) 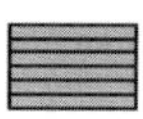 6 Days (5,25) = 365,25 Days a Year | |

2 days = 2 × 24 × 3600 sec = 172,800 sec for the 4-year period, i.e.

$$\frac{172,800}{4\text{yr}} = 43,200 \text{ sec/year},$$

$$\frac{43,200 \text{ sec/year}}{365.25 \text{ days/year}} = 118.2788 \text{ sec/day}.$$

Since one day has 72 × 20 minutes, it follows that each water-clock should have an error of the order of:

$$\frac{118.2788}{72} = \frac{1.6 \text{ sec}}{20 \text{ min}}.$$

This is well within the limits given in *Measuring the Solar System* [12].

It is reasonable that, if the measured time span (water-clock) is 20 minutes, then the corresponding error will refer not to the 1.5 min but to 20 min, as it is confirmed by the preceding ratio 1.6 sec/20 min.

Therefore, the observed difference of the two additional days in 4 years covers the inaccuracy of a water-clock measurement in relation with the motion of the stars, which gives the real duration of the year. From the three assumptions we made, which would fit to the principles of any solar calendar, we now come to examine the MCA of Herakleio Museum, as it is linearly shown in the figure. As it can be seen, the MCA's perimeter is decorated with 72 rose-like decorations, each one corresponding to the 20-min period. (According to Aristotle, the solar semi-god Heracles (Hercules), who is connected with the calendar, had 72 descendants [13].) The views of other authors also coincide with calendrical subdivisions, as follows from the explanation of the depictions (12 rose-like decorations for each part of the night or day, $3 \times 12 = 36$ decorations in total), since their product ($72 \times 20$) gives 1440 min, that correspond to the time duration of one day (24 hours). Next there is a strip in the shape of the capital Greek letter Π (Pi), which defines the sunlight hours at the Summer Solstice but also the "night hours" in relation with the "daily working hours". The days are represented with eight horizontal lines; they are surrounded by the Ten Dactyls. In the shape of the capital letter H we have the induced days per 4 years ($21 + 2 = 23$). To the right and to the left of the base there are four spirals ($2 + 2 = 4$) corresponding to the four weeks of the Minoan Calendar.

## 1.1 The Function of the Abacus

With the heliacal rising of Sirius, the water-clock started, and with the first 20-min time interval a pawn was placed on the first perimetric rose; the Priest assigned with that duty was moving it every 20 minutes on the next rose to the right or to the left. In any case, since the heliacal rising of Sirius was taking place sometime before the Sunrise, it follows that the first rose would be near the end/edge of the strip defining the evening hours. The question which rose was first was a matter of experience and correlation of the heliacal rising with the Sunrise. A full cycle covers all 72 roses and was giving a full 24-hour day; then a second pawn-indicator was placed on the first of the 8 lines representing the week of 8 days. And when a full 9-day week was covered, the white line completes the 9 days of a Minoan week (the 9th day is represented by the white line which surrounds the eight days), a third pawn was placed on the first 9-day interval at the lower side of the abacus. When all four 9-day intervals (36 days) had passed, this was one Dactyl, and then another pawn was placed on the first Dactyl. With this method, after 10 Dactyls a "full" year had passed, and this was marked with a placement of a fifth pawn on the first large rose to the left (out of the four roses representing the four years). This was the complete and detailed function of the Minoan Calendrical Abacus, which covered four full years as indicated by the four large roses at the lower part of the abacus. Therefore, this abacus offers a detailed representation of the measurement of time assisted only by two 20-minute water-clocks! Every four years, a difference of 21 days ($4 \times 5\frac{1}{4}$) appears, which from the structure of the Abacus and the historical data it can be shown that it was being added after the end of the fourth year, as a "lame" Dactyl (21 days long instead of the 36 days of the normal one). The Minoans, therefore, correcting this error every four years, started to celebrate a festival in the form of Games to honor Androgeos, son of Minos. The duration of these Games covered the difference without causing problems to the populace.

## 2 "Eneoros Minos"

The *Minotaur* was a Mythological monster, half man and half Bull (Taurus), restricted into the Labyrinth of Knossos. According to the ancient Greek writers he was called "Minoios Taurus", as belonging to King Minos, and "Knossius Taurus" (Euripides: Hercules 1327). Plutarch calls him "Mithoxer" (Theseus XV and Isocrates Orations 28). He resulted as the punishment of the perjurer Minos by Poseidon and the coition of his wife Pasiphae with the Holy Bull who emerged from the Sea, donated by the God to be sacrificed by Minos; instead, Minos kept the Bull in his herds. So the Minotaur was "locked" in the Labyrinth made by Daedalus and was being fed with seven (7) young men and seven (7) young women from Athens, being sent every nine (9) years to Crete by the Athenians as a tax of their subordination to the suzerain Minos (Diodorus of Sicily IV 61,3, Apollodorus III 203). Besides, Homer in his *Odyssey* (T, 179) gives to Minos the surname *Eneoros*.

It is evident that this story pertains to the pre-Minoan or the first Minoan Lunar Calendar, which had to be compatible with the newer one that was introduced as based on the Constellation Taurus at the Vernal Equinox; then the novelty of the Beginning of the Year at that Equinox was enacted, 21 to 22 March in modern dates. The whole Myth is composed by the Symbolism of concealing the Astronomical and Calendrical Knowledge.

1. The Bull (Taurus) as the Constellation defining the Vernal Equinox.
2. The Bull emerges from the Universal Ocean (Sea).
3. The Minoan Priesthood did not sacrifice the Bull (they did not change the existing Calendrical System).
4. But from the Bull and Minos' perjury resulted a new, "mixed" Calendar (MinoTaurus). The initial Minoan Calendar was based on the nine-day interval: 9 days × 4 "weeks" = 36 days each Month. The new calendar had the Seven-day interval (7 days). These two calendars had also a different length for the Year: The first one 360 days + 5.25 Induced and the second one 364 days + 1.25 Induced. The other difference was the length of the "weeks" (9 days for the first, 7 days for the second calendar). These two calendars co-align themselves every nine (9) years, because we have the following relations resulting into integer numbers. These relations exist for all the known Ancient Calendars as follows: The surname ENEOROS that we mentioned for Minos in Homer's

**Table 2** All calendars co-align every nine years.

| 9 years × 364 days = 3276 days + 11.25 Epagomenal | | | |
|---|---|---|---|
| Calendars | Days/Week | In 9 Years | Months |
| Minoan Calendar | 9 | 364 | 91 × 36 |
| Mayas | 18 | 182 | 91 × 18 × 2 |
| Patolli | 12 | 273 | 91 × 12 × 3 |
| Phaistos Disk | 6 | 546 | 91 × 6 × 6 |
| Lunar – Solar | 7 | 468 | 117 × 7 × 4 |

*Odyssey* means 9-hour-long, with the older and general meaning of the hour as a time period, repeating itself based on the Number NINE (ENNEA, 9).

5. The Mino-Taur, then, was "eating" every nine years Seven (7) Lads and Seven (7) Maidens sent by the Athenians from their aristocratic families to study and be incorporated in the Priesthood of the Minoans. And, as is mentioned by Strabo (B6, b10) and Herodotus (6, 7 and 6, 175), these young people under the leadership of the Minoan Votton and other Minoans sailed through the Adriatic Gulf and finally arrived in Macedonia by land; there they founded the Vottiaia or Vottian colony, where the Pelasgic, Minoan and Athenian tribes first resided, between the rivers Loudias and Axios (Herodotus 7, 123 and 7, 127).

After all of the above, one concludes that the Homeric texts have not been completely decoded and interpreted, even today.

## References


1. Homer *Iliad* and *Odyssey*.
2. Herodotus *History*.
3. Strabo Books 6, 10.
4. Euripides *Hercules*.
5. Isocrates *Orations*.
6. Plutarch *Theseus*.
7. Dimopoulou-Rethimniotaki, N., *The Archaeological Museum of Herakleion*, I. Latsis, 2005, pp. 332–333.
8. Evans, A., *The Palace of Minos at Knossos*, 1921–1929.
9. Herberger, C.F., *The Riddle of the Sphinx*, Vantage, New York, 1979.
10. Von Helmuth, Th.B., *Alt Kreta Kunst und Kunstgewerbe im Agaischen Kullturkreise*, Berlin, 1921.
11. Alexiou, S., *Minoikos politismos*, S. Alexiou Sons, Herakleion, p. 208.
12. *Measuring the Solar System/Motion Under Gravity: A Scientific Theory*, The Open University Editions, Section 1, 3 (Time) Table 2.
13. Rispen, J., *Great Greek Mythology*, Vol. II, p. 140 of the Greek edition.

# The Divine Fires of Creation: Homeric Hephaestos as a Comet/Meteor God

Amanda Laoupi

*National Technical University of Athens, Greece*

**Abstract.** Hephaestos belongs to the guardian-gods or "creators" of the Universe and functions as a pivotal force among the "proto-Hellenic" deities. This paper focuses, firstly, on the strong relationship between Hephaestos and the Pelasgian substratum of circum-Mediterranean region. The Pelasgian nuclei of prehistoric Attica remained active through the dual worship of Athena and Hephaestos and the close connection of the Athenian city-state with the island of Lemnos even in Classical Era. Furthermore, Kabeiroi, these primordial and mysterious daemons of NE Aegean, were introduced by the Pelasgians of 12th century BC, when removed from Boeotia to Samothrace, Imbros and Lemnos. According to another ancient tradition, they were children of Hephaestos and the daughter of sea-god Proteas. All the same, another striking complication arises. The Kabeirian Mysteries were celebrated yearly and were related to the element of fire. They lasted for nine (9) days, as a remembrance of the nine year period during which Hephaestos remained at the bottom of the sea.

This is the second axis of investigation. Although today, Vulcan is a name given to the near-Earth asteroid 2212, modern thinkers consider of Hephaestos as a symbol of earthen fires and related geological phenomena produced by volcanic activity. But Hephaestos' cult embraces challenging elements that require astronomical interpretation. Various ancient traditions include thought-provoking details:

(a) Lemnian labyrinth was famous in antiquity. Apart from being viewed as a symbol of our planetary system or as an astronomical map, labyrinth is also mystically connected to the protection from tsunamis, a serious side-effect of impact episodes.
(b) Ancient writers state that Lemnian earth was characterized by some special features, appearing with in impact events and with destruction levels all over the world of extraterrestrial origin.
(c) Pandora, one of the most fabulous works of the technician god was made by earth, water and divine fire. Furthermore, Pandora's box was used to compile all the positive and negative parameters of Life. When some invisible forces are released, then destruction strikes humanity. Respectively, impacts have always been seen both as a curse and as a blessing for life on Earth. Metallurgy changed the evolution of human history. Floods, epidemics and other disasters caused by extraterrestrial invaders did also.
(d) Hephaestos fell from Heaven, either on land (in the latitude of NE Aegean), or into deep sea (an underwater impact), where he remained invisible, working in his forge. Scientific research has shown that both arguments function logically. Furthermore, impact events can trigger increased volcanic activity, as the geo-archive of our planet has already revealed.
(e) If the Homeric Iliad is deciphered from the standpoint of Archaeoastronomy, Hephaestos is also related to the meteor swarm of Perseides. Finally, god's deformity and his reappearance in the latitude of Eastern Mediterranean may include the element of periodicity, may be as a comet.

*S.A. Paipetis (ed.), Science and Technology in Homeric Epics,* 325–340.

In fact, a great deal of information acquired from Iliad seems to refer to the extensive catastrophes of the beginning of 2nd millennium BC. Consequently, Hephaestos may function as a symbolic archetype of past impact events, being one of the pivotal figures within the gnostical system of the Pelasgians during 3rd and 2nd millennia BC. Phaethon's ride, the famous fall of Troy and 12th century catastrophes were added later, forming another mythological "layer" of information in the palimpsest of ancient astronomical knowledge in the circum-Mediterranean region.

## 1 Introduction

Dreams and myths bury the most intense memories under the conscious mind by suppressing and controlling anxiety. Various cultural personalities are characterized by different archetypes and phenotypes. That is why we encounter several identical kernels in the expression of the catastrophic events (e.g. the symbols of comets) along with many different stories about them. The use of common symbols make the social system operating correctly. Heroes and gods are known by many adjectives describing their traits. And … "the gods in Homer are great gods, because one thunderstorm does not make a great god, nor does one volcano …". Human disasters of the past gave birth to great myths [20]. Let us follow the path of Hephaestos.

Neoplatonists (a revival of Platonism of the third century AD) accepted Twelve Gods as a legacy from Plato. Generally speaking, Hestia (Vesta) represents earth, Poseidon (Neptune) water, Hera (Juno) air, and Hephaestos (Vulcan) fire. So, Zeus (Jupiter), Poseidon and Hephaestos belong to the Creators of the Universe, Hestia, Athena (Minerva) and Ares (Mars) to the Guards, Demeter (Ceres), Hera and Artemis (Diana) to the Life-givers and Hermes (Mercury), Aphrodite (Venus) and Apollo to the Uplifters. The creative and paternal gods make the universe, the life-givers give it life, uplifters harmonize it, and the guards preserve and protect it [27].[1]

## 2 The Fall on Earth

Crippled at birth, Hephaestos was hurled from Olympus (Heaven) by Hera, who was ashamed of his deformity (Homer Iliad, XVIII.136; Quintus Smyrnaeus Fall of Troy, 2.549). Another version of Greek myths wants him to be cast from heaven by Zeus himself, when the former tried to help Hera. Then the "cosmic invader" passed the "magic threshold" (possibly, earth's atmosphere), travelling "all day long" before landing at Lemnos at "about sunset". But most sources maintain that Hephaestus splashed in the sea, near Lemnos, and was washed ashore, where he lay broken until rescued by the Nereids, Thetis and Eurynome (Iliad, XVIII, 136 and 423–432; Homeric Hymn 3 to Pythian Apollo, 310; Apollodorus, 1.3.11ff.; Pausanias, 8.41.5). Secretly, Hephaestus lived with these goddesses in their underwater caves for nine

[1] http://cura.free.fr/decem/10kengil.html

years. He lived in their "mykhos", a Greek word meaning both innermost place and women's quarters in a house. This nine year hibernation holds a very strong symbolism reflecting a second womblike incubation that awoke his own creative energy. We must be very careful, though, because the word Eurynome was also used as an adjective for Artemis (Pausanias, 8.41.5)!

At this point we must make two crucial observations. Number nine, which was ritually repeated during ceremonies of Kabeirian Mysteries, covers probably an archaeoastronomical truth. One strong analogy is traced in the plasma model of the Plasma physicist Anthony L. Peratt [42].[2] He states that ca. in 4 kya BP, a giant plasma column was produced in the atmosphere of the Earth, so luminous that it was observed by human populations around the world. The early development of this column was transformed into a stack of 9 segments. During that episode, the magnetosphere of Earth was glowing as a semi-permanent aurora.

## 3 The Divine Works

Among the most prominent divine works was Pandora, when Zeus wanted to make an evil thing for men as the price of fire. Earth, water and divine fire were the ingredients of this magic creation. Her name was Pandora, a woman's name, "because each of the gods who have their homes on Olympus had given her a gift, to be a sorrow to men" (Hesiod Works and Days, 60; Hyginus Astronomica, 2.15). But there is another striking piece of information (Orpheus Argonautics, 972–977). Midea's mystical ritual invited the monstrous Hecate, this chthonic goddess, to appear from Hades agile and luminous, with her three heads and an iron face. Her name among the chthonian deities was Pandora!

Another allegory of cosmic invasions is the symbol of the necklace. Dragon's pearls are a worldwide symbol of impacts. The god forged a cursed necklace of Harmony as a gift for the girl at her marriage to Cadmus. The necklace was designed to curse her and all of her descendants, for Hephaestos had a grudge against this child born of Aphrodite's adulterous affair with Ares (Apollodorus 3.25; Diodorus Siculus The Library of History, 4.66.3; Statius Thebaid, 2.265; Pausanias, 9.41.1; Hyginous (Fabulae, 148). Nonnos (Dionysiaca, 5.562), characterizes this necklace as curious, consisting of many colours, as "breathing still of the furnace". Later on (Dionysiaca, 5.88), the description of this masterpiece includes serpents with coiling shape that spit poison from either mouth and of eagles, both alternative symbols of comets. The "meteoritic" symbol of the eagle is also repeated in the forge of Hephaestos concerning Prometheus' daily suffering, a incessant suffering of humankind by extra-terrestrial invaders (Hyginus Astronomica, 2.15).

The East-Asian dragons are almost invariably portrayed with a red sphere in their mouths, in front of their mouths, or (in Javanese art) on top of their heads.

[2] http://www.circlon.com/HTML/darkmatter.html; http:www.mythopedia.info/;
http://physun.physics.mcmaster.ca/~pgs/dark-matter3.html;
http://www.thunderbolts.info/tpod/2004/arch/041231predictions-rock-art.htm

This sphere is called by the Chinese "huoh chuh", meaning the "fire pearl" and also the "meteorite". In Greek mythology, the giant Typhon had serpents for legs and a body all winged or feathered, the feathers usually representing the flames of fire. Also his 100 heads emanated destructive heat (Hesiod Theogony, 820; Aeschylus Prometheus Bound, 356, 371; Apollodorus, 1.6.3).

On the other hand, one of the three cities of Troy, described by Homer, is referred to the period around 1800 BC, when the polar star Tuban (a Draco), according to the phenomenon of wobble of Earth's axis (Precession of the Equinoxes), gave its place to the star b Ursus minor in the heavens. The fall of that Troy was also symbolized by the retirement of the constellation Ursus major from the area of the celestial North Pole [72].

## 4 The Pelasgian Substratum: The Circum-Mediterranean Nuclei

### *4.1 Minoan Crete and Lemnos: The Labyrinth*

Homer (Odyssey, xix, 172–178) calls Crete a land of many peoples (Achaeans, great-hearted native Cretans, Cydonians, Dorians and goodly Pelasgians). The most intriguing connection between Crete and other Mediterranean areas (Egypt, Lemnos, Etruria) is the famous labyrinth (elaborate complex system of paths and tunnels). Apart from a great number of scientific works worldwide, trying to deciphering the symbol, one aspect seems fascinating. By the 5th century BC, the city of Knossos began to mint coins, the earliest of which shows the Minotaur on the obverse and a labyrinthine tetragamma (swastika) with a star or sun motif in the centre on the reverse. In time, the tetragamma gave way to the maze pattern and a human or bovine head replaced the central star [35].[3]

Pliny the Elder in his *Natural History* (XXXVI.13) speaks of a remarkable labyrinth in Lemnos, which has not been identified in modern times. Though called Lemnian Labyrinth in this section, Pliny previously refers to this Labyrinth as the temple built by Theodorus at Samos (34.83, 35.19, 82 & 36.90).

This misinterpretation though, derives from the fact that Samos was the old name of Samothrace, as the Homeric tradition passed into the verses of Apollonius' Argonautics (1.923) and beyond (Diodorus, 3.55.8). Diodorus' narration (5.47.1ff.) on the flood that affected NE Aegean, Asia Minor and Black Sea [47, 48]. The scientists date a major episode in 5600 BC, when the salty waters of the Aegean poured into the brackish waters of the Black Sea), correlates the island of Samothrace, the flood episode, the very ancient nuclei of Eastern Mediterranean cultural substratum and the fishermen's altars near the shores in an excellent and highly valuable framework of environmental information. In the ancient text, the flood is expressed by the term "labros", a Homeric word that means the impetuous waters of the sea or the rivers!

[3] http://www.athenapub.com/11labyr.htm

Recent archaeological, theological, and natural research, has brought forward new hypotheses about the origins of the labyrinth, and that the forces of an ancient tsunami (seaquakes) may play a special part in labyrinth history. There are remarkable examples of the labyrinth shape from a whole range of ancient and disparate cultures.[4] "In its earliest use, the labyrinth seems always to have been associated with death" [9, p. 10]. Since ancient times, people have used labyrinths to invoke the mercy of the Gods in their dealings with the sea. At the shores of Iceland and the Baltic Sea, there are still many old labyrinths, once used by fishing communities as indicators of wind direction and as shrines to the old sea-goddesses, who would protect the fishermen and return them home safe [51, 52]. Legends tell how labyrinths near lakes and sea sides have also been used as places to guide the souls of the deceased to the hereafter. Moreover, the name of Troy-town has been discovered in several locations in Northern Europe [21, 56].

Labyrinths appear in various countries throughout the world (India, England, Scotland, and the Hebrides), as a form of spiritual protection against "evil eye". This protective aspect was also reflected in the planning of prehistoric cities, in order to offer tactical protection from military invasion, as in the case of Troy, which was constructed in a maze-like configuration [9, 26, p. 413].

## *4.2 The Pelasgians and the Etruscans*

Early settlers of Lemnos are thought to be related to the Etruscans of Italy,[5] as evidenced by the burial rites of the pre-6th century BC and the inscriptions found in the island by archaeologists, with striking resemblance to those of the Etruscans (see 6th century inscription on a funerary stele known as Lemnos Stele (National Archaeological Museum, Athens).

The cult of Hephaestos was "Pelasgian". According to Herodotus (6.140.1) the "pre-Hellenic" population of the Lemnos island was Pelasgian, as the population of the Greek mainland before the flood of Deucalion (Thucydides, 1.3.2: the nation of Pelasgians). The Etruscans, said by Herodotus (1.94) to be Anatolian Lydians, arrived to Etruria before the Trojan War, were especial worshippers of Jupiter and lightning of all kinds (at least 30) due to a very consistent fear of lightning. Lightning prefers damp areas, underground waters, towers and hilltops, metallic substances and objects in the soil. In Etruscan mythology, Techulka, a smith-god and death-demon, hit his victims by a giant hammer, accompanied by a winged demon figured with snakes. Recent investigations have shown that their blood type is similar to the Urartu people of Lake Van in NE Anatolia [21]. It is also noteworthy that Homer (Iliad, V.480 & X.429) mentions that Pelasgians had once built a city, named Larissa, in the area of the Troad. For this reason, Larissans stood for the Trojans during the Trojan War.

[4] http://www.labyrinthos.net/centre.htm

[5] http://www.compmore.net/~tntr/lemstelea.html

## 5 Fire Festivities: The Kabeirian Mysteries

Among the mysteries of Panhellenic fame were those of Kabeiroi in Samothrace. The Kabeiroi were also worshipped in Lemnos and Thebes, where the sanctuary is 8 km west of Thebes.[6] From what is known, a non-Greek element is revealed in this cult, to which the experts have attributed a Hittite origin. The Lemnian Kabeiroi are related to the worship of Hephaestus and the Thyrrenian (probably Etruscan) past of the island. In Samothrace and Lemnos, the nine-day ceremony, during which all fires of the island were extinguished, took place in a grove near the temple of Hephaestos (the word orgies from the Greek *orgas*, i.e. area with high humidity, water and wood). The tribe called the Kabeirides was thought to be fathered from Hephaestos and a Thracian nymph Kabeiro, daughter of sea-god Proteus. Cadmus, Odysseus and Agamemnon were initiated to those Mysteries (Apollonius, 1.917ff).

Hephaestos' twin sons the Kabeiroi fought beside Dionysus' in his war against the Indians. Twice during the battle, Hephaestos intervened to carry his sons to safety, when the Indian River Hydaspes tried to drown them (Nonnos, 24.77, 27.120 & 325, 29.193, 30.42 & 36.5, 36.129). The motif of fire bearing Hephaestos, bringing destruction "with his blazing shower of deadly Lemnian flame" reminds us of (a) Gigantomachy when Hephaestos killed Mimas by throwing at him molten iron (Apollodorus, 1.37) or exhausted sank on the battlefield of Phlegra (Apollonius, 3.23), (b) Hera and the Trojan river Skamandros, and (c) Phaethon and river Po/Eridanus [7].

Within this conceptual framework, goddess Hera's symbolic substratum seems to be more related to the spirits of fire than purely to the moon, as Woods claim [72]. Hera "of the Pelasgians" (Apollonius, 1.15) had other characteristics and was distinguished from the moon goddess Hera, also appeared in the Homeric Epics. During the Trojan War, Hera appealed to her own dear son to rescue Achilles from the River-God Skamandros by pronouncing a few very important words, perhaps one of the most clear arguments for Hephaestos' cometary connection (Homer Iliad, XXI, 328).

## 6 Hephaestos and Athena: A Dual Celestial Archetype

Hesiod, as well as Roman sources, claims that Hera gave birth to Hephaestus parthenogenetically, without Zeus' participation, since she was angry at him for giving birth to Athena from his own head without first procreating with her. Thus, in Theogony (924–929), Hesiod highlights the analogy between Athena and Hephaestos (Apollodorus, 1.19; Cicero, 3.22). The Roman equivalent of Athena (Minerva) was Hephaestos (Vulcan). The sequence of the Twelve Gods appears in the Rustic Calendar, in Manilius and at the Altar at Gabii. Aries and Libra had Athena-Minerva and Hephaestus-Vulcan as their guardian gods. Aries symbolizes the head

[6] http://www.gla.ac.uk/archaeology/research/pgabstracts.html

from which Athena sprang. In the same conceptual framework, Hephaestos was treated as the creator of the constellations, a creative force in the Universe (Iliad I, 597–607) and the mythical fall on Lemnos as god's stay below the horizon, in the realm of Thetis (Heridanus constellation), where he created the constellations of the Southern Hemisphere [72]. Although in the verses of Odyssey (xviii, 283), the god forge is on the island of Lemnos, in the Iliad (XVIII, 369) it is located in heavens. Perseus constellation is more likely to connected with Hephaestos not because of its shape, but of its relation with the meteor swarms of Perseides, visible from the 25th of July to the 4th of August. These flames are also described as "burning the sky" (Iliad V, 4–8) [72].

Athena is a master of disguise, as Homer constantly points out. Pallas Athena represented the proto-planet Venus (Typhon = the cometary tail of proto-Venus), in her cometary behavior [70], and was worshipped among the peoples of Mediterranean [20, 61]. That Venus was later identified with goddess Aphrodite and planet Venus' dual appearance in the sky (evening = female, morning = male), was Aphrodite barbata (bearded), or the Cyprian goddess Aphrodite with a beard, a strong image of bisexuality (Pauly-Wissowa R.E.) Athena holds also her primordial androgynous image, as male, and bearded serpents were found on a pediment of the Archaic Athenian Acropolis [20]. Planet Venus is symbolized by the "crux ansata" (Egyptian ankh), a combined phallus and vulva. Consequently, as we can detect, two Heras, two Athenas and two Hephaestos in the Epics, we can also find the two Aphrodites, the Uranian/Selenian and the planetary Venus.

Athena "herself had no womb, for when she carried children, it was in a basket" [22]. In the Orphic Hymns (32.10-11) is clearly addressed by the words: "born both as male and female", "agile and luminous" and "dracaena".

## 7 The Archaeoastronomical Evidence

### *7.1 Hephaestos 2212*

A statistically significant number of Earth-crossing asteroids are part of the Taurid Complex of interplanetary objects. Another group is also identified, which appears aligned with (2212) Hephaestos. In addition, these two complexes may are of common origin [2, 3, 14, 58, 59, 60]. British astronomers Victor Clube and Bill Napier [12] built a strong argument, according to which the progenitor of comet Encke quite likely caused humanity a great deal of grief in the past. They list only one object, Hephaestos, as having once been part of the still active comet Encke. Hephaestos was discovered in 27 September 1978 (I.A.U.Circ. 3279) and is one of the largest Earth-orbit-crossing objects so far found. Its 9.5 km diameter (about the same as the hypothetical dinosaur slayer) is actually larger than comet Encke's estimated girth. This object has in an orbit very much like that of the Taurids, but rotated through about 90°. Probably some thousands of years ago, it split from the giant comet, whose debris now travel along the Taurid track [14]. Hephaestos represents a potential collision hazard, although not of immediately concern. It would strike Earth at just over 30 kilometres/second, with an impact energy of about 100 million TNT megatons! The giant comet break up hypothesis is consistent with the archaeostronomical, palaeoclimatic and geochemical evidence available. The last Ice Age could have been caused by the progenitor of comet Encke , part of which was the Tunguska meteorite, an interstellar object. It is estimated that further debris from the zodiacal cloud will intersect earth during the period AD 2000–2400.

## 7.2 The Three Cities of Troy in the Homeric Epics

"The Fall of a City" is a legendary symbol for a disaster in various cultures [20]. The Homeric Epics are a palimpsest of archaeoastronomical knowledge, as they are built on layers of information [21: "Homer describes at least one Trojan War"]. At least three "Trojan cities" are involved in the Epics. The latest, Troy VIIa, dated ca. at the beginning of the 12th century BC, was a victim of celestial conflagration (probably Phaethon's myth; the name of Phaethon was also used as an adjective among the inhabitants of Colchis, see Apollonius, 2.253–254).The strong signal around year 1159 BC, recorded on tree rings and ice-cores all over the world, as well as the total solar eclipse on April 10th, 1178 BC, visible in the geographical latitude of Eastern Mediterranean and described in the verses of Odyssey (xx, 350–356), produce a dating limit (ca. 1200 BC) for this event [31, 40]. Phaethon story implies that the disintegrating objects were travelling south of east.

The other, Troy VI, was under siege ca. 1312 BC, as new archaeoastronomical evidence of a total solar eclipse implies [28, 39] and Troy IIg was one of the scenes of "Hephaestos' rage". That Troy was also the witness of multi-regional catastrophes and collapse dated around 2200 BC (Egyptian Old Kingdom, Harappan Culture, Canaanite settlements, Malta and Akkad), and to cultural revivals, migrations and social reorganizations during the period 2000–1800 BC. The climatic upsets of the period 2200–1800 BC are interpreted by interdisciplinary studies.[7] Around 2200 BC, in Southern Asia, the Indian monsoons providing 80% of the Nile flow was deflected. At the same time, the famous historic flood of China was followed by aridification in 2000 BC, and the West China's cold event (2000–1500 BC). Similar phenomena of extended drought are registered near the sources of Nile, Tigris and Euphrates, Indus and Yellow rivers. The impact craters in Al Amarah marshes of Iraq are also dated to 2000 BC [36]. Finally, during the years 1900-1800 BC, the Harappan culture in the Indus valley collapsed.[8]

---

[7] www.iap.ac.cn/html/qikan/aas/aas2004/ 200402/040216.pdf
http://www.space-frontier.org/Projects/TheWatch/Archives/20011104middleeast.html;
http://www.unu.edu/env/govern/elnino/countryreports/inside/china/REFERENCES/
REFERENCES_txt.html

[8] http://www.grahamhancock.com/forum/BlackardA1.php

The impact signal of the 4kyr B.P. event throughout land and seas [25], shows some characteristics: (a) the co-occurrence in distant regions of flow-glass debris with similar petrographical and geochemical characteristics, and (b) the distinctive heated soil surface, both identified the distal dispersion of an impact ejecta. The particular configuration conditions, petrography and geochemistry of the distinguishing features are compared from sedimentary records in soils, ancient habitations, lakes and deep-sea cores in various regions of the Northern and Southern Hemispheres. The best-preserved record of ejecta dispersion (nearly intact signals) is observed in continental deposits at specific locations, where the impact-related surface was rapidly buried.

On the other hand, Sanskrit literature of ancient and medieval India is rich in information about environmental sciences. Fiery celestial body fallen on Earth, earthquakes, rise of sea-level, draught of rivers, lakes and wells, destruction from heaven, severe famine are some of the implications related to the Pleiades. These disasters should have taken place in the north-western part of India (23.5° N, 71.5° E), where the river Sarasvati joined the sea. The whole plain, now an arid area known as the Thar Desert, was once a very fertile plain crossed by this great river. In those days of Mohenjo-Daro and Harappa, the area was one of the richest places in the world. Renown scholars, planetarium softwares and astronomical calculations date the impact crater and the falling meteors around 1800–2200 BC [29, 64].

Finally, another piece of archaeoastronomical information has been deciphered in the Iliad, providing scientists with a strong argument on the chronological structure of the Homeric Epics [72]. On March 5th, 1953 BC, a conjunction of 5 planets (Jupiter, Saturn, Mars, Venus & Mercury) with the new moon was visible in the geographical latitude of Greece. This information, hidden in the verses of Iliad (I, 493–494), was referred to by Chinese astronomers of the time. Modern researchers, Kevin Pang of JPL (Jet Propulsion Laboratory) and John Bangert of the Naval Observatory have discovered that the conjunction was visible in the night sky for some days, in fact, during 26th of February their alignment was most perfectly observed for the last 6000 years! In fact, Pang found a passage in a 1st century BC text of Hong Fan Zhuan saying: "The Ancient Zhuanxu calendar (invented in about 2000 BC) began at dawn, in the beginning of spring, when the sun, new moon and five planets gathered in the constellation Yingshi (Pegasus)". This book was written by Liu Xiang who lived from 77 to 9 BC.

## 8 Comets and Their Symbolic Images

Physically, Hephaestus was a muscular man with a thick neck and hairy chest, who, because of a shortened, lame leg (lame = one-footed) and club foot (with feet facing backwards), supported himself with a crutch. Bearded, he most often dressed in a ragged sleeveless tunic and woolen hat. Most frequently, he was portrayed in art holding the tools of his trade, especially blacksmith's hammer and tongs. Sometimes, he was surrounded by the Kabeiroi (Herodotus, 3.37), the dwarf-like blacksmith servants of Mother Goddess, who helped him in his subterranean forge. The characteristics attributed to him remind of the coma, when comets are hit by the solar wind. The description of an ancient Greek painting by Philostratus the Elder (Imagines, 1.1), too, notices that Homer inspired the ancient artist in the scene of Skamandros and Hephaestos. And it is noteworthy that one of the two god's features is characterized by speed, an attribute not consistent to his malformation. But the adjective lame in ancient Greek may also be interpreted as strong-armed and ambidextrous.

Phaethon's western parallel, Quetzalcoatl (the feathered serpent), according to the Annals of Quauhtitlan, immolated himself on the shores of the eastern sea, and from his ashes rose birds with shining feathers (symbols of warrior souls mounting to the sun), while his heart became the Morning Star ... Tezcatlipoca, his antagonist defeating Quetzalcoatl in ball-play (a game directly symbolic of the movements of the heavenly orbs), cast him out of the land into the east, where he encountered the sun and was burned [4, vol. 3, p. 61; 31, 32].[9]

Another interesting aspect of this folk memory which might have shed some light on why the rolling cross motif is linked to birds is the image of a one-legged fowl. This is also a characteristic of the Chinese divine pheasant which was closely associated with the mythical, lame, raven-beak-nosed emperor Yu, who could transform himself into this pheasant or a bear. One of Yu's enemies, the Owl, who invented thunder and lightening was also one-footed [5, pp. 118–121, 150–151; 6, pp. 122–156].

A second aspect of comets, evident in ancient lore, involves shape-shifting. A comet is three dimensional and could appear as quite a different animal when viewed from a different angle (e.g. the mythical ability of Yu to transform into a pheasant or bear). Our ancestors' stories speak of weakening gods and fantastic births, because comets can also change spontaneously; a gas emitting area could become dust, or a piece of the comet could break away, creating another comet, perhaps initially more flamboyant than its parent. In Chinese lore, Ts'ang Chieh, the four-eyed legendary inventor of writing, was inspired to create written symbols from noticing the marks of birds' feet in the sand. His ancient style is known as "bird foot-prints writing" [34, p. 31].

According to the Comet/bird hypothesis, when a comet approaches so close to Earth, the jets of gas streaming from it, bent by the comet's rotation, became visible, looking like a tetragamma. This observation is drawn from an ancient Chinese manuscript that shows comet tail varieties [49]. Respectively, the swastika-like comet on the Han Dynasty silk comet atlas was labelled a "long tailed pheasant star".[10] So, many swastika and swastika-like motifs may have been representations of bird tracks, including many of those found by Schliemann in Troy [30].

Symbolic bird tracks, even if they had not been recognized as such, appear on objects unearthed by Heinrich Schliemann from Hissarlik in Asia Minor [55, pp. 334–353]. More than 700 owl-faced idols and vases were also collected from the third

[9] http://abob.libs.uga.edu/ bobk/bobk.html

[10] http://en.wikipedia.org/wiki/Han_Dynasty_silk_comet_atlas#Comet-inspired_motifs

city of Troy, as copies of the ancient Palladium, which was fabled to have fallen from heaven with joined feet. And Glaux is the little owl, Athene noctua, emblem of old and new Athens, reminding of the homeric "owl-eyed Athena".

# 9 The Geoarchaeological Evidence

## *9.1 Lemnian Earth and the Destruction Layers*

The "flame of Hephaestos" or his "red breath" (characterized as purest flame) was a leitmotif among ancients (Orphic Hymn 66 to Hephaestus; Homer Iliad, II, 426, IX, 467, XVII, 88 & XXIII, 33 and Odyssey xiv, 71; Hesiod Theogony, 864; Aristophanes Birds, 436; Quintus Smyrnaeus, 13.170,13.367 & 4.160; Suidas, s.v. "Hephaestos").

Although ancient writers mention Lemnian earth together with Keian, Cappadocian and Sinopic earths [43], all four being identified as red earths, Pliny's comment makes the difference. This earth (terra lemnia, rubricata or sigillata) resembles cinnabar (35.14), it had a pleasant taste, too, while Galen (13.246b) adds that "it differs from miltos because it doesn't leave a stain when handled". The same writer, during his visit to Hephaestias , analyzes the myth of Hephaestos and his relationship with Lemnos, saying that "the mythical hill, also known as Mosychlos, appeared to be burnt due to its colour and from the fact that nothing grows on it". Belon, during his journey in the 16th century, refers also to the yellow/white colours of the earth, equally explained by the presence of hydrothermally altered rocks [44].

Consequently, hydrocarbon evolution due to past volcanic activity may be one explanation. Destruction layers with hydrocarbon presence and other characteristics mentioned above (like cinnabar, with sweet taste, loosing its power with the time passing over or being periodically recharged) may be another evidence of past celestial events (combustion residues, chemical fusion). Troy IIg conflagration (first fall) produced an up to 6 m bed of ashes and a layer of calcinated debris up to 3 m high. Experts on wild fires claim that there was never seen "red ashes of wood in natural fires, because ash residue from the burning of a city is measured in inches, rather than feet". The mysterious melted copper and lead which covered a large area, according to Schliemann, might have originally been deposits that contributed to the attractiveness of the site for lightning discharges. After Schliemann's observations on this destruction layer of the "burnt city", the Cincinnati archaeologists, under the leadership of Carl Blegen, examined closely the ruins of the Burnt City-Level IIg by their code. The stratum of Troy IIg had an average thickness of more than 1m; it consisted mainly of ashes, charred matter, and burned debris. This deposit apparently extended uniformly over the entire site, eloquent evidence that the settlement perished in a vast conflagration from which no buildings escaped destruction. The catastrophe struck suddenly, without warning, giving the inhabitants little or no time to collect and save any belongings before they fled. Moreover, the Cincinnati team mention several places of the greenish-yellow discoloration (probably sulphur oxides). The calcinated debris of the old city was strong enough to become the foundation of the new city walls of Troy III [8, 19, 45, 53–55].

A cometary or planetary near-encounter results in falling gases, hydrocarbons, burning pitch and stones. Such events are unknown to modern experience but are indicated by ancient legends over many places and by various geological and biological phenomena [17–20, 65].

## *9.2 Impact Craters: Underwater or Terrestrial?*

In 1927 Franz Xaver Kugler, a Jesuit scholar who had devoted over thirty years to the study of cuneiform astronomical texts, published an essay entitled "The Sibylline Starwar and Phaethon in the Light of Natural History", asserting that a large impact event in the Mediterranean Sea inspired fire-from-above legends such as Phaethon's ride [31].[11]

Apart from the terrestrial impact craters all over the world, one underwater feature deserves special mention. Burckle crater is located in the central Indian Ocean on the edge of a fracture zone at 30.87° S, 61.36° E. The whole setting infers to a Shoemaker-Levy type impact of a comet, the fragmentation of which also produced two other large impact centres, one in NW Pacific and another in CE Pacific. The scientific team, examining the case, trying to co-estimate various astronomical, geoarchaeological and mythological data and information, suggests a specific calendrical date of around May 10, 2807 BC. Similar investigations correlate the major environmental and social upheavals of the past to crucial impact events [1].

Excessive influx of cometary debris (without impact events) may precipitate an enhanced zodiacal light, and provoke extreme aurora events by disturbing the geomagnetic field in our planet (http:www.mythopedia.info/). These spectacular instabilities in the plasma were remembered as dragons and waring gods. Of course, the phenomenon of an enhanced aurora may be triggered by an extreme solar weather, passage through a gigantic molecular cloud or disintegration of a giant comet in the inner solar system, even by a combination of these phenomena. Such visual experiences reformed the psychological, sociological and artistic aspects of the myths.

Respectively, the scenario of conflagration is enriched by other natural phenomena related to cosmic invaders, causing a "Tunguska type" events. Gas cloud [15], "swamp gas" (usually called that way even if it is generated from earthquake gassing fissures) that cause asymmetric areas of destruction[12] or "vacuum bomb" during high altitude explosions [37], can be added to the catalogue of impact phenomena [21], Moreover, mega-lightning and fire-storm (possibly due to interplanetary discharge or to bolide and lightning-fire shower) could be considered as the

[11] http://abob.libs.uga.edu/bobk/bobk.html

[12] Ion Nistor, http://www.tunguska101.org/hypothesis_uk.htm

main contributors to the destruction of Troy VIIa (Phaethon's time). Ancient writers often refer to the catastrophes of prosperous cities, like Sodom and Gommorah, and Bolsena, the richest town in Tuscany, by extreme thunderbolts [21].

## 10 Conclusions

There is a – till recently neglected – agent which could destroy civilization and cause earthly turbulence, extra-terrestrial encounters. Astronomical evidence indicates that our ancestors viewed a much more active sky than we do. Particularly, during the past twelve thousand years, such deliveries were not uncommon. Much evidence suggests that humanity witnessed, and was affected by, the break-up of a very large comet over this period. Along with the two luminaries, the solar deities (e.g. Apollo, Hercules, Helios) and the moon goddesses (Aphrodite Urania: Hecate-Hera- Artemis), who represented the female reproductive force, there was another sacred fire represented by cosmic "invaders" (e.g. Typhoon, Hephaestos, Phaethon) or other phenomena (e.g. Sirius, Saturn, Jupiter).

In summary, the strong parallelisms of symbols, words, images and allegories in the human knowledge indicate that the "Greek" Hephaestos was derived from the Pelasgian religious circum-Mediterranean substratum, reflecting some major celestial events dated back to the beginning of 2nd millennium BC. The Homeric Epics are a palimpsest of archaeoenvironmental knowledge, as they are built on layers of information.

Florence and Kenneth Wood, Alfred de Grazia and NASA first maintained that Vulcan may be connected to meteoritic phenomena. The present paper focuses on the arguments of this statement by collecting data from ancient writers, worldwide iconography and geoarchaeological, palaeoclimatic and archaeoastronomical evidence. Furthermore, the paper deals with the spatial and temporal itinerary of myth's elaboration. The psychological filtration is also taken under consideration, when comes to matters of disaster dealing within the socio-cultural framework of ancient societies.

To conclude, Disaster Archaeology finds a very powerful ally in the name of Catastrophist Mythology, which echoes real occurrences through various myths. The Homeric figure of Hephaestos and his puzzling mythical motif holds a prominent role in it.

## Acknowledgements

I wish to express my deepest gratitude to my mental companions in Disaster research, Professors Alfred de Grazia (one of world leading personalities in Disaster Studies), George Ferentinos (Marine Geology & Physical Oceanography, University of Patras), Stavros Papamarinopoulos (Applied Geophysics, Patras University) and Effie Photos-Jones (Department of Archaeology, University of Glasgow), who inspired me in various ways.

## References

1. Abbott, D.H., Masse, Br.W. and Bregerr, D. (2007). Burcle Abyssal Impact Crater: Did this impact produce a global deluge? in *Proceedings of International Conference on 'The Atlantis Hypothesis: Searching for a Lost Land'*, Island of Melos, Greece, 11–13 July 2005, St. Papamarinopoulos (Ed.), Heliotopos Publications, Athens, pp. 179–190.
2. Asher, D.J., Clube, S.V.M. and Steel, D.I. (1993). Asteroids in the Tauri complex, *Monthly Notices of the Royal Astronomical Society/MNRAS* **264**(1), 93.
3. Asher, D.J. and Izumi, K. (1998). Meteor observations in Japan: New implications for a Taurid meteoroid swarm, *Monthly Notices of the Royal Astronomical Society/MNRAS* **297**(1), 23–27.
4. Bancroft, H.H. (1886). *The Works of Hubert Howe Bancroft*, Vol. 3, The Native Races, The History Co., San Francisco.
5. Barnard, N. (Ed.) (1972). *Early Chinese Art and Its Possible Influence in the Pacific Basin*, Vol. 1, Intercultural Arts Press, New York.
6. Barnard, N. (1973). *The Chu Silk Manuscript. Translation and Commentary, Part 2*, Australian National University, Canberra.
7. Bate, H.N. (1918). *The Sibylline Oracles Books III–V*, MacMillan Company, New York.
8. Blegen, C. (1963). *Troy and the Trojans*, Thames and Hudson, London.
9. Bord, J. (1976). *Mazes and Labyrinths of the World*, E.P. Dutton, New York.
10. Carpenter, Rh. (1966). *Discontinuity in Greek Civilization*, The J.H. Gray Lectures for 1965, Clarendon Press, Cambridge.
11. Chang, S.H. (1990). *History and Legend*, University of Michigan Press, Ann Arbor.
12. Clube, S.V.M. and Napier, B. (1982). *The Cosmic Serpent*, Universe Books, U.K.
13. Clube, S.V.M. and Napier, W.M. (1987). The cometary break up hypothesis reexamined – A reply, *Monthly Notices of the Royal Astronomical Society/MNRAS* **225**(1), 55–58.
14. Clube, S.V.M. and Napier, B. (1990). *The Cosmic Winter*, Basil Blackwell Inc., Cambridge, MA.
15. de Grazia, Al. (1981). *Chaos and Creation*, Metron, Princeton/London/Bombay. Publications.
16. de Grazia, Al. (1983). *Homo Schizo I and II*, Metron Publications, Princeton, NJ.
17. de Grazia, Al. (1983). *God's Fire*, Metron Publications, Princeton, NJ.
18. de Grazia, Al. (1983). *The Lately Tortured Earth*, Metron Publications, Princeton, NJ.
19. de Grazia, Al. (1984). *The Burning of Troy*, Metron Publications,Princeton, NJ.
20. de Grazia, Al. (1984). *The Disastrous Love Affair of Moon and Mars*, Metron Publications, Princeton, NJ.
21. de Grazia, Al. (2005). *The Iron Age of Mars. Speculations on a Quantavolution*, Metron Publications/Eumetron, Naxos, Greece.
22. Deutsch, H. (1969). *A Psychoanalytic Study of the Myth of Dionysus and Apollo*, Int. I.U. Press, New York.
23. Diggle, J. (1970). *Euripides Phaethon*, Cambridge University Press, Cambridge.
24. Drews, R. (1992). *The end of the Bronze Age: Changes in Warfare and the Catastrophe ca. 1200 BC*, Princeton University Press, Princeton, NJ.
25. Evans, K.R., Horton, J.W. Jr., Thompson, M.F. and Warme G.E. (Eds.) (2005). *The Sedimentary Record of Meteorite Impact*, SEPM Research Conference, Springfield, Missouri, May 21–23, Southwest Missouri State University Printing Services.
26. Freitas, Lima de (1987). Labyrinth, in *The Encyclopedia of Religion*, M. Eliade (Ed.), 16 Vols. Macmillan, New York.
27. Gillman, K. (1996/1997). Twelve gods and seven planets, *Considerations* **XI**(4), 63–95.

28. Henriksson, G. (2006). The Trojan War dated by two solar eclipses, Paper presented at *International Conference on Archaeoastronomy – SEAC 2006 on 'Ancient Watching at Cosmic Space and Observation of Astronomical Phenomena'*, Island of Rhodes, Greece, 6–10 April 2006.
29. Iyengar, R.N. (2004). Profile of a natural disaster in ancient Sanskrit literature, *Indian Journal of History of Science, INSA* **39**(1), 11–49.
30. Kobres, B. (1992). Comets and the Bronze Age collapse, *Society for Interdisciplinary Studies* (Chronology and Catastrophism Workshop 1992) **1**, 6–10.
31. Kobres, B. (1995). The path of a comet and Phaethon's ride, *The World & I* **10**(2), 394–405.
32. Koss, N. (1979). The relationship of Hsi-Yu Chi and Feng-Shun Yen-I, *T'oung Pao* **LX**(4-5), 143–165.
33. Legge, J. (1960). *The Chinese Classics* (Vols. 3 & 4), Hong Kong University Press, Hong Kong.
34. MacCulloch, C.J.A. and Canon, J.A. (Eds.) (1928). *Mythology of All Races*, Vol. 8, J.C. Ferguson "Chinese Mythology", Marshall Jones Co., Boston, MA.
35. MacGillivray, A. (2003). Return to the labyrinth: A clew to the function of the Minoan palaces, *Athena Review* **3**(3), 62–66.
36. Master, Sh. (2002). Umm al Binni lake, a possible Holocene impact structure in the marshes of southern Iraq: Geological evidence for its age, and implications for Bronze-age Mesopotamia, in *Environmental Catastrophes and Recoveries in the Holocene*, Brunel University, Department of Geography & Earth Sciences, August 29–September 2, 2002.
37. Nikolsky, G.A., Tsinbal, M.N., Shnitke, V.E. and Shultz, E.O., Tunguska vacuum bomb of cosmic origin, http://www-th.bo.infn.it/tunguska/abstr3html
38. Nur, A. and Cline, E. (2000/2001). Poseidon's horses: Plate tectonics and earthquake storms in the late Bronze Age Aegean and Eastern Mediterranean, *JAS* **27**, 43–63.
39. Pang, K.D., Chou, H.H., Yau, K., Bangert, J.A. and Ahluwalia, D.A. (1989). Shang dynasty oracle bone eclipse records and the Earth's rotation rate in 1302 B.C., *Bulletin of the American Astronomic Society* **21**, 753.
40. Papamarinopoulos, St. (2007). Phaethon and Phaethousa. A shining comet passing in the 12th century B.C., in *Proceedings of International Conference on 'The Atlantis Hypothesis: Searching for a Lost Land'*, Island of Melos, Greece, 11–13 July 2005, St. Papamarinopoulos (Ed.), Heliotopos Publications, Athens, pp. 163–177.
41. Peiser, B.J., Palmer, Tr. and Baily, M.E. (Eds.) (1998). *Natural Catastrophes during Bronze Age Civilizations. Archaeological, Geological, Astronomical and Cultural Perspectives*, BAR International Series No. 728, Oxford, England.
42. Peratt, A.L. (1992). *Physics of the Plasma Universe*, Springer-Verlag, New York.
43. Photos-Jones, E., Cottier, A., Hall, A.J. and Mendoni, L. (1997). Keian miltos: The well-known iron oxides of antiquity, *ABSA* **92**, 359–371.
44. Hall, A.J. and Photos-Jones, E. (2008). Accessing past beliefs and practices: The case of the Lemnian earth, *Archaeometry Online*. Early Articles. doi: 10.1111/j.1475-4754-2007.00377.x.
45. Rapp, G. Jr. and Gifford, J.A. (1982). *Troy: The Archaeological Geology*, Supplementary Monograph 4, Princeton University Press, Princeton, NJ.
46. Rix, Z. (1977). Note on the Androgyne comet – I, *S.I.S.R.* **5**, 17–19.
47. Ryan, W. and Pittman, W. (2000). *Noah's Flood: The New Scientific Discoveries about the Event that Changed History*, Simon & Schuster, New York.
48. Ryan, W., Major, C., Lericolas, G. and Goldstein, S. (2003) Catastrophic flooding of the Black Sea, *Annual Review of Earth Planetary Science* **31**, 525–554.
49. Sagan, C. and Druyan, A. (1985). *Comet*, Random House, New York/Joseph, London.
50. Sallares, R. (1991). *The Ecology of the Ancient Greek World*, Cornell University Press, Ithaca, New York.
51. Saward, J. and Deb (1998). The labyrinth in Iceland, *Caerdroia* **29**, 58–60.
52. Saward, J. (2003). *Labyrinths and Mazes*, Gaia Books, London/Lark Books, New York.
53. Schaeffer, C.F.A. (1948). *Stratigraphie comparée et chronologic de l'Asie Occidentale*, Oxford University Press, London.

54. Schliemann, H. (1875). *Troy and Its Remains*, Murray, London.
55. Schliemann, H. (1881). *Ilios*, Harper & Brothers, New York.
56. Spanuth, J. (1975). *Atlantis of the North*, Sidgwick & Jackson, London.
57. Steel, D. (1995). *Rogue Asteroids and Doomsday Comets*, Wiley, New York.
58. Steel, D.I., Asher, D.J. and Clube, S.V.M. (1991). The structure and evolution of the Taurid complex, *Monthly Notices of the Royal Astronomical Society/MNRAS* **251**, 632–648.
59. Steel, D.I. and Asher, D.J. (1996). On the origin of Comet Encke, *Monthly Notices of the Royal Astronomical Society/MNRAS* **281**(3), 937–944.
60. Stohl, J. (1987). Meteor contribution by short-period comets. *Astronomy and Astrophysics* **187**(1–2), 933.
61. Talbott, D. (1994). The great Comet Venus, *AEON* **III**(5), 5–51.
62. Thomas, P., Chyba, C. and McKay, C. (Eds.) (1997). *Comets and the Origin and Evolution of Life*, Springer-Verlag, New York.
63. Thompson, W.I. (1981). *The Time Falling Bodies Take to Light*, St. Martins Press, New York.
64. Valdiya, K.S. (2002). *Saraswati, the River that Disappeared*, University Press, Hyderabad.
65. Velikovsky, I. (1950). *Worlds in Collision*, Mac Millan (April) and Doubleday Books (June), New York.
66. Velikovsky, I. (1955). *Earth in Upheaval*, Doubleday Books, New York.
67. Velikovsky, I. (1977). *Peoples of the Sea*, Doubleday Books, New York.
68. Wainwright, G.A. (1959). The Teresh, the Etruscans and Asia Minor. *Anatolian Studies* **9**, 197.
69. Ward, W.A. and Joukovsky, M.S. (Eds.) (1992). *The Crisis Years: The 12th century B.C.*, Kendall/Hunt Publishing Company, Dubuque, IA.:
70. Wallis, M.K. (1972). Comet-like interaction of Venus with the solar wind, *Cosmic Electrodynamics* **3** (April), 45–59.
71. Whipple, F.L. (1985). *The Mystery of Comets*, Smithsonian Institute Press, Washington, DC.
72. Wood, F. and Wood, K. (1991). *Homer's Secret Iliad*. The Star McCune Trust, USA.

# A Comet during the Trojan War?

S.P. Papamarinopoulos

*University of Patras, Greece*

This paper is dedicated to Victor Clube, Bill Napier, Alfred De Grazia and Bob Kobres. From their work I learnt to interpret the past.

**Abstract.** Plato, in Timaeos, describes the story of the sun's inadequate son who mimics his father. He is called Phaethon (shining) and caused hardships to several peoples on earth. In the past, Phaethon was associated with visible luminous electric phenomena and fires on the earth related to the volcanic explosion of Santorini. Plato regards Phaethon's story as a non-fabricated myth but on the contrary as true information from the past. In the text, time of the event is not given, but the general analysis of Timaeos' and Critias' passages in connection with Atlantis proves that the 12th century BC is the obvious century of the event. Plato claims it had happened thousands of years before Solon's 6th century BC. Herodotus, a century earlier, records that Egyptian history extends thousands of years before his time. However, centuries later, ancient Greek writers and Egyptologists illustrate the use principally of moon calendars by the Egyptian priesthood in all epochs. The priests, when recording the ancient history of their country used moon calendars whereas Pharaoh's high officials used solar calendar, when issuing governmental edicts. Dividing these thousands of years by the number of the full moons of the Metonic circle-year we come to the beginning of 12th century BC. A comet actually introduces a parallaxis in its orbit with respect to the orbits of the planets and their satellites as the myth describes. However, the luminous phenomena in volcanic eruptions do not. Homer, on the other hand, describes the same phenomenon with a variety of complex images as happening during the war between the Achaean Greeks and the Trojans. The comet appears as Athena coming from the west in the form of a shining "star", in other words, as Phaethousa. The female and male appearances of the same strange and rare phenomenon seen by different people in the Aegean can be explained as an optical illusion. Numerous traces of fires have been spotted in archaeological sites corresponding to 12th century BC in Bronze Age sites in East Mediterranean. They are assigned as results of earthquake activity. In fact, the seismic storm which occurred as a domino effect has left indisputable evidence at the above sites. The decisive scientific methodology may prove that, apart from fires caused by earthquakes, Phaethon could have also caused some fires. This may be proved by taking samples of burnt soil from 12th century BC horizon far away from palaces or settlements. The contact between falling burning cometary fragments and forests causes fires. Furthermore, its tail, consisting of ionized plasma, stardust and water particles, interacts with the atmosphere producing precipitation. For instance, Irish bog trees and Asia Minor oaks exhibit an anomalous precipitation event within their growing rings in 1159 BC and in the years after (see [2, 18] respectively). At exactly the same period, ice-cores present extreme high acidity peaks, whereas in Ireland, Greenland and Asia Minor a kind of climatic shift appears.

*S.A. Paipetis (ed.), Science and Technology in Homeric Epics,* 341–356.

## 1 Introduction

Plato, in *Timaeos*, explains that the story he intends to describe has the form of a myth, meaning a non-fabricated myth but a myth imprinted on the memory of the people in the past. It is therefore a truth, in other words logos. Let us examine more closely the case step by step by clarifying the differentiation between fabricated and non-fabricated myth (Tim 22.c.3–22.c.7):

τὸ γὰρ οὖν καὶ παρ' ὑμῖν λεγόμενον,
ὥς ποτε Φαέθων Ἡλίου παῖς
τὸ τοῦ πατρὸς ἅρμα ξεύξας
διὰ τὸ μὴ δυνατός εἶναι
κατὰ τὴν τοῦ πατρὸς ῥδόν ἐλαύνειν τα
τ' ἐπὶ γῆς συνέκαυσεν καὶ αυτὸς
κεραυνωθεὶς διεφθάρη.

For in truth the story that is told in your country as well as ours, how once upon a time Phaethon, son of Helios, yoked his father's chariot ,and, because he was unable to drive it along the course taken by his father, burnt up all that was upon the earth and himself perished by a thunder-bolt.

Plato describes an entity, a kind of object which mimics the sun, as if it were its relative in shining capacity but smaller in size, in other words its child. However, it results in being catastrophic due to its incapacity to light the earth with the same respect as the sun. One wonders what sort of an object this Phaethon could be causing hardships to peoples on earth. Could it possibly be a usually fabricated Platonic myth after all? Plato himself clarifies the concept of this myth in the following passage (Tim 22.c.7–22.d.3):

τοῦτο μύθου μὲν σχῆμα ἔχον λέγεται,
τὸ δὲ αληθές ἐστι τῶν περὶ γῆν
κατ' οὐρανόν ἰόντων παράλλαξις καὶ
διὰ μακρῶν χρόνων γιγνομένη τῶν επὶ
γῆς πυρὶ πολλῷ φθορά.

that story, as it is told, has the fashion of a legend, but the truth of it lies in the occurrence of shifting of the bodies in the heavens which move round the earth, and a destruction of the things on earth by fierce fire.

But the planets and their satellites do not perform any parallaxis of their orbits in the sky except for the comets which clearly exhibit irregular orbits with respect to the orbits previously mentioned. Plato clearly defines the concepts of the myths by saying (Resp 377.b.6–377.b.6):

μύθους πλασθέντας ἀκούειν τους παῖδας

fabricated myths for the children to listen.

He admits that he himself constructs fabricated myths in order to attract children's attention and then defines the case of the non-fabricated myths as follows (Tim 26.e.4–26.e.5):

τό τε μὴ πλασθέντα μῦθον ἀλλ' ἀληθινὸν λόγον εἶναι πάμμεγά που.

and the fact that it is no invented fable but genuine history is all important.

But Plato does not stop here. Moreover, in Critias, he also defines the concept of mythology (Criti 110.a.2–110.a.4):

τῶν ἐν τοῖς πρόσθεν καὶ πάλαι
ποτὲ γεγονότων ἠμέλουν.
μυθολογία γὰρ ἀναζήτησίς
τε τῶν παλαιῶν μετὰ σχολῆς
ἅμ' ἐπί τὰς πόλεις ἔρχεσθον,

and their talk was about them; and in consequence they paid no regard to the happenings of bygone ages. For legendary lore and the investigation of antiquity are visitants that come to cities.

In other words, Plato divides a myth into two parts. Firstly a fabricated myth is an attractive colourful sphere concealing a truth in its core constructed by him, whereas secondly a non-fabricated myth is also colourful but not constructed by him. The external surface of this sphere contains some of the initial observers' inventions and of other future users' added to it. This happens because it was transmitted from generation to generation throughout the centuries. Both the initial observers and the next users could not comprehend a fearful and rare past event. This event lies within the core of the non-fabricated myth. It is the traumatic memory of this event imprinted on the initial observers' memory which later on became a legend. This is the Phaethon case. Therefore it belongs to the second category of a non-fabricated myth. In both cases (fabricated and non-fabricated myth) Plato tells the truth. In the first category there is a philosophical truth whereas in the second there is a natural past event which can be identified scientifically and sometimes can be proved. Unfortunately Vida-Naquet [23] was unable to understand this difference, in spite of his 50-year effort in his study of Timaeos and Critias in connection with Athens and Atlantis.

Moreover, in Phaedrus, Plato defines science as follows (Phaedr. 277.b.5–277.c.3):

{ΣΩ.] Πρὶν ἀν τις τὸ τε ἀληθές
ἑκάστων εἰδῇ πέρι ὧν λέγει ἢ γράφει,
κατ' αὑτό τε πᾶν ὁρίζεσθαι δυνατός
γένηται, ὁρισάμενός τε πάλιν κατ' εἴδη
μέχρι τοῦ ἀτμήτου τέμνειν ἐπιστηθῇ,
περί τε ψυχῆς φύσεως διιδὼν κατά
ταὐτά, τὸ προσαρμόττον ἑκάστῃ φύσει
εἶδος ἀνευρίσκων, οὕτω τιθῇ καὶ διακοσμῇ τὸν λόγον,
ποικίλῃ μὲν ποικίλους ψυχῇ καὶ παναρμονίους διδούς
λόγους, ἁπλοῦς δὲ ἁπλῇ,

Socrates: First you must know the truth about the subject that you speak or write about, that is to say, you must be able to isolate it in definition, and having so defined it you must next understand how to divide it into kinds, until you reach the limit of division, secondly, you must have a corresponding discernment of the nature of the soul, discover the type of speech appropriate to each nature, and order and arrange your discourse accordingly, express the nature of the complex and simple soul with panharmonic and simple analogies.

Thus it is deduced that the philosopher defines both science and mythology with complete clarity. He further divides the myths into two categories: those which were genuine legends of past events and remained in people's collective unconscious and those which were fabricated by him for the purpose of attracting the young to discover by themselves the concealed philosophical truth in its "dramatic", so to speak, fabrication. Therefore Phaethon should be treated by future analysts scientifically bearing in mind what Plato meant in terms of philosophy, science and mythology. Without recognizing Plato's genuine interest in the past of the Greeks, their country and the world respectively, they could easily mislead themselves to erroneous misconceptions.

## 2 An Old Interpretation of Phaethon Phenomenon

Galanopoulos [12, 13] attempted to interpret the Phaethon phenomenon as a luminous image appearing during the explosion of the volcanic crater of Santorini in the Aegean Sea. He explained, in Figures 1 and 2, the platonic description as prehistoric information deduced from an already known, nowadays, electric phenomenon with luminous effects occurring before and during volcanic explosions. It has been repeatedly observed by different both scientific observers and laymen. Unfortunately such an image with a clear physical interpretation does not explain what Plato describes as parallaxis of the orbit of a phenomenon both round the earth and in the sky. The luminous image associated with a volcanic explosion cannot perform an orbit round the Earth and in other words perform a parallaxis mode with respect to the plane of the Earth's orbit.

## 3 A New Interpretation of Phaethon Phenomenon

As Clube and Napier [6] pointed out, fragments of the initial giant comet Enke, which first appeared 20000 years ago, approach the Earth every 1000 years or so, causing disasters. These years are 1200, 2300 and 3300 BC In this paper we examine the case of the 12th century BC disaster. Kobres [17] was the first to perform a simulation on legends associated with the 12th century BC round the world. He demonstrated a plausible scenario illustrating that Phaethon was in fact a comet recorded by prehistoric observers in the Aegean, Middle East, Egypt, India, China and Central America. Phaethon approached then the Earth very closely and interacted with it. Flaming particles from its head possibly induced fires in Europe,

(a)

(b)

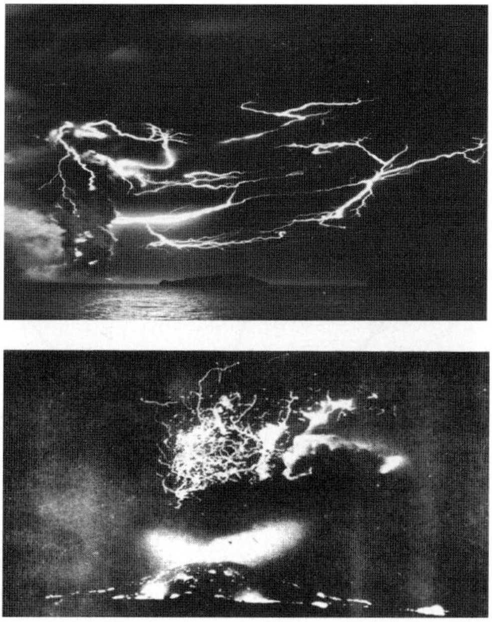

**Fig. 1** (a) A luminous electrical phenomenon during a volcanic explosion [12]. (b) Another electric luminous phenomenon resembling a flaming galloping rider. This feature does not exhibit parallaxis because it does not perform any orbit [13]. Reprinted with permission.

Anatolia and the Middle East. Ionized dust and water particles from its tail were released into the atmosphere and interacted chemically with it. Precipitating events occurred successively from this interaction. Apart from that, a late ancient Greek writer, Diodorus Siculus, mentions the elements of the earthquake and flood connecting them with the visual presentation of a comet in Peloponnesus. We do not have any means of dating the described event. One wonders if the air-vibrations of the passing cometary fragment induced some tremors on the ground or if an earthquake actually occurred, coinciding with the comet's passing in the area of the observers (Bibliotheca historica 15.p.1.32–15.p.1.33):

> Περὶ τοῦ γενομένου σεισμοῦ καὶ κατακλυσμοῦ περὶ τὴν Πελοπόννησον καὶ τῆς φανείσης ἐν ουρανῷ λαμπάδος.

> About the earthquake and the flood round Peloponnesus and the appearance of the torch in the sky.

A description involving earthquakes and fires but not earthquakes and floods, as the above, remind us what Clube and Napier [5] reported in their book *The Cosmic Serpent*. They found similarities between the Tunguska impact in Siberia in 1908 AD, a medieval account in 1296 AD of a stony meteorite swarm which fell in what is now Russian Federation (as recorded by a local chronicler) and the two battles of the gods as described by Hesiod in his *Theogony*. The last one, written

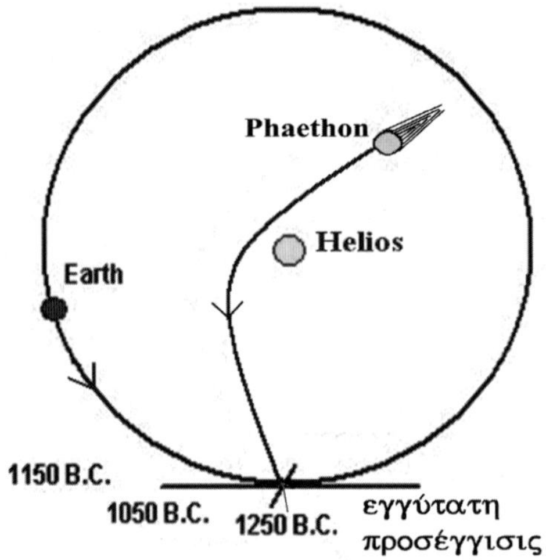


**Fig. 2** A comet exhibiting the phenomenon of parallaxis in its orbit as it goes round Earth. Reprinted with permission.

**Table 1** Characteristics of a comet's appearance as an extra close encounter to Earth.

| Tunguska (1908 AD) | Velikii (1296 AD) |
|---|---|
| Blinding ball of fire "darkened the Sun" | |
| Thick cloud of dust | Dark cloud |
| Intense thunder | Intense thunder |
| Column of fire | Clouds of fire arose and collided |
| Blast (flattened forest) Charred trees signed clothes | Great heat from lightning and thunder |
| Ground tremors Lightning (ceaseless) | Ground tremors |

about 800 BC, describes tales of vastly greater antiquity. The common characteristics of the events and the independent observations convince us that all these were not fabricated myths in connection with Hesiod's two stories. Whatever later added poetic trappings and misunderstandings of interpreters do not destroy the original kernel of the two Hesiodic stories. In Tables 1 and 2, Clube and Napier have tabulated the events of Tunguska in 1908 AD, in Velikii in Russian Federation in 1296 BC and the two Hesiodic cases of unknown time of occurrence. The reader may get some idea of what had happened, when Phaethon passed very close to the Earth in connection with fires.

**Table 2** Characteristics of a comet's appearance as an extra close encounter to Earth.

| Hesiod I conflict (?) | Hesiod II conflict (?) |
|---|---|
| Gleaming brilliance of thunderbolt and lightning | Fire from the monster |
| Dust, smoky thunderbolts smoky thunderbolts | Blazing thunderbolts |
| Intense thunder, great din Immense flame reached upper air | Thunder harshly Flame shot out |
| Hot blast winds Burned forest | Blast, hurricane winds Earth caught fire |
| Earthquakes Lightning (thick and fast) | Lightning |

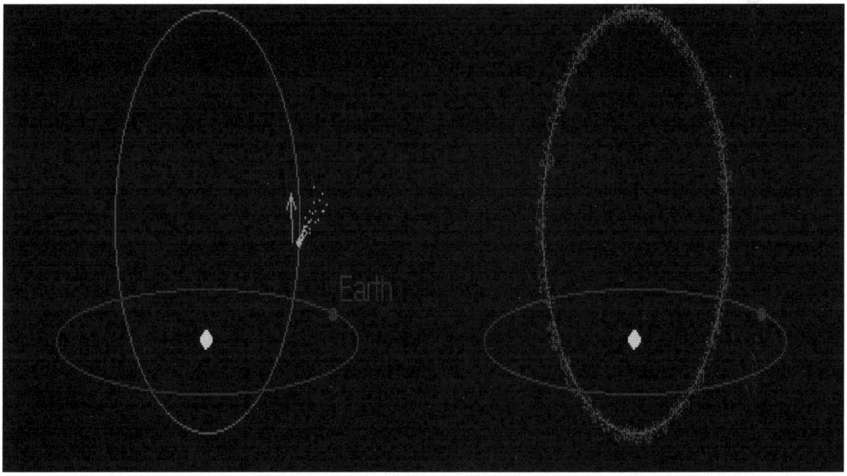


**Fig. 3** Plane of orbits, in vertical mode, offers a non-destructive meeting between a comet and the Earth.

The prehistoric observers saw Phaethon in a variety of forms. These various descriptions of the same object are the result of observers being in different pairs of longitude and latitude. It was also an optical illusion which occurs in the same pairs of longitude and latitude but occurring in different times that is just before sunset and immediately after dawn as Figure 6 illustrates. The words Phaethon (shining) meaning male and Phaethousa (shining) meaning female are the author's addition (Figure 6). This figure explains why Quatzacoatl in Figure 7, the celestial feathered serpent, appeared in paintings as the white god of South American Indians having a beard, whereas the Indians are neither white nor have beards due to genetic reasons. In Figure 7 the reader can see a drawing depicting Quetzalcoatl in the particular role

**Fig. 4** The results of the non-destructive meeting with a comet's orbit full of ionised dust with Earth are innocent and lovely to look shooting stars.

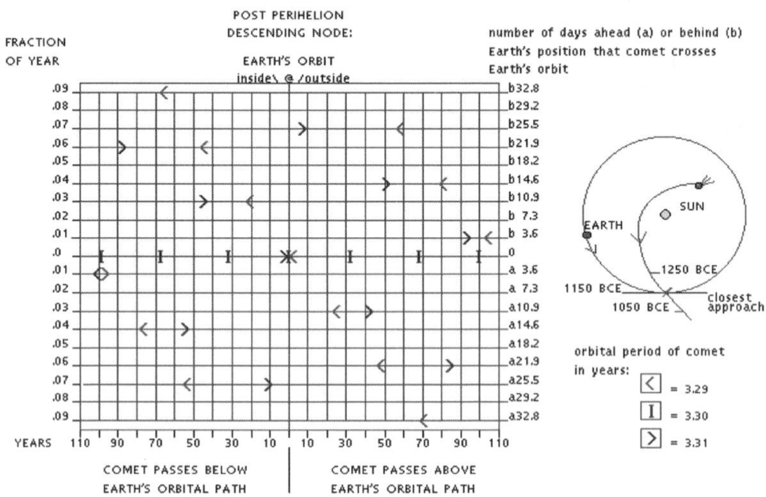


**Fig. 5** (a) Simulation parameters of Phaethon's path according to Kobres [17]. Reprinted with permission.

of Atlas. Apart from the previous reasons, a combination of religious beliefs and lack of understanding of a very unusual and rare celestial phenomenon made people interpret this event as a "double fighting of either heroes or animals or gods". Figure 3 shows a non-destructive "rendez-vous" between a comet and the Earth.

In Figure 4 shooting stars are presented in the blue sky. Figure 3 illustrates the intersection between the planes of the comet's and Earth's orbits at right angles without causing any fires on earth. However, in the 12th century BC the intersection of the planes of the same orbits was not serene. This allowed the Earth's atmosphere

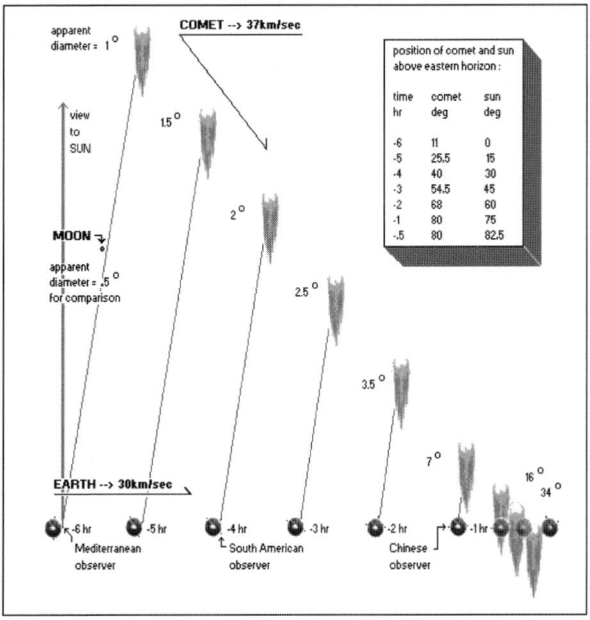


**Fig. 5** (b) Different observers round the world in Greece, Egypt, Israel, India, China and Central America saw the object successively in its last six hours [17]. Reprinted with permission.

to interact with the comet's head and tail. The complex coupling caused several problems to the populations on Earth at the time the "visit". Even now we can not fully comprehend of what is happening on Earth during such an event. King [16] published his observations about a cometary fragment which passed above Alabama on the morning of the 5th December of 1999 inducing fires in the nearby forests. The observers whose houses were in a distance of 50 m from the foci of fires said that they started exactly at the moment of the observation of the light of the passing fragment. The three different foci started to devour the forest simultaneously. King [16] wonders through what mechanism the fires were caused. Was electric induction between the passing body and the ground responsible for the simultaneous fires or the fired particles? The Alabama fragment did not induce any precipitation.

Figure 5a shows the simulation results of Phaethon's path, whereas Figure 5b illustrates what is happening during the last six hours from the initial observation of the object. Figure 5b explains how the prehistoric Greeks saw the object (comet), 1 hour before the end of the closest encounter, as a female war goddess, Athena, that is as Phaethousa (shining) coming to the region of Troy from the West. It explains why the Egyptians also saw the object hovering over the Nile's delta and associated it with a female war goddess, the lioness, that is wingless Sekhmet causing heat. In other cases Athena and Sekhmet have some other depictions in both countries. Phaethon's case can be proved or disproved in some known areas of catastrophes. It can be easily identified in archaeological sites in different countries on condi-

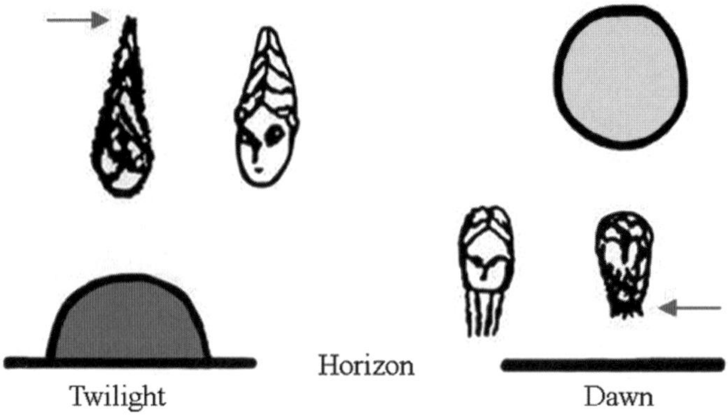


**Fig. 6** To the top comet Phaethon is mimicking the sun. Being an inadequate son, he does not simply light the Earth as the sun does, he induces fires on it. To the bottom an optical illusion of the comet conceived as female Phaethousa (shining) just before sunset and as male Phaethon (shining) just after dawn. The initial drawing was done by B. Kobres. They were modified slightly by the author. The words in Greek are added by the author. Reprinted with permission.

tion that the 12th century BC stratum is well defined. If there were forests in those sites, then a burnt horizon may still exist not in situ but accumulated in depressions close to them. Soil samples should be taken from such depressions and studied. The sampling sites should be away from the late Bronze Age settlements so as to avoid the implication of fires caused either by earthquake activity or advancing enemies. The mapping of the depressions will not be an easy procedure, since the possible fires due to Phaethon will not be expected to be found everywhere but only in the areas where fragmented fired tiny rocks from its disintegrating head would have fallen. The application of highly detailed micro-topographic, microgeological and geochemical mapping is the only way expected to offer hopeful results. The chronostratigraphic recording of past floods and the possible discrepancy of the pollen profile in these sites might offer a reliable list of hydrogeological and interesting bio-indices, which will further clarify the Phaethon's effect versus other slow advancing regular phenomena printed on the soil.

## 4 Attempts to Date Phaethon's Passing

Spedicato [22], Clube [5] and Kobres [17] offer three different datings for comet Phaethon's passing. They all associate it with the Exodus of the Hebrew people from Egypt and other environmental events. The first researcher suggests 1447 BC, whereas the second, Victor Clube and Bill Napier 1369 BC and the third 1159 BC as the year of Phaethon's visit respectively. Who could be correct? Studying Homer and especially the Odyssey, we found a particular passage which we present below for the reader because we suspect that there was a total solar eclipse hidden in it.

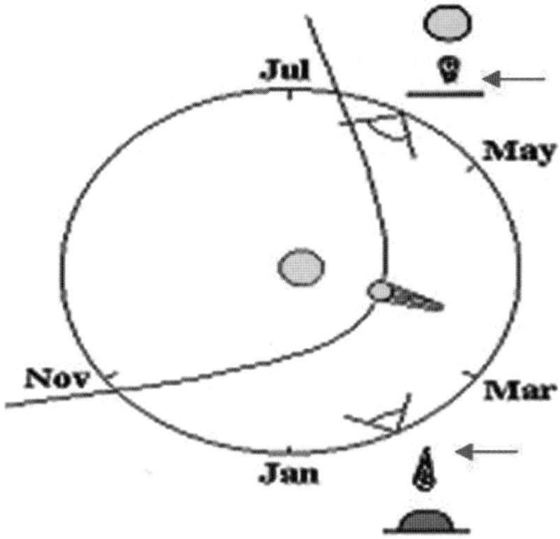


**Fig. 7** (a) The drawing and the text, in the top, is done by B. Kobres. Ishtar, the godess, is associated with different names and star systems in different times within a year. It was female or male as well due to the optical illusion in the same corresponding times within the year. It was confused, with planet Venus too. In fact it was a comet changing sex or names passing close to Venus and being associated with different star formations along its orbit during its passing close to Earth.

We found that such an event indeed happened on 16 April 1177 BC at 09.12, as Espenak [8] has proved. It is shown in Figure 8. Homer describes in Od. 20(350–356) a character Theoclymenus (whose name means godlike) who predicted in a dramatic way Odysseus' return to Ithaca, slaughtering of the suitors and in Od. 20.356–357 describing the following very characteristically:

ἠέλιος δ' οὐρανοῦ ἐξαπόλωλε, κακὴ δ' ἐπιδέδρομεν ἀχλύς.

and the sun has perished out of heaven and an evil mist covers all.

This is an exact example of a non-fabricated myth which has a colorful external surface full of invented dramatic events which attract attention to the young but in the center hides a real astronomical event, a total solar eclipse visible in Africa Asia, in Crete and other Aegean islands and Minor Asia that morning of 16th of April of 1177 BC. Due to the imposed Homeric meter and with the assistance of the recitation of the story, practiced even by illiteral people, the information remained alive up today in spite of the loss of written Linear B type of language in Greece during the turmoil of the 12th century BC.

Let us now come back to the comet. The second fall of Troy is generally accepted to have happened in 1184 BC. The first fall of Troy and the second (the legendary one) are assigned within the interval between 1250 and 1020 BC in the stratum called Troy VII. In accordance with Eratosthenes the war lasted between 1193 and 1184 BC. Taking into consideration Odysseus's ten-year length of adventures and

**Fig. 7** (b) God Quatzakoatl appears with a *beard whereas his worshippers did not have* any due to known genetic reasons. The working hypothesis of the author is that the "white Gods" in South America Quatzakoatl, Viracocha and others are possibly imitations of the comet. It was interpreted as such by the priesthood of the celestial phenomenon. Similarly certain shaved African chieftains and Pharaohs and beardless South American chieftains adding an artificial beard may exhibit not only a fertility symbol but power through distinction for the same reason as explained above.

also including the year of the solar eclipse in 1177 BC (the year he returned home), we define the end of the legendary war as 1186 BC. Zangger [25], on the contrary believes the war lasted just a few months and ended in 1186 BC. Accepting Homer's statements about the war duration and also counting inclusively from 1186 BC (end of war), we define as the beginning of the legendary Trojan war as the year 1195 BC. This means that if we had interpreted correctly Homer, the comet appeared as Athena Phaethousa (shining) in the above mentioned interval. Accepting Zangger's idea the comet consequently passed either in 1186 or few months later in 1187 BC. In Table 3 all data are tabulated.

## 5 Conclusions

Plato described in Timaeos a story which he himself calls a non-fabricated myth. His statement is accepted, as such, because the described phenomenon corresponds to our current astronomical understanding of a rare astrophysical phenomenon similar to a comet's passing. The object was observed by different peoples round the world in the 12th century BC. It produced legends, in other words non-fabricated myths, of the second platonic category, as Kobres [17] describes. The partial evidence, as Kobres believes, supporting the cometary working hypothesis comes from Baillie and Munro [2] and Hodell [10]. All these data do not fit with the analysis

**Table 3** Some of the violent events of the 12th century BC.

| | |
|---|---|
| 1. 1224 BC: | The first fall of Troy is 30 years earlier than the beginning of the legendary Trojan war. It was conducted with a small army in which Ajax's father was present. It did not become an epic. The evidence of the event is presented in Aphaea's temple in Aegina as Kakridis [15] has demonstrated. It is calculated by adding 30 years inclusively to the beginning of the legendary war of Troy which was 1195 BC |
| 2. 1225–1175 BC: | The seismic storm [19]. The beginning of the seismic storm almost coincides with the first fall of Troy by Ajax's father and his companions. |
| 3. 1208 BC: | The first appearance of the Sea Peoples in the fifth year of Pharaoh Meneptach [25]. |
| 4. 1184 BC: | The second fall of Troy, as Homer describes it, with Ajax himself participating in the events (Eratosthenes 3th century BC). On the basis of the solar eclipse of 1177 BC and Homer's statement about the duration of the Trojan war, (ten years) Eratosthenes' statement is shifted from 1184 to 1186 BC. |
| 5. 1190–1192 BC: | Acidity peaks found in Greenland. They were produced from volcanic eruptions in the Atlantic [26]. |
| 6. 1186–1195 BC: | The corrected length of the legendary Trojan war. |
| 7. 1177 BC: | The total solar eclipse happens a few days before Odysseus' return to Ithaca. We say few days or weeks taking into account that the eclipse was visible somewhere in the Aegean Sea. From the latter to Ithaca the boat could sail in the previously mentioned time easily. It was visible in Africa, Crete and other islands and Asia Minor. It is proved that the event is related to the darkening of the sun. This is highly unlikely to have happened firstly because usual atmospheric fog can not be seen as evil in the Aegean. Secondly if it had happened, it would have been incapable of blocking the sun completely and having lasted for a short time. As Homer says "and the sun has perished out of heaven and an evil mist covers all", that event lasted 04m 34s which proves its short duration. The characteristics of total darkness impressed the prehistoric observers, which was imprinted in their memory (collective unconscious) as an extraordinary event. It was unexpected, abrupt and was interpreted as something evil by them. Two more possibilities could be examined apart from the above. The first is a volcanic veil which could have produced the same effect as the darkening of sun, if a very strong volcanic explosion had occurred. As Zielinski et al. [26] reported, there occurred volcanic activity in 1192 and 1190 BC. Obtained material from drilled cores from Greenland exhibited acidity peaks produced by the falling of volcanic dust in the snow. There is a 50-year difference from the event of 1178 BC (total solar eclipse). The second possibility is the cometary veil. Homer describes the darkening of the sun in silence. He does not describe, as Hesiod and others do, darkness and events of the kind tabulated in Table 1b. Cometary veils, and associated events and their dissimilarities with solar eclipses in the prehistory and history of India are discussed by Ivengar [14]. The total solar eclipse offers a clear absolute dating of the above mentioned events independent of archaeological or archaeometric datings. We disagree with those who wonder, if the Trojan war ever occurred [4]) because the burnt stratum of Troy VII, and in particular the street 710, contains an arrow head. It presents a small but real evidence of a possible Achaean missile. |
| 8. 1176 BC: | The second assault of the Sea Peoples against Pharaoh Ramses the III [24], is close to Odysseus' return home! |
| 9. 1159 BC: | The passing of the comet producing precipitation events as Kobres suggests. |
| 10. 1159 BC: | Precipitation events are present in Irish oaks [2] and in Asia Minor oaks [18], respectively. They demonstrate anomalous ring growth in their trees of the same period. These events do not coincide with either the duration of the Trojan War (1186–1195) and the passing of "Athena" as shining comet or with the volcanic activity in the Atlantic. |

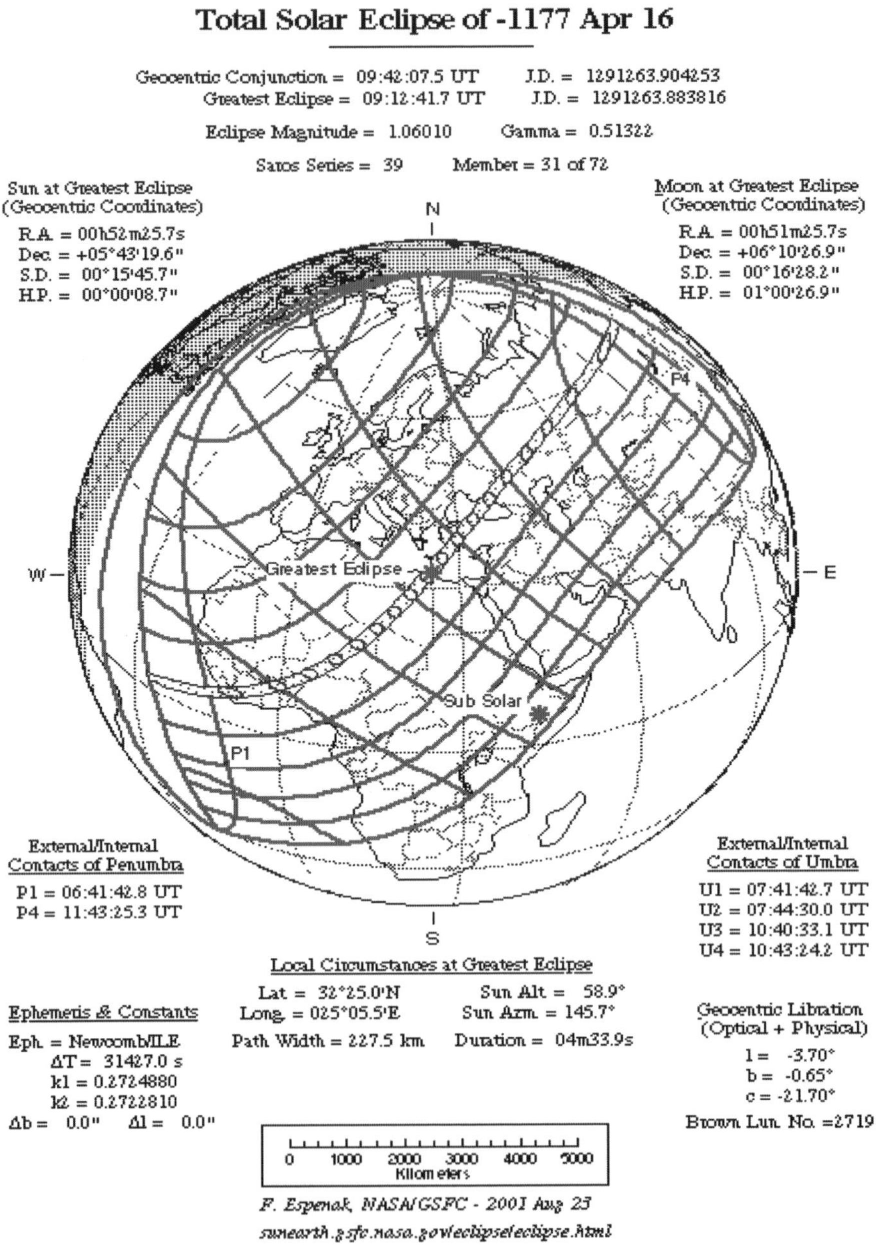


**Fig. 8** The path of the total solar eclipse. Its absolute date it signals the end of the ten years time interval of the adventures of Odysseus and defines absolutely the end of the legendary Trojan war. (The computer output was constructed by the NASA expert Fred Espenak.) Reprinted with permission.

presented in this paper. The climatic shift recorded in the oaks in Ireland and Asia Minor, is concordant with the abandonment of sites in Peloponnesus in Greece by the population during LHIII late phase as Betancourt [3] suggests. Therefore, this climatic shift could have been caused through some other mechanism rather than a comet. Although we do not have yet direct proof locked in the soil in connection with the comet's passing, we do not wish to abandon the search for this case. The complexity of the phenomenon producing either only floods with its tail's content, or just fire with its disintegrating flaming head in different parts of the world, illustrates our modern partial understanding of its function and requires multiple archaeometric and geoarchaeological attention to study it. A hope to detect exogenous carbon in a burnt stratum arises from the phenomenon of fulerenes. In the case of carbon originating from a comet, advanced laser spectroscopy may offer a solution, because this carbon locks noble gases within its structure which can be released and consequently detected if a carbon molecule is hit by a laser beam. Another hope arises from the mountainous lakes of the world. In these lakes events may be found from either past explosions or from fallings of comets in the oceans. Albot [1] has a brilliant idea which if it follows may offer results. A stratum produced by sea water in a mountainous lake can be easily recognised in its stratigraphy, studied and dated. Albot's idea is a new scientific horizon in the study of past comets. In the ancient Greek literature we have a Homeric window of 12th century BC, centralised on Troy, where the comet appears principally as a female flying entity which we call Phaethousa (shining) and a broader non-Homeric window, described and called Phaethon (shining) by Plato corresponding in the same century. In our symposium another theory connected with Phaethon was presented by Spedicato [22]. All Phaethon's theories are offered for further examination and comparison. This study is based on a previous study conducted by the author [21].

## Acknowledgements

Many thanks to Eric Wright for constructive criticism in some of the ideas presented in this paper. Special thanks to Bob Kobres who in many ways assisted me with writing this paper.

## Notes

The translation into English of the above mentioned passages from ancient Greek authors are based on the popular Loeb Classical Library. Phaethousa originates from Homer.

The author has done his utmost to trace the original copyrightholders of figures and illustrations in order to obtain official permission to reprint them in the present publication. In cases where this search was unsuccessful the author would be pleased to hear from copyright holders so that correct procedures can be followed.

## References

1. Albot, D. (2007), Detection of a sea-water stratum in the stratigraphy of mountainous lakes as events of past comet comings on earth, Personal Communication.
2. Baillie, M.G.I. and Munro, M.A.R. (1988), Irish tree rings, Santorini and volcanic dust veils, *Nature* **332**(24), March, 344–346.
3. Betancourt, P.P. (2000), The Aegean and the origin of the Sea Peoples, in *The Sea Peoples and Their world: A Reassessment*, E.D. Oren (Ed.), University of Pennsylvania Museum, 360 pp.
4. Bittlestone, R., Diggle, J. and Underhill, J. (2005), *Odysseus Unbound. The Search for Homer's Ithaca*, Cambridge, 598 pp.
5. Clube, V. and Napier, B. (1982), *The Cosmic Serpent*, Faber and Faber, 299 pp.
6. Clube,V. and Napier, B. (1990), *The Cosmic Winter*, Basil Blackwell Ltd., 307 pp.
7. Diodorus, *1st Century BC.*
8. Espenak, F. (2001), Total solar eclipse of 1177, April 16, NASA–GSFC.
9. Eratosthenes, *4th Century BC*, The Ancient Greek Literature.
10. Hodell, D.A. (1991), Reconstruction of Caribbean climate change over the past 10500 years, *Nature* **352**, 29 August, 790–793.
11. Homer, 8th century BC, *Iliad*, Ideotheatron-Georgiadis, Athens, 890 pp. [in Greek].
12. Galanopoulos, A. and Bacon, E. (1970). *Atlantis, The Truth behind the Legend*, Nelson, Third Edition, pp. 215 and 216.
13. Galanopoulos, A. (1971), *The Legend of Phaethon under the Light of Science, Santorini*, "Grafikes Technes", Papoulias et al., M.A. Danezis (Ed.), 396 pp. [in Greek].
14. Ivengar, R.N. (2006), Short communication on some comet observations in ancient India, *J. Geolol. Soc. India* **67**, 289–294.
15. Kakridis, I.Th. (1986), *Greek Mythology*, Vol. 4, Ekdotiki Athenon [in Greek].
16. King, D. (2003), The trans-Alabama superbolide of 5 December 1999, *EOS* **84**(27), 8 July, 253–257.
17. Kobres, B. (1995), The path of a comet and Phaethon's ride, *The World & I* **10**(2), 394–405.
18. Kuniholm, P.I. (1990), Archaeological evidence and non-evidence for climatic change, *Philosophical Transactions of the Royal Society of London A* **330**, 645–655.
19. Nur, A. and Cline, E.H. (2000), Poseidon's horses: Plate tectonics and earthquake storms in the Late Bronze Age Aegean and Eastern Mediterranean, *JAS* **27**, 43–63.
20. Plato, 4th century BC, *Timaeos and Critias, The Greeks*, Cactos, Athens, 330 pp. [in Greek].
21. Papamarinopoulos, St.P. (2007), Phaethon or Phaethousa: A shining comet passing in the 12th century BC?, in *Proceedings of International Conference 'The Atlantis Hypothesis: Searching for a Lost Land'*, Melos Island, 11–13 July 2005 (in print).
22. Spedicato, E. (2008), A super Tunguska event c. 1447 BC: a scenario for the Phaethon explosion, the Indo-Aryan migration and Exodus events, in *Proceedings of International Symposium on 'Science and Technology in Homeric Epics'*, Ancient Olympia, 27–30 August 2006, S.A. Paipetis (Ed.), Springer, Dordrecht (this volume).
23. Vidal-Naquet, P. (2005), *L'Atlantide*, Les Belles Lettres, 198 pp.
24. Wachsmann, S. (2000), *To the Sea of the Philistines*, University Museum Monograph 108, University Museum Symposium Series 11, The Sea Peoples and Their World: A Reassessment, E.D. Oren (Ed.), The University Museum, University of Pennsylvania, 360 pp.
25. Zangger, E. (1995), Who were the sea peoples?, *Aramco World* **46**(3), 20–31.
26. Zielinski, P. et al. (1994), Record of volcanism since 7000 BC from the GISP2 Greenland ice core and implications for the volcano-climate system, *Science* **264**, 948–952.

# Homeric Calendar and Helios Charioteer

Maria K. Papathanassiou

*University of Athens, Greece*

**Abstract.** Although no calendar is mentioned in the Homeric epics, there are references to years and months, which suggest the existence of an early lunisolar calendar. This hypothesis is further supported by the total number of the cattle of the Sun in Homer's *Odyssey*, which seems to be a poetic reference to the most ancient Greek calendar, namely the period of two lunar years. There are also traces of the tripartite division of the month in three decades of days used throughout the Greek world until Imperial Roman times. Literary evidence is supported by archaeological evidence: The circular disc of the Sun and the crescent of the Moon are displayed in scenes of ritual offerings on golden rings from Mycenae and Tiryns (15th cent. BC), while month-names and every year offerings to divinities are recorded on Linear B tablets from Knossos (Crete) and Pylos (Peloponnese). Linear B tablets also record a large number of chariots used by Minoans and Mycenaeans. It seems then very likely that later Helios's iconography as a warrior charioteer and his symbol as an "all-seeing eye" can be traced back to Mycenaean world.

## 1 Homeric and Mycenaean Calendars

A research in the Homeric Epics shows that there is no reference to any calendar. In spite this, general information regarding years and months, without mention of month-names, gives evidence of some early form of a lunisolar calendar. The Homeric year seams to be a tropical (seasonal) year, as it goes wheeling around (περιτελλομένων ἐνιαυτῶν, *Iliad*, II, 551; XXIII 833 or περιπλομένων ἐνιαυτῶν, *Odyssey*, X, 467; XI, 294) and the same seasons return: "but when the year had passed in the waning of moons and the long days had come round" (Od. X, 469–470, ἀλλ' ὅτε δὴ ῥ' ἐνιαυτὸς ἔην, περὶ δ' ἔτραπον ὧραι,/ μηνῶν φθινόντων, περὶ δ' ἤματα μακρὰ τελέσθη (XIX, 153). The reference to "the waning of moons" (μηνῶν φθινόντων) is an indication that the month (lunation) was measured from one new moon to the next one. This, in relation to "the wheeling around of the seasons" (περὶ δ' ἔτραπον ὧραι) of a whole year (ἀλλ' ὅτε δὴ ῥ' ἐνιαυτὸς ἔην), gives evidence of a tropical (seasonal) year measured by lunations. A reason for it could be the very ancient Greek custom, the sacrifices offered to gods to be held at ex-

*S.A. Paipetis (ed.), Science and Technology in Homeric Epics,* 357–368.

actly determined dates in a solar (seasonal) year composed of lunar months.[1] As the length of the solar year ($365\frac{1}{4}$ days) is incommensurable to the length of the lunar month ($29\frac{1}{2}$ days), this implies that from very early times the Greeks tried to find the best calendar system, namely that which could keep pace with the seasons of the solar year. Here we can recognize the traces of the tripartite division of the month in three decades of days – used throughout the Greek world until the Roman Imperial times and the official adoption of the planetary week in 321 AD by the Emperor Constantine – as we can deduce from the Homeric epics (Od. XIV, 161–162):

> Ulysses will return in this same year (τοῦ δ ' αὐτοῦ λυκάβαντος);
> with this waning moon or the next waxing moon
> (τοῦ μὲν φθίνοντος μηνός, τοῦ δ' ἱσταμένοιο)
> he will be here (...)

In later times Hesiod either numbers the days in the period of a waxing and of a waning month or he also numbers the days consecutively through (the "29th" τρισεινάδα, *Works and Days*, 814), and he speaks of the "middle" days of the month. The close links between calendar and cult are shown in the festival held by the apparition of the crescent of the new moon (Od. XIX, 306; XX, 156). Later, in classical times the names of the months were generally derived from a festival, which was celebrated in the given month.

Let us now come to the archaeological evidence. On a famous golden ring (Athens, National Archaeological Museum) found in the grave circle A in Mycenae, the goddess of vegetation sits under a tree and receives the offering of sacred lilies from two women and a young girl; in the background we can distinguish a double axe, symbol of Minoan civilization. A sinusoidal double line separates the upper part of the scene depicted, which symbolizes the sky, from its lower part symbolizing the earthly environment. In the upper part we can clearly see the circular disc of the Sun with its numerous tiny rays and the crescent of the Moon (Figure 1).

On another famous ring, this from Tiryns (Athens, National Archaeological Museum), a sitting goddess receives liquid offerings from four daemons. Here, also a sinusoidal line separates the upper part of the scene symbolizing the sky and in which the Sun and the Moon are depicted. But here the Sun and the Moon are among four ears of wheat while this area is plenty of small "grains" (very likely grains of wheat) like those on the garment of the goddess (Figure 2).

Is not it a clear indication of the close links between celestial and terrestrial phenomena, between the annual motion of the Sun in the sky and the alternation of the seasons on earth, and consequently the changes in vegetation? Moreover, the faster motion of the Moon offers a means for the measurement of time, counting days of lunation.

Let us now come to the literary evidence. The decipherment of the Linear B Minoan script and the reading of the Linear B tablets, written before c. 1200 BC, found in the palaces of Knossos (Crete) and Pylos (western Peloponnese), offer im-

[1] Geminus, *Introduction to the Phaenomena*, ch. VIII (On months), 7.

**Fig. 1** Golden ring from the grave circle A in Mycenae (early 15th century BC). We can distinguish the symbols of the Sun (a circular disc with many tiny rays) and the crescent of the Moon (Athens, National Archaeological Museum).

**Fig. 2** Golden ring from Tiryns (early 15th century BC), with the symbol of the Sun (like a six-spoked wheel) and the Moon (crescent) among ears of wheat (Athens, National Archaeological Museum).

portant information as far as the calendar in Mycenaean times is concerned.[2] There are many references regarding the year (we-to wetos ἔτος "year"). For example, a "year-by-year contribution" (do-so-mo we-te-i-we-te-i, δοσμὸς ἀπὸ ἔτος εἰς ἔτος), "this year" (to-to we-tos; za-we-te adv. tsawetes), "annual" (e-ni-ja-u-si-jo Eniausioi

[2] The calendar data cited and the references to the tablets of Knossos and Pylos are taken from the work of Michael Ventris and John Chadwick [3].

ἐνιαύσιος), "the next year" (a2-te-ro we-to di-do-si, hateron wetos didonsi), "from last year" (pe-ru-si-nwa).[3]

As far as the month is concerned, the word *me-no* (μηνὸς) "of the month" recurs at Knossos on three tablets, but the month-names are unfortunately lost. Strong evidence that from the earliest times Greek months were lunar – it was the Romans who eventually divorced the month from the moon – is the sign of the crescent moon, which was employed on the tablets to denote month. At Knossos, eleven tablets contain month-names, which very probably formed part of a ritual calendar, specifying or recording offerings of oil sent to a limited number of places, priests and divinities.[4] Among these divinities are included "all gods" (pa-si-te-o-i = pansi theoi "to all gods") and *Paiawon* (pa-ja-wo-[ne, pa-ja-wo = Homeric Παιήων *Iliad*, V, 401, Doric Παιάν (*ΠαιάϜων, perhaps from the verb παίω "strike"), healer god, later identified with Apollo. Dedications "to the priestess of the winds" (a-ne-mo i-je-re-ja, 'Ανέμων ἱερείᾳ) – one locates her at the town of *U-ta-no* – show an early cult of the winds. In the Homeric Epics we are given names of some of the winds and Aiolos, their lord, plays an important role in the *Odyssey*. Cults of the winds were preserved in some Greek areas in later times.[5]

The evidence for month-names from Knossos is as follows:

- ?]de-u-ki-jo-jo me-no (Δευκίου μηνός) "in the month of Deukios". According to Promponas, "very probably it means the month of γλεύκους, exactly as the name πετμεζᾶς in modern Greek denotes the month of September in the district of Epirus" [10, p. 252].
- a-ma-ko-to me-no "in the month of A-ma-ko-to", Haimakto? (αἱμακτός, Euripides).
- ra-pa-to me-no, "in the month Lapatos". The name of this month survives in the homonymous Arcadian month, as is shown by an inscription from Orchomenos (μηνὸς Λαπάτω).
- wo-de-wi-jo me-no, "the month of Wo-de-wi-jo", wordewios "month of roses". wo-de-ri-jo-jo me-no ϜορδηϜίου μηνός, 'Ροδηίου μηνός, from 'Ροδήιος μήν, "month of roses". Promponas explains the etymology of the name of Eordaia (a district of Macedonia) in relation to the name of this Mycenaean month, and refers to the related passage of Herodote (*Urania* (VIII), 138), that in Macedonia

> near the place called 'the Gardens of Midas, son of Gordias.' In these gardens there are roses which grow of themselves, so sweet that no others can come near them, and with blossoms that have as many as sixty petals apiece.

(φύεται αὐτόματα ῥόδα, ἓν ἕκαστον ἔχον ἑξήκοντα φύλλα, ὀδμῇ τε ὑπερφέροντα τῶν ἄλλων) [6, pp. 50–52]. Promponas says also that "the month

---

[3] Ibid., 278, 279, 294, 295, 372, 465, 536, 537, 543, 568, 571, 587, 591, 593, 594.

[4] Ibid., 303–312.

[5] Promponas [8, pp. 144–145]: "Acts of worship related to winds were performed in Icaria until recently".

of May is named τριανταφυλλᾶς (= "of roses") in Thrace".[6] As far as the content of this tablet is concerned, it is very plausible that this tablet was a record of the unfavourable days (ἡμέραι ἀποφράδες or *dies nefasti*) of the first or second half of a Knossos month [3, p. 311]. The expression o-u-te-mi (οὐ θέμις, οὐχὶ θέμις) was also used in classical times.

- ka-ra-e-ri-jo me-no "in the month Ka-ra-e-ri-jo"; from κραῖρα "head". It should be noted that there is an Ephesian month Κλαριών and that the following words ]-jo and pa-ja-ni-jo may also be the names of months.
- di-wi-jo-jo me-no, Diwioio menos. In classical times the month Dios (Δῖος), named after the festival of Dia (Δῖα) celebrated during it, was the first month of the Macedonian calendar and the fourth of the Aetolian calendar. We find it later in Gaza, Ascalon and Bithynia [1, pp. 20, 48]. According to Promponas, as the names of the ancient months were generally derived from a festival, which was celebrated in the given month, the evidence of month *Dios* in Macedonia shows that there was a very well organized religious life earlier than 1900 BC [6, pp. 26, 57].
- e-me-si-jo, e-me-si-jo-jo. Previously were taken as month name, but this is not in keeping with the context.

From Pylos we have the following month-names [3, pp. 277, 286, 287, 458, 478, 480]:

- ki-ri-ti-jo-jo Krithioio "in the month of Krithios".
- pa-ki-ja-ni-jo-jo me-no = P. . . anioio menos "in the month of P. . . " for a month-name, very probably derived from a place, as it is the case of Attic month Μουνυχιών.
- po-ro-wi-to-jo. The name of this month is derived from πλόος<*πλοFος, *πλόFιστος (= the most suitable for sailing); consequently, the month was called *Plowi(s)tos*, which would mean "the month of sailing". The Greeks were scared of navigation in the stormy winter months and this month could correspond to our September, as according to Hesiod (*Works and Days*, 663–665)[7]

  Fifty days after the [summer] solstice,
  when the season of wearisome heat is come to an end,
  is the right time for me to go sailing . . .

  or at the end of March, when spring begins (ἄλλος δ' εἰαρινὸς πέλεται πλόος ἀνθρώποισιν, l. 678). According to Promponas it is very likely that this month relates to 'Ηροσούρια, a spring festival honoring the favorable winds for sailing [9, p. 91].

The calendars of Knossos and Pylos palaces may of course have been somewhat differing. None of the names identified shows the typical Attic-Ionic month formation -ιών, which is probably a comparatively late secondary development. Epigraphic

[6] Promponas [10, p. 253].

[7] ἤματα πεντήκοντα μετὰ τροπὰς ἠελίοιο,/ ἐς τέλος ἐλθόντος θέρεος, καματώδεος ὥρης,/ ὡραῖος πέλεται θνητοῖς πλόος . . .

**Fig. 3** Clay tablet from Knossos palace (KNSc230) depicting a chariot with four-spoked wheels (Herakleion Museum).

evidence of many month-names in Knossos and Pylos tablets shows the existence of early local calendars in Mycenaean times as well as their continuity as seen in the numerous local calendars in Classical and Hellenistic times. Therefore, it seems very likely, that the absence of any reference to any special calendar in the Homeric epics is intentional. Homer composed his rhapsodic poems to be sung by all Greeks; for this reason he did not mention any special month-name, referring thus in a special local Greek calendar.

## 2 Helios Charioteer

On the golden ring from Tiryns the Sun is depicted as a circle with six radii, alike to a six-spoked wheel. This reminds us that in Greek Mythology the Sun (Helios) as well as the Moon (Selene) crossed the sky riding in their chariots.[8] The Mycenaean chariot is depicted on many Knossos tablets, as for example KN Sc230 in the Herakleion Museum (Figure 3), or it is denoted by the symbol ⊕.

It is also frequently seen on other artistic objects, as for example on the Pylos fresco depicting two warriors – one of them riding in a chariot (Figure 4), and numerous scenes on pottery, like the sarcophagus from Episkopi (Hierapetra, Archaeological Collection) (Figure 5). The chariot consists of a light body, possibly with a fixed axle on which are mounted the two four-spoked wheels, like that on another golden ring from the grave circle A at Mycenae (Athens, Archaeological Museum) (Figure 6). But on the cylindrical side of an ivory pyxis from Engomi-Alasia (Cyprus, now in the British Museum, London), a warrior rides in his chariot which has six-spoked wheels (Figure 7) – an image very close to the symbolic image of the Sun on the golden ring from Tiryns.

Some details of the Mycenaean chariot and its fittings emerge from the inventories at Knossos. The chariot bodies are indexed separately from the wheels, very likely because they were kept in an outlying building. At Pylos no tablets listing chariots have been found but their existence can be inferred from the inventories of

[8] *Homeric Hymn* XXXI (to Helios (Sun)), l. 8ff., and XXXII (to Selene (Moon)), l. 9ff.

**Fig. 4** Fresco from Pylos palace depicting two warriors; one warrior rides his chariot with four-spoked wheels (13th century BC).

wheels. References to pairs of wheels "with ivory borders" ([e-re-]pa-te de-de-me-na ἐλε]φάντει δεδεμένα), "bound with bronze" (ka-ko-de-ta, ka-ko de-de-me-no) or "bound with silver" (a-ku-ro de-de-me-no ἀργύρῳ δεδεμένω) show that there were luxury chariots (see [7, pp. 81–83] and [8, pp. 193–194, 197–207]). This view is further supported by Homer's detailed description of the very luxury chariot of goddess Hebe, which was decorated with gold, silver, copper and iron, and had eight-spoked wheels, which the goddess fitted on before she rides in her chariot (*Iliad*, V 722f.):[9]

> Hebe with all speed fitted on the eight-spoked wheels of bronze
> that were on either side of the iron axle-tree.
> The felloes of the wheels were of gold, imperishable, and over these there was
> a tire of bronze, wondrous to behold.
> The naves of the wheels were silver, turning round the axle upon either side.
> The car itself was made with plaited bands of gold and silver,
> and it had a double top-rail running all round it.
> From the body of the car there went a pole of silver,
> on to the end of which she bound the golden yoke, with the bands of
> gold that were to go under the necks of the horses.

Generally, it is considered that in areas with a physical environment such as Messenia or Crete, chariots can hardly have served other than as means of transport and perhaps as a prestige symbol. But from the saved tablets it seems very

[9] The English translation of the passages of the Homeric epics and Hesiod is taken from the Internet site of *OMACL* (Hesiod, the Homeric Hymns and Homerica), *Online Medieval and Classical Library*, Relaese #8. The *Iliad* and the *Odyssey* are translated by Samuel Butler.

**Fig. 5** Sarcophagus from Episkopi (Hierapetra, Crete, 13th century BC), depicting a scene of chariot races (low left); the chariot has four-spoked wheels (Hierapetra, Archaeological Collection).

likely that both Knossos and Pylos could put in the field a great number of luxury vehicles. We can safely conclude that Pylos had a chariot force of some sort, probably around 200 chariots; but this evidence suggests that it was on a smaller scale than that at Knossos, with a force of around 550 chariots. One can only speculate the reason; possibly the smaller size of the kingdom of Pylos has something to do with it, but the absence of the chariot records must leave the true size of the force at Pylos in doubt. Let us now see the description of the chariot of the Sun (Helios) as given in his *Homeric hymn* (XXXI, 8f.):

> tireless Helios who is like the deathless gods. As he rides in his chariot, he shines upon men and deathless gods, and piercingly he gazes with his eyes from his golden helmet. Bright rays beam dazzlingly from him, and his bright locks streaming form the temples of his head gracefully enclose his far-seen face: a rich, fine-spun garment glows upon his body and flutters in the wind: and stallions carry him. Then, when he has stayed his golden-yoked chariot and horses, he rests there upon the highest point of heaven, until he marvelously drives them down again through heaven to Ocean.

This description of Helios as a warrior charioteer reminds us the warrior who rides in his chariot on the above mentioned Pylos fresco. Therefore, it seems very likely

**Fig. 6** Golden ring from the grave circle A in Mycenae (16th century BC) with a scene of hunting; the chariot has four-spoked wheels (Athens, National Archaeological Museum).

**Fig. 7** Part of the cylindrical side of an ivory pyxis from Engomi–Alasia (Cyprus), dating in the Late Bronze Age (1550–1050 BC) and depicting a scene of hunting; the chariot has six-spoked wheels. (London, British Museum).

that the Sun's iconography in later times, presenting him as a heavenly charioteer, has its origin in the extensive use of chariots in Mycenaean times.

## 3 The Cattle of the Sun

A great lot of the sufferings of Ulysses and his companions were the result of the slaughter of some animals belonging to the cattle of the Sun, when they arrived at the Thrinacian island (= Sicily), although they had been warned accordingly, that if they dare to do it, they will be in great unhappiness (Od. XII, 127f.):

> You will now come to the Thrinacian island, and here you will
> see many herds of cattle and flocks of sheep belonging to the sun-god
> seven herds of cattle and seven flocks of sheep, with fifty head in each
> flock. They do not breed, nor do they become fewer in number, and
> they are tended by the goddesses Phaethusa and Lampetie, who are
> children of the sun-god Hyperion by Neaera. Their mother when she
> had borne them and had done suckling them sent them to the
> Thrinacian island, which was a long way off, to live there and look
> after their father's flocks and herds.

The number seven is closely linked with Apollo, as the seventh day of each month was sacred to Apollo throughout the Greek world. Hesiod (*Works and Days*, 770–771) establishes that this was his birthday:

> To begin with, the first, the fourth, and the seventh
> – on which Leto bare Apollo with the blade of gold –
> each is a holy day.

Especially at ancient Athens many festivals sacred to Apollo were celebrated on the seventh day of the lunar month. For example, in Hekatombaion 7, there was the festival of Hekatombaia; in Metageitnion 7, the inhabitans of Salamis sacrificed to Apollo Patroos, Leto, Artemis, and Athena. In Boedromion 7 there was the festival of Boedromia; In Pyanopsion 7, that of Pyanopsia; in Gamelion 7 the Erkhians sacrificed to Apollo Delphinios and Apollo Lykeios. In Thargelion 7, there was the festival of Thargelia. It is also noteworthy that only Apollo's festivals occurred on or included the seventh day of the month. The festival of no other deity was allowed to impinge upon this day [4, p. 19]. Consequently, in Greek cult the number seven has nothing to do with the phases of the Moon and the week, which were introduced by the Romans, who were influenced by religious rites of the Near and Middle East peoples.

Multiplying, "seven herds of cattle and seven flocks of sheep, with fifty head in each flock", we have $7 \times 50 = 350$ cattle and $7 \times 50 = 350$ sheep, their sum being $350 + 350 = 700$ – a fixed total number as "they do not breed, nor do they become fewer in number". What does it mean? It seems very likely that Homer had wished to refer poetically to a calendar used in his time. The number 350 is produced by Apollo's sacred number 7 multiplied by the number *fifty* (another number linked with Greek Mythology, for example the *fifty* Sea-nymphs, *Nereids*). There is no other integer number than fifty (as multiplier of seven) that gives the best approximation to the 354 days of a lunar year with its twelve synodic months, namely lunations (29.5 days × 12 lunations = 354 days). On this ground we could further suggest that the number 700 (= 7 × 50 + 7 × 50) is a poetic reference to the most ancient Greek calendar system, namely the long ago established period of two lunar years (Διετηρίς) (354 × 2 = 708 days). As the number of days of two lunar years (708 days) is smaller than that of two solar years (730 days),[10] and their difference amounts in 22 days, at the end of the period of two lunar years a thirteenth (intercalary) month should be added, so that in the third year the lunisolar calendar keeps pace with the solar (seasonal) year. On the other hand, if we calculate "poetically" like Homer, a total sum of 700 (= 7 × 50 + 7 × 50) days, the difference can be covered by a full lunar month.

## 4 Helios Who Sees Everything

According to Greek Mythology, Helios "who sees and gives ear to everything" (Od. XI, 107) revealed to lamenting Demeter what god has rapt her daughter, Persephone (*Homeric Hymn* II (to Demeter), 22–26, 62–63, 69–71):

> But no one ... heard her voice, ... only tender-hearted Hecate ...
> heard the girl from her cave, and the lord Helios ...
> So they [Demeter and Hekate] came to Helios, who is
> watchman of both gods and men, and stood in front of his horses:
> and the bright goddess enquired of him: ...
> Helios, you – for with your beams you
> look down from the bright upper air over all the earth and sea –
> tell me truly of my dear child, if you have seen her anywhere.

Among Minoan hieroglyphs there is one depicted as a human eye signifying "supervision". The evolution of this hieroglyph in Linear B script is a circle with a point in its center. This symbol, used to denote the Sun in the astronomical and astrological Byzantine MSS, as well as the gold in the alchemical Byzantine MSS, is still in use. Moreover it is used as a mystical symbol of the divine power emanating from the central point towards the circumference. In Christian cult, the comparison of Jesus to Helios as "the Sun of Justice" implies the attribution of the qualities of the Sun-Helios to him; consequently, Helios-Jesus is the divine eye, "God's eye", which is "All-seeing", like Helios, sometimes depicted above the central door of the "iconostas" in Christian churches.

[10] Here we refer to the integral part of the number of days of a solar year (365), as practically the additional fraction (c. 1/4 day) is not taken into account.

## Acknowledgements

I am grateful to the National Archaeological Museum at Athens, the Herakleion Museum, the 24th (Hag. Nikolaos, Crete) and the 38th (Kalamata, Peloponnese) Ephorias of Prehistoric and Classical Antiquities of Greece, and the British Museum at London, for permission to use the above pictures in this publication.

## References

1. Bickerman, E.J., *Chronology of the Ancient World*, Thames and Hudson, London 1968 (rev. ed. 1980).
2. Chadwick, J., *The Mycenaean World*, Cambridge University Press, 1976.
3. Ventris, M. and Chadwick, J., *Documents in Mycenaean Greek*, 2nd ed., Cambridge University Press, 1973.
4. Mikalson, J.D., *The Sacred and Civil Calendar of the Athenian Year*, Princeton University Press, 1975.
5. *OMACL* (Hesiod, the Homeric Hymns and Homerica), *Online Medieval and Classical Library*, Release #8.
6. Promponas, I., *The Affinity of Macedonian and Mycenaean Dialects and the Proto-Hellenic Origin of Macedonians*, Athens, 1973 [in Greek].
7. Promponas, I., *A Brief Introduction to Mycenaean Philology*, Athens, 1977.
8. Promponas, I., *Dictionary of Mycenaean Greek*, Vol. 1 (ἀ-βέλεμνον), Athens, 1978.
9. Promponas, I., *Mycenaean Epic Poetry Based on Mycenaean Texts and the Homeric Epics*, Athens, 1980 [in Greek].
10. Promponas, I., The ancient Macedonian Menologion and its importance for the nationality of Macedonians, in *Proceedings of 6th Conference on 'Mount Olympos through the Centuries'*, Elasson, September 1993, Larissa, 1994, pp. 249–254 [in Greek].
11. Sali, T., *Dictionary of Mycenaean Technical Terms*, Athens, 1996 [in Greek].

# Homer and Orosius: A Key to Explain Deucalion's Flood, Exodus and Other Tales

Emilio Spedicato

*University of Bergamo, Italy*

**Abstract.** In a previous paper (Spedicato, 2005), taking the hint from a passage in Paulus Orosius, we have explained Deucalion's Flood and the "passage" of the Red Sea by Moses in terms of a super Tunguska type explosion over northern Germany of a body known in the Greek tradition as Phaethon. In this paper, we observe how an enigmatic passage in Homer's Odyssey about Lampos and Phaethon sheds additional light on the above events.

## 1 Orosius as a Key to Understanding Deucalion's Flood and Exodus

In the little-read *Histories against Pagans*, written at the beginning of the 5th century by Paulus Orosius, a friend of St Augustin, we find, in Book 1, 8–10, the following passage:

> 810 years before the foundation of Rome Amphithion was king in Athens. At his times a flood destroyed most of the people of Thessaly. Only a few could save themselves on the mountains, in particular on the Parnassus which was under the jurisdiction of Deucalion ... Plato states that at that time Ethiopia was affected by many terrible diseases, which almost destroyed the whole population ... at that time Father Liberus conquered India shedding lot of blood and killing many people ... against a nation that was never hostile to others and lived peacefully ...
>
> Pompeus [Trogus] and Cornelius [Tacitus] state that 805 years before the foundation of Rome, terrible disasters and diseases affected the Egyptians ... they expelled Moses ... he stole the sacred objects of the Egyptians ... Cornelius Tacitus refers that ... a pestilence deforming the bodies developed in Egypt at the time of king Boccoris, who after consulting the oracle of Ammon was ordered to deport out of Egypt certain people disliked by the gods ... Moses, one among the expelled people, admonished his people to confide only in him, as a leader sent by the celestial power ...

*S.A. Paipetis (ed.), Science and Technology in Homeric Epics,* 369–374.

> There were extreme heats, long lasting and unbearable; it was unusually hot in Ethiopia and the Scythians were unable to bear the hot weather.
>
> This was also the reason why some people, not willing to recognize God's absolute power, have invented the ridiculous fable of Phaethon, in order to provide an explanation that is however lacking of basis.

The above passage has been exploited by us, together with Septuaginta, Josephus Flavius' *Antiquities of the Jews*, the *Legends of the Jews* as collected by Ginzberg, other material in classic Greek and Latin writers, in Philo and in the Fathers of the Church, to explain the above quoted events as follows.

Contrary to Orosius' opinion, who saw miracles where rare natural events actually occurred, Phaethon was indeed the cause of the events cited. We interpret Phaethon as a super Tunguska type object that after wild evolutions enters the upper layers of the atmosphere. Coming from south-east it breaks up over Arabia probably, the event being described in the Bible as the *glory of God* dividing in two. A main part continued in the direction north-west, touching again the atmosphere over eastern Mediterranean and sending a heat wave that burnt buildings in Crete. After rebounding over the Parnassus, where Deucalion was probably staying at the Delphi sanctuary, it re-entered the atmosphere over the Balkans sending again a heat wave that burned the Central Europe forests. It finally exploded over river Eider, the Eridanus of the classical tradition, the explosion causing a huge earthquake and sending a powerful hot wind to all directions.

The earthquake reached after a few minutes the Sinai peninsula, where Moses was stranded in the flat area of Pi-Hahirot, the present Nuweiba. His trail to the north went between the sea and the coastal mountains and was blocked by a previous earth-fall or rock-fall. The effects of the earthquake are described in the following psalm 113–114:

> When Israel left Egypt
> ...
> The sea saw and withdrew
> The Jordan inverted his course
> The mountains jumped as rams,
> The hills as the lambs of the flock.

After a few hours, during the night, the hot wind came, reduced in speed and temperature from the point of explosion but lasting possibly a few hours. The wind flew almost exactly in the direction of the Red Sea, which is long about 2500 km, and sent its waters to accumulate around the narrow southern exit, the Bab el Mandeb (with flooding of part of Yemen, where interruption of constructions is documented for about three hundred years). In the northern part of the Red Sea, and in particular in the Aqaba Gulf, the level of the waters lowered, this happening during the night. How much it lowered cannot be said presently, but could be calculated in principle, by making assumptions on the energy and the elevation of the body that exploded over Eridanus. The lowering of the waters allowed Moses to walk over the dried sea bed bypassing the point where the road was closed by the rock-fall; beyond this point the road continued in the direction of higher land. *Notice that nowhere does Septuaginta say that Moses passed from one coast of Sinai to the other*. Moses passed during the night, visibility being provided by the reflection of sunlight over the huge cloud of dust sent by Phaethon explosion thousand of kilometres high in the sky, a phenomenon that albeit in minor scale happened after the famous Tunguska explosion in Siberia in 1908. The Egyptians (who were trying to retrieve from Moses the gold he had stolen at the Pani sanctuary of Baal Seefon, *Lord Shiva*, present Ras Muhammad, and who reached Pi-Hahirot probably by boat from a port in present Safaji Jezirat . . . ) began following the Hebrew over the seabed at sunrise. They were destroyed by the waters that returned after the wind had stopped.

The Deucalion flood relates to the events in the Mediterranean after the explosion over Eridanus. The wind flowing over the Adriatic pushes the waters south, where they are deviated eastwards by the Italian coast and especially by the Gargano peninsula. Therefore we have a large influx of waters into the Patras-Corinth-Alkuonessi gulf. Some of the waters move over the low lying Corinth isthmus flooding Attica and destroying Athens. Others enter the Amphissa plane and move up the slope of the Parnassus. They probably did not reach Delphi, about 800 m high, but were close providing to Deucalion dramatic evidence of the flood.

In Crete the fire that was started by the passage of Phaethon in the high atmosphere was followed by the earthquake caused by the explosion and maybe an hour later by the Mediterranean waters pushed south by the wind associated with the explosion. Our scenario therefore explains naturally the three elements archaeologically related to the destruction of the Minoan civilization (fire, earthquake and flood), independently of the effects of a supposed Santorini explosion.

In our paper we claimed that the volcanoes that erupted during this period with catastrophic consequences were the about 300 volcanoes in Dancalia, the possible eruption of Santorini being unimportant for the general scenario. We are now glad to notice that C14 dating of a whole olive tree found in Santorini and presented in a Science issue of February 2006 has changed the date of the great eruption significantly, setting it to the period 1600–1630 BC. Our dating for Exodus is exactly the Biblical dating, 1447 BC. Since Bible states that the Hebrews had been in Egypt 210 years before Exodus, and we can reasonably take as beginning of their stay the arrival of Joseph, then the new dating for Santorini may well correlate with the seven year food shortage that led to the establishment of Jacob and his sons in Egypt after their call by Joseph. In other terms the Santorini explosion appears to have nothing to do with Exodus but may be correlated with the seven years of drought associated with Joseph.

The flood due to the Phaethon explosion affected the whole Mediterranean and, of course, the Baltic, the Northern Sea and parts of the Atlantic. Here we only comment that the flat coasts of Egypt and Libya must have been invaded for many km by the rising sea. The delta must have been severely beaten. Therefore Egypt, already suffering from the Ten Plagues and an incoming invasion from the East (with Velikovsky we accept the identity of the Hyksos with the Amalek, who must have been the Amu often coming from Turan, beyond the Amu Darya . . . ) was further destroyed and took centuries to regain the ancient greatness.

## 2 Before the Phaethon Explosion: The Key in Homer

Orosius speaks of climatic events, e.g. unbearable heats, leading to local difficult living conditions and to events as the Liberus war against the Indians that can be seen as an expedition to another land to escape local difficult conditions. In our work [1] we have proposed that the Amu-Amalek-Hyksos move against Egypt was a transfer of population from Turan, to avoid an ivasion by strong people from the north (the army of Liberus or Dionisius, locally knows as Sindhy/Hindhi, i.e. the *Lions*, coming from north of the *Syr Daria*, *sea/river of the Lion*, the border line still existing at Alexander's time, as stated by Curtius Rufus, indcated by a line of trees and stones). The Sindhi could not be resisted because they had superior military technology (including iron). The Liberus army crossed the Amu Darya and after bloody battles with the people of present Afghanistan in the Kush mountains (meaning *mountains of the kill*; the name changed into Hindukush, or *mountains of the killing of the Hindhi*, after the above battles) entered India and fought along the river Idaspe the local king, a fight that gave origin to the late but still useful poem *Dionisiaca* by Nonnus of Panopolis.

In [2] Vinci has proposed that the Homer epic should be set originally in the Baltic area, with the Trojans in southern Finland and the Danai/Teucri/Myceneans/ Achaeans in Denmark and southern Sweden. Vinci's arguments are geographic and climatologic. He claims that after some probably dramatic climatic change these people moved south, to the Mediterranean, where they renamed local places with the names of their original places. In our scenario the migration from the Baltic to the Mediterranean can be set at the time just before the Phaethon explosion.

After writing paper [1], additional search provided confirmation of our scenario, including:

- A statement in Nonnus' *Dionisiaca* that the waters of the river Ydaspes (possibly but not certainly the Indus), near which there was fighting, took the colour of wine. In ancient times wine was usually red, hence the statement appears to show that the reddening of the waters of the Nile during one of the Ten Plagues took place also in India, albeit with less intensity and without poisoning effects; this phenomenon can be explained by the cloud of material emitted by the many volcanoes in Dancalia reaching the Hindukush/Karakorum (Himalaya and then been washed by the rains into the Indian rivers having their source there).
- A statement in Iustinus' *Epitome* of Pompeus Trogus that at the beginning of the annual flooding the waters of the Nile arrived with a red colour, lasting a couple of days. This can be interpreted as the volcanoes of Dancalia still erupting quite often, albeit not with the intensity at the time of Exodus.
- When the material emitted by the eruption of Krakatoa in 1983 was first studied, around 1929, it was found that there was a layer of pumice with a strong blood red colour; it was supposed that it was due to interaction of the standard pumice with the seawater that entered the crater when it collapsed. We can therefore hypothesize that the blood red colour of the Nile, or the Idaspes, was due to material emitted from the Dancalian volcanoes that somehow interacted with

seawater (possibly the Dancalian depression at that time was partially filled with seawater). It should be noted that dating of the Dancalian eruptions in the last 12,000 has never been performed, as communicated to me by prof. Abate of Florence University, head of the Italian team that does geological exploration in that area.

An important step of our recent work was reading again the works of Homer, to look if there was any interesting material for our scenario. We read the Hymns, the Iliad and the Odyssey (this one for the fourth time in our life!). We noticed that there are no references to Deucalion flood or to the explosion of Phaethon. This suggests that the stories relate to a time before Exodus, hence before 1447 BC. This is supported also by the fact that in the Dionisiaca there are frequent references to the events described by Homer, which would set again their dating before 1447, on the basis of the statement in Orosius that Dionisius invasion was about the same time as Exodus.

We read Odyssey as the last work. When we were very close to the end we found the following verse, book 23, line 243:

> Lampos and Phaethon, the two horses that carry Aurora

We give the following preliminary interpretation of the above enigmatic verse:

(a) Lampos and Phaethon were small bodies, two little satellites of Earth (as Phobos and Deimos are now small satellites of Mars).
(b) Due to their small size they were visible only in special light conditions: not during the night or the day, but at sunrise, when the air is generally cooler and drier, the sky is still partly dark, and they were illuminated by the rising sun, whose light was not yet too strong. We suspect they appeared in the sky a little before the Sun and could therefore be interpreted as carrying Aurora.
(c) The above association with sunrise explains why Phaethon was called *son of the Sun*.
(d) For some reason, possibly to be found in further search of ancient literature, including northern Europe sources, the orbit of Lampos and Phaethon were perturbed and they moved closer to Earth, in an orbit that led to their destruction and catastrophic events on Earth.
(e) As far as it is known to this author, the classic literature does not explain the end of Lampos, while the end of Phaethon, the *glory of God*, of the Bible, is known as briefly described about. A reason for this lack of information may be that Lampos terminated earlier its life as a mini satellite, crashing over Earth far away from the European and Mediterranean region, weeks before the end of Phaethon. The multiple eruption of the Dancalian volcanoes that we have considered suggest that it crashed in Africa or south Arabia, triggering the eruptions. We are presently unable to present a crater that for size and age could be associated with Lampos.
(f) Alternatively, the orbits of Lampos and Phaethon may have become unstable and were slung to outer space. Finally, they might very well be two *companion asteroids*, such as the recently discovered 3753 Cruithne and

2002 AA29, eventually may have driven away in a way similar as above (http://www.astro.uwo.ca/~wiegert/3753/3753.html).

So, if we are correct, the events associated with Exodus and the other related facts, including worldwide migrations, the demise of the north American civilization that exploited the pure copper mines in Isle Royal of Lake Superior, recently dated at circa 1500 BC, should be explained not only in terms of Phaethon, as Orosius said, but also of Lampos. Of this, we are for ever indebted to the great Homer, source not only of poetical material of immense beauty, but of inexhaustible information on the ancient civilization.

## Acknowledgement

This work was partially supported by Fondi Accademici 1996.

## References

1. Spedicato, E., A super-Tunguska event circa 1447 BC: A scenario for the Phaethon explosion, the Indo-Aryan migration and the Exodus events, Report 1-2005 Miscellanea, University of Bergamo, 2005. Also in *Proceedings of the 2005 Milos Conference on Atlantis*.
2. Vinci, F., *Omero nel Baltico*, Palombi, 1998.

# Part 10

# SEAFARING

# Homer at Sea (ἰχθυόεντα κέλευθα)[1]

M.T. Wright

*Imperial College, London, U.K.*

**Abstract.** The problems inherent in drawing on Homeric epic as an historical source are here reviewed with specific reference to the light that this material may shed on ancient seafaring. Some passages from *The Iliad* and *The Odyssey* are discussed as case studies.

## 1 Introduction

History begins with Herodotos, and he begins with the Phoenicians (see [1 and 7: I, 1]):

> ... Φοίνικ[ε]ς ... ναυτιλίῃσι μακρῇσι ἐπιθέσθαι, ἀπαγινέοντας δὲ φορτία Αἰγύπτιά τε καὶ ᾿Ασσύρια τῇ τε ἄλλῃ χώρῃ ἐσαπικνέεσθαι καὶ δὴ καὶ ἐς Ἄργος.
>
> ... Phoenicians ... took to making long trading voyages. Loaded with Egyptian and Assyrian goods, they called at various places along the coast, including Argos, ....

Trading, he continues, led on to woman-snatching, and that led to war; and so we come to the abduction of Helen, to the Trojan War and to Homer.

History cannot be conjured out of the air, but we have so little material. For the more ancient Mycenaean palace culture we have the tantalizing fragments of tablets inscribed in Linear B. These show that boatbuilding was then already an established occupation [12, p. 123]. Tablets from Pylos tell too, with foreboding, of the "watchers guarding the coastal area" [13, p. 40], waiting for an enemy who would come by sea. All the known palaces were sacked between about 1200 and 1150; and then the curtain fell, leaving Greece in the so-called Dark Age [15, Ch. I, *passim*].

The destruction of Troy VIIa, widely agreed to have been Homeric Troy, is dated to just a little earlier: between about 1250 and 1200 [15, Ch. I, *passim*]. Scholarship

[1] See [4], III, 177.

*S.A. Paipetis (ed.), Science and Technology in Homeric Epics,* 377–384.

dates the Homeric epics to some 500 years later: the Iliad to the late 8th century and the Odyssey to perhaps a generation later (see Introduction in [9]). In truth, we know very little about the genesis of these poems, but ever since they led Heinrich Schliemann to dig at Hissarlik and discover Troy, few people have doubted that they had a factual basis. If so, perhaps we can use it in writing history.

## 2 Problems of Historiography

Homer was no historian, but a poet working with semi-legendary and mythical material. His heroic past was perhaps poorly remembered, altered to suit circumstances and artistic effect, and shot through with anachronism. It may have been based on fact, but it has been well said that its historical value lies less in Homer's attempt to describe an actual past than in the picture which, in the course of this attempt, he cannot help giving us of the life and manners of his own day (see Introduction in [10, p. 11]).

As historians of science and technology, we know all too well that writers have rarely taken a close interest in these activities, or recorded them with care; but we must make what we can of what they write. Much of Homer's material seems to be drawn from the practical experience of everyday life, but we must be careful in using literature, even (or especially) such great literature, for purposes for which it was not intended.

Firstly, there are the questions of chronology and anachronism. By the time the epics were written down Greece was firmly in the Iron Age. The events recalled lay way back in the Bronze Age, and there seems to be a conscious effort to set a period atmosphere. For instance, swords are bronze and spears have bronze heads (in all but one specific instance). So too are the metallic parts of armour, except for Achilles's heaven-sent replacements. On the other hand, we notice knives, axes and ploughshares of iron, and what is written about the hot stick fizzing in the eye of Polyphemus is clear evidence that the poet had witnessed the quenching of steel (see [4 and 10: V, 391–393]). What are we to make of the sea-borne trade in iron, the excuse for Athene's arrival in Ithaca (see [4 and 10: I, 182–184])? Was it up-to-the-minute stuff, as befits a goddess, or was it simply anachronistic? We cannot be sure.

Secondly, while Homer often aims to engage his audience and his readership by describing the artefacts and circumstances of ordinary experience, sometimes he transports us into a scene that is beyond human expectation, a world in which human agency may be supported or supplanted by the mysterious and the divine. Intermediate between these, perhaps, are scenes in which the remarkable nature of events or the high status of characters is signified by the execution of near-incredible feats or the use of exotic or surprising materials.

For these reasons we must be careful in interpreting anything Homer says as historical evidence. In his accounts of seafaring, however, a natural and man-made material environment is described that seems, for the most part, very real – no matter that divine agency is often invoked – provided that we have the wit and the imagination to understand his words.

## 3 The Form of the Ship

What were Homer's ships like? Casson considers the evidence in detail, and perhaps he builds a little too speculatively on it [14, Ch. 4 *passim*]. We agree, however, in equating the ships described with those depicted on a number of "geometric" vases. The norm was the long ship (μακρὰ ναῦς), designed to be rowed but commonly also carrying sailing gear. The "catalogue of ships" (see [3 and 9: II, 484–759]) indicates that the pentecontor (πεντηκόντορος) was the usual size. Homer makes a point of stating that the Boiotian ships held 120 men each (see [3 and 9: II, 510]). In the opinion of Thucydides this remark indicated that ships with two banks of oars had already been invented [6, 1.10.4], but there seems no necessity to suppose that all the men rowed at once.

The pentecontor was a general-purpose craft, used for trading and for the sort of piratical raid that lies, as a barely-mentioned fact of life, behind the Iliad. Ample rowing power was essential. Although Gillmer et al. suggest that the comparable ships depicted in the "House of the Admiral" on Thera, dateable to before c. 1510 B.C., already had an adequate sailing capability [18], such long narrow ships cannot be well-adapted to sailing.

A hull with 25 rowing benches is about as long as can be built, as a simple structure. There was space in the boat's bottom, beneath the oarsmen, for freight (or booty), probably with little or no decking. The pentecontor remained as a type recognized by Herodotos in the 5th century, although he evidently looked back on the Phocaeans' use of them for trade as quaint (see [1 and 7: I, 163]).

Just twice we read of smaller boats, of 20 oars. Such boats carried Telemachos on his errand to Pylos (see [4 and 10: II, 212ff.]), and Odysseus when he returned Chryseïs to her father (see [3 and 9: I, 309ff.]. In both contexts the small, agile boat makes sense: θοὰ ναῦς, a "fast ship": a cutter. What, though, are we to make of the statement that Odysseus took on board one hundred oxen? Even though the Achaeans were in a dire emergency and a true hecatomb was indicated, no single ship of that time could have carried such a herd! We cannot take all Homer's statements at face value.

## 4 Ship Construction

How were boats and ships constructed? We turn to the passage in which Odysseus makes himself a water-craft for his escape from Calypso (see [4 and 10: V, 234 ff.]:

δῶκέν οἱ πέλεκυν μέγαν, ἄρμενον ἐν παλάμῃσι,
χάλκεον, ἀμφοτέρωθεν ἀκαχμένον· αὐτὰρ ἐν αὐτῷ
στειλειὸν περικαλλὲς ἐλάινον, εὖ ἐναρηρός·
δῶκε δ' ἔπειτα σκέπαρον ἐύξοον· ἦρχε δ' ὁδοῖο
νήσου ἐπ' ἐσχατιῆς, ὅθι δένδρεα μακρὰ πεφύκει,
κλήθρη τ' αἴγειρός τ', ἐλάτη τ' ἦν οὐρανομήκης,
αὖα πάλαι, περίκηλα, τά οἱ πλώοιεν ἐλαφρῶς.
αὐτὰρ ἐπεὶ δὴ δεῖξ' ὅθι δένδρεα μακρὰ πεφύκει,
ἡ μὲν ἔβη πρὸς δῶμα Καλυψώ, δῖα θεάων,
αὐτὰρ ὁ τάμνετο δοῦρα· θοῶς δέ οἱ ἤνυτο ἔργον.
εἴκοσι δ' ἔκβαλε πάντα, πελέκκησεν δ' ἄρα χαλκῷ,
ξέσσε δ' ἐπισταμένως καὶ ἐπὶ στάθμην ἴθυνεν.
τόφρα δ' ἔνεικε τέρετρα Καλυψώ, δῖα θεάων·
τέτρηνεν δ' ἄρα πάντα καὶ ἥρμοσεν ἀλλήλοισι,
γόμφοισιν δ' ἄρα τήν γε καὶ ἁρμονίῃσιν ἄρασσεν.
ὅσσον τίς τ' ἔδαφος νηὸς τορνώσεται ἀνὴρ
φορτίδος εὐρείης, εὖ εἰδὼς τεκτοσυνάων,
τόσσον ἔπ' εὐρεῖαν σχεδίην ποιήσατ' Ὀδυσσεύς.

First she gave him a great axe of bronze. Its double blade was sharpened well, and the shapely handle of olive-wood fixed firmly in its head was fitted to his grip. Next she handed him an adze of polished metal; and then led the way for him to the farthest point of the island, where the trees grew tall, alders and poplars and firs that shot up to the sky, all withered timber that had long since lost its sap and would make buoyant material for his boat. When she had shown him the place where the trees were tallest the gracious goddess left for home, and Odysseus began to cut the timber down. He made short work of the task. Twenty trees in all he felled, and lopped their branches with his axe; then trimmed them in a workmanlike manner and trued them to the line. Presently Calypso brought him augers. With these he drilled through all his planks, cut them to fit across each other, and fixed this flooring together by means of dowels driven through the interlocking joints, giving the same width to his boat as a skilled shipwright would choose in designing the hull for a broad-bottomed trading vessel.

Many commentators suppose that Odysseus built a raft. Rieu's English translation is even altered accordingly: "[Odysseus] cut them [the timbers] to fit across each other, and fixed this flooring together …". These words portray a flat rectangular grid, but of course the Greek text does not say this: the timbers are simply fitted to one another, and the definite article τήν rendered here as "this flooring", stands for "the boat". This interpretation stems from the prejudice that a raft is the best that one man could do in four days.

However, πολύτροπος Ὀδυσσεύς (polytropos Odysseus) is a hero. He can perform superhuman feats of endurance and strength; and he has the know-how: *polytropos* can mean "skilled in many trades". He does many things more surprising than rapid-action boatbuilding. Let us look, then, at what Homer tells us about the boatbuilding itself:

> [The planks] were bored, fitted to one another, and driven together with joining-pieces and pegs.

Casson was the first to point out that this description fits with the mounting archaeological evidence that, throughout antiquity, Mediterranean boats were commonly made in an astonishingly elaborate way [17]. Starting with a keel, the shipwright built the hull shell-first by fastening shaped planks edge-to-edge with joining pieces (false tenons or tongues, ἁρμονίαι) which were then bored through and pinned together using wooden pegs (treenails, γόμφοι).

Mark, however, argues that the description is of a different type of shell-first construction, in which the planks are sewn or lashed together [20]. He draws attention to several literary allusions. In his Latin version Pacuvius is explicit: Odysseus sews the boat together; and the Roman encyclopaedists Varro and Pliny both suggested that the Achaean boats at Troy were sewn, on the basis of the line:

καὶ δὴ δοῦρα σέσηπε νεῶν καὶ σπάρτα λέλυνται·

... the planks of our ships have rotted away and the cords are parted.[2]

These are weak arguments. We are not told that the cords were part of the hull; they might have been rigging. Mark makes much of the selection of tools that we are told about: axe, adze and auger, but no chisel; but neither is there any mention of the mallet for driving the planks together, nor about cordage, nor about many other necessary details.

On this rather flimsy basis Mark argues that the elaborate construction with tenon joints, already attested as early as the 14th century B.C. [16], dropped out of use during the post-Mycenaean Dark Age in Greece in favour of a less labour-intensive sewn construction; but that it was re-established later to survive – as the archaeological record indicates – as the main way of building substantial vessels through to Roman imperial times at least. The textual evidence will not however bear this weight of interpretation; and none of the few relics of sewn construction found in the Mediterranean has been ascribed to the Homeric period. It seems as likely that ships with tenon joints and boats with sewn joints coexisted for some time, as Tzalas concludes [21]. Tenon joints would suit thick planks, and therefore larger, heavier-built vessels, while sewn construction would be more appropriate for light craft.

The various timbers named in Homeric epic have been reviewed by Kokkini et al. [19]. The three named in this passage are of rather low mechanical strength, and poplar is rapidly destroyed by seawater. All three are however light and particularly easily worked, making their selection for the present purpose rational.

Two points seem not previously to have been noticed. The description of the trees, αὖα πάλαι, περίκηλα refers to the practice of ring-barking the standing tree so that the wood is seasoned before it is felled; but, as Theophrastos says, the shipwright requires green timber that can be bent [5, 4.2.8]. Then, in Homer's words, στάθμην ἴθυνεν, Odysseus prepares straight timbers; but the most obvious characteristic of building a boat by either of the shell-first methods described is that the planks are hewed into curves. Here, I think, we notice a limitation of Homer's usefulness to us as a witness of craft practice: perhaps the poet has seen wood being prepared for house-carpentry or joinery and has, unthinking, transferred what he saw to the boatyard. How much reliance, then, can we place on other details of procedure? It seems unwise to take them literally.

[2] See [3, II, 135]; the translation is supplied by Mark [20].

What are we to make of the comment that follows (see [4 and 10: V, 249–251])?

> ... giving the same width to his boat as a skilled shipwright would choose in designing the hull for a broad-bottomed trading vessel.

I think this is the only place in which Homer describes a broad ship. It is important, because the underlying factor is the intended method of propulsion: a broad-bottomed ship – convenient, as Homer implies, for carrying goods – cannot be rowed efficiently. We conclude that, by Homer's time, there existed efficient sailing rigs. Of course the long ship was sailed, at least in running or reaching, but there was manpower to work against wind and current, and to save the ship on a lee shore. Sailing, with its far lower manning level, was appropriate for a merchantman on economic grounds; but only once the ship could be fitted with a rig that made it sufficiently manageable in unfavourable winds.

Homer cannot, however, possibly mean that Odysseus made a boat as big as a merchantman; what we understand is that he made his boat with the *proportions* of a merchantman. In other words, he was making a sailing-dinghy, not a rowing-boat.

## 5 Navigation

As Odysseus sails away, we turn naturally to the question of navigation. There is absolutely no reason to suppose that the Homeric seaman had any instrumental aids at all. The experienced sailor can tell much from the set of the waves, the birds, the floating seaweed and so on. He can tell direction – and in very general terms latitude – by observing the sky, and this is what Calypso instructs Odysseus to do (see [4 and 11: V, 271–277]:

> ... οὐδέ οἱ ὕπνος ἐπὶ βλεφάροισιν ἔπιπτε
> Πληιάδας τ' ἐσορῶντι καὶ ὀψὲ δύοντα Βοώτην
> Ἄρκτον θ', ἣν καὶ ἅμαξαν ἐπίκλησιν καλέουσιν,
> ἥ τ' αὐτοῦ στρέφεται καί τ' Ὠρίωνα δοκεύει,
> οἴη δ' ἄμμορός ἐστι λοετρῶν Ὠκεανοῖο·
> τὴν γὰρ δή μιν ἄνωγε Καλυψὼ, δῖα θεάων,
> ποντοπορευέμεναι ἐπ' ἀριστερὰ χειρὸς ἔχοντα.

> ... nor did sleep ever descend on his eyelids
> as he kept his eye on the Pleiades and late-setting Boötes,
> and the Bear, to whom men give also the name of the Wagon,
> who turns about a fixed place and looks at Orion,
> and she alone is never plunged in the wash of Ocean.
> For so Kalypso, bright among goddesses, had told him
> to make his way over the sea, keeping the Bear on his left hand.

The use of the circumpolar constellations in keeping one's direction is sound; Odysseus is sailing East. "Pleiades and late-setting Boötes" seems to indicate the time of year: Spring or early Summer, when the Pleiades rise and Boötes sets just before dawn. In this it recalls the star-calendar used extensively by Hesiod and later writers [2, 8], and perhaps it hints at a forgotten element of Homer's original material according to which Odysseus had a choice as to when to set out.

Despite Odysseus's use of the night sky, it was probably the norm to put in to shore at night, and it is clear that in general one found one's way around by keeping in sight of land. Here, emphasizing their anxiety of sailing over the horizon, we have Nestor and others debating how to make the crossing of the Aegean from Lesbos to Euboea (see [4 and 10: III, 168–178]):

> ὀψὲ δὲ δὴ μετὰ νῶι κίε ξανθὸς Μενέλαος,
> ἐν Λέσβῳ δ᾽ ἔκιχεν δολιχὸν πλόον ὁρμαίνοντας,
> ἢ καθύπερθε Χίοιο νεοίμεθα παιπαλοέσσης,
> νήσου ἔπι Ψυρίης, αὐτὴν ἐπ᾽ ἀριστέρ᾽ ἔχοντες,
> ἦ ὑπένερθε Χίοιο, παρ᾽ ἠνεμόεντα Μίμαντα.
> ᾐτέομεν δὲ θεὸν φῆναι τέρας· αὐτὰρ ὅ γ᾽ ἡμῖν
> δεῖξε, καὶ ἠνώγει πέλαγος μέσον εἰς Εὔβοιαν
> τέμνειν, ὄφρα τάχιστα ὑπὲκ κακότητα φύγοιμεν.
> ὦρτο δ᾽ ἐπὶ λιγὺς οὖρος ἀήμεναι· αἱ δὲ μάλ᾽ ὦκα
> ἰχθυόεντα κέλευθα διέδραμον, ἐς δὲ Γεραιστὸν
> ἐννύχιαι κατάγοντο· …

> … and late in our wake red-haired Menelaus followed too. He caught us up in Lesbos, where we were hesitating whether to choose the long passage outside the rugged coast of Chios and by way of Psyria, keeping that island to our left, or to sail inside Chios past the windy heights of Mimas. In this dilemma we prayed for a sign, and heaven made it clear that we should cut straight across the open sea to Euboea to get out of harm's way as quickly as possible. A whistling wind blew up, and our ships made splendid running down the highways of the fish, reaching Geraestus in the night.

… and the successful landfall called for a handsome sacrifice in thanks to Poseidon. Were we in any doubt that the question of navigation must have been a crucial one for the Homeric sailor, note that it is in this particular that the Phaeacean ships ran as by magic [4, VIII, 557–563].

## 6 Conclusion

As maritime archaeology and studies in ancient technology advance, our improved understanding of early material culture enhances our perception of the world of the epic; equally, used with imagination but contained within limits that the reader must be careful to observe, Homer's poetry sheds light on the world in which the poet lived and on that from which his material was drawn. Homeric epic and the artefactual record are complementary.

## References

1. Herodoti, *Historiae*, C. Hude (Ed.), Oxford University Press, Oxford, 1926.
2. Hesiodi, *Theogonia &c.*, F. Solmsen (Ed.), Oxford University Press, Oxford, 1970.
3. Homer, *The Iliad*, D.B. Monro (Ed.), Oxford University Press, Oxford, 1884.
4. Homer, *The Odyssey*, W.W. Merry (Ed.), Oxford University Press, Oxford, 1870.
5. Theophrastus, *Enquiry into Plants*, Loeb Classical Library, Harvard, 1970.
6. Thucydidis, *Historiae*, H.S. Jones (Ed.), Oxford University Press, Oxford, 1898.
7. Herodotus, *The Histories*, trans. A. de Selincourt, Penguin, London, 1954.
8. Hesiod, *Theogony, Works and Days, &c.*, trans. D. Wender, Penguin, London, 1973.
9. Homer, *The Iliad*, trans. M. Hammond, Penguin, London, 1987.
10. Homer, *The Odyssey*, trans. E.V. Rieu, Penguin, London, 1946.
11. Homer, *The Odyssey*, trans. R. Lattimore, Harper & Row, New York, 1965.
12. Chadwick, J., *Documents in Mycenaean Greek*, Cambridge, 1973.
13. Chadwick, J., *Linear B and Related Scripts*, British Museum Press, London, 1987.
14. Casson, L., *Ships and Seamanship in the Ancient World*, Princeton, University Press, 1971.
15. Murray, O., *Early Greece*, 2nd edition, Fontana, London, 1993.
16. Bass, G.F., The construction of a seagoing vessel of the Late Bronze Age, in *Proceedings of 1st International Symposium on Ship Construction in Antiquity*, Piraeus, 1985, p. 25.
17. Casson, L., Odysseus' boat (Od. V, 244–257), *American Journal of Archaeology* **85**, 1964, 61–64.
18. Gillmer, Th., Marcoyannis, G. and Roukis, Ch., The Thera ship, in *Extraordinary Machines and Structures in Antiquity*, S.A. Paipetis (Ed.), Peri Technon, Patras, 2003, pp. 231–242.
19. Kokkini, S., Krigas, N. and Dardioti, A., Natural environment and Homeric man: Plants and constructions, in *Ancient Greek Technology*, Thessaloniki, EMAET et al., 1997, pp. 233–238 [in Greek].
20. Mark, S.F., Odyssey 5.234–53 and Homeric ship construction: A reappraisal, *American Journal of Archaeology* **95**, 1991, 441–445.
21. Tzalas, Ch., The shipbuilding methods of the Greeks during pre-classical, classical and Hellenistic times, in *Ancient Greek Technology*, Thessaloniki, EMAET et al., 1997, pp. 507–515 [in Greek].

# The Redness of Ulysses' Ships

Thomas Th. Katsaros

*University of Aegean, Rhodes, Greece*

**Abstract.** The ships of all warlords mentioned in the Homeric Epics were black except for 12 red-cheek vessels of Ulysses. Pitch was the common way to protect wood from seawater. The Iliad and the Odyssey provide full information of it. A similar reference can be found in the text of Genesis (6, 14) concerning Noah's Ark. Many scholars have studied the relevant verses of Homer (Il. 2.637, Od. 9.125), however, none of them seems to have understood the specific meaning of the structure mentioned. One of the first to investigate *miltos* from Kea Island was X. Landerer in mid-19th century. The view of Landerer was right, but his conclusion was uncertain. Others followed, yet the problem remains unsolved until present, e.g. why the conqueror of Troy had 12 ships only, while all other kings possessed 40 ships in the average. On the other hand, the ships of Ulysses were red-painted and not black-painted as the others. A technological question arises from this description. The redness of Ulysses' Ships was a matter of aesthetics or an application of Mycenaean maritime technology? Was red colour something "precious" or highly expensive and this was the reason that the king of Ithaca had twelve of this kind only? We will try to specify the substance of this material by experimental means and draw conclusions.

## 1 Introduction

Pliny the Elder [1][1] was the first out of many scholars, who attempted to make a note on the Homeric text about the redness of Ulysses' ships:

> ... from the era of Trojan War they appreciated the colour of rubrica so much, as Homer mentioned, how glorified the ships which were painted with this, although his mentions on painting and pigments were rare.[2]

Homer exactly referred that were painted with red colour the cheeks of the 12 ships of Ulysses. The problem is what space of the ship is the cheek. In Greek language is easy to understand the meaning of this through etymology.

The Homeric term is *miltopareas nēas*, i.e. 12 ships with red cheeks. Many things can be hidden under this verse, such as what kind of red is *miltos* and what place is

[1] 35, XXXIII, xxxi (7), 49.

[2] XXX, 115.

*S.A. Paipetis (ed.), Science and Technology in Homeric Epics,* 385–389.

the cheek of a sea vessel? From the depictions of ancient Greek ships we can see the eye shape design in front of the vessel. So a tentative interpretation is that the Homeric warships, like Ulysses' ships, had this eye shaped design and the part under it was called *cheek*. We note here that another famous ship from Greek antiquity Argo of Jason had no only a face as prow figure, but had also the utility of speech.

Many scholars attempted to solve this question both the characterization of miltos itself as chemical substance and the specific place of its application on the Ulysses' ships. Stefanides [12] notes that under the name of *miltos*, many things can be perceived, such as cinnabar, minium, ruddle, rubrica etc. Kordellas [10] provides a description of *miltos*, but his interpretation was rather unsuccessful. Orlandos[13] maintains that *miltos* is red lead oxide (minium). Pliny the Elder mentioned colours and their use for painting by the ancients [1].[3]

This paper is a preliminary work on characterization of *miltos*, both as raw material and on the other hand as a term itself. According to this the redness of Ulysses' ships was a matter of aesthetics and an application of Mycenaean Maritime Technology.

## 2 Experimental

We tried to prove that the use of *miltos* was a case of maritime Mycenaean technology and in accordance to this hypothesis, we made an experiment with *red earth* from the Island of Kea and *Hematite* from the mines of *Laurion*. Both places were well known during the Mycenaean era [8, 9].

We cut 12 pieces of pine wood in the same dimensions ($10 \times 5 \times 1$ cm) and make 12 samples. After, we cover the 10 of these, with Beeswax by heating. The other 2 we put uncovered with their surface clear. The 10 samples separated in two groups, the first with the application of *red hematite* and the other with application of *minium*. We immersed all these samples in seawater for seven (7) days, corresponding to an ancient sea trip of 7 day duration (Figure 1).

After exposition, the results were observed and the following conclusions were drawn: The two samples had a surface of sea plants on the whole area of the sample, the red paint wood had stability in shape and had no growth on its surface, no kind of mollusks no sea plants. The colour remained unchanged. The sample without paint layer started to putrefy. But we have another point; samples with hematite had a different behavior against sea water environment than the other with minium on its surface. This was expected, since iron oxides in sea water environment are unstable and only lead possesses high stability. The observation of the traditional method to protect sea vessels in the Aegean Sea until our days is by using minium before last paint layer. At first, ship carpenters burned surface of the cheek of vessel and then applied minium with a brush.

[3] 35, XII–XXXII.

**Fig. 1** The 12 samples of the experiment: Upper line: the 2 on the left are without colour cover, the 2 samples in the middle are without application of wax as medium of the colour, the 2 samples on the right are with a film of wax.Lower line: the 2 samples on the left are covered with hematite, the 2 in the middle are covered with a film of wax, and the 2 last on the right with minium plus wax in hot.

# 3 Discussion

The experimental process was based on texts by Pliny the Elder [1], Theophrastus [2], Xenophon [3], Herodotus [4], Dioscurides [5], and above all of Homeric Poems. In the Odyssey, Homer mentioned that Ulysses sailed for seven days and then arrived to *Laestrygonians' Harbor* (Od. 7.48). This is very important, being in full contrast with common belief, that people in antiquity used to sail in the sea during the day only. From this piece of information, the duration of our experiment was determined to the seven days.

Another matter was the specific substance of *miltos*. It was hematite or something such as a lead-rich material like minium? We follow the text of Pliny the Elder, who mentioned that *ancient people used to paint their war ships with cerussa* [10], which is the same reference to Herodotus' on the description of the Samians' Fleet [4]. Herodotus said that the ancient painted their ships with red colour. So we have two authors who mentioned that the ancients did the same thing, e.g. painted their ships red.

On the *cerusa* of Pliny the Elder is well known for the confusion with *cerusa usta*, which is minium artificial. Here, one must note that Pliny, at the beginning of his career was a ship painter, an information transmitted by Protogenes. Pliny was well aware of painting methods and techniques, such application of colours on wood, since this was his first job. Theophrastus praised the quality of miltos from Kea Island [2], but Xenophon made a note on Athenian monopoly of raw materials from all Athenians' allies areas [3]. Moreover, a stone inscription of the Acropolis of Athens, stated the rules of monopoly on Kean miltos. The punishment for violators was death [6], a fact demonstrating the importance of raw materials from the mines of Kea for the Athenians.

**Fig. 2** Mosaic from Lisbon 3rd century AD: Ulysses and the Sirens. This depiction corresponds to the Homeric verse describing Ulysses' ships. The hue of the red is the colour of minium (red lead), and this supports our hypothesis.

The specific material chemical characterization was given by the author et al. [7], and the determination of the mineral form of minium and masicot together with hematite. The hypothesis of Landerer on miltos from Kea was something relevant, because he observed in mid-19th century on Vourkari Harbor a huge stone of Litharge, but after this he gave up research and rejected the identification of Kean miltos as lead-rich earth [11]. Dioscurides tell of the use of wax on the surface of the ships through a precious description of a recipe for Zopissa [5]. Dioscurides described the mixture of wax with resin in hot, to produce a material for covering of the cheek of the sea vessel. In accordance to Dioscurides (I, 72):

> Sopissan called the resin with wax from the ships, by others named apohyma because of its solubility in seawater, some others (called) with this name the resin of pine tree.

Now, the question is whether Homeric sailors had these materials in use. Homer himself makes a note for the presence of wax on Ulysses' ship, when describing the approach to the Sirens. The most impressive example is a mosaic in Lisbon, Portugal, of Roman times representing the scene (Figure 2). It is a naïve and provincial portrayal in the mosaic of the Frigidarium of the baths of the Luso-Roman villa of Santa Vitória do Amerxial (Mon forte), dating from 3rd century AD, National Archaeological Museum of Portugal, Lisbon.

In this mosaic the ship has a characteristic red colour such as that described by Homer. In Ulysses' times, sailors use to put wax on ships abundantly, to protect wood from seawater. In accordance to our experimental study, the use of wax was a well known technique in the Mycenaean era.

## 4 Conclusions

1. The substance of miltos from Kea Island is *lead-rich red earth.*
2. *Miltoparea* means the front area and the whole of the bottom area of the vessel.
3. Red colour was to protect wood from seawater and keep the vessel water-tight.
4. The *encaustic technique* (with beeswax) along with the method of painting was most probably initiated in the Homeric Era.
5. It is likely that the cost of *miltos* was very high and this may be the reason that the King of Ithaca could afford twelve ships of this kind only.
6. The application of *miltos* ensured high-speed sailing of the ship.
7. The toxicity of lead took prevented growth of mollusks and other sea plants.

## Acknowledgements

This paper is part of the 03ED141 research project, implemented within the framework of the "Reinforcement Programme of Human Research Manpower" (PENED) and co-financed by National and Community Funds (25% of the Greek Ministry of Development, General Secretariat for Research and Technology and 75% of EU-European Social Fund).

## References

1. Pliny the Elder, *Historia Naturalis*, Books (35) XXXV, (30), XXX.
2. Theophrastus of Eresus, *On Stones (De Lapidus)*, VIII, 52.
3. Xenophon, *Atheniensium respublica*, 2, 12.
4. Herodotus, *Histories*, Thalea 56–58.
5. Dioscurides, *De re Medica I*, 72.
6. Todd, M. (reprint edition 1985), *Greek Historical Inscriptions*, p. 185.
7. Katsaros, Th. and Basiakos, Y. (2002) The colours of Theophrastus, Aristoteles Univ. of Thessaloniki and Paul Getty, in *Proceedings International Conference: Color in Ancient Greece*, M. Tiverios (Ed.), Thessaloniki, p. 201.
8. Cascey (1964), *Hagia Irini of Kea.*
9. Konophagos, C. (1980), *Ancient Laurion* [in Greek].
10. Kordellas, A. (1886), *Chromatology* [in Greek].
11. Landerer, X. (1854), On Miltos from the Island of Kea, *Archaeol. Eph.* **1854**(5), 1330 [in Greek].
12. Stephanides, M. (1896), *Mineralogy of Theophrastus* [in Greek].
13. Orlandos, A. (1986), *Lexicon of Ancient Architectural Terms*, p. 179.

# Ambrosia, Nectar and Elaion in the Homeric Poems

Doukaina G. Zanni

*University of Patras, Greece*
*Democritus University of Thrace (Komotini), Greece*

**Abstract.** Ambrosia and nectar, supposed to be the divine food consumed by the Greek gods, appears in the Homeric Poems as ointment, used as cosmetic to clean and care for the bodies of gods and mortals, as well as a substance preventing decomposition of dead bodies. Ambrosia was used both externally by application on the skin, as well as internally, along with nectar, in embalming processes. Although it is not possible to describe the composition and properties of ambrosia more specifically, in Homeric Poems, all references to ambrosia (noun and adjective) together with *elaion*, put the question of the early knowledge of the properties of *elaion* in general. From the several uses of *elaion* described in Homeric scenes, it is clear that the specific qualities attributed to *elaion* relate it with the immortal gods, not only symbolically, but also practically, through "divine" characteristics it could transmit to the mortals. The early knowledge of its qualities, at the time, definitively empirical, is confirmed by recent scientific research and explains many of the uses of various kinds of *elaion* during the past decades and, in some cases, even today.

## 1 Ambrosia, Nector and Elaion

Ambrosia and nectar are commonly known as the divine food consumed by the ancient Greek gods. Nevertheless, ambrosia appears in the Homeric Poems in a twofold way: used as an ointment used as a cosmetic for the cleaning and care of the bodies of gods and mortals, and as a substance preventing decomposition of dead bodies. The former was possible by application on the skin and the latter by its internal use, together with nectar, in embalming.[1] It is not possible to describe the properties of ambrosia, since they are mentioned in a mythic-poetic way, and it was not the poet's intention to provide specific descriptions, however, all references to ambrosia, together with *elaion*, in the Homeric Poems, put the question of the early knowledge of the qualities of *elaion* in general.

[1] Nectar and ambrosia symbolize the early and the natural death respectively, while their consumption as food, functions protectively against death. On the Indo-European representation of death and pain and its characteristics which are common also in prehistoric Greece, refer to [23]. On the various functions of ambrosia (food and ointment) and their connection with the question of otherness between divine and human elements and the anthropomorphism of gods, see [4].

*S.A. Paipetis (ed.), Science and Technology in Homeric Epics,* 391–399.

The question is connected to that of the evidence of medical knowledge, in general, provided in the Homeric Poems, e.g. knowledge of human body and its care, as well as diseases and methods of treatment, which are given mainly through the descriptions of lethal wounds of Homeric heroes.[2] The main objection against using Homeric Poems as a source for medical information, consists in the fact that, although the epic myth represents Mycenaean civilization, the poet inserts elements of his own time. This fact makes the distinction between the medical information, reflecting the poet's time, and those actually referring to his heroes' time, difficult.[3] Nonetheless, this difficulty seems less complicated, given, on one hand, the oral tradition of the epics and, on the other, the fact that the poet's descriptions illustrate what the poet's audience would have expected or taken for granted in the late 8th century BC.[4] Recent linguistic research reveals a striking familiarity between Mycenaean texts and the Homeric Poems.[5]

Although the consumption of ambrosia and nectar as food by the gods is not explicitly stated in the Homeric Poems, it is suggested as such, in comparison to the basic human food and drink, e.g. bread and wine.[6] The references of the term *amvrosia* in the Homeric Poems do not allow its connection with a specific kind of food, although it is implied that it derives from some kind of plant.[7] Since humans are maintained in life through food and drink and death is often caused by bleeding, the immortality of the gods is indicated, on one hand, by their (immortal) food, and, on the other hand, by *ichor*, the fluid running through their veins, which is distinguished from blood.[8] The word *amvrosia* probably derives from the privative a- and the word related to the adjective *vrotos*, meaning "mortal".[9] Etymologically the word derives from the Indo-European stem *mer, meaning "death", found in

---

[2] On medicine in the Homeric Poems, see [20, 22, 25]. There have been various opinions expressed on the question if Homer had actually any knowledge of anatomy and the methods of healing lethal wounds. This question resides in the fact that there can be found realistic as well as imaginary elements in the poetic descriptions of warfare. See [10] and II 24–29 in [43]. Also, see [35, 40–42].

[3] See II 19 in [43] and 19 in [26].

[4] See [30, p. 37].

[5] According to these linguistic studies, the Mycenaean overtone found in many Homeric verses is due to the fact that Homeric gods are represented in resemblance to the Mycenean kings. See [37, p. 79].

[6] On the suggestion that ambrosia and nectar are the divine food, see Il. 5.339–342. Achilles regained his strength after the consumption of ambrosia and nectar offered to him by Athena in Il. 19.347–8. See Eust., *Commentarii ad Homeri Iliadem* 2.86.4. On ambrosia as solid food, see Eust., *Commentarii ad Homeri Iliadem* 1.247.22 και 25, 2.200.24. On ambrosia as substance which prevents death, similar to *amtra*, which was, according to the Indian mythology, the food of the gods, see [17, 357-7].

[7] On ambrosia as food for horses see Il. 5.775–7. According to Kirk [19, pp. 256–257], this use of ambrosia leads to the conclusion that ambrosia was a substance produced by plants similar to bread and cereals. On the plant amvrosia (or votrys or artemisia), see Dioscour., *Materia medica* III 14.

[8] The idea of the consumption of ambrosia and nectar by the gods contrasts with the ancient concept of the gods consuming the offerings of the humans in Mesopotamia. See [19, pp. 198–199].

[9] Eust., *Commentarii ad Homeri Iliadem* 1.247.27 and 1.248.1.

the ancient Greek adjective *mortos*. The adjective *amvrosios* or *amvrotos* describes nouns like the blood of the gods, clothes, sandals, hair, the night and the sleep, and means "divine", "immortal" and "giving life". The same adjective is used for *elaion*.[10]

The red and strongly fragrant nectar is the divine drink, similar to wine.[11] Among many etymological interpretations suggested for the term nectar, the most valid ones associate it with the etymon *nek-*, found in the Greek word *nekys-nekros* and the Latin verb *necare*.[12] Among the most recent etymological interpretations of the word, it has been suggested that nectar derives from the Egyptian word *ntry*, which is connected to nitro (sodium carbonate).[13] This etymological interpretation is convenient, as it will be more thoroughly studied below, for the connection of ambrosia and nectar to substances used by the mortals in embalming. The adjective *nectareos* means "fragrant", "aromatic with essential oils", symbolically, also, "divine", and appears in the Iliad when describing clothes.[14]

Apart from its nutritional value as the divine food, ambrosia appears in the Homeric Poems with an external use. In the famous scene of Hera, making her toilet, before her meeting with Zeus, Il. 14.170–1, the goddess cleans up her body using *amvrosia* and then anoints her skin with *elaion* (ἀμβροσίῃ μὲν πρῶτον ἀπὸ χροὸς ἱμερόεντος / λύματα πάντα κάθηρεν). According to recent views, the *amvrosia* used in this case is similar to *kallos amvrosion*, with which Athena anoints Penelope's face, Od. 18.192–4, and which appears to be a cosmetic ointment used, also, by Aphrodite (κάλλεϊ μέν οἱ πρῶτα προσώπατα καλὰ κάθηρεν / ἀμβροσίῳ οἵῳ περ ἐϋστέφανος Κυθέρεια / χρίεται).[15] In the Homeric Hymn of Aphrodite (63), the goddess cleans up her body and anoints it with *amvroton elaion* (λοῦσαν καὶ χρῖσαν ἐλαίῳ ἀμβρότῳ),[16] a use which could be compared with that of *amvrosia*, in the rhapsodies of Iliad and Odyssey mentioned above. Similar to the use of *amvrosia*, as a divine cleansing and cosmetic ointment, is the use by humans of the

---

[10] On the adjectives *amvrosios* and *amvrotos* describing various nouns see Il. 2.19 (the sleep), 2.57, 10.41, 14.78, 18.265–66, 24.363, Od. 4.429 and 574, 7.283, 10.404, 15.8 (the night), Il. 5.338, 14.172–3 and 14.178 (Aphrodite's veil, cf. Hymn. Aphrod. 184), 16.670 (the clothes), 5.339 (the *ichor*), 14.175–7 (Hera's hair), 23.186–7 (oil), 23.340–1, Od. 1.95–6, 5.44-5 (the sandals). The adjectives *amvrosios* and *amvrotos* are equivalent to the adjective *avrotos* in the literary tradition. On etymology and meaning of the adjective, see [13, p. 293].

[11] On nectar as a substance similar to wine, see [17, p. 358]. On the red color of nectar which is explicitly stated in T 38, and as a motif is found also in Horace, see [33].

[12] The Sanskrit writing of the term is *naśyati*. The etymology of the word explains the use of nectar, in the literary tradition, as remedy drink. See [9, pp. 153–154].

[13] On the word *nectar* and its etymological association with sodium carbonate, see [7]. On nectar as the scent sent out at the libation, as a substance used for the heat of the altar and as aromatic wine, see [24].

[14] On the adjective *nectareos* in the Iliad describing Helen's cloth (Il. 3.385) and Achilles' tunic, which was his mother's gift (Il. 18.25), meaning "fragrant", "aromatic with essential oils", see [18, p. 501] and [19, p. 265]. On the adjective with the symbolic meaning "divine", see [7].

[15] On the association of ambrosia with *kallos amvrosion*, see [17, p. 358]. According to Russo [36, p. 188], the term *kallei* is considered by the poet to be a material substance, and is a semantic *hapax*.

[16] On the *amvroton elaion* in the Homeric Hymn of Aphrodite, see [8, p. 438].

aromatic *elaion* for the body, after being cleaned with warm water. This use is attested in Il. 10.572ff., where Ulysses and Diomedes bathe after the battle and spread *elaion* over their body, as well as in many scenes of the Odyssey, in which the care taken of Ulysses includes his anointing with *elaion*.[17] In Il. 14.171–2, *amvrosia* is used as ointment consisting oil for the preservation of clothes (ἀλείψατο δὲ λίπ' ἐλαίῳ / ἀμβροσίῳ ἑδανῷ). The use of *elaion* as an ointment for linen clothes, to maintain their silkiness and sleekness after wash, is also confirmed by the archaeological evidence as a Mycenaean practice.[18]

This use of *elaion*, for the preparation of pharmaceutical and cosmetic ointments and perfumes also used in religious rites, was widely known in Egypt.[19] In Egyptian papyri, there are plenty of references to all kinds of *elaion* (olive oil, castor oil, flax oil, rose oil, cedar oil, safflower oil, sesame oil, seed oil, and almond oil).[20] The most commonly used sort of oil in archaic Greece was olive oil.[21] The external use of *elaion* was confirmed, in times posterior to Homer's, also, in other sectors. Suggestively, we mention the use of *elaion* by athletes for the protection of their skin, from the protracted exposure under the sun and the excessive perspiration during the games. *Elaion* and fat, in general, were known to increase the elasticity of the skin. Information on its various uses in medical treatment is drawn from the Hippocratic Corpus.[22]

The use of *elaion* appears in the Iliad not only for cosmetic reasons but also in the scenes of the care of a dead body. Similar was the use of *aleiphar*, which, according to recent interpretation of the term, consisted of or contained either plant or animal fat.[23] In Il. 18.350–1, Myrmidones anoint Patroclus' dead body with *elaion* and fill his wounds with *aleiphar* (λοῦσάν τε καὶ ἤλειψαν λίπ' ἐλαίῳ / ἐν δ' ὠτειλὰς πλῆσαν ἀλείφατος ἐννεώροιο). Similarly, in Il. 24.587, Hector's dead body is bathed and anointed with *elaion* (λοῦσαν καὶ χρῖσαν ἐλαίῳ). Aphrodite persistently anoints Hector's dead body with *amvrosion* rose oil, in Il. 23.186–7 (ῥοδόεντι δὲ χρῖεν ἐλαίῳ / ἀμβροσίῳ), in order to prevent the scrapes which are about to be caused on it, by its being dragged behind Achilles' chariot, while Apollo, in the following verses (Il. 23.188–191), covers the body with a black cloud, in order to prevent, during the same procedure, its desiccation by the sun. This is the only case in the Iliad, where there is an explicit indication of a specific sort of *elaion*. The use

[17] On the use of *elaion* in the Odyssey, see Od. 4.49, 4.252, 8.454, 10.449ff., 17.88, 19.320.

[18] See [17, pp. 358–359], where also the discussion about the *hapax* ἑδανῷ, meaning "cloth" appears.

[19] See [20]. On the use of pharmaceutical plants and perfumes in Egypt, see [1, 2, 11, 27, 29, 47].

[20] On the meaning of the term *elaion* in the papyri, see [5, 38, 39].

[21] On the production of olive oil, its use and the olive presses in ancient Greece, see [31, 34].

[22] On the pharmaceutical uses of *elaion* for various illnesses in the Hippocratic Corpus, see [16, pp. 247–248] (s.v. ἔλαιον). On the different sorts of *elaion*, their qualities and uses see also Dioscour., *Materia medica* I 30ff.

[23] With one exception (Od. 3.408), all references to *aleiphar* in the Homeric Poems are connected with the care of the dead body: Il. 18.351, 23.170–2 and Od. 24.44–5, 24.67, 24.73. On its composition, see [22, p. 161] and note 425 for earlier bibliography. On *aleiphar* as "aromatic oil used as an ointment", "myros" and its preparation from oil boiled with aromatic essences, according to the Linear B tablets, see [37, pp. 78–80.

of this sort of oil can be explained by the fact that rose oil had a strong fragrance, which would repel the dogs sticking around the body,[24] and, perhaps, cover, at the same time, the malodor caused by the incipient decomposition. The same care is taken of the dead body of Sarpedon by Apollo, which is cleaned in the river waters and then anointed with *amvrosia*, in Il. 16.669–70 (λοῦσον ποταμοῖο ῥοῇσι / χρῖσόν τ' ἀμβροσίῃ).

More specific is the use of ambrosia and nectar by Thetis, in Il. 19.37–9. Ambrosia and nectar are introduced in Patroclus' nostrils, to prevent decomposition of his dead body (Πατρόκλῳ δ' αὖτ' ἀμβροσίην καὶ νέκταρ ἐρυθρὸν / στάξε κατὰ ῥινῶν ἵνα οἱ χρὼς ἔμπεδος εἴη). This act could be regarded as symbolizing the immortality given to Patroclus by the goddess, through the use of the immortal substances, *amvrosia* and *nectar*, which is similar to that of *amvrosia*, in the Hesiodic List, for the regeneration and immortality of Iphigenia's body, after her sacrifice.[25] In this Homeric scene, it is possible to perceive an embalming technique. More evidence for this interpretation is provided by the use of the verb *tarchyein*, for the description of the care taken of the dead body, in Il. 7.85. The verb *tarchyein* is associated with the verb *taricheuein* and the noun *tarichos*, which means "mummy" and "dried fish".[26] It has been suggested recently, that Theocritus was referring to this Homeric scene, when describing the deification of Berenice by Aphrodite (Edyl. 15.106–8), in which the infusion of *amvrosia* in the queen's bosom was reflecting the Ptolemaic embalming practices.[27]

The embalming techniques are known from the papyri, the mummies and the objects found in the Egyptian graves.[28] According to Herodotus' description (2.86–8), who is recording the customs of his days, one of the three embalming techniques entailed the use of cedar oil, which was injected rectally in the body.[29] The *elaion* was prevented from returning out of the body, for a period of about seventy days, as long as the embalming procedure would last, the body being preserved in *litron*.[30] On the last day, the *elaion* was allowed to escape, along with all the dissolved viscera. The *litron*, in which the body was being preserved, had eaten away the flesh, leaving

---

[24] On this special function of rose oil, see [22, p. 161]. On this sort of oil coming from Cyprus, see [9, pp. 135–136].

[25] Hesiodic List 23a. 22–3 M.-W. Cf. Hymn. Dem. 237. On the connection, see [8, p. 438].

[26] On the connection of the terms, see [19, p. 398]. See the modern conservation of fish in oil.

[27] On the connection of Il. 19.37–9 and 23.184–190 with Theocritus' Eidyllion (15.106–8), see [15, pp. 133–134].

[28] On the use of oil in the Hellenistic and Imperial periods in Egypt, according to the information drawn from the papyri, attesting all different kinds of oil (olive oil, radish oil, sesame oil, safflower oil), see [28, 29].

[29] This technique was chosen by the middle social classes. The cost of each technique depended on the method used for the embalment. The first method, which was the most expensive and was chosen by the upper social classes, entailed an incision made for the extraction of the viscera and the brain.

[30] This phase of the embalming process was the same for all the dead.

the body nothing but dry, hard skin and bones, while the cedar oil had removed the humidity from the inside of the body, thus preventing decomposition.[31]

The Herodotean litron is identified with the Egyptian natural soda, containing sodium carbonate, sodium chloride, sodium sulphate, with additions of clay and calcium carbonate.[32] Some scholars claimed that the substance used for the embalmment was nothing but natural salt.[33] Should it not be something so simple, natron or other substances containing natron, were, in any case, used for the embalming procedures. If we accept, on one hand, the etymology of the word "nectar", as deriving from the Egyptian "ntry"-niter, and the allusion to the embalming of Patroclus' body with ambrosia and nectar, on the other, we could come to the conclusion that *amvrosia* is identified with *elaion* and *nectar* with natron, both used as embalming substances.

These qualities of *elaion*, which, by Homer's time, were known, undoubtedly, only through practical experience, are now, also, scientifically substantiated, through the chemical analysis of its components, which explain many of the uses of various sorts of *elaion* during the past decades and, in some cases, even nowadays. It has been generally established that oil application interrupts oxygen exposure, thereby preventing skin-dehydration and blocking the growth of the aerobic micro- organisms, involved in decomposition procedures. More specifically, olive oil was used, in the past decades, as soothing agent in several dermatoses (eczema, psora), afforded relief in burns, and was used in the preparation of plasters as well. Flax oil served as topical antiphlogistic or analgesic, and was used in the preparation of emollient applications for skin burns and ointments.[34]

Rose oil, used by Aphrodite to protect Hector's dead body, and rose water are both used in pharmaceutics and cosmetics as odor correctives.[35] According to recent studies concerning the general action of ethereal oils, these agents are useful for the treatment of dermatoses caused by excessive skin dryness or wetness.[36] The aromatic oil, which was used for the care of Patroclus' dead body, is identified with myhrr, an oil resin used as incense and as aromatic. Like all other resins, it has local antiseptic and irritant effects and was also used as styptic agent.[37] As far as ancient internal use of *elaion* in embalming, this is explained by the scientifically confirmed

[31] The body was then covered with a kind of gum and wrapped tightly in linen bandages. The organs, which had been liquidized, were kept in the canopic jars and were buried with the mummy.

[32] The term *litron* used by Herodotus is the earlier word for *niter*, i.e. the sodium carbonate or nitric potash. This substance, when mixed with oil, served by early times for the preparation of soap. For the meaning and function of *litron* in the Herodotian description, see [14, p. 210]. On the qualities of natron, which were known in antiquity, see Dioscour., *Materia medica* V 113.

[33] On the discussion, see I 43 and note 37 in [43]. Similar, though more brief, and indicating the confused knowledge of the author is the description of the same embalming procedure by Diodorus (I.91).

[34] On the qualities and use of olive oil, see [32, p. 548]. On flax oil, see [32, p. 313]. On castor oil, see [6, p. 144], [32, p. 443]. Cf. [45, p. 95]. On castor oil, sesame oil and safflower oil, see [3, 46].

[35] On rose oil as an odor corrective, see [32, p. 148]. Almond oil is also employed in the preparation of aromas, see [45, p. 93].

[36] On the ethereal oils in general, see [12].

[37] On the Myrrh, see [6, p. 216], [32, p. 356]. Cf. [45, p. 265].

lytic and cathartic qualities of some of its several kinds. Olive oil was used earlier as laxative or diluent. Almond oil was employed in the preparation of enemas, and its systemic administration had a laxative effect. Castor oil at the specific dose of 15 ml served as laxative as well.[38]

Apart from the scientific explanation of the use of *elaion* and the information on the level of the scientific knowledge reflected in the Homeric Poems, the use of *elaion* has another semantic dimension. The special treatment of the dead body, through the embalmment, suggested in Il. 18.37–9, seems to reflect a custom, according to which, the body of those who died away from home, had to be preserved till its proper sepulture.[39] Anyway, *elaion* has a practical value, although its meaning, in the Homeric Poems, is also symbolic, similar to that born by *elaion* in the Archaic times, in religious rites and games.[40] The symbolism consists in the connection of *elaion* with the gods, who preserve the dead body, in the same way they assure, on other occasions, the treatment of the weak.[41]

The various uses of *elaion*, mentioned above, reveal the early, definitively empirical knowledge of its qualities, known even before Homer's era, which are now confirmed in recent scientific research. In the Homeric scenes examined above, *amvrosia* (noun and adjective) and *elaion* are being used almost constantly in the same phrase in various combinations, to indicate either the revitalization through the application of the substances on the skin for cosmetic reasons, or the divine immortal element, as reflected in their use throughout embalming, together with nectar. From these uses of *elaion*, in the Homeric Poems, it is clear that there were specific qualities acknowledged to *elaion*, which related it with the immortal gods, not only symbolically, but also practically, through the "divine" qualities it could transmit to the mortals.

## References

1. Aufrère, S., Nature et emploi des parfums et onguents liturgiques, in M.-C. Grasse (dir.), *L'Égypte. Parfums d'histoire*, M.-C. Grasse (Ed.), Paris, 2003, pp. 116–143.
2. Aufrère, S., Nature et emploi des parfums et onguents liturgiques. Recettes, in *L'Égypte. Parfums d'histoire*, M.-C. Grasse (Ed.), Paris, 2003, pp. 144–151.
3. Ashri, A., Poetiray, P. and Pineda, C.R., *Sesame and Safflower: Status and Potentials*, Food and Agriculture Organization of the United Nations, 1985.
4. Ballabriga, A., La nourriture des dieux et la parfum des déesses: à propos d'"Iliade" XIV, 170–172, *Métis* **12**, 1997, 119–127.

[38] On the lytic and cathartic qualities of olive oil, see [32, p. 548]. For almond oil, see [45, p. 93]. For castor oil, see [6, p. 144], [32, p. 443]. Cf. [45, p. 95].

[39] In the Hellenistic uses of the term *tarchyein*, the meaning of the official sepulture is more evident than that of the special treatment of the dead. See [18, p. 398].

[40] On the use of olive oil reflected in the representations of the Attic pottery, see [34, pp. 113–115].

[41] The use of *elaion* by the gods, in the scenes cited above, corresponds perfectly to the main characteristic of the Homeric medicine, i.e. the belief, that health or illness are given by the gods. See [43], II 22; and [26], 20. This is also reflected in the adjectives describing the gods-doctors. On the subject see [26], 22; and [22, pp. 88ff].

5. Bagnall, R.S., Vegetable seed oil is sesame oil, *Chronique d'Égypte* **75**, 2000, 133–135.
6. Claus, E.P., Tyler, V.E. and Brady, L.R., *Pharmacognosy*, London, 1970.
7. Griffith, D.R., Nektar und nitron, *Glotta* **72**, 1994, 20–23.
8. Edwards, M.W., *Homer's Iliad, Text and Commentary*, Vol. 5 (Books 17–20), Transl. into Greek by M. Kesar, A. Regakos (Ed.), Thessaloniki, 2003.
9. Faure, P., *Parfums et aromates de l'Antiquité*, Paris, 1987.
10. Friedrich, W.-H., *Verwundung und Tod. Homerische Darstellungsweisen*, Göttingen, 1956.
11. Germer, R., *Untersuchung über Arzneimittelpflanzen im Alten Ägypten*, Dissertation, Hamburg, 1979.
12. Hadji-Minaglou, F. and Bolcato, O., The potential role of specific essential oils in the replacement of dermatological drugs (strong, medium, and weak) in the treatment of acute dry or weeping dermatics, *International Journal of Aromatherapy* **15**(2), 2005, 66–73.
13. Hainsworth, B., *Homer's Iliad, Text and Commentary*, Vol. 3 (Books 9–12), Transl. into Greek by M. Kesar, A. Regakos (Ed.), Thessaloniki, 2004.
14. How, W.W. and Wells, J., *A Commentary on Herodotus in Two volumes*, Vol. I (Books I–IV), Oxford, 1912 (reprinted 1989).
15. Hunter, R., *Theocritus and the Archaeology of Greek Poetry*, Cambridge, 1996.
16. Index Hippocraticus Thesauri Linguae Graecae, Göttingen, 1986.
17. Janko, R., *Homer's Iliad, Text and Commentary*, Vol. 4 (Books 13–16), Transl. into Greek by E. Hameti, A. Regakos (Ed.), Thessaloniki, 2003
18. Kirk, G.S., *Homer's Iliad, Text and Commentary*, Vol. 1 (Books 1–4), Transl. into Greek by H. Tsirigakis, D. Jacob–A. Regakos (Eds.), Thessaloniki, 2003.
19. Kirk G.S., *Homer's Iliad, Text and Commentary*, Vol. 2 (Books 5–8), Transl. into Greek by F. Philippou, A. Regakos (Ed.), Thessaloniki, 2003.
20. Köhner, O., *Die ärztlichen Kenntnisse in Ilias und Odyssee*, München, 1929.
21. Koura, B., Die "7-Heiligen Öle" und andere Öl- und Fettnamen. Eine lexikographische Untersuchung zu den Beziehungen von Ölen, Fetten und Salben bei den Alten Ägyptern von den Frühzeit bis zum Anfang der Ptolemäerzeit (von 3000 v. Chr. – ca. 305 v. Chr.), *Aegyptiaca Monasteriensia* 2, Aachen, 1999.
22. Laser, S., Medizin und Körperpflege, *Archaeologia Homerica*, fasc. S, Göttingen, 1983.
23. Lazzeroni, R., Il nettare e l'ambrosia, Su alcune rappresentazioni indoeuropee della morte, *SSL* **28**, 1988, 177–179.
24. Levin, S., The etymology of νέκταρ. Exotic scents in early Greece, *SMEA* **13**, 1971, 31–50.
25. Lorenz, R., *Beiträge zur Hygiene bei Homer*, München, 1976.
26. Lypourlis, D., *Hippocratic Medicine*, Thessaloniki, 1972 [in Greek].
27. Manniche, L., *Sacred Luxuries. Fragrance, Aromatherapy, and Cosmetics in Ancient Egypt*, Ithaca-New York, 1999.
28. Mayerson, Ph., Qualitative distinctions for elaion (oil) and psomion (bread), *BASP* **39**(1–4), 2002, 101–109.
29. Mossakowska, M., Les huiles utilisés pour l'éclairage en Égypte (d'après les papyrus grecs), *JJP* **24**, 1994, 109–131.
30. Nutton, V., *Ancient Medicine*, London, 2004.
31. Phaklaris, P.B. and Stamatopoulou, B.G., Olive and olive tree in Ancient Greece, in *Proceedings of International Conference on Olive Tree and Olive Oil from Antiquity till Present*, Athens, 1–2 October 1999, Ekat. Polimerou-Kamilaki, P. Kamilakis-L. Karapidaki, Athens, 2003, pp. 33–47 [in Greek].
32. Phokas, G.K., *Special Pharmacognosy*, Thessaloniki, 1975.
33. Pulleyn, S., Horace Odes 3.3.12: "purpureo bibet ore nectar", *Mnemos. Ser. 4* **50**(4), 1997, 482–484.
34. Raftopoulou, E., Selling and using olive oil and scented oil in Archaic Greece, in *Olive Tree and Olive Oil*, Three-Day Workshop, Calamata, 7–9 May 1993, Culture and Technology Foundation, ETBA, ΕΛΑΪΣ Α.Ε., 1996, pp. 110–117 [in Greek].
35. Robertson, C., Wounds and wounding in the Iliad, *AHB* **16**, 2002, 103–110.
36. Russo, J., Fernandez-Galiano M. and Heubeck A., *Homer's Odyssey, Text and Commentary*, Vol. 3 (Books 17–24), Translation into Greek by F. Philippou, A. Regakos (Ed.), Athens, 2005.

37. Sali, T., *Dictionary of Mycenaean Technical Terms*, Athens, 1996 [in Greek].
38. Sandy, D.B., Oil specification in the papyri: What is elaion?, in *Atti del XVII Congresso Internazionale di Papirologia*, III, Napoli, 1984, pp. 1317-1323.
39. Sandy, D.B., The production and use of vegetable oils in Ptolemaic Egypt, *BASP*, Suppl. 6, Atlanta, 1989.
40. Saunders, K.B., The wounds in Iliad 13–16, *CQ* **49**, 1999, 345–363.
41. Saunders, K.B., A note on the strange death of Mydon in Iliad 5, *Symbolae Osloenses* **75**, 2000, 24–33.
42. Saunders, K.B., Fröhlich's table of Homeric wounds, *CQ* **54**(1), 2004, 1–17.
43. Sigerist, H.E., *History of Medicine, I–II*, New York, 1961–1967.
44. Singer, Ch. and Ashworth Underwood, A., *A Short History of Medicine*, Oxford, 1962.
45. Souseles, Chr., *Pharmacognosy*, Thessaloniki, 2000 [in Greek].
46. Weiss, E.A., *Castor, Sesame and Safflower*, Barnes and Noble, New York, 1971.
47. Wozny, D. and Simones, I. (Eds.), *Parfums et cosmétiques dans l'Égypte ancienne*, Le Caire–Paris–Marseille, 2002.

# Dietary Habits in Homer

Sophia P. Christopoulou

*Achaia Prefecture, Patras, Greece*

**Abstract.** Although differing by region, dietary habits are closely interconnected with society, since they change very slowly. In fact, they pass on from one generation to the next almost unchanged. At the same time, diet plays an important role in the life of people, since they affect numerous vocational and social activities. Like a mirror, they reflect the economic, social, moral and intellectual status of the various social classes; as well as people's religious beliefs. Food is related to daily routine and entertainment, as can be inferred from the significance of *symposia*[1] in the ancient world. The Greek gastronomical identity is rooted far deep in time, dating back to prehistoric ages, as excavations have shown. Fire signaled the beginning of cooking. Since then, lots of people, devoted their time and inventiveness, to fascinate the spirit, taste, smell and sight of those who followed. Our ancestors have created gastronomic records early in time, a real treasure for the ensuing generations.

## 1 Introduction

According to recent research [12, pp. 18–19], Minoans and Mycenaeans of Bronze Age, whether rich or poor,[2] followed diets rich in animal proteins. Studies of ancient pottery proved that our ancestors were fond of stews containing (a) vegetables and lots of olive oil or fish oil,[3] (b) vegetables (possibly sweet courgettes of Crete),

[1] *Symposium* = syn + pino (drinking together), *tsimpoussi* in Modern Greek. There were many kinds of symposia: *eilapinai*, *eranica*, *koina*, *fyletica*, *fratrica*, *syssitia* (Spartans), *andreia* (Cretans). A symposium was divided in two parts: the *deipnon* where hosts and guests would eat without much discussion and *potos* (drinking), which lasted longer, hosts and guests would drink having to choose among a great variety of nuts, fruits, cheese, salted fish, desserts (*tragimata*), and they would talk.

[2] According to excavation finds among rich and poor burials in the Late Minoan cemetery of Armenoi, Crete, both rich and poor had a lot of meat in their diet. Also, in the period between 1390 and 1190 BC, the inhabitants of Armenoi were not eating fish [12, p. 19].

[3] Neolithic cave of Gerani, Rethymnon, situated on the north coast of Crete, c. 6000–3800 BC, [12, p. 80].

*S.A. Paipetis (ed.), Science and Technology in Homeric Epics,* 401–412.

fruit waxes and meats[4] (lamb, beef or goat), (c) meat (pork), (d) cereals and beans, (e) honey[5] and (f) chickpeas, oil and meat.[6]

In Mycenae, the material found in the East Basement Granary (c. 1100 BC) [11, pp. 129–130]

> comprised five main components: Emmer wheat, einkorn wheat, barley, bitter vetch and fava beans. Finds such as these can suggest some components of diet, when the evidence is related to ancient written sources and modern rural practices. All three kinds of grains, two types of wheat and barley, are most likely to have been used as the base for porridge of some kind. Barley could also have been used as a thickener for soups and stews or as the basis for beer. Fava beans could also have been used in stews. It is also likely that these dishes were made more attractive by the addition of wild herbs, still popular in Greece today.

The list of flavorings on the so-called "Spice" tablets, which were found in the House of the Sphinxes in Mycenae, includes: cardamom, celery, coriander, cumin, fennel, mint and sesame [11, pp. 129–130].

As excavation research and ancient texts reveal, ancient Greeks included pulses (lentils, split peas, chickpeas, and broad beans), wheat, barley, olives, grapes, fish and meat – especially game – in their diet.

## 2 Homer and the Habits of Achaean Heroes

In Homer's *Iliad* and *Odyssey*, a whole world is pictured. It is not the real Mycenaean world of 1300 BC, with its labyrinthine palaces and its clay tablets, nor is it quite the Greek world of 800 or 700 BC, in which Homer is supposed to have lived. It is, however, a world so convincing in every detail, and so familiar to all readers of the epics, that has to be part of any reconstruction of Greek life during a long period.

The *Iliad* tells the story of the Achaean warriors' quarrels outside the walls of Troy, where they stayed for ten years; sometimes raiding, sometimes hunting. Their life is described in every detail: they ate roasted meat and drank strong, sweet red wine, brought there by seamen, peoples of the sea, who would dock their ships, set up a market on shore and accept captive slaves and livestock in exchange for their vintage. In the *Odyssey* the poet tells of Odysseus' ten-year wanderings on his way home to Ithaca.

Food and drink were at the center of Homeric life. Lots of food and trays with all sorts of meats are offered to a stranger generously (Odyssey I, 140–142):

---

[4] Settlement of Apodoulou, Rethymnon, (c. 1900–1700 BC), Minoan Kydonia (1480–1425 BC), [12, p. 85].

[5] Three different cooking pots from Mycenaean Thebes, c. 1380–1130 BC, [12, p. 116].

[6] In a cooking jar at Midea-Argolida c. 1340–1100 BC, [12, p. 122].

A maid poured water from a beautiful gold jug
over the visitor's hands into a silver bowl,
and drew up a carved table. An aged housekeeper
had put out bread, adding many relishes,
and encouraged them to taste all that was in house.
A waiter set out for them bronze trays of all sorts of meats,
and gold drinking-cups. A servant often passed by to pour wine for them.
(Odyssey I, 136–143)

Around the central fireplace of a Homeric household, women, children and servants clustered. Men sat on stools form wall to wall or by the fire, each with a well-scrubbed table before him. The table had to be clean, for the diners used no plates. Joints of roast meat – beef, mutton or pork – were served; bread was taken around piled high in baskets, while a wine-waiter was always at hand. A singer sang stories of gods and heroes, and thus, could earn a well-chosen cut of meat, in reward for his moving tale. That very scene was what Odysseus saw, when, disguised as a mysterious beggar, he stood at the doorway of his own smoke-filled hall, watching the carousal of the young men were competing in claiming his faithful Penelope.

In reference with food and bender, it is nowadays certain that sacrifice was an important element of the etiquette, an equivalent of today's prayers [17]. Beasts were slaughtered and offers were made to the Gods on the altar, which stood within the courtyard, right across the main hall. Pieces of meat were then carried inside the manor and were roasted on jacks right over the big fire of the fireplace. The entrails were roasted as well, inside the stomach of a young animal, filled with blood and suet. Those were roasted and not boiled. Nutrition was based on some sort of bread, made by women from ground barley and wheat.

At the time Homer's epics were written, Greece was relatively small, filled with isolated farms and small cities, constantly threatened by war and piracy. No big city-states existed yet; each small city had its own defense walls and each farm its own bushy fence.

## 3 The Fruits of the Earth

Agriculture comprised a significant resource for the Greeks, even since the Homeric era [13], for the field, the property, was the basis and the essence of true ownership.[7] Ancient writers acknowledge the great power of agriculture and its positive effect on prosperity and virtue of those cultivating the soil with their own hands it was considered beneficent and honest labor.[8] Odysseus names the Cyclopes "lawless" and "inhuman", since they "neither plant nor plough" (Odyssey X, 108). Agriculture was not only considered an excellent resource: "... for he had large estates of rich corn-growing land, with much orchard ground as well ..." (Iliad X, 122–123), but

[7] For the important role agriculture played in the ancient Greek political society, see [2, 4].

[8] Hesiod extols agriculture in his poem *Works and Days* (in Greek).

also a sacred activity, according to the religious beliefs of the people, that Demeter was patron of agriculture. Demeter, also known as Sito [1], contributed to the spread of grains.[9]

## 3.1 Cereals

The two most important types of cereals in Homeric Greece were: (a) barley, whose grains were called either *alfiton* – usually roasted grains grounded into coarse meal, namely porridge – or *krimnon* (*kri*) – crudely crashed barley, and (b) white-wheat (*zeiae*), the most common variety of wheat. Another variety of wheat, related to tifi, the bulrush, spread throughout eastern and central Europe in the end of the prehistoric era. Porridge made of barley was named alfita. Relevant to this are the words of Telemachus, when he addresses Menelaus, comparing Sparta with his own homeland, Ithaca:

> ... for you have much flat ground in your kingdom
> where lotus thrives, as also meadowsweet
> and wheat and rye,
> and barley with white and spreading ears.
> (Odyssey, IV, 602–604)

An excellent description of Achilles' shield, crafted by Hephaestus, as cited in the Iliad, depicts scenes from the agrarian life (the ploughing and the harvesting):

> He wrought also a fair fallow field,
> large and thrice ploughed already.
> Many men were working at the plough within it,
> turning their oxen to and fro, furrow after furrow.
> Each time that they turned on reaching the headland a man would come up to them and give them a cup of wine [...]
> He wrought also a field of harvest corn, and the reapers
> were reaping with sharp sickles in their hands.
> Swathe after swathe fell to the ground in a straight line behind them,
> and the binders bound them in bands of twisted straw ....
> (Ilyad XVIII, 541-6/550-3)

As mentioned in the Odyssey, harvesting of the crops was usually done in May or June and mowing was performed, not with a long, but with a short scythe. Wheat, barley and peas were hulled on a threshing-floor under the feet of oxen (Iliad XX, 495–497); the winnowing was preferably done on a windy day:

> ... or as one who yokes broad-browed oxen
> that they may tread barley in a threshing-floor –
> and it is soon bruised small under the feet of the lowing cattle – even ...

[9] For the protection of agriculture by Demeter, see [4, p. 245] and [10, pp. 6–7].

Beside the preparation of bread, barley-meal was also used to make various kinds of porridges. In the Iliad (XVI, 628–641), when Patroclos visits Nestor's tend, he finds Hecamede preparing a porridge:

> First she set for them a fair
> and well-made table that had feet of cyanus; on it there was a vessel of bronze and an onion
> to give relish to the drink, with honey and cakes of barley-meal. [...]
> In this the woman, as fair as a goddess,
> mixed them a mess with Pramnian wine; she grated goat's milk cheese into it
> with a bronze grater, threw in a handful of white barley-meal,
> and having thus prepared the mess she bade them drink it.

The above-mentioned scene probably describes the preparation of *kykeon* (derivative of the ancient Greek verb *kykan* "to stir, to thicken by stirring"), a type of porridge, known in the classical era as well, very often appearing in early Greek poetry. *Kykeon* combined the gifts of Dionysus and Demeter, for it was a mixture of strong wine, unseasoned honey, cheese and barley-meal. It is the same mess that Circe offers Odysseus' companions, using *cheese*, *honey*, *meal*, and *Pramnian wine*, before she turned them into pigs (Odyssey X, 234–235).

## *3.2 Fruits and Vegetables*

Though Odyssey abounds in standard descriptions of meals, the consumption of fruit is never mentioned. These blanks are filled in from posterior information. Juicy fruit were served as appetizers, while sweet fruit, dry fruit and nuts were served as side dishes, something to munch on, while drinking wine at the symposia [8, pp. 129–131]. They were also common food for travelers and workers in the fields, though the relative sources were scarce. Nevertheless, when Odysseus finds himself to be a castaway on the island of Scheria, he comes to a standstill, mesmerized by the view of king Alcinous' neat vegetable garden and the orchards with fruit-bearing trees; pears, pomegranates, apples, figs, olives and of course vineyards:

> Outside the yard is a big orchard on both sides of the gates,
> of four acres, and a hedge runs along its side of it.
> There, tall trees spread their leaves, pears and pomegranates
> and shiny-fruited apples and sweet figs and leafy olives.
> Their fruit never fails or falls short, winter or summer, all the
> year; as the West Wind blows it fertilizes some and ripens others.
> Pear upon pear grows old and apple upon apple,
> grapes upon grapes and fig upon fig.
> (Odyssey VIII, 112–121)

A relevant reference is found in the Odyssey (XXIV, 340–344), when Odysseus goes to his father's orchard, in order to reveal him his true identity. Laertes has built a house there, surrounded by huts for the slaves, a big garden and a vineyard. Odysseus reminds the old man of the trees that belonged to him as a child: 13 pears, 10 apples, 40 figs and 20 rows of grapevines cultivated on terraces.

Thus, the Odyssey provides information about the cultivation of orchads, though little prehistoric evidence can be found in Greece. Most of the fruit, of which there is evidence that existed in Classical Greece, belong to common species of the region, according to findings of prehistoric archeology. Many of those species have been dramatically improved at some point in history, in the late years of the prehistoric period. Several improvements were achieved in the region of Greece, yet there have been some exterior manipulations. Numerous paleobotanists agree that certain species, which disappeared from Europe during the last ice period, survived in northeastern Middle East and re-inhabited Greece and the Balkans at the close of the prehistoric period [8, pp. 129–131]. Apple-tree, for example, which is mentioned in the Odyssey, was probably domesticated in the Middle East. The only early archaeological findings of seeds from cultivated apples come from Iraq and date back to before 2000 BC, while similar discoveries were made later on in Israel. True domestication must have been achieved later, as a result of the development of engrafting techniques and vegetal propagation [19, pp. 162–166].

According to paleobotanists, fruit-bearing trees mentioned in Alcinous' orchard and similar ones in Laertes' garden are nothing else but earlier, indigenous and probably not yet domesticated varieties of those species. Also, Odyssey provides indications of the existence of other fruit, such as the *kranon*, the acorn, which was edible, though mainly used as hog feed on the island of Circe (Odyssey X, 242); and the lotus (Odyssey IV, 603).

Some kinds of acorns were eaten, such as the fruit of the oak (*akylus*) and the beech masts. These acorns constituted the basis of the nutrition for some of the poorest agrarian populations. Circe fed Odysseus' companions acorns and beech masts after she had turned them into pigs (Odyssey X, 242).

Homer mentions also a variety of vegetables and pulses, such as broad beans, chickpeas and onions. However, the plant that marked ancient Greek horticulture was the leek (*prason*), and its impact was so big that Homer used the word *praseae* when referring to vegetable plots. The word *praseae*, as used in the Odyssey (VII, 127–128 and XXIV, 247) "... in the furthest part of the ground there are beautifully arranged beds/ that are in bloom all the year round", probably had the same implications with the word green – that is green vegetables; consequently ancients Greeks were involved in gardening. Thus, a wild variety of leek, the broadleaf wild leek (*Allium ampeloprasum*) must definitely have been used at the time, as it did by the generations that followed.

## *3.3 Wine*

The time of the harvest, the treading of grapes, the production of must, and the tasting of fresh wine were significant dates on the agrarian and religious calendar, therefore, were often depicted on pots and reliefs. Homer, in a fascinating way, reproduces a scene of agrarian life, depicted on Achilles' shield (Iliad XVIII, 561–567):

> ... he wrought also a vineyard, golden and fair to see,
> and the vines were loaded with grapes. The bunches overhead were black,
> but the vines were trained on poles of silver ...
> there was only one path to it,
> and by this the vintagers went when they would gather the vintage.
> Youths and maidens all blithe and full of glee, carried the luscious fruit in plaited baskets ...

The vines were trimmed in the beginning of spring and the stubble and the grass, leftovers of the summer season and the wine harvest in September, was collected to be used to manufacture mattresses. The grapes were dried in the sun for ten days and left in the shade for another five, before they were treaded in wine presses; and the must was poured into jars.

Some vinicultural areas are mentioned as well; Priam states in the Iliad (III, 184) that he has been in "... Phrygia with the many vineyards ...", while Hecamedes (Iliad, XI, 639) prepares a mess with "Pramnian wine". This wine probably was named after a mountain of Icaria; Galen describes it as "strong red wine"; thus, its name denoted its qualities rather than its origin. Circe (Odyssey XII, 19) offers Odyssey some "red, sparkling wine".

The importance of grain and wine becomes even more evident from the scene where Telemachus, in his treasury, shortly before his departure for Pylos, decides on the supplis he will take with him (Odyssey II, 340–41): "... well-ripened wine, unblended/ and fit for a god to drink, were ranged against the wall ...", and orders Euryclea:

> ... draw me off some of the best wine you have, after what you are keeping for my father's own drinking ...
> Let me have twelve jars, and see that they all have lids;
> also fill me some well-sewn leathern bags with barley meal – about twenty measures in all.

The ability to put in storage large quantities of wine for a long period of time, signified wealth, power and noble descendance. In another part of the Odyssey (III, 391–392), it is stated that the jars with wine were "unsealed and opened after eleven years".

Wine was stored within large amphorae, which facilitated its transportation. The Iliad embeds a scene (VII, 472–475), where the transportation and trading of wine, newly arrived from Lemnos, is described:

> From this supply the Argives bought their wine,
> some with bronze, some with iron,
> some with hides, some with whole heifers,
> and some again with captives.

Since the 8th century BC, wine import from different areas led to the acknowledgement and appreciation of several famous types of wine, such as the "Pramnian wine" or the wine of Maron (Odyssey IX, 196–198, 208–211), by which Odysseus managed to get Polyphymus drunk:

> ... I also took a goatskin of sweet
> black wine
> which had been given me by Maron,
> Apollo son of Euanthes, who was priest of Apollo the patron god of Ismarus,
> ...
> when he drank it
> he mixed twenty parts of water to one of wine,
> and yet the fragrance from the mixing-bowl was so exquisite
> that it was impossible to refrain from drinking.

The inference that can be drawn from Homer's epics is that the most important vinicultural areas of the time were the islands of northern Aegean Sea and the coastline of Ionia – Aeolis. Evidence from 5th and 4th centuries BC supports this belief. It seems that Homeric wine was red, gritty and sweet or smooth.

### *3.4 Olive Oil*

Along with wine and cereals, it was considered the wealth of prehistoric Greek settlements, long before the word oil was even inscribed on Linear B plates (e-ra-wo/ *elaiwon* = *elaion* – oil) [6, pp. 177–178], or mentioned by Homer. It is seems though, that it was not used in cooking, its use was rather external, as a drink-offering, in cosmetics (it was applied to the skin); and as fuel (oil lamps).

## 4 Dairy Products and the Consumption of Meat

The first written reference tp dairy products can be traced back in early ancient Greece. In Odyssey (IV, 87–89), Menelaus speaks of Africa and states that:

> Every one in that country, whether master or man,
> has plenty of cheese, meat, and good milk,
> for the ewes yield all the year round.

Further on, in the same epic, Odysseus, king of Ithaca, on his voyage back home after the end of the Trojan War, arrives at the island of Cyclopes and encounters Polyphymus, who lived a cave and pastured sheep. Ovine cheese, curd and milk come up for the first time at this very point:

His cheese-racks were loaded with cheeses
and he had more lambs and kids than his pens could hold ...
... as for his dairy, all the vessels, bowls, and milk pails into which he milked,
were swimming with whey.

A second reference can be found in the Iliad (XI, 628–641), when Patroclos, on a visit to Nestor's tend, is exceptionally received by Hecamede, who prepares a mixture:

In this
the woman, as fair as a goddess, mixed them a mess
with Pramnian wine; she grated goat's milk cheese into it
with a bronze grater,
threw in a handful of white barley-meal,
and having thus prepared the mess she bade them drink it.

It seems that milk was curdled into cheese with the juice of the fig-tree (Iliad V, 902–903).

Affluence acquired from animal husbandry was a typical criterion of social stratification. Odysseus was called by Eumaeus, the swineherd, "a man of great wealth", for he possessed 12 herds of cattle, 12 flocks of sheep, 12 flocks of goats and 12 droves of pigs in the mainland, as well as 11 herds of goats on the island of Ithaca (Odyssey XIV, 100–104). When talking about meat, Homeric heroes mean flesh of domestic animals, since game, though delicious and dainty could not be obtained on a regular basis [8, pp. 99–103]. Among domestic animals, sheep, hogs, goats and cattle compose the main source of food. Three of the above species were in the centre of attention in numerous scenes throughout Iliad and Odyssey. That dealt with sacrifice and celebration.

The role of goat meat was only complementary, according to Homer (Odyssey II, 300 and 213–214) and Hesiod (*Works and Days*, 590). Although ancient Greeks ate and sacrificed goats, these animals were primarily used for the production of cheese. The meat of the ram, which Odysseus' companions offered him after he blinded Polyphymus, is the food of the feast that day, yet Jove must be the first to get a share: "Thus through the livelong day to the going down of the sun / we feasted our fill on meat and drink ..." (Odyssey IX, 556–558).

Like sheep, hogs were offered as sacrifice as well. On his master's land, Eumaeus (Odyssey XIV, 14–20)

... had built twelve sties
near one another for the sows to lie in.
There were fifty pigs wallowing in each sty,
all of them breeding sows; but the boars slept outside and were much fewer in number, for the suitors kept on eating them,
and the swineherd had to send them the best he had continually.
There were three hundred and sixty boar pigs ...

Eumaeus sacrifices one of those pigs for his guest, Odysseus in disguise, the workers and himself to eat (Odyssey XIV, 413–456).

Technical terms defining age, gender, best parts of the animal and ideal time for consumption were used for all of the above-mentioned species. The entrails of the animals were eaten as well, as cited at least twice throughout Odyssey:

> Then he (Peisistratus) gave them their portions of the inward meats . . .
> (III, 40)

> They sacrificed the sheep, goats,
> pigs, and the heifer,
> and when the inward meats were cooked they served them round.
> They mixed the wine in the mixing-bowls.
> (XX, 250–255)

The sacrifice of a heifer is described in full detail, as it is the theme of a whole scene in the Odyssey (III, 437–63). The ritual begins with the gilding of the animal's horns, and goes on with the participants washing their hands, sprinkling of barley-meal on victim's head, the lethal stroke with the axe, whining of the women, decapitation of the animal and collection of its blood. Then, parts of the thigh bones are cut off and wrapped into fat (probably, animal's caul), are offered to the god and are burnt, while wine is poured on top of them. Shortly after (III, 461–472):

> When the thighs were burnt and they had tasted the inward meats,
> they cut the rest of the meat up small, put the pieces on the spits and toasted them over the fire. [. . . ]
> When the outer meats were done they drew them off the spits
> and sat down to dinner where they were waited upon by some worthy henchmen,
> who kept pouring them out their wine in cups of gold.

In Odyssey, Telemachus and Peisistratus, son of Nestor, arrive at Menelaus' manor, in Sparta on a chariot. After welcome, they take a moment to stare at the palace and then (IV, 51): ". . . the two took their seats by the side of Menelaus", while shortly after (IV, 55–58):

> An upper servant brought them bread,
> and offered them many good things of what there was in the house,
> while the carver fetched them plates of all manner of meats and set cups of gold by their side.

Menelaus honors them by offering them a piece of fat roast loin, his own portion of meat.

Thus, throughout the Epics, the basic food is meat roasted – on jacks. Whether sheep or goat, fat hog or wild boar, ox or heifer, Homeric people indulged themselves with roasted meat, on special occasions. They cut it in pieces, basted oil on them or wrapped them in caul, salted the pieces and finally roasted them. Afterwards, they shared the meat into portions and served it along with barley bread. Meals were necessarily accompanied with wine, diluted in craters.

## 5 Seafood

Although Achaeans and Argives traveled afar, it seems that they were not particularly keen on eating sophisticated seafood. References on seafood consumption are rare. However, there is a part of the Iliad (XVI, 406–408) which speaks of fishermen, sitting on jagged rocks and throwing their lines into the sea; using rods, with arched hooks, made of horns or copper, and leaden plummets to make the hooks sink [17]:

> ... as one who sits at the end of some jutting rock
> and draws a strong fish out of the sea
> with a hook and a line ...

Oysters are also mentioned in the following verses (Iliad, XIV 746–748):

> If we had been at sea
> this fellow would have dived from the ship's side and brought up
> as many oysters as the whole crew could stomach ...

## 6 Conclusions

All the facts mentioned above lead to the conclusion that people, during the homeric era, based their diet on meat (connected to ritual sacrifices), bread and various porridges, made of barley meal. These are, at least, the most frequently mentioned nutriments. It does not mean that other kinds of food were not consumed – such as fish, poultry, fresh and dry fruit, olives, nuts and vegetables – yet they are not cited throughout Homer's work. In addition, good wine, especially sweet and red, was drunk in large quantities.

It is important to note that the diet of the rich differed from the one of the poorer, lower classes. Opportunities to eat better were given to the latter on account of a normal devotional system, which was, among other reasons, applied in order to maintain social balance.

Greece was, however, an unusual, for the prehistoric ages, land: commerce was largely carried on via the sea, on occasional markets that lacked regularity in the flow of various products [5]. The Aegean Sea and the Ionian islands were parts of a wider Mediterranean commercial network ever since the 1st millennium. A special food market, however, did not exist, yet the trading of food was an essential part of commerce altogether [18].

On the way towards their final destination, away from the regions they were produced, products had to pass through the markets of small islands and maritime cities. At the same time, products, from the surrounding countryside, arrived at the markets of the mainland cities almost on a daily basis. Exclusive production of certain goods became imperative, due to the diversity of ground and the limited production potential of the largest part of the Greek region. However, the inhabitants of the islands and the coastal cities in the Aegean and the Ionian Sea had a hold over maritime commerce throughout the geographically isolated eastern part of the Mediterranean Sea, as well as the passes that linked it with the western part.

## References

1. Athenaeus, *Deipnosophistes*, (transl.) Th.G. Mavropoulos, Athens, 1992.
2. Venizelos Th., *About the Ancient Greeks*, Athens, 1873 [in Greek].
3. Botsford and Robinson, *Ancient Greek History*, for Prof. Donald Kagan, (transl.) S.E. Tsintonis (Ed.), MIET, Athens, 1977 [in Greek].
4. Voulodemos, Ch., *Essay on Ancient Greeks' Private Life*, Vol. 1, Odessa, 1875 [in Greek].
5. Braund, D., Fish from the Black Sea: Classical Byzantium and the Greekness of trade, in *Food in Antiquity*, J. Wilkins et al. (Ed.), Exeter, 1995.
6. Christides, A.Ph., *History of the Greek Language*, Centre of the Greek Language, Institute of New Greek Studies (M. Triandafyllides Foundation), 2001 [in Greek].
7. Dalby, A. and Grainger S., *Cooking in Antiquity*, Papademas, Athens, 2002 [in Greek].
8. Dalby, A., *Sirens Dinners: The History of the Nutrition and Gastronomy in Greece*, (transl.) Hel. Patrikiou, University Press in Crete, Herakleion, 2000 [in Greek].
9. Demopoulos, P.N., *Ancient Greeks' Public and Private Life*, Athens, 1933 [in Greek].
10. Evangelides, J., *Dissertation about the Wheat and the Meal, Namely about the Ancient Greek Diet*, Erlangen, 1890 [in Greek].
11. French, B.E., Archaeological evidence of food in Mycenae, in *Minoans and Mycenaeans Flavours of Their Time*, Ministry of Culture, Hellenic Cultural Heritage SA, National Achaeological Museum, Kapon, Athens, 1999 [in Greek].
12. Martlew, H. and Tzedakis, G. (Eds.), Archaeology meets science, in *Minoan and Mycenaeans Flavours of Their Time*, Ministry of Culture, Hellenic Cultural Heritage S.A., Culture Olympiad, Kapon, Athens, 2001.
13. Micha-Lampaki, A., *Ancient Greeks' Diet According to Ancient Comedy Writers*, Dr. Thesis, Athens, 1984 (in Greek)
14. Homer, *Odyssey*, (transl.) N. Kazantzakis and J.Th. Kakrides, Athens, 1979 [in Greek].
15. Homer, *Iliad*, (transl. P. Giannakopoulos), Kaktos, Athens, 1992 [in Greek].
16. Homer, *Odyssey*, (transl. P. Giannakopoulos), Athens, 1992 [in Greek].
17. Quennel, M. and Quennel, C.H.B., *Homeric Poems and Their Era*, (transl.) S. Markianos, Athens, 1965 [in Greek].
18. Sherratt, S. and Sherratt, A., The growth of the Mediterranean economy in the early first millennium BC, in *World Archaeology* **24**, 1993, 361–378.
19. Zohary and Hopf, *Domestication of Plants in the Old World*, 2nd edition, Oxford, 1993.

# Part 12

# GEOGRAPHY

# Trojan Plain and Homeric Topography

P. Malfas

*Athens, Greece*

**Abstract.** Substantial evidence of the Homeric topography composes the Trojan landscape and maps out this particular area at the time of the Trojan War, in such a way that it is impossible to associate Homer's Iliad with a "Trojan Bay", facing the castle. The setting of the Iliad takes place in the entire plain, from Troy up to the Hellespont, from the city walls up to Straits and the Sigean promontory (Kum Kale). In this way, there must have been in-between the battlefield, the confluence of the two rivers, Scamander and Simoes (5.774), the swelling of the plain (Kumkioi) (*throsmos*), the Herculean Wall ("ἀμφίχυτον") and, of course, the extensive camp of the Achaeans with its defensive wall, at a large distance from the coast of the Hellespont.

## 1 Introduction

Strabo, the 1st century BC geographer, disputed the identification of Ilion of his time with the Homeric Troy, because, by taking into consideration the alluviums of the river Scamander, as well as the (occasional) proximity of the coastline, he estimated that at the time of the Trojan War – 12 centuries before his time – there would be no plain in front of the hillock of the new city, but the sea, and in this way Iliad's descriptions would not be applicable.

As we know, the excavations contradicted that ancient geographer, since, beneath the ruins of Ilion, the walls of Troy were discovered, while the specialists of the time, Schliemann's advisors, made an opposite evaluation as to the prehistory of the plain.

In the last decades, geological surveys conducted in the frame of new archaeological research, aimed at solving the old above-mentioned "topographical issue of Troy". Geologists John Kraft, Ilhan Kayan and George Rapp, who undertook this investigation (1977), were confident at the beginning and reached conclusions similar to Strabo's, when they observed further to the first boreholes holocenic marine depositions dating 10,000–7,500 years earlier (B.P.) in the plain's stratigraphy. They proceeded even to the publication of geographical reconstructions with successive coastlines of a sea bay, which was gradually restricted from the southern side of the valley up to the coast of to date, without succeeding in dating the alluviums at all (Figure 1a).

*S.A. Paipetis (ed.), Science and Technology in Homeric Epics,* 415–431.

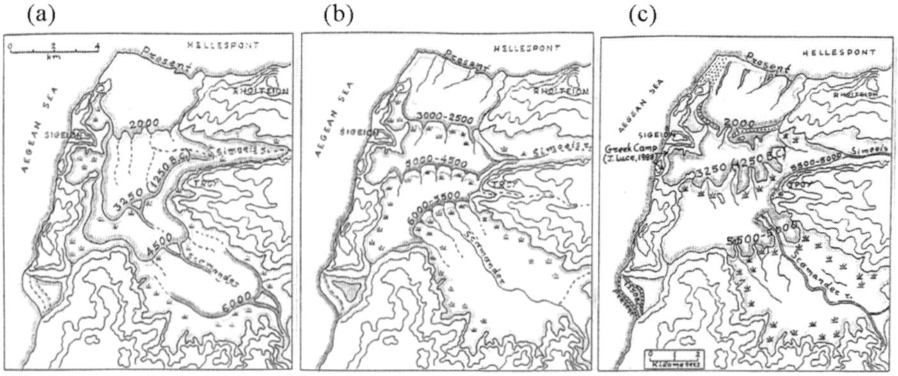


**Fig. 1** Paleogeographic reconstructions of Scamander valley with successive coastlines of the delta for several eras from 6,000 B.P. to date: John Kraft et al. (left) [8], Ilhan Kayan et al. (middle) [10], John Kraft et al. [15] (right). (Copies of the original, which have been drawn by the author in such a way as to facilitate comparisons.)

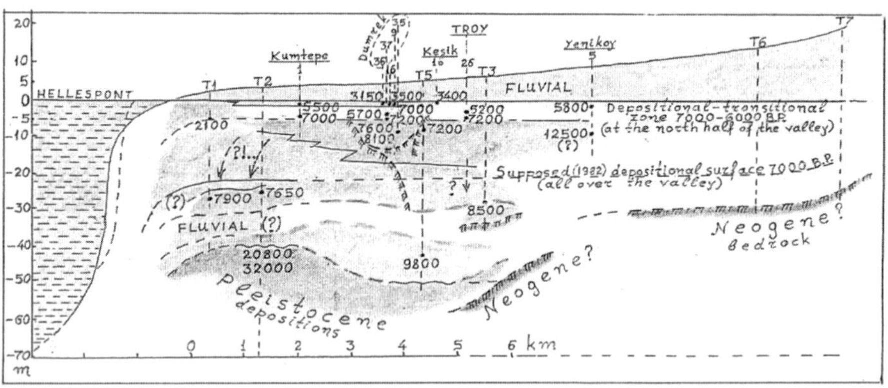


**Fig. 2** Stratigraphic figure depicting a north-south section along the flood plain of the Scamandrean valley with boreholes (vertical lines), sediment datings (numbers in years B.P.) and marine and fluvial depositions. (Drawn by the author, based on the respective diagrams, side cross sections and topographical maps, included in geologists' reports – [8, 10, 11, 12, 17]).

The research was repeated later on and proved a much different state of things. Further to a large number of boreholes and adequate datings of sediments, it was proven that the sea bay in Troy was subsequent to an earlier alluvium of the basin, which had been completed around 5,000 B.P. (3,000 BC) (Figure 3). It had been created as a result of an "invasion" of the sea into the valley, due to subversive geophysical phenomena, which took place in the area towards the end of the Bronze Age (Figure 4), while it was closed by the new alluviums in the centuries AD and after Strabo.[1]

[1] The views above have been elaborated thoroughly in a study which was published in 1998 (see References).

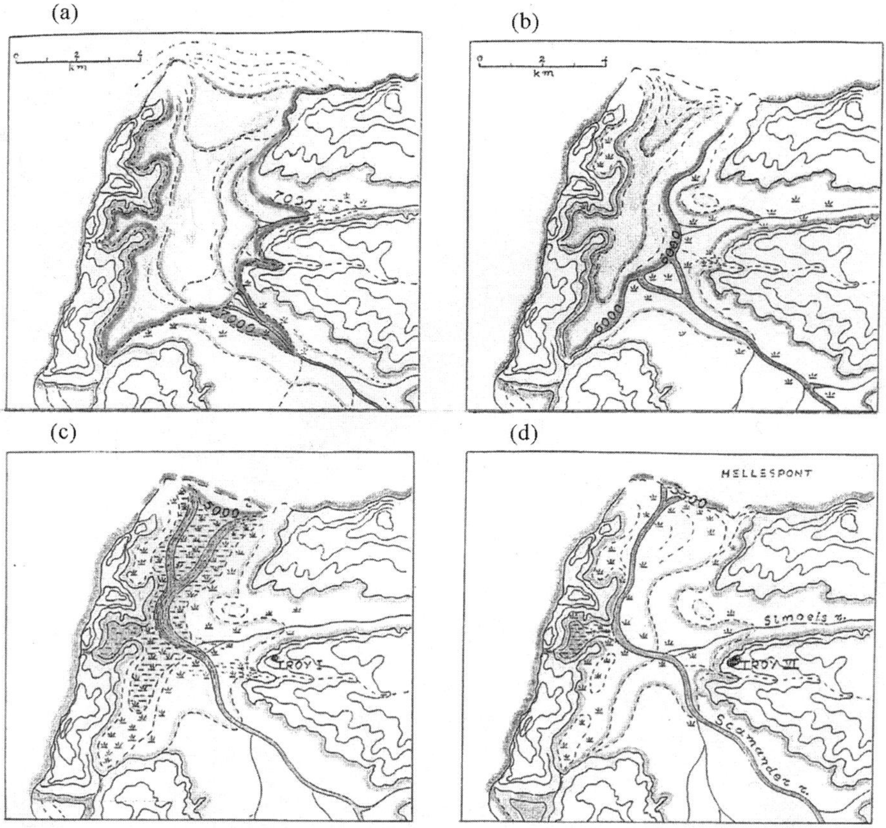


**Fig. 3** Approximate phases of the regression of the sea from Scamander valley and the corresponding development of Trojan plain, based on geological research data, as interpreted by the author (initial development): (a) 7,000 B.P., (b) 6,000 B.P., (c) 5,000 B.P. (Troy I), (d) 3,500 B.P. (Troy VI).

Nevertheless, the said geologists avoided disproving their initial premature conclusions. Now, they opt to appear reserved and they are content with variations of the so-called "Trojan Bay" (Figures 1b, c) and with a declining evolution – unnaturally slow – in the northern half of the valley from 6,000 B.P. (4,000 BC) up to date. They achieve this by making use of limited information.

We could say that an objective reason, compelling them to maintain the old theory may be the difficulty of dating the "invasion" of the sea and its chronological correlation with the Trojan War (before or after?). We believe this, because we know very well that they suspected, at least, a collapse of the bottom of the Scamandrean basin, but, as they wrote "there is insufficient evidence" [10, p. 230].

Another discouraging factor for refuting their initial conclusions might have been also the ... embellishment of the "Trojan Bay" with elements from the Homeric topography, applied unreservedly – as well as in vain – very early (1984), by Professor

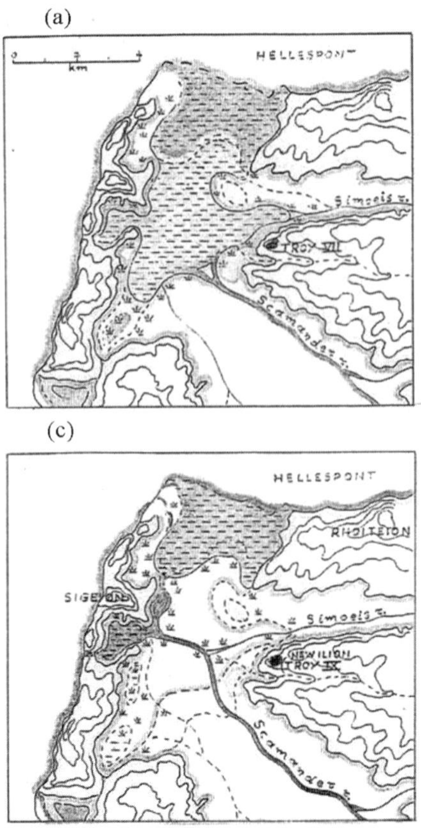


**Fig. 4** The geography of the vicinity of Troy at several periods since the new penetration of the sea into the valley (3500/3000 B.P.) until today, according to the author (earlier development): (a) The initial limits of the penetration (hypothetically on the 7 m level curve). (b) In Homer's time (2750 B.P.), based on the analogy of the topographical data in Strabo's time. (c) In Strabo's time (2000 B.P.) based on his description. (d) As it is today.

John Luce, philologist, well-known researcher of Homer. We say this, because the geologists involved claim lately [16], by referring to a more recent adaptation of Luce (1998), that "the geomorphology and the sedimentology of the plain complement Homer's Iliad".

However, as shown in Figure 5, a bay next to Troy and subsequently an extensive delta far south – deliberately omitted in the reconstruction – leave no space for the development of Iliad's war activities. What is positive is that the said geologists admit that "there are many questions … and more work remains to be done" [15, p. 377]!

For this reason it is time that we examined closely, which are the actual requirements of the Epic and if Homer can cast aside the doubts about the chronological correlation of the geophysical disaster in the area of Troy with Iliad's actions.

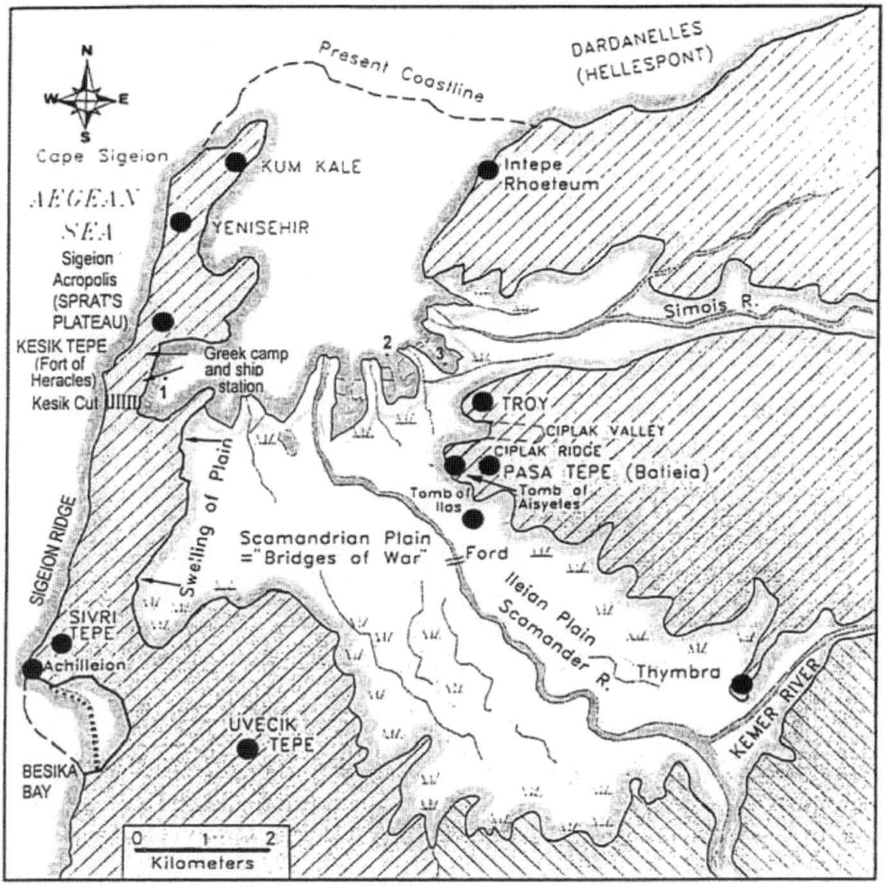


**Fig. 5** The "Trojan Bay" as supposed to be at the time of Trojan War [15], with elements of the Homeric topography, adapted by J. Luce. Reprinted by permission of J. Luce.

## 2 Characteristics and Orientation of the Achaean Camp

Before we refer to certain characteristics of the Homeric topography, which compose Troy's landscape and map out the area, where the events of the Iliad take place, it is necessary to try to sketch out a general outline, resulting from various indications in the Epic, concerning especially the area occupied and organised with defensive installations by the Achaeans during the siege of Troy.

The camp of the Achaeans was set up between two capes, along a wide coastal zone, which extended to the battlefield (14.30–36). Many thousands of men were quartered in this camp, divided in army units and by birth place. Hundreds of ships, which were drawn ashore, were arrayed in three lines, and fastened on poles in the ground.

For the protection of the camp, a long wooden wall was built, with stone foundations, at a much farther distance from the sandy coast. In front of the wall, a defensive ditch was dug and stakes were thrusted in its bottom.

Conflicts were taking place also within the camp, between the walls, which was seized at a moment during the war by the Trojans (Book 12 – "Τειχομαχία") and the third line of the ships, a fact which indicates the great distance between the wall and the landing coast. It was actually a large bridgehead.

With regards to the currently disputed orientation of the camp in relation to the location of Troy, we agree with the opinion of former researchers, Ch. Maclaren, H. Schliemann, W. Leaf (Figure 6) et al., further to our thorough evaluation of all information. More specifically, we place the camp on the location, which, we believe, the poet means, namely to the North of Troy on the flat coastal zone between the two capes, at the exit of Scamander valley. A zone, which, also during prehistoric times constituted the extension of the former Trojan plain towards Hellespont, as it roughly today is. This is proved by the stratigraphy of the plain, as we have already mentioned. Besides, due to their morphology, the other coasts could not afford to accommodate such a large camp.

"Strabo's Bay", which during its initial formation covered almost the entire initial field to the North of Troy, at a time, which cannot be defined at present, but much earlier than Strabo's time, could be the reason for doubting the existence of the northern Trojan field and subsequently a reason for doubting the orientation of the Achaean camp in relation to Troy, since, according to geological indications, the time of the invasion of the sea into the valley seems to point widely around (± 250 years) the time of the Trojan War (circa 1250 BC).

However, all doubts can be cast aside, regardless of the vagueness of the geological data related to the time of this radical geophysical disaster (3500–3000 B.P. or 1500–1000 BC), because, according to the landscape, which is composed by the Iliad's indications and especially based on the poet's reference to a natural disaster which took place there later on (Il. 12.13–34), the bay must have been formed definitely after and not before the great war.

The above conclusions are justified in our study that follows.

## 3 Key Elements of Homeric Topography

Let us begin reversely from the environs of Troy. The first important element in the Homeric topography is that the two rivers of the Iliad, the Scamander (Karamenderes) and the Simoeis (Dumrek) joined their streams near the battlefield (Il. 5.773–774):

*ἀλλ' ὅτε δὴ Τροίην ἶξον ποταμὼ τε ῥέοντε,*
*ἧχι ῥοὰς Σιμόεις συμβάλλετον ἠδὲ Σκάμανδρος,*

But when they were come to the land of Troy and the
two flowing rivers, where the Simois and Scamander join their streams ...

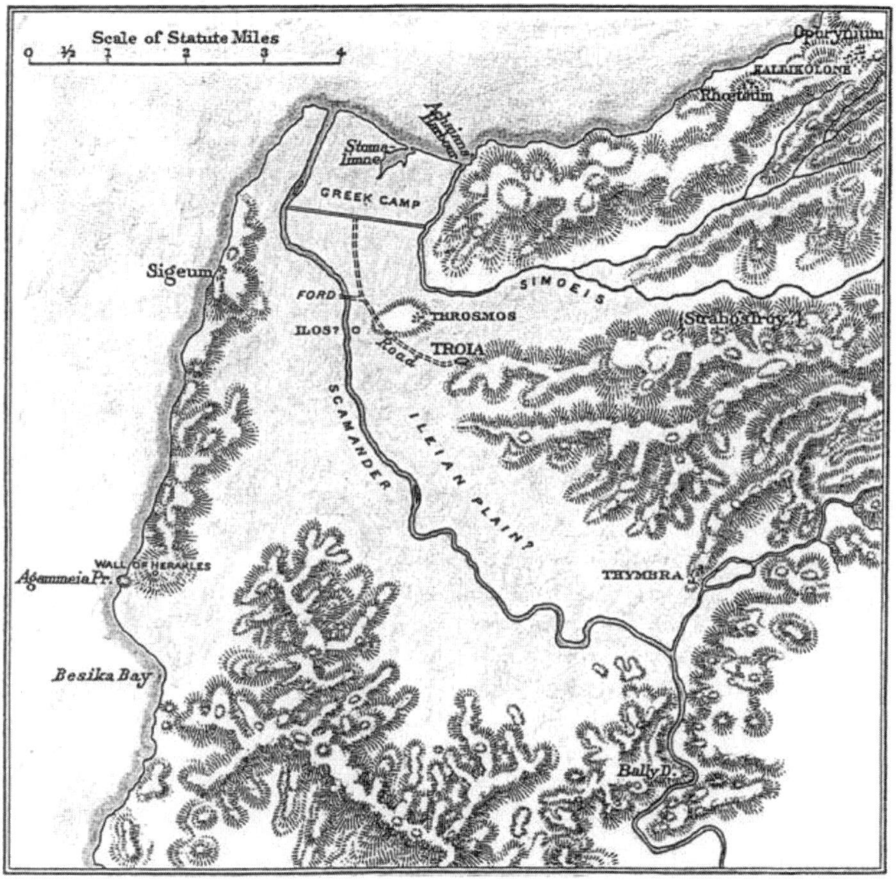


**Fig. 6** The plain of Troy according to the Iliad [7].

The southern part of the plain, where the two rivers could join together – a joining point changeable over the centuries – is the vicinity of Troy, in its western area, since the lower flow of the smaller river is found entirely in the northern part of Troy's ridge. This presupposes the existence of dry land, common delta and mouth, much northern of Troy, and, at least, it excludes the existence of a sea bay with a coastline in front of the foot of the ridge, as it is claimed by the latest version of the "Trojan Bay" [15] and depicted in Figure 1c.

The second element in the Iliad – an actually distinguishable feature today in the field, as it has remain unchanged throughout the centuries – is more indicative not only of the location of the Achaean's camp, but also of the approximate location of the coastline in the Hellespont, because its correlation in the area, where the events occurred, gives us a sense of the entire extent of the field. This feature is the "swelling of the plain" (Il. 10.160, 20.3, etc. "θρωσμός πεδίοιο").

The conflicts must have swayed back and forth, in relation to the *throsmos* and therefore we estimate that it was situated towards the middle of the battlefield. It was used alternatively as a starting line of operation for the Trojans, or as a fall-back and regrouping site or even as a line of containment. Consequently, we believe that it was closer to Troy than to Achaeans' camp.

A sandy low and wide hill, situated 1.5 km NW of the hillock of Troy, with the current name of Kumkioi was identified by the former researchers (H. Schliemann, W. Leaf et al.) with the *throsmos* (Figure 6),[2] and this was done very accurately, as there is no other land elevation, extending far enough into the plain, capable of satisfying such a military purpose. The only other similar hill (Kalafatli), situated to the South of the ruins, is found at a close distance from the wall of the lower city (600 m). As it is located at 1/10 of the distance from the (steep) Aegean coast, one cannot possibly imagine that conflicts could take place back and forth in relation to a territorial point so close to the city.

Consequently, the wide sand hill of Kumkioi sets explicitly the orientation of the Achaean camp north of Troy, while, as it is closer to the castle than to the camp, it indicates also the approximate location of the Achaeans' wall. Namely it must have been situated far from the city at a distance twice as much as the sand hill's distance (1.5 km) and even more, meaning at 3–4 km from the Troy castle.

By large leaps and safe orientation, we have already moved almost to the boundaries of the camp and very close to the exit of the valley (Figure 7).

For reasons already mentioned, the size of the camp must have been very large. With a most likely arch-shaped sandy coast between the two capes, the opening of the valley's mouth and a depth of 1 km (at least) in the bridgehead, added to the distance of the wall from the city (3–4 km), we may also determine the approximate location of the alluviated coast in the Hellespont at that time, i.e. 4–5 km from Troy, namely where it is located today as well, and as, roughly speaking, it is proved by the geological findings.

As we can see, only few fundamental characteristics of the Homeric topography contradict the existence of any sea bay, small or large, in the valley at the time of the War. Nevertheless, let us proceed even further.

The wings of the camp were occupied on the one side – the side facing also the "wine-dark sea" (Il. 23.143: "... ἰδών ἐπι οἴνοπα πόντον") – by the forces of Achilles, while on the other side there were the forces of Ajax.

Within or very close to one of the wings, i.e. close to one of the capes, Achilles founded a great tomb ("μέγα ἠρίον") to honor Patroclus, his close friend and second-in-command, who was killed in battle, a tomb destined for himself as well. As it can be deduced by both of Homeric Epics, the location chosen was neither on

---

[2] During the presence of the sea into the plain, there must have been a large coastal sand area. It is possible that its formation relegates to an older geological period, or it could be attributed to prevailing conditions similar not only at the time of the initial holocenic regression of the sea, but also during the earlier one (sea and river currents, strong winds etc) This is not indicated on the topographic maps, but only in W. Leaf's reconstruction (Figure 6), and also in few recent geological maps. Therefore it constituted the key element of the identification of the field, which was carried out by the author in July 1998.

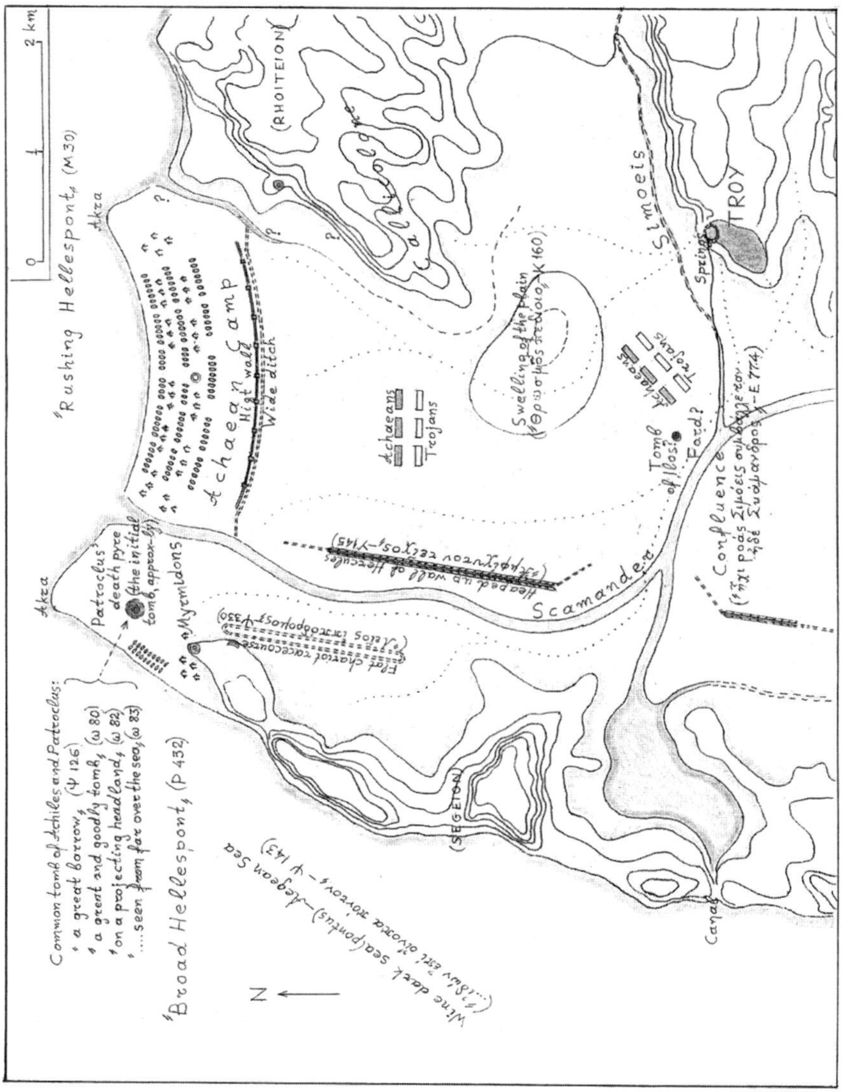


**Fig 7** The plain of Troy according to the Iliad as interpreted by the author.

a hill nor on a ridge but at a low "projecting" coast ("ἀκτῇ ἔπι προὐχούσῃ"), so that the tomb could not be confused with other territorial elevations in the landscape and could be visible from great distance, "from far over the sea" ("ἐκ ποντόφιν") to those who were heading to the Hellespont. This is clearly mentioned in the Odyssey 24.80–83:

*ἀμφ' αὐτοῖσι δ' ἔπειτα μέγαν καὶ ἀμύμονα τύμβον*
*χεύαμεν Ἀργείων ἱερός στρατός αἰχμητάων*
*ἀκτῇ ἔπι προὐχούσῃ, ἐπὶ πλατεῖ Ἑλλησπόντῳ,*
*ὥς κεν τηλεφανής ἐκ ποντόφιν ἀνδράσιν εἴη*

And over them we heaped up a great and goodly tomb,
we the mighty host of Argive spearmen
on a projecting headland by the broad Hellespont,
that is might be seen from far over the sea …

The above evaluation can also be deduced indirectly from Il. 23.120–126:

*... τὰς (δρυς υψικόμους) μὲν ἔπειτα διαπλήσσοντες Ἀχαιοί*
*ἔκδεον ἡμιόνων· ταὶ δὲ χθόνα ποσσὶ δατεῦντο*
*ἐλδόμεναι πεδίοιο διὰ ῥωπήϊα πυκνά.*
*πάντες δ' ὑλοτόμοι φιτροὺς φέρον ὣς γὰρ ἀνώγει*
*Μηριόνης θεράπων ἀγαπήνορος Ἰδομενῆος.*
*κὰδ δ' ἄρ' ἐπ' ἀκτῆς βάλλον ἐπισχερῶ, ἐνθ' ἄρ' Ἀχιλλεύς*
*φράσσατο Πατρόκλῳ μέγα ἠρίον ἠδέ οἷ αὐτῷ.*

Then the Achaeans split the trunks asunder
and bound them behind the mules, and these
tore up the earth with their feet as they hasted
toward the plain through the thick underbrush
And all the woodcutters, bare logs …
Then down upon the shore they cast them,
man after man, where Achilles planned a great barrow
for Patroclus and for himself.

Only the northern low and flat extension of the ridge of Sigeion which ends to a cape (Kumkale, where there is an Ottoman castle and a settlement) can actually correspond to the points discussed above, meaning the topographic characteristics of the tomb area, drawn by Homer's verses, and no other point, low or high, of this billowy ridge.[3]

From the northern slopes of the ridge the mules must have descended, loaded with wood, passing "through the thick underbrush" and hasting "toward the plain" ("ἐλδόμεναι πεδίοιο"). And the woodcutters, as well, after having cut timber on this ridge-spur ("κνημούς") of the "many-fountained Ida", they were carrying branches

[3] On the ridge of Sigeion there are other tombs to be found, and also in the other ridges around the plain, dating variably, some from historic times mainly, but there are also few dating from prehistoric times.

in their hands for the death-pyre and they were heading down to the area of the tomb for the funeral ceremony (libations, sacrifices, burning of the dead).

Next to the area of the ceremony there was the area of the subsequent sports games, which Achilles organised to honor his departed friend (Il. Bk. 23 – Ἆθλα ἐπὶ Πατρόκλῳ). The area included also a flat chariot race course – λεῖος ιππόδρομος – which was found on a broad field with appropriate road construction and the necessary signalling (Il. 23 326–33 and 358–66).

The chariot race was attended by the Achaeans' leaders, who watched the scene from elevated places (Il. 23.451) in front of the starting point, and all the area of the games must have been most likely protected against the strong south-western winds (eastern foot of Sigeion).

It is well understood that the area of the games was located close to the camp and it was at an edge of Scamander field (Il. 23.365). The charioteers, beginning the race, galloped along the field, leaving the camp behind them in a distance (Il. 23.365: "... νόσφι νεῶν ...") and they approached the camp reversely after turning at the given point, to finish at the starting point.

Based on the above, we are able to define the approximate location of the chariot racecourse at the western zone of the northern Trojan field, as it is depicted in the map of Figure 7. Maclaren had made exactly the same assessment in 1863 [5, p. 213].

The common tomb of Achilles and Patroclus was completed as a "great barrow" after Achilles' fall (Od. 24.80–83), but must have been destroyed later on, either due to natural causes, or due to man's intervention for the establishment of settlements and defence works in the passing of the centuries, as it was situated in such an exposed but also contested area.

Nevertheless, we are aware of the fact that the Aeolian settlers, and subsequently lords of the territory, in honor of their heroes and in memory of the events erected, a separate tomb for Achilles and a separate tomb for Patroclus in the area of one of the two capes, and likewise a tomb for Ajax, in the area of the opposite cape, while according to Strabo's testimony, his contemporary Ilians paid tribute to them every year [4, par. 32 (C.596)]. Based on the geographer's descriptions – and on information by other writers – the tombs were identified by the researchers of the 18th and the 19th century with those still salvaged at the slopes – and at the foot, one in the west – of the opposite ridges.[4]

It is not difficult to imagine what the Aeolians meant to align with the tombs, despite the broader "Strabo's Bay", which was located in-between in the area ever since their first era (11th and 10th century BC), and extended to the visible – in those centuries – ruins of Troy.

---

[4] It is not certain that Strabo refers to these tombs exactly, at least with regards to those located in the western side, at Sigeion. I believe that it is likely that the tombs of the Aeolians would have the same fate with the initial tomb (the common tomb of the two heroes), two thousand years after Strabo if they were built in the same area, meaning in the low and flat zone ("projecting" coast and distinguishable "from far over the sea"), which was more exposed to the elements of nature and to human activities.

To complete the map of the ancient Trojan plain we need also to align the course of the Scamander main river bed – of that time – according to the few indications appearing in the Epic. This course could not be exactly the same as today's course, while it is possible that it was regulated by the Trojans by means of flood controls and irrigation works, and also, perhaps, even with works deviating the course of its water. At this point, we would like to seize the opportunity to note that the waters of the Scamander must have been enriched by other carstic springs, which were gushing at that time out the heart of the Troy's ridge, from the foot of the castle, and this is the reason that they are called "Scamander's springs" (Il. 22.147–148). Apparently at a certain time, these springs dried and therefore they have constituted one of the riddles of Homeric topography for the contemporary research until recently, which were clarified by the archaeological pickaxe.[5]

The references related to the river are clearer towards the interior, but they become gradually limited in the exterior of the field of action.

More specifically, every time that the conflicts moved closer to the city, the appearance of the river Scamander is noticeable and its riverbed seems very close, as for example during the pursuit of the Trojans by Achilles along the riverbanks and out of Troy's walls (Il. Bk. 21 – "Μάχη Παραποτάμιος"). When the conflicts are brought back in the middle of the field, the riverbed appears distant (Il. 5.35 and 7.490), or instead of the river there is mention of a construction work. We mean the "heaped-up wall" ("ἀμφίχυτον τεῖχος") of Hercules, the "high wall that the Trojans and Pallas Athene had built for him" (Il. 20.145–146). Finally, no reference is made to the river when conflicts and other activities are described, to take place close to or at the limits (at the wall) or within the Achaean camp.

The last negative indication may be or may be not coincidental. Perhaps, at this point, the river is omitted as insignificant, or perhaps its main riverbed did not influence the activities in the vicinity and within the bridgehead (reduced flow, low riverbanks, improvised bridging, etc). It is very likely that a large amount of Scamander's water was channelled into the lagoon, at that time, of the western side of the plain – right opposite Troy – and through an artificial canal or tunnel there, at the very narrow neck (Kesik) exiting in the Aegean Sea.

Based on the above indications of the Epic and on certain stratigraphic information we align roughly the main riverbed of the Scamander or Ksanthos, as shown in Figure 7 with its course separated in western and northern direction, as one of the various similar versions that may be considered. Also, we note a hypothetical location of the *amphichiton* wall; hypothetical in relation to the riverbed but in accordance with the Homeric topography regarding the rises of *Callicolone*, meaning on the left and on the right side of the battlefield (Il. 20.145 & 151). By considering the *amphichiton* wall as a flood control work, we imagine it similar in form and construction to the Mycenaean work of Kopais in Boeotia,[6] adapted of course to the requirements of the Trojan field.

[5] St. Troica – 11, 2001, pp. 36–39. (M. Karfmann).

[6] *History*, Ekdhotike Athinon, Vol. A', pp. 311–313 [in Greek].

## 4 The Location of the Camp and Homer's Testimony

Apart from the fundamental and to a certain degree currently acknowledged Iliad's topographic elements, which we pointed out earlier, there are also other indications, support the orientation of the Achaean camp to the North of Troy indirectly and which are omitted here due to space restrictions.

Nevertheless, there is a very specific reference made by the poet, who in the beginning of Bk. 12, by narrating a natural disaster which destroyed the Achaean installations, after the War, he defines their location "close to the rushing ("ἀγάρροον") Hellespont" (Il. 12.30), meaning at the Straits, north of Troy, where flows a continuous stream, coming from the Black Sea. If we actually consider that from one side (the one occupied by the Myrmidons), the camp was facing at the same time also the "broad Hellespont" (north-east Aegean Sea) then its location in the Iliad becomes even more specific, as in verses 17.432–433:

*τὼ δ' οὔτ' ἂψ ἐπὶ νῆας ἐπὶ πλατὺν Ἑλλήσποντον*
*ἠθελέτην ἰέναι οὔτ' ἐς πόλεμον μετ' Ἀχαιούς,*

yet neither back to the ships to the broad Hellespont were the
twain minded to go, not yet into the battle amid the Achaeans

referring to Achilles' renowned horses, which had been tied up in the battle field and weeped in front of Patroclus, the deceased charioteer, refusing to return to their base (and go to battle).

However, the reference "to the Hellespont" is the least which Homer makes. The most important part is hidden under the poetic veil of the specific narration, which, due to its utterly mythical resemblance, has not been assessed geographically by the researchers up to now. Still, closer examination of the said verses, indisputably prove the specific location, covered by the meaning of "close to the rushing Hellespont" (disputed today)[7] and of course decisive for the deployment of the Achaean camp along Hellespont Straits. In other words, it was there that the continuous stream can only be "rushing" and not in the coast of the Aegean Sea, in which it flows after its exit from the Straits.

We refer to the first 35 verses of Bk. 12 ("Τειχομαχία"), describing the destruction of the wall of the Achaean camp and its sinking into the sea by the Gods after the War. On the one hand, Zeus appears, causing a long lasting flood and on the other Apollo turns the course of all large rivers of Troy against the wall of the camp, while in addition Earth-shaker Poseidon uproots the wall with his trident and lets it sink into the sea.

Truly, in this destruction, apart from Scamander and Simoeis, it seems utterly unreal to call forth other rivers located at a great distance, are directed to other coasts and are completely irrelevant with Trojan plain (Rhesus, Heptaporus, Caresus, Rhodius, Granicus and Aesepus) even though all of them (Il. 12.20–21) spring from Ida's central massif.

[7] It is claimed that the current, even of reduced intensity, keeps flow for several kilometres, even out of the Straits.

**Fig. 8** Troad's peninsula according to Strabo's description.

Yet, if we examine the map of the large peninsula of Troad closer (Figure 8) – which has the shape of a circle's quadrant – we will realise in contrary that it is completely natural to have the indirect participation of the rest of the peninsula rivers in this destruction, since their waters, which flow into the Propontis and meet the continuous stream originating from the Black Sea, follow the same direction and soon reach the Straits of the Hellespont jointly, to end up into the Aegean Sea. It is, therefore, a natural sequel to have all these waters of the rivers contributing to the sudden swelling of the stream in cases of heavy downpours in the area before its exit into the Aegean Sea, especially when it is also possible to be stemmed by the strong south-western winds [5, p. 63].

Consequently, the poet grounds his description on the geographical pecularities of the area and not on magic powers.

Meanwhile, the river Scamander with his tributaries accumulates all the rest of the waters of the flood, from the central plateau of Troad, and leads them also to the Hellespont, where its natural exit is.

The converging point of the flood, coming from the (upper and lower) basin of Scamander, and the flood coming down from Hellespont, is of course none other than the exit of the plain of the river into the Straits. A gigantic mass of water, suddenly accumulated over the sensitive and exposed outer zone of the alluvial Trojan plain, is possible to overturn the natural balances and develop into a large overflow also within the valley in case of a large scale flood.

Therefore, Homer, becomes on the one hand even more specific for the point of destruction – in the exit of the valley "along the strong stream of the Hellespont" ("... παρ' ἀγάρροον 'Ελλήσποντον") – in other words the location of the Achaean camp, while, on the other hand, he invokes a state of flood, which is likely to occur and cause destructive consequences in the northern Trojan plain.

Since there is now the involvement of the Earth-shaker Poseidon in the disaster, who uproots and sinks the walls, according to Il. 12.27–30:

*αὐτὸς δ' ἐννοσίγαιος ἔχων χείρεσσι τρίαιναν*
*ἡγεῖτ, ἐκ δ' ἄρα πάντα θεμείλια κύμασι πέμπε*
*φιτρῶν καὶ λάων, τὰ θέσαν μογέοντες Ἀχαιοί,*
..............................................................................................
*λεῖα δ' ἐποίησεν παρ' ἀγάρροον Ἑλλήσποντον,*

And the Shaker of Earth, bearing his trident in his hands
was himself the leader, and swept forth upon the waves all the foundations
. . . . . . . . . . . . . . . . . . . . . . . . . . . . . . . . . . . . . . . . . . . . . . . . . . . . . . . . .
and made all smooth along the strong stream of the Hellespont.

This means that the destructive consequences of this phenomenon were not only those of a temporary flood, but much greater, more radical, to a point of utterly altering and deforming the landscape of Troy. The fact that this information is not used to enrich a fictional scenario of destruction is confirmed now by the stratigraphy of the northern part of the plain, in which, apart from the fact that there is proof for the existence of a former plain – created prior to Strabo's Bay – there are also indications of collapse of the outer, in relation to Hellespont, zone. It was at this exact point, where the authors of the respective diagram (Kraft et al.) had expressed a few questions in 1982 and we added afterwards in our copy a few exclamation marks (Figure 2), after taking into account also the non-geological facts and following its comparison with other diagrams, which were published in subsequent years.

In conclusion, the correlation of the fact of the sinking in the Homeric narration with the image of collapse, which is revealed in the stratigraphy of the specific area, on the one hand proves the indisputable location of the point of destruction – i.e. the camp – right there, as well as the meaning "along the strong stream of the Hellespont", and on the other hand proves that the poet does not only use simply the natural characteristic of the area to enrich a "scenario" of his own, but also he tells of a fact which has actually taken place; a natural disaster, which did not only concern the Achaean premises – which disappeared forever – as of course the settlement itself (Troy VI/1750–1250 BC) – which was soon rebuilt (Troy VII/1200–1000 BC) – but also overturned and altered the prehistoric landscape of Troy for many centuries, until it was restored gradually by the new alluviums of the "divine" Scamander. At this point is where the difference lies between the Homeric topography and Strabo's topography, i.e. the reason that, for many centuries, Troy was sought after elsewhere.

Besides the above, the poet's testimony covers also the geological indications' inadequacy (1500–1000 BC) to respond to the critical question whether this radical alteration in the landscape of Troy took place before or after the "traditional" – in other words variable – dating of the Trojan War (circa 1250 BC).[8] Since, as the poet clarifies from the beginning of this extract, the destruction, which he describes subsequently, occurred some generations after the War and definitely when "the city of Priam was sacked" (Il. 12.13–16), a fact which Homer emphasizes also when he closes the said narrative (Il. 12.34):

*Ὣς ἄρ ἔμελλον ὄπισθε Ποσειδάων καὶ Ἀπόλλων θησέμεναι*

Thus were Poseidon and Apollo to do in the aftertime

## 5 Conclusion

Lastly, by taking into account, on the one hand the topographical facts of the Iliad and the specific indications on the other, both geographical and temporal, provided by Homer, we arrive at the conclusion that it is impossible for Homeric topography to be related to any sea bay, neither the Troy of prosperity, nor that of Priam, but only the Troy of decay, first, and, of desolation afterwards, as well as Ilion of historical times.

Consequently there is no reason to strive to adjust Iliad's war scenery to the marshy environment of a closed sea bay, or of a river delta extending in front of Troy, nor to seek a new location for the Achaeans' camp.

Interpretations of this kind, conciliatory to the "Trojan Bay", drawn by the geologists of Troy (Kraft et al.), such as those of J. Luce, to which we referred at the beginning, are refuted by the Epic itself. In conclusion, we ought to acknowledge that the former researchers Charles Maclaren, Henry Schliemann, Walter Leaf etc, who established the interpretation of Homeric topography – at least in its fundamental points – by casting aside "Strabo's Bay", are once again justified.

By giving credit to the conviction of the majority of the ancient world regarding the actual location of Troy and not to Estiaia's, Demetrius' and Strabo's arguments, they succeeded in re-establishing the location of Homer's descriptions on the reconstructed by the "divine" Scamander Trojan Field.

---

[8] This is the average of various datings regarding the conquest of Troy, which derive from ancient sources. (*Greek Mythology*, Ekdhotike Athinon, Vol. 5, p. 159, in Greek).

## References


1. Page, T., Capps, E., Rouse, W., Post, L. and Warmington, E. (Eds.), *Homer's Iliad*, The Loeb Classical Library, Harvard University Press, 1960. English translation by A.T. Murray.
2. Page, T., Capps, E., Rouse, W., Post, L. and Warmington, E. (Eds.), *Homer's Odyssey*, The Loeb Classical Library, Harvard University Press, 1960. English translation by A.T. Murray.
3. *Homer's Iliad*, Papyrus Publ., Athens, 1970 [in Greek].
4. Strabo, *Geographica*, Bk. 13 *Troad – Aeolian Land*, Cactos Publ., Athens, 1994.
5. Maclaren, C., *The Plain of Troy Described*, 1863.
6. Schliemann, H., *Ilios: The City and Country of Trojans*, 1880, pp. 78–95.
7. Leaf, W., *Troy: A Study in Homeric Geography*, 1912, pp. 24–52 & 384–389.
8. Kraft, J., Kayan, I. and Erol, O., *Troy. The Archaeological Geology*, G. Rapp and J. Gifford (Eds.), Princeton University, 1982.
9. Luce, J., The Homeric topography of the Trojan Plain reconsidered, *Oxford Journal of Archaeology*, 1984, 31–43.
10. Kayan, I., The Troia Bay and supposed harbour sites in the Bronze Age, *Studia Troica* **5**, 1995, 211–235.
11. Kayan, I., Holocene statigraphy of the Lower Karanederes – Dumrek Plain and archaeological material in the alluvial sediments to the north of the Troia Ridge, *St. Troica* **6**, 1996, 234–249.
12. Kayan, I., Geomorphological evolution of the Ciplak Valley and archaeological material in the alluvial sediments to the south of the Lower City of Troia, *St. Troica* **7**, 1997, 487–507.
13. Luce, J., *Homer's Landscapes. Troy and Ithaca Revisited*, Yale University, 1998.
14. Malfas, P., *"Troad: In Search of the Achaean Camp". A Study for Topographical-Geological Problem of Troy*, Private Publication, Athens, 1990 [in Greek].
15. Kraft, J., Kayan, I., Bruckner, H. and Rapp, G., Sedimentary facies patterns and the interpretation of paleogeographies of Ancient Troia, in *Troia and the Troad*, G. Wagner, E. Pernicka and H.P. Werpmann (Eds.), Heidelberg, 2002, pp. 361–377 (ch. 24).
16. Kraft, J., Rapp, G., Kayan, I. and Luce, J., Harbour areas at ancient Troy. Sedimentology and geomorphology complement Homer's Iliad, *Geological Society of America*, February, 2003, 163–166.
17. Gobel, I., Statir, M., Kadereit, A., Wagner, G., Kayan, I., Stratigraphy, geochemistry and geochronometry of sedimentary archives around Hissarlik Hill – A pilot study, in *Troia and the Throad*, G. Wagnner, E. Pernicka and H.-P. Werpmann (Eds.), Heiderlberg, 2002, pp. 341–359 (ch. 3).

# Part 13

# GENERAL INTEREST

# Mêtis and the Artificial

K.P. Anagnostopoulos and S. Chelidoni

*Democritus University of Thrace, Greece*

**Abstract.** Banished from the field of true knowledge, as defined by classical philosophy, *mêtis* does nonetheless pervade Greek mythology and social practice for a millennium, disappearing for good from both oral and written discourse after the 5th century AD. Homer's epics represent a dual source that permits the reconstruction of the ancient Greek *mêtis*. Indeed, they reveal some of its aspects by narrating, on the one hand, the activities of cunning gods, among who, besides Zeus himself, we find prominent Athena and Hephaestus, who, by no fortuitous coincidence, are also the gods of technology; and, on the other hand, mostly human practices, and among them those of a human being cunning *par excellence*, namely the *polumetis* and *poikiloboulos* Odysseus. In this paper, *mêtis* is considered to cover all cognitive processes that are necessary for man to face adverse or confrontational situations against powerful adversaries in unstable and complex environments, about which there is neither documented knowledge nor the possibility of rigorous analysis. In modern terms, it corresponds to design – as a cognitive process employed by all those devising courses of action aimed at changing existing situations into preferred ones.

## 1 Introduction

Having ancient philosophy as its source of origin and its starting point, Western thought evolved based on two forms of knowledge, science and craft (know-how), which are quite often complemented by the Aristotelian *phronēsis*, or practical intelligence, as a distinct category. Starting from the "first available witness", Homer, and getting at Oppian, a thousand years later, M. Dètienne and J.-P. Vernant suggest that there is another distinct way of intelligence, bearing the name of Zeus' first wife, running through the Ancient Greek world [4, p. 12]:

> There is no doubt that *mêtis* is a type of intelligence and of thought, a way of knowing; it implies a complex but very coherent body of mental attitudes and intellectual behaviour which combine flair, wisdom, forethought, subtlety of mind, deception, resourcefulness, vigilance, opportunism, various skills, and experience acquired over the years. It is applied to situations which are transient, shifting, disconcerting and ambiguous, situations which do not lend themselves to precise measurement, exact calculation or rigorous logic.

*S.A. Paipetis (ed.), Science and Technology in Homeric Epics,* 435–442.

This unique thought, fully action-oriented, lateral and conjectural, can be detected in all major professions and activities in antiquity. Moving quite often along the thin line that separates wit from deviousness and perspicacity from breach of the game rules, *mêtis* is considered the mental mechanism *par excellence* that allows adaptation in a changing and complex environment and/or opposition to stronger adversaries in adverse or confrontational relationships – whether this "adversary" is man or nature, which, unlike human beings, does not develop strategic behaviour.[1]

Banished from the field of true knowledge, as defined by classic philosophy (the word is not to be found among the 1,600 most used words in the 5th century BC [13]), *mêtis* does nonetheless pervade Greek mythology and social practice for a millennium, disappearing for good from both oral and written discourse after the 5th century AD. Divine *mêtis*, first of all, as captured mainly in Zeus, but also in Athena and Hephaestus who, by no fortuitous coincidence, are also the gods of technology.[2] Animal *mêtis*, then, encountered in many animals, for example the octopus and the fox, the partridge and the crab.[3] But mostly, human *mêtis*, as revealed in Homer's epics, and in particular in the person of an extremely cunning man, Odysseus, who is *polumetis*, *poikilometis*, *aiolometis*, *polutropos*, *polumechanos* and *poikiloboulos*, i.e. full of resources that allow him to get out of any trouble (*aporia*), while being at the same time skilful in constructing daidala and a very competent shipbuilder.

## 2 Mêtis Models

*Mêtis* constitutes, first of all, a praise to human ingenuity and creativity. In this sense, it marks the primacy of the human mind over brute force as a means of prevailing both in human conflicts and in struggles against nature. For example, Nestor advises his son Antilochus that "the woodman does more by skill than by brute force; by skill the pilot guides his storm-tossed barque over the sea, and so by skill one driver

[1] The fall into discredit of *mêtis* is undoubtedly also due to the fact that it is often associated, if not identified, with deceit, lying, deviousness and breaking the rules of the game, that one can recognize in Antilochus' tricks; in the violent pulling by Erechtheus of Skelmis' chariot reins during the chariot race narrated by Nonnus of Panopolis [4, pp. 239–240]; in Clytemnestra's feminine guile (*gunaikoboulus* ... *metidas*), in the tricks of the *aiolometis* (swiftly cunning) and *doloplokos* (plotting) Aphrodite (*oi exapatas tes goeteeias* – the deceptions of charm). Regarding the latent misogynism in *mêtis*, see [5, pp. 210–211].

[2] Not all gods are endowed with *mêtis*. Apollo, Mars and Poseidon, unlike Hermes and Aphrodite, are not included among the cunning (*metioentes*) gods. On the other hand, the *agkulometes* (possessing twisting *mêtis*), *dolophron* (intriguer) and *poikiloboulos* (full of inventive ploys) Titan Prometheus helped Zeus with his *mechanai* in his battle against Kronos.

[3] Plenty of animals are depicted as cunning (*metioenta*), e.g. the cuttle-fish (*dolometis sepia*), the *aioloboulos* and *poikilophron* fox, the partridge pretending to be lame (*perdikos skelos*) and the *poluplokon* (having many coils) octopus. Cunning animals, firstly, behave in a smart manner, providing models for specific types of intelligence (e.g. the octopus, for the sophist's many coils (*periplokai*) and the politician's many bonds (*poluplokon*)). Secondly, they prove that for *mêtis* to be defeated more *mêtis* is required; that the hunter and the fisherman, dealing with cunning creatures, will only achieve their goal if they are able to exhibit more *mêtis*.

can beat another" [Il., XXIII, 315–318]. The gods are praising the ingenious, but lame and slow, Hephaestus and are laughing at the very fast but coarse Ares, whom the ironsmith caught using a trick (*dolos*), after he made an unbreakable chain-link net around the bed on which Mars was lying with his wife, Aphrodite [Od., VIII]. Odysseus defeated the Cyclops using deception, not force (*dolo oude biephin*) [Od., X, 408], and cleverness (sophie) proves to be more valuable than inflexibility (*atropie*) [4, p. 55]. The object, the daidalon, devised and constructed by the *demiourgos* (creator) aided by *mêtis*, is a "marvel to the eye", an expression which, for the spectators, initially means amazement mixed with various other emotions and, then, the beginning of aesthetic contemplation [5, p. 85]. Similarly, when Thetis goes to Hephaestus to ask him to make her son's armour, she finds the lame god fixing wheels to tripods, so that they might go of their own selves to the assemblies of the gods, and come back again – "marvels indeed to the eye" [Il., XVIII, 372–376].[4]

Force may often be effective, but *mêtis* is the condition for both conquering and preserving power. Mêtis was a Titan and she held a prominent position in Theogony, although her position became quite marginal, even non-existent, in the religious cult of ancient Greeks. Zeus prevailed in his battle against Kronos and the Titans thanks to Mêtis' cunningness; he then married her (his first marriage), but, fearing that Mêtis would in the end turn against him, he swallowed her. Thus, stability was achieved in the universe, a status quo, an absolute rule, because the now *metieta*, *metioeis* Zeus was able "to conceive in advance all tricks that men, gods or unknown monsters might concoct in the future" [4, p. 354]. Order was not ensured so much thanks to his thunderbolts, but mostly because of Zeus' thorough cunningness. From this point on, *mêtis* would be used to delimit the powers and allocate knowledge among the gods.

Two gods are particularly endowed with *mêtis*: Hephaestus and Athena. The *poluphron* Hephaestus [Il. XXI, 355, 367] is the god of ironwork, who fully controls the power of fire, while his craft is "the product of his wise thought" [8, p. 88]. A daughter of Mêtis and Zeus, Athena is an expert in weaving and carpentry, protector of daidala specialist Phereclus and protector of the crafts of weaving and carpentry in Athens [4, p. 69]. Athena not only devised and delivered the bridle to Bellerophon, not only taught carpenters how to make chariots and how to build ships, she also taught man the craft of reining, i.e. the craft of steering a chariot, and the craft of steering a ship [4, p. 274].

In addition to gods, many human beings also seem to be endowed with *mêtis*: for example, Nestor, the good advisor [Il., XIV, 107: *ameinona metin*], whose reciprocated affinity with Odysseus is based on their mutually acknowledged perspicacity. Kleoboulos calls his daughter *Eumetis*, because of her cleverness in making up and

[4] In Homer, the term *techne* refers both to the creators' (*demiourgoi*) technical expertise – ironsmiths and carpenters – and to the sorcery of Hephaestus or the spells of Proteus, although both imply the same *mêtis* [14, p. 61]. In the Homeric world, the creator's social prestige is greater than that of the classic times' craftsman. Creators are organized in guilds and, due to the competition among them, produce works that cause amazement and admiration. In classic times, *mêtis* inspires other persons, too, e.g. the sophist who spellbinds others by the ploys of *logos* and not anymore by technical intelligence [14, pp. 81–84].

solving riddles. Palamedes, whose manual dexterity (*palamen*) is associated to an extremely prolific spirit (he is considered, inter alia, the inventor of some alphabet letters and games) is the only one who can annihilate Odysseus [5, p. 101]. Daedalus is also a man of *mêtis* [5, pp. 91, 98ff.]: son of Metion or Eupalamus – names that designate a man of *mêtis* and manual dexterity, respectively – and Mêtiadhousa, i.e. she who delights in *mêtis*, he is a master of crafts, a constructor of daidala, the inventor of sculpture and architectural design, an excellent engineer, the constructor of the Labyrinth.

Above them all, however, there stands Odysseus, who excels in all manner of stratagems and subtle cunning (*pantoious te dolous kai medea pukna*) [Il., III, 202] and whom Homer qualifies 81 times using the epithet polymetis (he who has multiple cunningness) [5, p. 87]. Like his protector Athena, who laughs with his deeds as he reaches the shores of Ithaca, Odysseus is at the same time a capable mason and ship helmsman [4, pp. 276–277; 5, pp. 62–63], and a shipbuilder who also built his bed and bridal bedchamber [Od., Books V and XXIII]. Odysseus is the one who will annihilate the Cyclops by means of deception, fooling him using the name of outis (a pun made by Homer referring to *mêtis*), transforming himself successively, according to Homer's metaphor, in a carpenter and an ironsmith when plucking out the Cyclops' eye bulb [Od., IX, 384–394, 8, p. 15].

## 3 Design and Action

Modern authors approach *mêtis* primarily from a "sociological" perspective, i.e. emphasizing the possibilities it offers the weak to encounter stronger "adversaries". In this sense, *mêtis* encompasses the practical, local intelligence which people possess and use, often against central plans emanating from "social engineering" perceptions.[5]

In this paper, we suggest a "psychological" approach, in the sense that *mêtis* covers all cognitive processes that are necessary for man to face adverse or confrontational situations against powerful adversaries in unstable and complex environments, about which there is neither structured knowledge nor the possibility of

[5] M. de Certeau suggests the distinction between the so-called "strategies" – a creation of institutional and ruling structures – and "tactics" (*mêtis*' field *par excellence*), involving the individuals' efforts to ensure themselves a bearable life or to turn to their advantage the rules and products defined by strategies [3]. In the same spirit, J. Scott focuses on the so-called high modernism in particular, as captured in large-scale projects, especially in the 20th century – from Lenin to Le Corbusier and to Julius Nyerere; from the 19th century deforestation in Germany, through the Soviet Union collectivisation, to the violent urbanisation in Tanzania and the construction of Brasilia – to underline the advantages of *mêtis* as practical, local knowledge and expertise as compared to knowledge proceeding from bureaucrats and scientists of contemporary large-scale bureaucratic capitalism [10]. At a more regulatory level, but moving in the same direction, J. Stiglitz praises the importance of *mêtis* and tacit knowledge that could increase the efficiency of World Bank policies [12]. At the micro-level of organizations, the concept of *mêtis* is used as opposed to notions that perceive the organization through "representations", i.e. a sense-making activity which defines the context for action [2, 8].

rigorous analysis. In this sense, *mêtis* coincides with the mental mechanisms that allow adaptation in a rapidly changing and complex environment, i.e. human inventiveness and ingenuity. In this context, *mêtis* encompasses, on the one hand, the design of action, i.e. devising a course in order to achieve a specific, delimited goal, and, on the other hand, the implementation of the plan prepared.

The starting point for this approach is the dual nature of human (material) artifacts, as defined by H. Simon. On the one hand, artifacts are objects that obey natural laws, just like all natural objects. On the other hand, they are purposeful objects, in the sense that they correspond to the goals and pursuits of man, who acts upon world with a purpose – an action explained in terms of reasons rather than causes. It is this relationship between how things ought to be and how things are that renders particularly difficult the development of a science of artifacts and their design.[6] In any case, for an artifact to be characterized as such three conditions must be present: the goal or purpose, the artifact's character, and the environment in which the artifact functions. "Description of an artifice in terms of its organization and functioning – its interface between inner and outer environments – is a major objective of invention and design activity" [11, ch. 1]. By analogy, the assumption for the acting human being is that he represents an adaptive system and, consequently, the cognitive processes that are involved in problem solving can be examined based on the objectives and the environment defined by the problem.[7]

"The cunning (*metioeis*) man ... shall embody the main types of human being in ancient Greek society" (charioteer, politician, fisherman, ironsmith, orator, weaver, navigator, hunter, sophist, carpenter, strategist, physician), while "governance and construction are two types of action ... However, for ancient Greeks, these activities are to a great extent similar" [4, pp. 356, 274]. As can be deduced from these statements and as already mentioned, it is quite often that one of these activities is used to clarify another, for example carpentry may be used to clarify navigation. This fact can be interpreted, firstly, in a narrow sense, as being simply a metaphorical use of the former to better understand the latter; and, secondly, in a larger sense, meaning that all these activities involve design processes governed by the same principles. This second approach, which we advocate here, is also supported by H. Simon [11, ch. 5]:

> Everyone designs who devises courses of actions aimed at changing existing situations into preferred ones. The intellectual activity that produces material artifacts is no different fundamentally from the one that prescribes remedies for a sick patient or the one that devises a new sales plan for a company or a social welfare policy for a state.

---

[6] Plato's opinion on the work of both the craftsman and the Creator (*Demiourgos*), as expressed in *Timaios*, is similar but not identical [9, pp. 78–79]. In addition, it is in step with Aristotle's opinion about *techne* as production of works [1, 1094a].

[7] Cf. "... there are only a few 'intrinsic' characteristics of the inner environment of thinking beings that limit the adaptation of thought to the shape of the problem environment. All else in thinking and problem-solving behaviour is artificial – is learned and is subject to improvement through the invention of improved designs and their storage in memory" [11, ch. 3].

In the course of design, alternative actions that shall lead to the objective to be achieved must be identified, but these are not usually given; they have to be constructed. The general mechanism applied in this case is the means-ends mechanism and the finding of "good enough" actions (*satisficing*). The capacity of every goal-seeking system to achieve its targets depends on establishing correlations, whether simple or complicated, between specific changes in world situations and the specific actions provoking these changes. In this mechanism, where an objective is not operational, it is broken down into operational sub-objectives, whereas actions are broken down into sub-actions.

Design means studying in advance, i.e. devising and evaluating actions. The Homeric human being is fully aware of the fact that action has to be designed; that, like in hunting or fishing, one weaves, plots, constructs a cunning action or a trick (*metin plekein/huphainein, tektainesthai metin*) [4, p. 64]. Thus, Antilochus carefully studied his enterprise in advance, demonstrating prudence and foresight [4, p. 35]. In other instances, like the steering of a ship, decisions have to be made in real time – something that some people might call today (maybe a bit hastily) use of tacit knowledge. Either way, when implementing a plan or in unforeseen situations (e.g. rough sea, hunting), sureness of eye (*eustochia*) and sharpwittedness (*agchinoia*) are required.

As already mentioned, *mêtis* is differentiated from force. It is equally differentiated from blind, random search for solutions. The ambiguous cosmogonic powers Tuche – denoting both change and mobility and favourable outcome and success – and Kairos – denoting the best opportunity – may rule over human affairs, but successful action is based on the art of anticipation and, according to Plato, "bad weather is better faced with the intelligence of steering rather than without it" [4, pp. 255ff.]. Excluding force and the random search for solutions, man is lost in a labyrinth, is deprived of resources, a situation he can escape from by devising *mechanai* (manoeuvres, tricks), and *poroi* (a path, a way out, a trick). In modern terms, *heuristics* are used or devised, i.e. simple empirical rules for solving complex problems, where there is neither structured knowledge nor full information.[8] In describing an action characterised by *mêtis* the following verbs are involved: presume (*tekmairomai*), which suggests tracing a course, with the aid of signs and by keeping one's gaze constantly fixed on the pursued end of the journey; intend (*stochazomai*); and straighten (*ithunein*), used for both the carpenter and the shipbuilder (*epi stath-*

[8] Another modern example is provided by the algorithms for an NP-hard combinatorial optimisation problem. "Brute force" would amount to a "quixotic" enumeration of all solutions; the opposite extreme would be random, "blind" search in the "labyrinth" of the solutions space; heuristic search would correspond to *mêtis*. Cf. also F. Frontisi-Ducroux's observation that Daedalus' Labyrinth is "much more than a construction with a complex design, it is a 'problem to be solved', the demonstration in space of the concept of lack of resources (*aporia*), the unsolved problem" [5, p. 255]. We would also like to note the debate that has arisen among experts in artificial intelligence following the victory of Deep Blue over Kasparov. A great part, if not the majority, of them claimed that Deep Blue, an unquestionable technological achievement, only used, after all, "brute computing force" (according to information provided by IBM engineers, it was able to examine 200 million moves per second), lacking actual intelligence capabilities.

*men ithunein*) and for the ship's helmsman and the charioteer headed to their goal without any swerving [4, pp. 89, 278].

## 4 Conclusion

From the concept of *mêtis*, pervading the entire Ancient Greek world, a multifarious behaviour emerges which, despite some differences, seems to obey a single logic. *Mêtis* encompasses the cognitive functions activated in deliberate action, whether this regards design and construction of (material) artifacts or solving riddles, navigating, hunting and fishing, rhetoric and political action – it is, indeed, interesting to note that behind such a heterogeneous set of activities the Ancient Greek world saw only one moving principle. *Mêtis* is the condition for success when the balance of forces is unfavourable, because the adversary is stronger and/or intelligent, rendering a frontal conflict pointless. Remaining consistently focused on his specific objective, without resorting to brute force (either deliberately or out of weakness) and without leaving anything to chance, man must devise a oblique, conjectural and contingent course of action, that will lead him to the achievement of his goal. His behaviour is determined, firstly, by lack of structured knowledge (a "theory") about the complexity and the speed at which the environment within which he deploys his action is changing; secondly, by his aim, which is not of a strategic, long-term nature, but focuses instead on solving a specific problem.

M. Dètienne and J.-P. Vernant suggest that the banishment of *mêtis* is due to a large extent to choices made in Classical Greece. Should our approach be convincing, i.e. that the ancient Greek *mêtis* encompasses everything covered nowadays by the term "design", and given that no satisfactory design theory has been developed to date, one might add that it is possible that the obvious difficulties involved in a systematic approach of such a multifaceted phenomenon have also had a discouraging effect. The estimate that it is rather improbable to find worthy common properties in so many diverse activities is perhaps not unfounded. Nonetheless, it would be equally reasonable to assume that *mêtis* contains a few, simple cognitive mechanisms, which prove to be very productive and effective in a great variety of complex situations.

## References


1. Aristotle (1993). *Ethica Nicomacheia*, Kaktos Publishers, Athens.
2. Baumard, P. (1999). *Tacit Knowledge in Organizations*, Sage, London.
3. de Certeau, M. (1990). *L'Invention du Quotidien*, 1. Arts de Faire, Gallimard, Paris.
4. Dètienne, M. and Vernant, J.-P. (1993). *Les Ruses de l'Intelligence: La Mètis des Grecs*, Flammarion, Paris, 1974 (Greek edition I. Zacharopoulos Publishers, Athens).
5. Frontisi-Ducroux, F. (2002). *Dédale: Mythologie de l'Artisan en Grèce Ancienne*, Editions La Découverte & Syros, Paris, 2000 (Greek edition, Olkos Publishers, Athens).

6. Homer (2006). *Odyssey*. Modern Greek translation by D.N. Maronitis, Institute for Modern Greek Studies (Manolis Triantafyllidis Foundation), Thessaloniki.
7. Homer (1955). *Iliad*. Modern Greek translation by N. Kazantzakis, I.Th. Kakridis. Athens.
8. Letiche, H. and Statler, M. (2005). Evoking metis: Questioning the logics of change, responsiveness, meaning and action in organizations, *Culture and Organization* **11**(1), 1–16.
9. Plato (1997). *Timaeus*. Modern Greek translation by V. Kalfas, Polis Publications, Athens.
10. Scott, J. (1999). *Seeing Like a State: How Certain Schemes to Improve the Human Condition Have Failed*, Yale University Press, New Haven.
11. Simon, H. (2006). *The Sciences of the Artificial*, MIT Press.
12. Stiglitz, J. (1999). *Scan Globally, Reinvent Locally: Knowledge Infrastructure and the Localization of Knowledge*, Keynote Address, First Global Development Network Conference, Bonn, Germany.
13. Vallet, O. (1987). Notes bibliographiques: M. Dètienne et J.-P.Vernant. Les ruses de l'intelligence: La mètis des grecs, *Revue Française de Science Politique* **37**(2), 261–263.
14. Vernant, J.-P. (1989). *Mythe et Pensée en Grèce Ancienne*, Editions La Découverte, Paris.

# Interpreting the Representations on the Shield of Achilles

D. Kalligeropoulos and S. Vasileiadou

*Department of Automation, TEI Peiraeus, Greece*

**Abstract.** In Book XVIII of the Iliad, Homer describes the construction of Achilles' shield by Hephaestus, god of technology. This shield is the symbol of Achilles, the central hero of the Homeric epic. At the same time, the shield constitutes a totality, wider than the epic itself. In a poetic and descriptive way, it sums up the whole of human knowledge in the Homeric era. The shield itself is a universal model of the world, human and natural, a model of the very structure of the physical world. In fact, a model of the universe, which deeply affected the perception of nature by the Presocratic philosophers – Anaximander, Anaxagoras, Empedocles, and later on, Plato and Aristotle. Moreover, it is a model of the social structure of the Greek cities, which also influenced classical philosophers. At the same time, the shield constitutes an introduction to the Platonic philosophical method of dichotomy and to the Aristotelian digital analysis of the world into categories. In this way, Homer paves the way to the ensuing developments in science and technology.

## 1 Introduction

The Homeric Iliad is a polemic epic. It is the description of a war, uniting the Greeks and setting a landmark for the beginning of their History. Iliad includes the first known record of the Greeks' cities, of their leaders, gods and heroes. Iliad summarizes the then historical knowledge into the form of total History. And as an epic, it expresses the values and the visions of the Greeks.

It is not accidental that in this epic the theoretical and technical thought of the time is summed up. It is in the same epic that Homer dares to express his own conception of the world and to pave the way to the scientific thought of the physical philosophers of the archaic period.

And as this epic influenced for centuries education, art and philosophy of the Greeks, we may now say that it also influenced their science and technology.

*S.A. Paipetis (ed.), Science and Technology in Homeric Epics,* 443–450.

## 2 The Homeric Shield

In Book XVIII of the Iliad, the so-called *Oplopoiea* (= *weapon manufacture*), Homer presents Hephaestus' workroom, where the coppersmith god constructs weapons for Achilles, the central hero of the epic. Among these weapons, the poet chose the shield to describe thoroughly, transforming it thus into a shield-poem, a shield-symbol of the whole epic. Homer depicts on it the whole world – natural and human, earth, sky and sea, heavenly bodies and fixed stars, the human social relations, cities and activities, and the outer limit of the world – Oceanus.

From the remarkable poetic description of the mythical Achilles' shield two questions arise:

(a) Is there any relation between this description and Homeric technology? In other words, has such a shield really existed? Or, can it be reconstructed?
(b) What does the shield symbolize? Has any scientific interpretation? Which is its relation to the ensuing scientific developments?

## 3 The Shield's Representations

A rational structure is inherent in the poetic description of the Homeric shield. The investigation of this structure leads to the analysis and hierarchy of the partial components. And after that, it follows the classification of the components into categories and species.

The description begins with the physical world. The Earth is placed in the middle of the world, following the geocentric conception of the universe. Around the Earth the perceptible elements of the world are placed, i.e., the Sky, the Sea, the Sun, and the Moon, as well as the heavenly vault with the fixed stars, i.e., Polar Bear, Orion, Pleiades, and the Hyads.

The description goes on with human activities and creations. First of all, it reports the social institution of human Republic. It distinguishes two types of human cities: the city in peace with a marriage, market, and law court, and the city in war with armies, siege, and battle.

The description continues with economic structures and human works. Firstly, it refers to agricultural works, such as ploughing, reaping, and grape-harvest. Stock breeding, e.g. the pasture of bulls and sheep, appear next. Dance and the rest of hard work appear last.

The whole description is completed with the return to the boundary of the natural world. The outermost frontier of the world is conceived as a great river, the unlimited sea, the Oceanus.

This analytical description cites the world's elements, both natural and human, in a linear form, i.e. the one by the other. However, it is possible that these elements be placed in a rectangular or even better in a cyclic form – suiting also to the form of the shield.

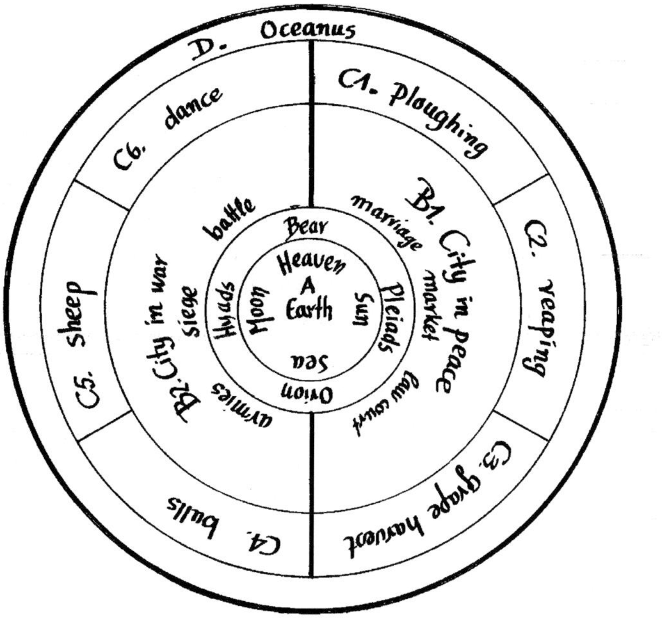


**Fig. 1** The land-planning representation of the shield of Achilles.

In such a polar diagram, the cyclical shield would have Earth in the middle, in a central circle the perceptible natural world, in an outer ring the heavenly vault, and with the general categories of human cities and activities following in distinct rings. The partial categories would be placed in circular form into the rings. And all these would be surrounded by the outermost ring, i.e. the end limit of the world, Oceanus.

This polar earth-planning of Homeric shield's elements allows for the transfer from the poetic description to the holistic representation, to synthesis, to the design and construction of the shield.

Such a subjective representation of the shield is the one we cite here. Even if such a shield never existed, it could at least be designed and constructed.

## 4 Interpreting Shield's Representations

A new question now arises: What was, and what did symbolize this shield? And what was its influence on the scientific theories to follow?

First of all, the shield is the framework where the epic of Iliad takes place. And though shield's description is part of the Homeric epic, the epic itself is enclosed in the shield, in the category referring to war and city siege.

By means of this shield, Homer summarizes the existing or under formulation theory on the structure of the natural world and human society. Consequently, it is

**Fig. 2** A complete representation of Achilles' shield.

a model of the whole world – natural and human. Besides, the Homeric mythical conception allows the combination of natural and human world into a united whole, expressing coexistence, interaction and analogy between natural and human world.

On the other hand, by separating these two worlds, two representative models arise: the synthetic Homeric model of world's structure and the corresponding synthetic model of human social and economic life. In the former, a globe map appears for the first time, with the earth in the middle and the Ocean around it. The morphology of the earth is represented with the sky above the earth and the sea beneath it. A notion of the planetary system is indicated, where the sun and the moon turn around the earth and the fixed stars of the heavenly vault are placed at their physical positions, i.e., the Polar Bear in the north, Orion in the south, Pleiades in the east, and Hyads in the west.

In the latter, the human life model, the city, is located at the central position of the new social structure. Its functions are distributed harmonically in peaceful and

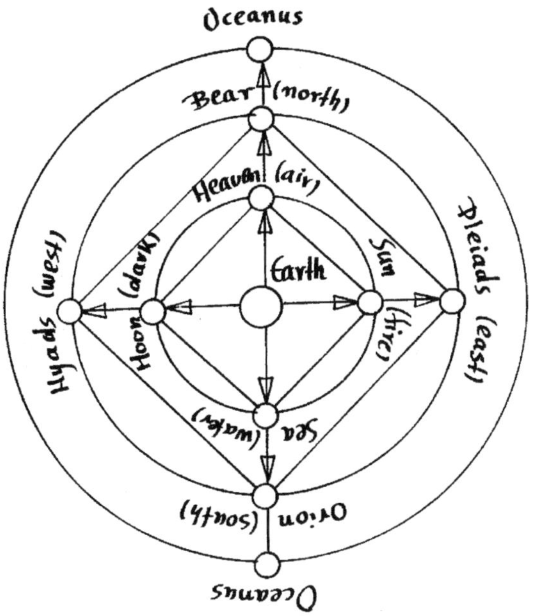


**Fig. 3** Synthetic model of the natural world according to the Homeric description.

polemic periods. Social institutions as marriage, market, and justice emerge. The polemic activities are sorted: army, siege, battle. The repetitive economic activities of the year are placed circularly, in the form of a peculiar calendar that starts with ploughing and ends with dance and the rest.

Both of these models have a crucial effect on after-Homer philosophers and thinkers. On the philosophers who tried to formulate models of the form and the structure of the universe, as well as on those who tried to analyze the structure of the human city and society. Let us examine the first of them.

Anaximander (611–546 BC) represents a model of the universe that has earth in the centre, surrounded by Oceanus and in the centre of the earth is the Mediterranean Sea, "Μεσόγειος" in Greek, meaning "in the middle of the earth" ("στο μέσον της γης").

Anaxagoras (500–428 BC) considers that in the center of the universe is Mind – "Νους" and introduces the idea of opposite qualities, inherent in a unity but separated because of Mind's rotation. In this way, the dense is separated from the rare, the hot from the cold, the bright from the dark and the dry from the moist. Then cold, dense, dark, and moist come together and solidify in the form of Earth, whereas hot, rare, bright, and dry become Aether or Heaven, encircling Earth.

Empedocles (495–435 BC) is who defines all four fundamental elements, earth, water, air and fire, as constituting the basic components of the world.

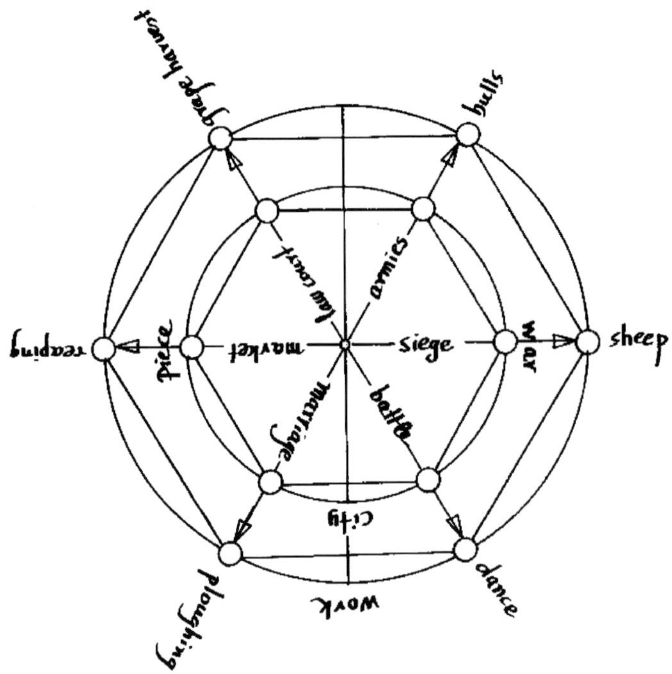


**Fig. 4** Synthetic model of human social and economic life according to Homeric descriptions.

The planetary system of Plato has the earth at the centre. Around the earth, different whirls carry the planets and their rims determine the orbits of Moon, Sun, Venus and Mercury.

The universe, according to Aristotle, has also earth at the centre. The terrestrial sphere around the earth consists of water, air and fire. The rotation of moon takes place into the area of the terrestrial sphere, whereas, outside it, planets rotate in circular orbits. The celestial sphere of the fixed stars surrounds everything.

In this way, the Homeric conception of the world evolved into a structured astronomical model of the universe until the classical years and afterwards.

It is worth mentioning that the analysis of the structure of the Homeric shield into categories and species refers to the Platonic method of dichotomy. Based on the principle that "the one is divided into two", Plato created a genealogical tree, digital or binary, in modern terms, for every general category of ideas. By means of binary decisions, someone goes from the general category (genus) to the partial subcategories and finally to the species that constitute the primary idea.

Therefore, Homer, starting from the whole, the shield, goes to the binary *nature-man*. In the sequence, he divides nature to the binary earth-world, the world to the binary *perceptible world-celestial vault* and so on. He ends up to all of the humans and the naturals, the species, which constitute the whole. In this way, Homer introduces the concept of logic, as well as the process of creating new ideas and concepts.

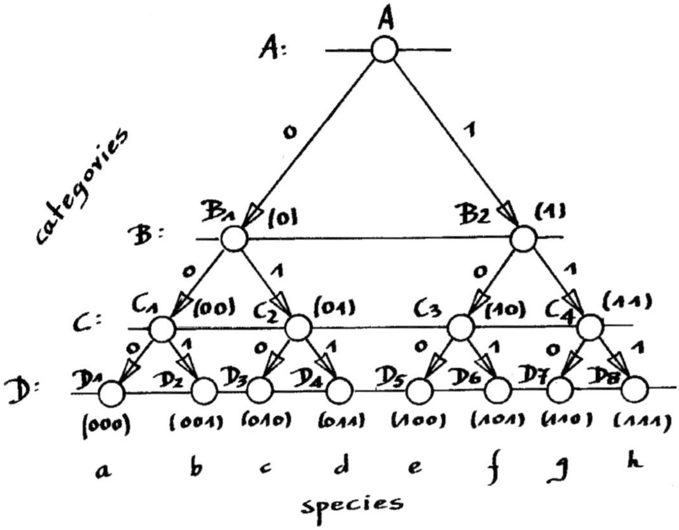


**Fig. 5** The genealogical tree of the method of dichotomy, starting from the primary category A (genus) end ending to the final category D (species).

# 5 Summary and Conclusion

- In the Iliad Homer, from a poetic viewpoint, tries to present his own conception for human and natural things. To this end, he chooses Achilles' shield manufactured by Hephaestus, god of technology.
- Concerning the content of Achilles' shield, it could be considered as a model of the world, both human and natural. The natural world, with the earth at the center, consists of the primary elements later introduced on by Empedocles. Moreover, the structure of the natural world is similar to the structure later introduced by Presocratic and classical philosophers.
- The human world, with Man at the center, is split into the social institution of the city and the economic institution of the work.
- The Homeric method approaches the method of dichotomy, by which Plato analyses primary ideas-categories into their species.
- The shield's description includes elements and notions of science and technology, whose development took place in the centuries to follow.

# References

1. Kalligeropoulos, D. and Vasileiadou, S., *History of Technology and of Automata*, Synchroni Ekdotiki, Athens, 2005 [in Greek].

2. Kalligeropoulos, D., *Myth and History of Ancient Greek Technology and of Automata*, Vol. A, Technology in Ancient Greek Myths – Mythological Automata, Kastaniotis, Athens, 1999 [in Greek].
3. Vasileiadou, S., *Evolution of System Modelling and Control Concepts in Ancient Greece*, PhD Thesis, City University London, 2002.
4. Homer, *Iliad*, transl. Nikos Kazantzakis and Ioannis Kakridis, Estia, Athens, 1985 [in Greek].
5. Homer, *Odyssey*, transl. Nikos Kazantzakis and Ioannis Kakridis, Estia, Athens, 1986 [in Greek].

# Homer and the So-Called Homeric Questions

A. Tziropoulou-Efstathiou

*"Helleniki Agoghi", School of Ancient Greek, Athens, Greece*

**Abstract.** Euclous, the very ancient soothsayer, also called Empyrivitis, had predicted that Homer "will be immortal and ageless for ever ...". Indeed, humanity never ceased dealing with Homer. Even nowadays, in our materialistic society, Homer becomes more and more the object of studies, research, discussions, conferences, and symposia. From Japan to Scandinavia or Australia, Homeric poems editions succeed one another in ever increasing numbers and selling indices. For many centuries a great deal of researchers and commentators are dealing not only with the translation and commentary of the Iliad and Odyssey but also with the various so called Homeric questions, namely: Has Homer ever existed? What does the word Homer mean? Was "Homer" one or more persons hidden behind this name? Were there two poets, one who composed the Iliad and another who composed Odyssey or the same person composed both? If Homer was a real person, when did he live? Where was he born? Was Homer blind? Have Homer epic poems been written or have they been disseminated verbally? What is the Homeric Ithaca? Did Ulysses ever sail out to the ocean? We shall try to answer these questions briefly, based on ancient Greek literature, ancient commentators and of course on specific references of the Homeric poems themselves.

## 1 Introduction

Euklous,[1] the very ancient soothsayer, among many other prophecies, had foreseen the military campaign of the Persians against Greece, as well as the imperishable glory of Homer, who "... alone and first, will sing the 'evils' of broader Greece". Indeed, it was a great evil for Greece, the devastating civil Trojan war, between the Helladites (main land Greeks), Achaians, Danates, Argiates, in short, All Greeks, and the Trojans-Darnites of Greek origin from Peloponnesus, since "... and the nation of the Trojans is Greek once set forward from Peloponnesus" (Rom. Arch. $\alpha'$ 61 a. f.). The prophecy concludes by foretelling that: "... he will become immortal and ageless for ever".

Indeed, humanity has never stopped researching Homer. Even today, in our clearly materialistic era, all the more Homer becomes the object of study, discus-

[1] Euclous, an ancient Cyprian bard and soothsayer, who, according to Pausanias (x. 12. §6, 14. §3, 24), lived before the time of Homer, who, as he predicted, was to spring from Cyprus. Pausanias quotes some lines professing to be the bard's prophecy on this event.

*S.A. Paipetis (ed.), Science and Technology in Homeric Epics,* 451–467.

sions, research, quest and admiration. For centuries now, a multitude of researchers and scholars are concerned with not only the interpretation and the annotation of the verses of the Iliad and the Odyssey, but also with several so-called "Homeric Questions":

- Did a poet Homer really exist?
- What does his name mean?
- Could it be that there was not "one" Homer, but several poets who took cover under the name Homer?
- Could there be two poets, of which one composed the Iliad and the other composed the Odyssey?
- Finally, if there really was a poet Homer, when did he live?
- Where was he born?
- Was he blind or not?
- Where his epics written or oral?
- What is the Homeric Ithaca?
- Did Ulysses sail to the ocean?

To the most basic question, *whether there was a poet Homer*, the answer is categorical: *yes, Homer did exist*. All ancient writers refer to Homer without any trace of doubt as to his very existence. Most of the times they simply call him "the poet" and not "poet Homer", attributing to him the due honor and the indisputable primacy: "The foremost of poets Homer" (Diod. 37.1)

And they cite Homeric verses in their works; "as the poet say ...". In the Platonic Dialogues alone, 175 Homeric quotations have been counted. And although Plato, "exiles" poets from his ideal Republic, still he writes for Homer: "truly this poet has educated Greece" (Republic 606). Aristotle (Poetics) also writes about Homer: "he is worthy of much praise". And Heraclites: "He is the wisest of all Greeks". Euripides calls him "excellent and most divine", while Aeschylus states that his tragedies are but the breadcrumbs of Homer's table. And Aristophanes: "the divine Homer morally tutored classes, virtues, shields of men" (Frogs 1035). Pindar calls him "sweet talking". Athenaeus "the king of poets".

Also Strabo, a methodical and precisionist geographer, states that he attributes utmost importance to the information rendered by Homer for "the transpired Ilion war", as well as for the different locations: "... Homer is the master of geographical experience, ... in all of inhabited land and sea ..." (A, 2).

Strabo, just as many others, also provides with information of the poet been buried in Ios Island. Historian Plinius also places Homer's grave in Ios "... Ios Homeri sepulcrum veneranda ...". Souedas refers this extant sepulchral: "Here the earth covers the holy caput of divine Homer, dignifier of hero men".

Even the top-most historian Thucydides (3.104) cites Homer's testimonies: "Homer indeed states that these are ...". Conclusively, "there is no one who has not heard of Homer".

From these few excerpts, one can draw the conclusion that, in ancient times, no one ever doubted the existence – birth and demise – of Homer. The "doubts" on this subject were fabricated in later times, as part of a wider effort, with more or less known, more or less dubious aims. Still, by crossing-out Homer and placing "two or more epic poets" in his place, they in fact admit that several poets equal to Homer existed, since the *Iliad* and the *Odyssey* maintain their undiminished value.

In recent years, from different directions, efforts have been made to alter his nationality. The Turks project him as part of their "cultural heritage". These are known ... Even Vuk Draskovic, a Serb politician, in his book *The Russian Consul* (Euroekdhōtiki Publications) writes: "... the epics of Iliad and Odyssey were written in a language resembling very much Serbian ... should we naturalize Homer as a Serb?"

Even in ancient times, many Greek cities disputed, strife, each one of them claiming the honor to be Homer's birth place. The known ancient Epigram mentions: *Seven cities argue for Homer's wise root, Smyrna, Chios, Colophon, Ithaca, Pylos, Argos, Athens.*

To *Ithaca* and *Pylos* we will refer in the sequence (as well as to Argos, also located in Peloponnese, as Pylos is). We should notice that Athens was simply the birthplace of Pronapides,[2] i.e. Homer's teacher. This is why, in the *Scholia* to Aeleon Aristides (181, 18-TLG) we read:

> And the poet uses attic dialect; even from the first verse he says Ριηληιάδεω (i.e. son of Peleus). This is Attic, just as is Μενέλεω (i.e. Menelaus). At most parts he uses attic dialect.

Smyrna and Colophon are excluded, "from where Homer is thought to be" (comments as above). Homer was not born in Ionia because, as will be confirmed further down, he lived prior to its colonization. As for Chios, G. Mistriotis in the *Lectures about Homer* writes that Homer was not born in Chios, but simply "resided in Chios", obviously for a short time. This is exactly why Aristotle points out: "All people honor the wise ones ... and Chios honors Homer, who is *not* (its) *citizen*" (Rhetorics 1398α).

Proclus (Chrestomathy B) writes on the subject: "It is not easy to declare who Homer's parents were or what his birth place was, as neither did he refer to those, nor did the historians studying him agree upon it" (Stray Papyrus 3).

The erroneous assumption that the Iliad and the Odyssey are creations of two or more lesser epic poets, has been definitively rejected. Computer studies have proved that these texts have been written by one and the same person. Just as Aristotle has stated: "Homer ... has composed the Odyssey, as well as the Iliad" (Poetics 1451α).

Already, since Alexandrian times, the so-called "separatists" existed, such as Xenon and Hellanicus. However, the majority of scholars and commentators considered both epics as works of Homer. The most significant supporter of this view was Aristophanes of Byzantium.

Mistriotis (*About Homer and Homeric Epics*, p. 99) states:

> In general, it is believed that the Iliad was created first and the Odyssey later. The poet, in his primes, created the saga of Achilles, the brave hero and, in his old age, that of wise Ulysses.

---

[2] "Pronapidis the professor of Homer" (Galin. Allil. Isiodus Theog. 364.28)

## 2 What Does "Homer" Mean?

Those insisting on the theory that no poet named Homer existed, invoke his name: Omiros, from, ομού + αραρίσκω, meaning a person who has assembled several rhapsodies created by others into one single work ["ομηρεύσαι, παρά το άρω = αρμόζω ταις φωναίς αλλήλαις ομοή αραρυίαις"].

Still the question arises: why reject the idea that the same poet arranged the individual chapters of the whole work? G. Mistriotis notes: "If one seriously reads the Odyssey in depth, he will find links everywhere, which connect the different parts of the legend" (*History of the Homeric Epics*).

Just exactly as Aristotle sums up in his *Poetics*:"... from numerous acts ... these poems in an *excellent manner* he assembled ...". Accordingly, Homer, with his unrivalled skill, was the one who composed, assembled "the different parts of the myth" into the two unequalled Epics.

He also assembled all the values of Hellenism: "virtues, shields of men", heroism, sacrifice, friendship, spousal love, eros and longing for one's country, wisdom and every knowledge: Martial arts: "he was the first to write about tactical theories", Geography: "Homer was knowledgeable of Geography" (*Ecateus*, Testimonies Frg. 11B), Astronomy, Navigation ... As proven from within the verses of the Odyssey (see e.g. 5.296), Meteorology (e.g. 10.190), Medicine, Psychology, History, Rhetorics, Technology ... "just exactly as Homer contrived the creations of Hephaestus to be self movable", and many more, Etymology ... He assembled all the types of the Greek dialects: "Homer the all wise, the sea of words" (Vita. 8).

He bequeathed elements of culture, hospitality, attire, hygiene, customs. Moreover, as already known, Homer's works bear the *seeds of drama*: "Tragedy and her father Homer".

The name *Homer* expresses, signifies, codifies, includes the ability of a great Poet. His name is not real, something very much customary in ancient times. Plato's name, for example, was Aristocles. He was renamed Plato not only due to his broad chest and his wide forehead, but also for the broadness of his thinking ("because of his words' broadness he was so renamed, previously known as Aristocles"). Aristotle's successor in the Lyceum, Tyrtamus, was renamed Theophrastos (God-spoken) "because of his exquisite expression". Stysichoros, the poet, previously called Tissias "was called Stysichoros, as he was the first to form a chorus accompanied by a guitar". Sapfo (Psapfa), by her name, reveals her attributes: (ψαύω, ψάλλω, ψαύστης → ψάλτης), e.g. a Chanter.

Plutarch notes: "It seems that nicknames cover a lot of attributes". And in the *Life of Coriolanus* (ch. 11) he sums up: "The Greeks used to name (somebody) by the way (they) behaved ...".

All persons in Prehistory, or Mythology, bore names indicating their character and actions, since "the myth must be easily recollected". And the most easily recollected is "that by which one is called": Kypselos, Theseus, Hercules, Oedipus, Electra, Ion, Iphigenia, Aegisthus, Epaphus, Ulysses, Helen, Menelaus, Agamemnon, Penelope, Calypso, Circe, Achilles, Ajax, Protesilaus, etc.

As for the tradition concerning Homer's blindness, it is clear that it was incorrect. Mistriotis notes ironically: "They blinded Homer by wrong etymology: he who cannot see!" Of course, he may have lost his sight at his late years. Homer was not blind. The "Bios" by Herodotus writes (30) that "his mother bore Homer not blind but having sight". This is attested by his marvelous, precise and detailed descriptions. The exquisite Homeric pictures of the moving crowds assimilated to the movement of the sea, or the sowed fields when the wind strikes them. The unseen nods of the heroes at crucial moments. The movement of the helmets' tufts, the movements of the fire, the flight of the bee ... The scientific descriptions of the wounds. The references to colours and shadings, from *the wine-coloured sea and the bright yellow dawn*, to the spots on animal skins. Various verbs with thin notional shadings regarding the meaning of the verb *to see*: distinguish, examine, observe, check in fear (... λεύσσω, λάω, παππαίνω, οπιπεύω, οψείω, ανγάζω, δενδίλω, επιλλίζω, παρατηρώ, υπόδρα) and many more.

> Those who declared him blind are not of swift minds. Because what that man had seen, have never been seen by any other human.

G. Mistriotis comments:

> He cannot be considered to suffer from congenital blindness (if he had been blind at all), since he portrays the perceptible world so vividly, which not only requires open eyes, but also high perceptiveness.

The description of Homer's statue, according to an ancient Epigram, concludes as follows: "He supported himself with both hands on a cane, exactly as he used to when he was alive ...".

Therefore, the question arises: When did Homer live? Many ancient writers have dealt with the matter: "Regarding Homer's poetry, his parentage and the time he lived ...".

Modern conventional dating places him around middle 9th century BC, with any earlier possibility dismissed right away. There is a persistent trend to place him in the 9th century BC, i.e. at the historical beginning of Greek civilization.

The "scientific" and also the domineering establishment stands on guard to characterize any controversialist with well known adjectives. The strange thing is that this also occurred in ancient times, according to a testimony of Pausanias, but for different reasons.

Pausanias, referring to the antiquity not only of Homer but also of Hesiod, writes (*Boeotian 30,3*) that although he himself knows that both topmost poets are very ancient, he avoids to elaborate on the subject in order not to have the "others" turn against him, as well as the poetic establishment of his time.

The same occurs with dating of Trojan War, which is placed conventionally around 1280 BC. Still, a persistent and impartial investigator could justly express questions and arguments for all so-called conventional chronologies, by carefully studying the ancient texts.

Eratosthenes (3rd century BC), third director of famous Alexandria Library, in his work *Chronographies*, estimates that Homer lived only a century after the Trojan War at the most.

In Proclus' *Chrestomathy* we read: "The historians around Crates (FR. ED.W.40) place him at the Trojan times".

And in the comments on the Iliad (*Eustathius* 4, 27): "Since Homer's birth, six hundred and twenty two years passed to Xerxes' campaign ..." (i.e. $1102 - 622 = 480$, the year of Xerxes' passing through Greece).

Similarly, in Plutarch's On Homer (Plutarchi de Homero testimonia, Π 3), it is mentioned that:

> Just as there is confusion about his origin, there is also about the time he was born. The historians around Aristarchus say that he was born around the time of the Ionian colonization, therefore he is younger than Heraclids Descent by sixty years ...

(that is: Descent of the Heraclids: $1200 - 60 = 1140$; note that it is testified that the Descent of Heraclids occurred at the time of Tisamenus, son of Orestes and grandson of Agamemnon). And continues: "The historians around Krates say that he was born even before the Heraclids (Dorians) Descent".

Also, Xenophon suspects that Homer is even more ancient, as "he says, the man was born during the Heraklids' Descent".

The extant fragments (Fragmenta) – References to Diodorus Siculus (TLG VII,2) quote: "Diodorus proves that he [Homer] died prior to the Descent of Heraklids".

Indeed, Homer has no knowledge of the Descent of Dorians, that is the Descent (= return) of Hercules' descendants, who were banished by Eurystheus from their ancestral place, formerly Dorian Peloponnese.

Homer has no knowledge of the colonization of Ionia. Also, we must not overlook the fact that Homer does not know the name *Nile* and refers to the river with its very ancient name "Egypt". He ignores place names, such as Thessaly, Peloponnese (calls it Argos), Macedonia, Larissa, Megara, Farsala. And, most important, Homer is not aware of the destruction of Cretan civilization.

He refers to the 100 flourishing Cretan cities in the time of Trojan War. Modern science dates the explosion of Thera volcano, which destroyed Minoan civilization, in 1620 BC. And it seems that ancient people were equally ignorant, since they used to calculate time by generations of heroes, so they estimated that Trojan War took place one generation after the Argonautics Campaign, without an exact dating of the ever ancient Argonauts. Accordingly, it remains unknown when exactly Homer lived, but of course much earlier than 8th century BC, where he is, rather thoughtlessly, placed.

As Vilamowic in his *Introduction to the Iliad* aptly notes,

> even in ancient times, it made an impression that the Homeric heroes did not wear wreaths, something that was necessary at later times in every religious activity and feast.

Still, the habit of the Greeks to put wreaths on their head in every celebration comes from very old. Athenaeus, in the *Deipnosophists* informs us that "Aeschylus in Prometheus Unbound clearly states that, to honor Prometheus, we put the wreath on the head, counteracting that bond".

This is exactly why Athenagoras, Bishop of Paramythia, in his study *Chaos*[3] summarizes:

> It is wrongly believed that poet Homer, who lived only 400 year before Herodotus, as he naively thought, but he preceded Herodotus in time almost by a millennium.

## 3 On Homer's Parents

Information of the names of great Poet's parents are also dark and confused.

> Regarding his parents much dispute exists amongst all. Hellanicus (6) and Cleanthes (592 ARN) on the one hand, say it was Maion, while Eugaeon (F.H.G.II 16) says it was Meles. Callicles says it was Demasagoras, while Democritus of Troezenia says his name was Daemon, a merchant, and there are those (who say), *Telemachus, son of Ulysses.*

Suedas Lexicon[4] only accepts the assumptions regarding Maion or Telemachus of Ulysses: "Homer the poet, was son of Maion or Metias, and, according to others, he was *son of Telemachos and Polykaste*, daughter of Nestor".

As for his mother, we cite the versions of Dublin Papyrus[5]:

> ... as for his mother's name, others say it was Metis, others Kritheis, others Themiste, others Yrnetho, others Calliope the Muse, and some others *Polykasti, daughter of Nestor* ...

We notice that all possible names (both for his father and mother) have special meaning, signifying that "this Poet, so gracious, so joyful, so great" originated from parents wise, sagacious, brilliant, prudent ...

> Maeon (= Μαίων): from maio, mao, mo = desire, request to learn, from which *Moa*, *Mosa*, *Muse*, *mathesis* = learning are derived.
> Metios (= Μήτιος) or Μητίας (= Metias): the wise, the prudent.
> Melis (= Μέλης): relevant to music = melody.
> Daemon (= Δαήμων): experienced, knowledgeable.
> Thameras (= Θαμήρας): dense, condensed, dense = wise.

---

[3] Bishop Athenagoras of Paramythia and Philiates Αθηναγόρας Μητροπολίτης Παραμυθιας και Φιλιατών (1869–1942): Chaos = God (ΧΑΟΣ = ΘΕΟΣ), Eleusis Publ., Athens, 2001.

[4] http://en.wikipedia.org/wiki/Suda#References

[5] Greek Papyri from Dublin, ed. B.C. McGing, Bonn, 1995. (Pap. Texte Abh. XLII). Nos. 1–34. [RH]

Menemachus (= Μενέμαχος): he who persists in the battle, the brave.
Metis (= Μήutauις): wisdom, prudence.
Critheis (= Κρηθηίς): or Crathais, Cratais, the potent.
Themiste (= Θεμίτη): Themis, goddess of justice – themistes = laws.
Yrnitho (= Υρνηθώ): ύραξ (yrax) = mix + νήθω (netho) = twine (from the verb υφαίνω (= yfaino) = to weave – the writer's "pen" = "ύφος" (yfos = style)
Calliope: the Muse of epic poetry. Clearly symbolical.

The common denominator of all these indicative names, remains the couple *Telemachus* and *Polykaste*. And indeed, most Lexicons refer to Polykaste as the wife or Telemachus. Eustathius of Thessaloniki, it is said, he referred to their wedding: "They were married ... and similarly *Telemachus wedded Polykaste*, daughter of Nestor" (*Menandrus*, Epid. 408.27).

It is worth mentioning the prophecy given to Emperor Adrian, when he questioned Pythia[6] about Homer:

> You ask me of the unknown ancestry and fatherland of the immortal bard. He is from Ithaca, *Telemachus is his father, and Epikaste, daughter of Nestor, his mother*, who gave birth, of all mortals, to the most wise of men ...
>
> We must very much believe in these and due to – the value – of both, the one who received the information, as well as the one passing it on, in this way the poet (Homer) in a magnificent way, through his epics, glorified his ancestor (Ulysses).

The romance between the two young people flourished "when Telemachus, driven by Athena, went to Pylos to hear of Ulysses' fate from Nestor". There, he was welcomed, but not with the full formality of hospitality, since: "Polykasti, Nestor's youngest daughter, takes Telemachus and bathes him. And after bath she anoints him with oil, and with a beautiful robe she dresses him ... ".

We are reminded that amongst the seven cities arguing "for the wise root of Homer" were *Ithaca* and *Pylos*. If this is true, the assumption that Homer is almost contemporary to the Trojan War is confirmed, and that the respective facts were passed on to him, by first or at the most second hand, by his grandfather Ulysses and his grandfather Nestor, or by his father Telemachus. (Note that Pan. Kavadias detects "a great similarity between archaeological findings in Cephalonia and Pylos".)

In *Comments Vetera on Homer's Odyssey* (93.464.1) we read: "bathing someone is not a virgin's work".

The fact that Polykasti herself bathes Telemachus, seems to be even stranger, if we compare it to the respective scene with shipwrecked Ulysses to be handed over by Nausicaa to her slaves to be bathed. Still, Ulysses declines, as he is embarrassed to appear naked in front of the girls (even though they are slaves and not the princess Nausicaa herself) and tells them: "Slaves, I will bath and anoint myself ... because I am ashamed to bare myself in front of girls".

[6] http://www.answers.com/topic/delphic-oracle

After Pylos, when Telemachus and the son of Nestor visited Menelaus in Sparta, they were led to "bath tabs" beautifully sculptured and the slave girls bathed them, anointed them with oil, and dressed them in robes.

In Circe's palace, although Ulysses had already been in bed with her, again, the slave girls are the ones to wash and anoint Ulysses' shoulders and then dress him in "beautiful robe and coat".

Therefore, it is apparent that Homer, who writes nothing accidentally, means to say something, to "pass" it to us disguised through the bathing of Telemachus by Polykaste, in a way though, that one must be predisposed to understand.

With Polykaste herself bathing Telemachus and anointing him with oil, and dressing him and admiring him, he places inside the saga (in a hidden, "encoded" way) his "identity", the descent of Homer: "Telemachus the father, and the Nestorian Epikaste – or Polykaste – the mother".

From Homer one must learn

There are other "insinuations", "deliberate brush-strokes" within the two epics, full records of which have been published [4], however, there is no space and time here to elaborate further. Generally, Homer adorns Ulysses and Nestor with most of the virtues, deepest wisdom and the greatest merits. As for the adorning adjectives, Ulysses holds the lion's share ...

Accordingly, one is justified to notice: "The poet ingeniously glorifies his ancestor through the epics ... " i.e. the poet, in an ingenuous manner, has glorified his grandfather through the epics, with all praises.

And something more, unprecedented: goddess Athena compares Ulysses' wisdom to her own:

> we both know what is to our benefit, since *you of all humans*, are the first in will and oration, *and for my wisdom* and great knowledge, *I am reputed in heaven among all immortals.*

In the main and most crucial episodes, resourceful Ulysses is the triumphant protagonist. Again, there is no time and space here to elaborate,

The fact that "the poet ingenuously glorifies his ancestor through the epics", is not the only remarkable point. Equally the poet is rushing to justify Ulysses for any carelessness, negligence or mistake, as this can already be noticed easily in the very first verses of the *Odyssey*. The poet-grandson also diligently conceals all errors and sins of his grandfather, especially those regarding the wise hero Palamedes and Ajax. This is also a very interesting chapter, with notes from ancient writers, but we shall not elaborate.

## 4 Delivery Oral or in Writing?

One must also examine *how were the epics delivered: orally or in writing*? The prevailing dogmatic version is that the Iliad and the Odyssey were passed on from generation to generation orally, until they were recorded in writing in Peisistratos' time (6th century BC). Any other assumption is rejected.

However, true science is based on research and not on dogmas. Immovable and rigid dogmas are unscientific, to put it mildly. For example, 60 years ago, encyclopedias wrote that the atom cannot be split, and yet it was split. that the Mycenaean-Minoan script expresses a language of Semitic origin, yet in 1952 M. Ventris proved that the language was Greek. Professor J. Highet, in his book *The Classical Tradition* [1] writes: "... such a poem as the Iliad, is impossible to be delivered without being written ...", while Schadewaldt [3] notes:

> The Iliad, as all great epics, was a product of a later epic era, rich in possibilities, when script was definitely known in Greece ... from the moment that the dogma for the delayed use of writing ceased to apply ...

The dogma of the "oral" delivery of the Homeric epics is partially based on the known Epigram referring to Peisistratos, the ruler of Athens, stating that: Peisistratus – the one who had put together Homer, who previously was sporadically sung.

In fact, this "proof" means nothing. The Epigram mentions "put together" and not "recorded", meaning to collect, to gather. Indeed, this is what Peisistratus did. He sent Onomakriton, the poet, to collect the scattered written rhapsodies of Homer. Souedas is very clear:

> The poet Homer, *wrote* the Iliad not as a whole, neither in the sequence as it is, but every rhapsody was written and presented during his tours, upon departure from each city, he gave the written text of the rhapsodies, for meals, as a hospitality gift. Later it was composed and compiled by many, especially by Peisistratus, the tyrant of Athens.

This is also confirmed by meticulous Cicero, who in his work *De Oratore* (111,137) writes:

> Pisistratus, primus Homeri *libros*, confusos ante, sic disposuisse ...ut nunc habemus,
>
> Peisistratos first compiled – arranged – the *books* of Homer, which were previously muddled, and now we have them.

Cicero, precise, meticulous and knowledgeable, who came to Greece to study and perfect his rhetoric skills and the art of speaking literally, by conviction uses the word *libros* (= books) of Homer, and not odes, epics, songs. He knows very well that Peisistratos did nothing but collecting the *scattered but already written Rhapsodies*.

There were other publications of the Homeric epics that preceded those of Peisistratos. Diogenis Laertius, for example, in his *Life of Solon* (quot. 57) notes: "It was rather Solon than Peisistratus, who cast light on Homer".

There is another relevant testimony of Plutarch, who in *Life of Lycurgus* (4) writes:

> Lycurgus, sailed towards Asia ... there he was, met with the poems of Homer for the first time, as it seems, maintained by the descendents of Creophylus ...
>
> ... Lycurgus of Lacedaemon was the first to bring the entire poetry of Homer to Greece.

G. Mistriotis writes (*Lecture about Homer*, p. 111):

> The availability of a script is supported by the inscriptions quoted by Herodotus, as well as the (inscription) discovered in Nubia, which was engraved by Greek mercenaries at the time of Psammetichus (the Pharaoh). Based on all that, it was given to us, that Lycurgus who got Homeric epics from the descendents of Creophylus, brought them to Sparta. If writing was unknown, how did he bring them?

The deceitful information regarding the recording of the Epics for the first time by the envoys of Peisistratos, was given by the Jewish writer (a completely untrustworthy historian) Joseph Ben Mathia, known as Flavius Josephus, in his effort to project the antiquity of his nation, as he himself confesses: "... I taught everyone of our antiquity" (Apinos A-3) ... Ben Mathia is dominated by the obsession to "prove" the "superiority" of his race over the Greeks, concerning civilization and culture. So he writes:

> It is true that the Greeks have no older ancient writing than the recording of the Homeric poetry; he seems to be born after the Trojan times; it is also said that he did not leave his poetry in writing, but it was maintained orally through songs, and later on recorded. (Apinos A-15)

The lies and fabrications of Josephus collapse by the contents of an ancient Greek Inscription (Hellenic Anthology 16, 292–293 TLG) signed by the famous poet Anyti of Tegea, also called "the Female Homer". In this Inscription, Anyti expresses her admiration for divine Homer who "wrote", "etched" the two epics, the Iliadian War and Ulysses' repatriation (... γραψάμενος σελίδας ... σελίδεσσιν χαράξας ...).[7]

However, if so, the next question arises: which letters, *what script did Homer use*? Diodorus Siculus (c 67.5) replies: "He used Pelasgian characters of Orpheus and Pronapides, Homer's teacher" ("...χρήσασθαι τοις Πελασγικοίς γράμμασι τον Ορφέα, και Προναπίδην τον Ομήρου διδάσκαλ ον ..."). *And since Pronapides Athenian, teacher of Homer, wrote in Pelasgian script*, e.g. Early Greek script, Homer also used the Pelasgian characters, the most ancient Greek script, probably, co-existing with Linear scripts. Exactly as, "in a 4th cent. BC inscription from Lassithi (Crete), on cinder stone, alphabetic and linear scripts co-exist" (Marinatos,

[7] Note that the word "σελίς" = page (comes from ύλη (= matter = forest, wood) means *wooden board*. Literally, "σελίς" is the "wooden plate on which letters were written by etching". "Γράμματα δέ λέγεται, διά το γραμμαίς καί ξυσμοίς τυπ ούσθαι" (= *letters are called, those that are imprinted by line and etching*.)

1958). The idea that Homer used syllabic scripture must be rejected, since dactylic hexameter cannot be recorded in Linear B script:

'Ανδρα μοι | έννεπε | Μούσα πο | ός μάλα | πολλά πλάγχθη ...

In Linear script:

Α ΝΑ ΔΑ ΡΑ ΜΟ Ε ΝΕ ΠΕ ΜΟ ΣΑ

ΠΟ ΛΥ ΤΟ ΡΟ ΠΟ Ο ΜΑ ΛΑ ΠΟ Λa

ΠΑ ΛΑ ΓΕ ΧΕ ΘΕ ...(!)

Already since the Trojan War (at least), there was an alphabetic script, as can be deduced from the collection of testimonies about the History Index, registered in TGL "Ibycus" (1a, 49T.1): "*A history Index, wrote a Gazette of Trojan decor*. He wrote about the abduction of Helen and about Menelaus and the entire Iliadic case ... etc. etc.".

The words "γράφω = *graphō*, to write" and "γράμμα = *gramma*, letter" are sound-made and very ancient words. If there were no "script", then the word that defined it in the first place would not exist either. Homer uses the verb "write" literally, referring to the known letter of Proitus: "writing on a folded board (ἐν πίνακι πτυκτῷ)". Dionysius of Alicarnassus (About comp. 14) explains: "... elements and letters – and letters means that they are lines ...". They named them letters because they are revealed, acknowledged in lines.

And the ability to know, to understand and to read those lines was exactly the ability to "γράμματα επίστασθαι" (= to know the letters well) and γράμματα φθέγγεσθαι" (= to utter the letters). And Homer "knew the letters well".[8]

## 5 Ithaca

Which is *Homeric Ithaca*? Is it really today's homonymous island? Others consider the island of Leukas to be the Homeric Ithaca (amongst them and the German Dörpfeld), others believe it to be Cephalonia, while others speak of an island that disappeared, and some look for Ithaca beyond the Greek seas, see for example Pantazis [2]: "Often, in these efforts of locating, the motive of projecting the native land of the writer has played its own, not insignificant role".

There is also still vague information that during the Middle Ages there were maps showing today's Ithaca without the homonymous channel, literally "attached" to the opposite shore of Cephalonia as an inseparable part of it before the rift that broke it away from the body of the island of Cephalus.

Akousilaus of Boeotia clarifies that "Ithacus and Neritus, from the lineage of Zeus, dwelled in Cephalonia" (ex. 43). After the 11th century, the Italians used to call today's Ithaca *Anticephalonia* or *Cephalonia piccola* (= small Cephalonia).

[8] To know letters = to be knowledgeable, well educated.

Let us state our view right away. Ithaca is obviously a symbol. To determine its exact meaning, an etymological analysis of the name will be attempted: ίθι + άγε → Ιθάγη → Ιθάκη (the middle palatal γ, after the aspirated dental θ, is transformed into the corresponding fine κ (at this point, as always, the musicality of language prevails) and *Ithagi* becomes *Ithaki*, i.e. be off and go, depart to return.

Everyone has his own *Ithaca*, *the place* where one must return, repatriate, to enjoy the *tasteful day* (= the day of return, the sweetest in the life of the wandering Greek hero) and not to remain without return. "Ithaca" is the place where everyone starts from and where everyone must return to. "... I always long to return home and my eyes to see the day of return" (Od. 5.219), "I have never seen anywhere sweeter than my homeland in the world" (Od. 10.27). As stated, everyone has his own Ithaca – where he must return. "If death should find him in a foreign land", his companions must arrange for the repatriation of his relics, see Il. 8.335: "to burn the dead ... so that each of us shall bring back their relics to their children when we return to our fatherland". "ίθι άγε" (go and return) ... applies both to the living and to the dead. Everyone to his *Ithaca*.[9]

## 6 Ulysses' "Ithaca" Is Not Lefkas But Cephalonia

During the time of Ulysses, Lefkas was not an island. It is known that

> ... during the Homeric era, Lefkas was jointed with Acarnania by an isthmus and probably belonged to the mainland, as it was not an island.

However, Homer, "the leader of geographical experience", according to Strabo, obviously calls his native land an *island*, as in Od. 13.95, and defines it as "αμφίαλον" (amphialon), the adjective meaning "surrounded by the sea".

In this *Homeric Catalogue* (characterized as "the whole truth"), meticulous Homer officially records (Od. 13.631): "... Ulysses was leading the Cephalonians *the generous* ...". Also, his father "Laertes ruled the Cephalonians" (Od. 24.378). He was a King, not of symbolic "Ithaca", but of very much real Cephalonia. Besides, Strabo mentions that "Σάμον δε την νυν Κεφαλληνίαν ... Ου γάρ ευκρινώς αποδίδωσιν ο ποιητής, ούτε της Κεφαλληνίας ούτε περί της Ιθάκης και των άλλων πλησίον τόπων, ώστε και οι εξηγούμενοι διαφέρονται (= έχουν διαφορές) και οι ιστορούντες" (C. 453).

However, at the gathering of the Greeks in Aulis, where precise record was necessary, the poet mentions that *Ulysses was leader of Cephalonians* (not of Ithacans). Just as leader of Athenians was Menestheus (v. 552), Thoas of Aetolians (638), Idomeneus of Cretans (645), Tlepolemus the young leader of Rhodeans (654) etc. We notice that, in this historical enumeration, *Ulysses is explicitly mentioned as the leader of Cephalonians*. Exactly as in Apollodorus' *Historical Library*:

---

[9] See also C. Kavafis' *Ithaca*: "Always keep Ithaca in your mind, to arrive there is your destination, ... By now you must have known what Ithacas mean". In addition, see also Bittlestone [5].

And the army gathered in Aulis.
And those who camped at Troy were the following:
Boeotians ... Orchomenians ... Phocaeans ...
......
Menestheus of Athenians ...
Ajax Telamonian of Salaminians ...
Diomedes Tydeus of Argeans ...
Mycenae, Agamemnon son of Atreas and Aeropeae ...
Menelaus son of Atreas of Lacedaemonians ...
Nestor of Pyleans ...
Ulysses of Cephalonians ...
Idomeneus of Cretans ...
Achilles of Myrmidons ...

and so on.

Also, in Book 4 of the Iliad, when, after Menelaus' injury, Agamemnon prepares the troops for war, he finds Athenian Menestheus standing with the Athenians around him (Il. 4.328) and nearby cunning Ulysses stood with the lines of Cephalonians around ... (Il. 4.329).

Also, all tragic poets, refer to Ulysses as the king of the Cephalonians, as for example, Euripides in his Tragedy Cyclops, ex. 92: "Χαίρ' ω ξένε, όστις δ' ει φράσον, πάτραν τε σήν. – Οδυσσεύς: "Ithacos Ulysses, King of the land of Cephalonians". Also, Sophocles in his *Philoctetes*: "the two leaders and the King of Cephalonians" (v. 264), "oh stranger Cephalonian" (v. 790). In a retrieved extract of Hellanicus (see *TLG Hellanicus Frg*) we read: "Cephalonians under Ulysses and Laertes, while the island is called Taphos by Mentes and the Cephalonians Taphians".

It is necessary to refer to the well known verses of the Odyssey (Od. 10.19) where Ulysses himself, while revealing to Alcinous – King of Phaeacians – his true identity, he describes his island, his "Ithaca", as follows: "This (my island) lies in the sea at a low place, at the western extremity, while others lie further towards the East and the Sun". These verses are practically "photographing" Cephalonia.

In the Comments on the Odyssey (*TLG Scholia vetera q.21*) we read:

> Ithaca: Od. 10.25 – is called most supreme towards darkness, that is most western of all, and this does not agree with the position of Ithaca ...

Note that Ulysses, coming from the West, perceives the boundaries of his homeland from afar (Od. 10.28): "Nine days we sailed ceaselessly, night and day, and on the tenth the homeland began to appear".

## 7 Ulysses' Sailing Out to the Ocean

Strabo writes (A, C, 38): "On Menelaus' wandering ... οι μέν περίπλουν των δια Γαδείρων μέχρι της Ινδικής εισάγουσιν ... therefore, neither Ulysses' wandering was impossible". Dealing with Ulysses' exit to the ocean in detail, with his endless, remote wanderings outside Mediterranean, lies beyond the limits of the present work. We simply note two points regarding *Scheria* and the *Herculean Columns*, i.e. Gibraltar straits.

*Scheria*: Wrongly and quite inaccurately the island of Corfu has been considered as Scheria. Even the name itself signifies that it is not an island (Scheria from *hersos* = wasteland and quite a vast place) must be very remote. If Scheria *were* Corfu, Ulysses would not need the Phaeacians to transport him, to carry him home. He himself could build a raft "by cutting and curving wood" as he used to (Od. 5.243), and within a few hours he would have sailed to his "Ithaca", through the sea he knew so well, the one that "separates" and "unites" the two islands. *Homer himself makes clear that Phaeacians have no close neighbors: "no one is near them"* (Od. 6.279). *Moreover, Scheria is constantly mentioned by Homer as "land" and not as island.* From this remote land, Nausicaa promises Ulysses that her father Alcinous, would send him back to his homeland.

Because of all these, and many other strange things (like the details of the preparation for departure by the super-automated ships of the Phaeacians) the ancient commentators remark: "Scheria was the name of the land of Phaeacians, not *Korkyra* (Corfu), which was outside the – land – we dwell" (*Scholia vetera* Od. 6.4). Also "that clearly displaced far away from here the land of Phaeacians exists. It is not Korkyra" (Sch. P.T. 6, 204). Strabo notes: "What is said about the Phaeacians, that we live far away at the wavy seas uttermost, and no mortal comes in touch with us. All these appear to take place in the Atlantic".

*Hercules' Columns*: Diodorus states (D. 18): "Hercules passed to the ocean towards Gaders, and placed one column on each of the continents".

*Telepylus* (Tele-Gate): There are some verses (Od. 10.80–99), having passed unobserved (mainly due to bad translation-interpretation of the text) telling of Telepylus. First, the meaning of Telepylus: τῆλε = (tele) far, πύλη = gate, opening, muzzle, entrance (entrance to a country), θύρα (thyra = door): into the house, πύλη (pyli = gate): into the city (even today, entrance and exits of ports and airports are called *gates*). I believe that the Homeric Telepylus is today's Gibraltar straits, the ancient columns of Bryareus, later named Herculean Columns, Γάδειρα (Gadeira) → Cadiz ....[10]

Now, let us examine *Gadeira*. The name is not foreign. It is Greek and shows exactly what it was before the opening of the straits by Hercules. Γά-δειρα: γα = γη (*ghe* = earth), δειρή ή δειρα (*deira* = neck, as in necklace). Exactly what "isthmus" is. Hercules dug between the continents opening a canal, so that the Atlantic would meet the Mediterranean (Diod. IV, 18, 2.4).

Accordingly, Homer states (Od. 180-99): Ulysses has departed for the island of Aeolus: "Six days we sailed ceaselessly, night and day, and on the seventh we arrived to the steep city of Lamos". *Lamos*: the same kind of association as with *isthmus*

---

[10] Straits (πορ-θμός = por-thmos): a narrow sea canal, separating two lands and joining two seas. From the verb πείρω (peiro) = to pass, to cross the sea. *Isthmus* (ισ-θμός = is-thmos): a narrow strip of land, joining two lands and separating two seas. Isthmus is also the neck of a human (probably by assimilation). That is why a *necklace* is also called *isthmion*.

and *ga-deira* (*laimos* (= neck) – *lamia* = gaps, *Lamos* = son of Neptune). Also, *into the Telepylus the Laestrygonian* [*Lae-s-try-gonia*: *lae* (*laimos* = neck), *s* (*kinesis* = movement), *try* (*trypo* = to drill), *gonia* (from *gignomai* = to become)]. Meaning the neck land that was cut, opened, became by drilling.

> Entering there, one shepherd calls the other shepherd.

*Poimen* = shepherd does not mean just "herdsman". It also means guard, and leader. The verb ελαύνω (*elavno*) = enter is used mainly on the sea, and that is why in the mouth of the people it became *lamno* = to row.

> and the other (shepherd = guard?, navigator?) excited, obeys.

This means that the one who sails in calls, and the other who sails out obeys (the nautical signals).

> There, any man who has no sleep could earn, receive, enjoy, two salaries ...
> those of a shepherd

*boukolos* = herdsman, not used here with the strict meaning of the word, *kello* (= to moor) here metaphorically, *voe* = shout + kello = moor. And all the passages-poroi are "vosporoi" as "passages of shout" (Bosporus).

> As for the other salary, by shepherding, and also collecting *agryfa* (= shining)
> *mela* (=apples)

*Agryfa* = bright, shining (from *argyros* = silver), *mela* = means sheep, goat, ox. Poetically it means a silver coin baring the picture of a sheep, goat, ox, used as passage fare. Therefore, the "brilliant apples = *agryfa mela*" are not "white sheep" as interpreted by several translations of the text, but money, silver coins, commonly "whites". *aspra* (= whites) is money, wealth, property, fortune, cash. Thus, the popular saying "the whites (= *aspra*) bring the stars (= *astra*) down".

> Because the nautical routes are near the night and the day

*kelefthos* = is basically a nautical route (boatswain, command, direct). "It refers to the length of day and the shortness of night, meaning that just as the night begins its way and immediately afterwards the day's way begins ..." (Pandazidis Lexicon). This means that ships and boats went in and out all day and night and there was a great need for guards, fare collectors and navigators.

> There when we arrived at the famous port
> That was stone, rock, steep[11]

*tetychike* = as constructed (of the verb τεύχω (= *teucho*) = to construct), therefore man-made, not natural. *diamberes* = through and through, across the entire width, "through the passage". *amfoterothen* = from both sides, from the front and the back.

[11] See: "the steep rocks of Gibraltar". Also: "mount of the Iberics it is ... with great height and erect ... it stands on the right side of those sailing from our see to the outside ..." (Strabo C, 140).

coasts and piers opposite each other in the entrance stood out.

"And opposite each other stood the Herculean columns".

But the entrance was narrow.
There everyone had their agile ships,
And on both sides there was white, shinny, slightly frothed the sea.
I alone kept my black ship towards the exit
There next to the end
After tying the ropes on a rock
And having climbed at the steep at the look-out post I stood.
And there works of animals or works of men were visible (wilderness ... no cultivation)
And only haze, vapour, fog we could see raising from the ground ...

When later, the savage residents will attack them, Ulysses, as already having tied his ship at the end of the pier, he unties it and *goes out into the ocean* ... "going through an Oceanic passage".

And while the damaged my companions in the port
I pulled the sharp knife from my thigh
and cut the stern ropes of my black ship ...
Joyfully thus escaped the suspended rocks
my ship to the sea; all the rest were gone.
......
*From that place we further sailed ...*

... *full speed ahead*, outside Herculean Columns ... Thus, "Ulysses' outing to the Ocean" began.

## References


1. Highet, J., *The Classical Tradition*, Oxford University Press, 1985.
2. Pantazis, D.E., *Homeric Geography and Homeric Era: Homerization of Ancient Greece and the Mecenaean Problem*, Kastaniotis Publ., Athens, 1996, p. 31 [in Greek].
3. Schadewaldt, W., *Aufsätze und Auslegungen zur homerischen Frage* (From Homer's world and work: The Homeric question), transl. into Greek, Ph. Kakrides, Educational Institution of National Bank of Greece.
4. Tziropoulou-Efstathiou, A., *Homer, Telemachus' Son, Odysseides*, Georgiadis Publ., Athens, 2003 [in Greek].
5. Bittlestone, R., *Odysseus Unbound. The Search for Homer's Ithaca*, Cambridge University Press, 2005.

# Atlantis in Homer and Other Authors Prior to Plato*

Stavros P. Papamarinopoulos

*University of Patras, Greece*

This paper is dedicated to Catherine Foliot, an excellent scholar.

**Abstract.** It is demonstrated that ancient sources, *prior to Plato*, mentioned a *sacred* circular entity somewhere West of Gibraltar. Homer in the 8th, Hesiod in the 7th BC, Pindar and Hellanicus in the 5th centuries respectively also wrote about it. We do not expect to read the name *Atlantis* in their texts, since it was invented by Plato as he clearly declares. We know that in the Euro-African region, west of Gibraltar, both in North West Africa and South West Iberia there are several submerged craters which are products of diapyrism. Some of them in Cadiz's bay are large. At least a small one is visible in Andalusia. We explore the case of ancient Erytheia mentioned in early ancient Greek texts as a vanished red *island*. We propose that in Andalusia's palaeocoast of 1200 BC, a nature-made *circularity* existed. In the past, it served once as a religious, cultural site for the nearby maritime prehistoric populations. It was associated with Ibero-Mauretian and Tartessian culture. It was understood, due to its circular shape, by the visiting prehistoric Greeks, as Poseidon's work. It vanished together with Achaeans and *Atlantes* after earthquakes and floods. Atlantis' giant *island* is Germano-Celto-Iberia, since the concept of the island in prehistory in either Greek or Egyptian was not that of the 5th century BC, because Herodotus added the word *peninsula* for the first time. The mythological part of Iberia reflects Heracles' activities. In other words, it reflects the collective unconscious of the prehistoric and historic Greeks in that region of the world. We note the first archaeological finding of an Achaean shard close to Guadalquivir's estuary. Plato's *Atlantes* is a remote strange and deformed incomplete echo of Sea Peoples' second assault against East Mediterranean countries. The giant continent West of Gibraltar was known to the Egyptians prior to 6th century BC, because of the trade of unique products existing only in South America. In other words, Plato's story about *Atlantis* begins to be comprehensible and gradually proved.

## 1 Introduction

Many analysts have entirely rejected *Timaeos* and *Critias* as a possible source of useful historical information. They firmly claim that it was Plato himself who invented the *Atlantis* episode. We express a different opinion. We need to re-examine the literature, *prior to Plato*, and demonstrate that there are other sources describing

* The present paper is the author's synthesis based on several papers published in different proceedings of previous international symposia on the subject of Ancient Greek Technology and Atlantis, respectively, organized by the University of Patras at Olympia in 2001, by EMAET in Athens in 2005 and by Heliotopos on the island of Melos in 2005.

*S.A. Paipetis (ed.), Science and Technology in Homeric Epics,* 469–508.

Atlantis circular entity as a nature-made formation. All of these sources describe in a fragmented and symbolic way a religious and cultural, nature-made, *circularity* associated with the rich in metals prehistoric Ibero-Mauretian and Tartessian world, known to the Achaean Greeks even before the end of their period in the 12th century BC.

## 2 Ancient Greek Sources about Atlantis Prior to Plato

We start with Paul Jordan [11], who refers to Hellanicus, a 5th century BC writer, and interprets the few extant excerpts from his texts in an unscientific way. According to him, Hellanicus, who lived some decades prior to Plato, did not mention anything about Plato's Atlantis in his lost book. Jordan [11] believes that Hellanicus described briefly Atlas' daughters, the so-called *Atlantides*. Let us view the text (Fragmenta 1a, 4, F.20.2):

[Ζε]υς μίογεται τῶν [δε γίγνεται Λακεδαίμων]

In HARPOKR.SUID. it is mentioned:

Ὁμηρίδαι˙ ... Ὁμηρίδαι γένος ἐν Χίῳ, ὅπερ Ἀκουσίλαος ἐν γ (2F)

2) Ἑλλάνικος ἐν τῇ Ἀτλαντιάδι ἀπό τοῦ ποιητοῦ φησίν ὠνομάσθαι. Σέλευκος δε ἐν β Περί Βίων (IV) ἁμαρτάνειν φησί Κράτητα νομίζοντα ἐν ταῖς Ἱεροπιοιίαις.

In the above excerpt, Hellanicus is mentioned by his name and so is *Atlantiás*, the title of his lost book. Jordan [11] states that Hellanicus simply mentions Atlantides, meaning the constellation of the Pleiades. Let us carefully examine Hellanicus' main text (Fragmenta 1a, 4, F.19a.15):

συνελθεῖϝ˙ Ταυγέτην Διί ὧν γενέσθαι Λακεδαίμοϝα˙ Μαῖαν Διί, ἀφ' ὧν Ἑρμῆς Ηλέκτραν Διί, ὧν Δάρδανος Ἀλκυόνην Ποσειδῶνι, ὧν 'Υγριούς Στερόπην Ἄρει, ὧν Οἰνόμαος Κελαινώ Ποσειδῶνι και αὐτήν συγγενέσθαι, ὧν Λύκος Μερόπην δὲ Σισύφῳ θνητῷ θνητῷ ὄντι, ὧν Γλαῦκος, διὸ καί ἀμαυράν εἶναι.

Here the names of Atlantides are presented, their associations with the Gods and their earthly localities. For instance, Taygete is associated with Zeus and from this association Lakedaemon is born. Another one, called Kelaino, whose name means the one *who is dark in appearance* or darkish in the skin mates with Poseidon and their erotic offspring, Lykos, comes to life. Poseidon places him in the blessed islands where he becomes immortal (Fragmenta 1a,4,F.19b.4).

λανθάνων ἐν σπῆ τ[ῶν] δὲ γίγνεται Ἑρμ[ῆσ] φιλήτης, ὅτι αὐτῆ φιλησί[μωσ] συνεκοιμ[ᾶτο] καὶ γ[ίγνεται θε]ῶν κῆρυχ ἀγήρπ[αοσ] καὶ ἀθάνατος Κ[ε]λαινοῖ δὲ μίσγαεται Ποσειδέων τῶν δὲ γίγνεται Λύκος, ὅν ὁ πατήρ κατοικίζει ἐν μακάρος νήσοις, καὶ ποιεῖ ἀθάνατον Τηυγέτῃ.

Luce [17] was the first to locate Hellanicus' particular phrase presented below, to have an exceptional similarity to Plato's phrase (firstly that of Hellanicus' and second the one of Plato):

Fragmenta 1a,4,F.19b.4 – 1a,4,F.19b.6
Κελαινοῖ δὲ μίσγεται Ποσειδέων τῶν δὲ γίγνεται Λύκος, ὅν ὁ πατήρ κατοικίζει ἐν μακάρων νήσοις, καὶ ποιεῖ ἀθάνατον.

Poseidon mated with Kelaeno and their son Lykos was settled by his father in the isles of the blest in order to become immortal.

Plato in Critias, wrote something similar to this (Criti 113.d.4–113.d.5).

αὐτῆς δὲ ἐπιθυμίαν Ποσειδῶν ἐλθὼν συμμείγνυται.

and Poseidon, being smitten with desire for her, wedded her.

We note that Plato immediately after this phrase describes the building of Atlantis by Poseidon. According to Hellanicus, Poseidon had intercourse with a woman who is either dark in appearance or her skin is of a darkish type. Then he places the offspring of their love in the islands of the blessed. The islands of the blessed are most likely West of Gibraltar, as most ancient Greek authors suggest. According to Plato's version, Poseidon has intercourse with Cleito. One of his descendants is called Azaes whose skin is darkish, so to speak. Gill [7] interprets Azaes as a *parched, dark-skinned person*. One can now see the double similarity between the two passages. Jordan [11] missed the point entirely, because he saw in that *only* the Atlantide Kelaeno as a star in the constellation of the Pleiades.

Then, in accordance with Plato, Poseidon also builds his palace West of Gibraltar. The common elements between the two versions are the erotic act, which produces a *circle* in the woman's body implying the Earth. The darkish skin in both stories is very relevant in that Euro-African part of the world. In other words, a nature-made circularity in the Earth, situated somewhere west of Gibraltar, interpreted by the prehistoric Greeks as Poseidon's act became the core of Atlantis' case. Unfortunately, Atlantias, Hellanicus' book, is missing.

Let us now consider what was written by Pindar who lived some decades before Plato. Pindar describes another sacred wedding. *Somebody*, whose name is missing, marries a woman, a Nereid, related to the sea as *islander* women in Plato's and Hellanicus' cases related in their respective stories. Pindar says (N 4.65–4.70):

ἔγαμεν ὑψιθρόνων μίαν Νηρεΐδων, εἶδεν δ' εὔκυκλον ἕδραν, τὰν οὐρανοῦ βασιλῆες πόντου – τ' ἐφεζόμενοι δῶρα καὶ κράτος εξέφαναν ἐγγενές. Γαδείρων τὸ πρὸς ζόφον οὐ περατόν ἀπότρεπε αὖτις Εὐρώπαν ποτί χέρσον ἔντεα ϝαός

He married one of the lofty-throned Nereids. He beheld the line *circle of seats*, on which the lords of the sky and sea sat and revealed to him their gifts and his race's power. That which lies to the west of Gadeira cannot be crossed; Turn back again the ship's tackle to the mainland of Europe.

The poet describes a *circular* base where kings of sea and heaven were offering gifts to visitors and speaking to them about their riches. Pindar in the following phrase defines the region of this rich race of kings. It is somewhere round Gadeira. Pindar, however, offers another fragment describing both Poseidon's act and a particular catastrophe. The similarity to Plato's particular phrase in *Timaeos* in connection with Atlantis was spotted by Luce [17]. The textual similarities in both ancient Greek texts are printed in italics (Frg Paian.52d.41–52d.44):

> Διός 'Εννοσίδαν τε βαρ[ύ]κτυπον, χθόνα τοίποτε καὶ στρατόν αθρόον πέμφαν κεραυνῷ τριόδοντί τε ἐς τον βαθύν Τάρταρον
>
> With Zeus and I fear loud-rumbling Earth-shaker. With their thunderbolt and trident they once sent *the land and all the people into deep Tartaros.*

The loss of a land with a whole army on it due to an earthquake is associated with the position of *Tartaros*. The location of Tartaros is most likely in Iberia in where *Tartessos* was also placed. The root tar means west of the dying sun in accordance with Graves [8]. Therefore Tar-tar-os is associated with the West and the repetition of the root tar-tar indicates not simply West but far West from the standpoint of the Greeks. Additionally, Hesiod associates Tartara with the end of the known world. Once in that locality Poseidon acted lethally when he dropped a land and an army into the Earth's depths. This tradition of the loss of an army somewhere in a far west region existed prior to Plato's time and bears striking resemblance to a *Timaeos* passage following below (Tim 25.c.6–25.d.3):

> ὑστέρω δὲ χρόνῳ σεισμῶν ἐξαισίων καὶ κατακλυσμῶν γενομένων, ἡμέρας καὶ νυκτός χαλεπῆς επελθούσης, το τε παρ' ὑμῖν μάχιμοϝ πᾶν ἀθρόον ἔδυ κατά γῆς, ἥ τε Ατλαντίς νῆσος ὡσαύτος κατά τῆς θαλάττης δῦσα ἠφανίσθη
>
> But at a *later time* there occurred portentous earthquakes and floods, and one grievous day and night befell them, *when the whole body of your warriors was swallowed up by the earth*, and the island of Atlantis in the manner was swallowed up the sea and vanished.

What Plato implied with the words *at a later time* was Atlantis' catastrophe after Athens' earthquake. One can see in italics, both in the Greek and the English texts, the similarity of the event involving an army swallowed up by the earth after an earthquake. This means that Pindar describes, as Plato does, an earthquake and the vanishing of an army into a *chasm* produced by it. What is noted here is the similarity between Pindar and Plato. Poseidon's act somewhere in the West and the wording are common elements in both descriptions. Let us come to 7th century BC and examine Hesiod's texts (Fragmenta 372.10–372.11):

> 'Ιόνιον δ' ἀνά κῦμα φέρον Γαδειρόθεν ἄξει Βρέττιος ἢ Καμπανός ἢ ἐξ ἀγαθοίο Τάραντος

Firstly Gadeira is mentioned and then a sea route connecting it with Taras in Italy and the Ionian Sea in Greece. The direction of the sea route is obvious and there is no

**Fig. 1** The distribution of Heracles Pillars (after Zhirov [43]).

confusion whatsoever with Gadeira of the East Mediterranean. This is an answer to attempts to demonstrate that Solon of 6th century BC did not know which pair from the available eight of Heracles Pillars was the one associated with Atlantis' position. According to them, Atlantis had to be sought within the Mediterranean Sea. But, on the contrary, the search must be directed within the Atlantic Ocean. Figure 1 shows the distribution of all pairs of Pillars of Heracles. The pair with number 1 is the one Plato meant. Hesiod was the first to give the answer that the Greeks of 7th century BC knew: Cadiz, west of Gibraltar. There are also publications corroborating the fact that some contact between Achaean Greeks and Iberians existed even before 7th century BC. Martin de La Cruz [20], Mommden et al. [21], and Podzuweit [31] studied Mycenaean pottery found in archaeological excavations in the effluence of Guadalkivir river.

This poor, so far, archaeological evidence illustrates that some sort of trade existed between the Achaean Greeks and the Atlantic Ocean Tartessians, even before the end of their period in 12th century BC in Iberia. On the other hand Hesiod describes what Poseidon's son, Chryssaor, achieved mimicking his father (Th 287–290):

> Χρυσάωρ δ' ἔτεκε τρικέφαλον Γηρυονῆα μιχθεὶς Καλλιρόῃ κουρῃ κλυτοῦ Ὠκεανοῖο· τὸν μὲν ἄρ' ἐξενάριξε βίη Ἡρακληείη βουσί πάρ' εἰλιπόδεσσι περιρρήτῳ εἰν' Ἐρυθείῃ.
>
> But Chryssaor was joined in love to Callirrhoe, the daughter of the glorious Ocean, and begot three-headed Geryones. Him mighty Heracles slew in sea-girt Erytheia.

Poseidon's son, Chryssaor, had an intercourse with the Ocean's daughter Callirrhoe on the land of Erytheia and gave birth to Geryones. Before attempting to interpret what Geryones might be, let us remind, that what Hesiod writes is very similar to what Plato describes. Poseidon in Plato's version is replaced by Chryssaor, that is Poseidon's son, in Hesiod's version. They both do the same as Hellanicus and Pindar describe. It is an erotic act at Erytheia, somewhere west of Gibraltar. The latter is geographically defined by Stesichoros of 7th centure BC to be close to

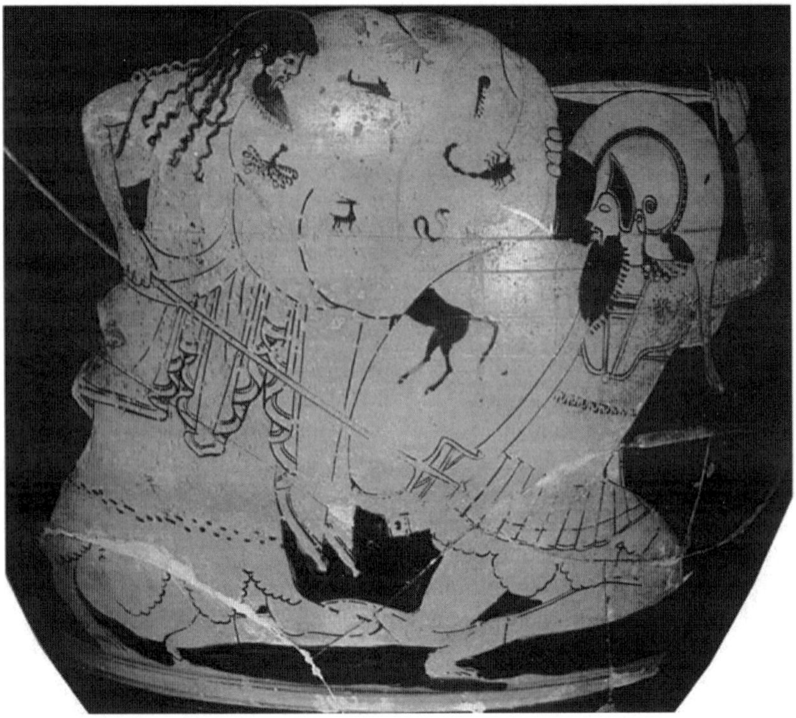

**Fig. 2** Poseidon (left) fights with Polyvotis (right). Poseidon had just grabbed a piece of rock from Kos island and he is ready to hurl it against his *enemy*.

Gadeira. But it is time to leave Hesiod temporarily and show what the impression of the Greeks about Poseidon's act was in producing *circularities* in the Aegean Sea. In Figure 2 Poseidon fights against a warrior *Polybotis*, whose name means *many bulls*. Polyvotis' name reminds us of Geryones' association with bulls in Erytheia region, somewhere west of Gibraltar. Poseidon hits Polyvotis firstly with his trident and then by hurling a huge rock at him. The result of this action is a *circular chasm* in the island of Nisyros, which is the volcanic crater shown in Figure 3. Polyvotis, in other words, is a circular chasm. Is it possible that Geryones could be something like that? In order to explore this question we need to leave the Aegean Sea and go west of Gibraltar.

Let us now come back west of Gibraltar and follow what Poseidon himself normally does, apart from producing earthquakes and inducing floods. Hesiod writes (Th 726–726):

χάλκεος ἄκμων ἐκ γαίης κατιών, δεκάτῃ κ' ἐς τάρταρον ἵκοι. τὸν περὶ χάλκεον ἕρκος ἐλήλαται·

round it runs a fence of bronze.

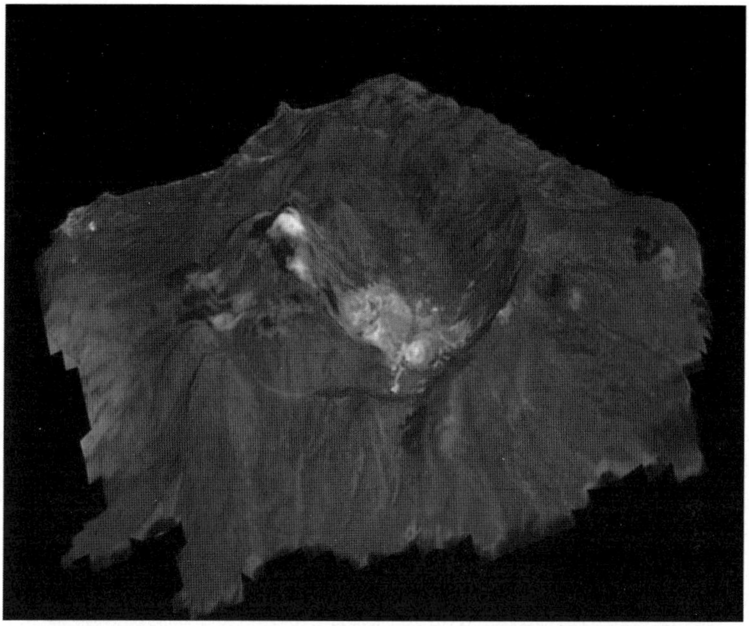

**Fig. 3** A huge rock has already *fallen* upon Polyvotis and created Nisyros' crater. Polyvotis and Geryones remain *buried* in the Aegean Sea and the Atlantic Ocean respectively. Poseidon is responsible as a warrior and a lover respectively. This was Achaeans' common understanding of large circular holes irrespective of their real generating mechanism of our modern science.

This *circular fence of bronze* is a characteristic example of Poseidon's building activity, created around a chasm. Furthermore, Poseidon does again something similar as Hesiod writes (Th 732–733):

> θύρας δ' ἐπέθηκε Ποσειδέων χαλκείας, τεῖχος δ' ἐπελήλαται ἀμφοτέρωθεν.
>
> for Poseidon fixed gates of bronze upon it and a wall runs all round it on every side.

As one may observe, Hesiod keeps insisting on describing Poseidon's capacity, e.g. to build circular walls round Earth's chasms. Before we interpret what Geryones could be, we leave Hesiod and deal with a writer of 1st century AD, who also mentions Geryones. He will assist us to interpret Geryones' nature and define his locality.

## 3 Red Tears

We present here some passages from Philostratus and focus attention on some particular details, connecting Geryones with a region west of Gibraltar suggesting Geryones' nature and his immediate environment. Then we will comprehend what Hesiod describes. The first passage (VA 5.2.6–5.4.1).

> The Ocean is pushed by underwater exhalations coming from the many crevices that are situated both below and around the earth and goes forwards and retreats again as the breath like exhalation dies away. This is corroborated by sick persons in the Gadeira region. During the time when the waters are at their height, souls do not leave the dying, something which would not occur unless the exhalation was coming towards land. As for what is said about the moon's appearance when it is new, full, and waning, I know that effects the Ocean which follows the moon phases by sinking and rising in sympathy with it. 5.3. Day follows night and night follows day in the Celtic region, with the darkness and the light retreating gradually, as here, around Gadeira and through the Pillars they say the alteration strikes the eyes suddenly, like lightning. They also say that the Islands of the Blessed lie at the extremity of Africa, rising near the deserted promontory.

In the above passage, Philostratus clearly presents the region round Gadeira (Celto-Iberia) and connects it with the phenomenon of the tides existing in the marine vicinity of Gadeira. He attempts some explanation of how the tides are caused by involving the appearance of the moon and its effect on Ocean's waters.

Let us study another passage, adding more information of the region west of Gibraltar to which the *Athenians* were surprisingly related. We remind that Plato also associated the prehistoric *Athenians* with *Atlantes*, somewhere west of Gibraltar. Philostratus' further text follows (VA 5.4.9 to VA 5.5.9):

> καὶ μὴν καὶ Ελληνικούς ἒναί φασι τὰ Γάδειρα καὶ παιδεύεσθαι τὸν ἡμεδαπόν τρόποϝ˙ ἀσπάζεσθαι ψοῦν 'Αθηναίους 'Ελλήνων μάλιστα καὶ Μενεσθεῖ τῷ Αθηναίῳ θύειν καὶ Θεμιστοκλέα δὲ τὸν ναύμαχον σοφίας τε καὶ ανδρείας ἀγασθέντες χαλκοῦν ἵδρυνται ἒννουν καὶ ὥσπερ χρησμῷ ἐφιστάντα. 'Ιδεῖν καὶ δένδρα φασίν ἐνταῦτα, οἷα οὐχ ἑτέρωθι τῆς γῆς. καὶ Γηρυόνεια μὲν καλεῖσθαι αὐτά, δύο δὲ εἶδος ἕτερον, λείβεσθαι δὲ αἵματι, καθάπερ τῷ χρυσῷ τὴν 'Ηλιάδα αἴγειρον, ἡ δὲ νῆσος, ἐν ᾗ τό ἱερόν, ἔσταιμὲν οπόση ὁ νεώς, πετρῶδες δὲ αὐτής οὐδέν, ἀλλά βαλβῖδι ξεστῇ εἴκασται.
>
> They say that the Gadeirans are Hellenised, and enjoy our kind of culture. Certainly they welcome Athenians more than anyone, sacrifice to Menestheus the Athenian, and admire Themistocles the navalfighter for his intelligence and courage, so that they have erected a bronze statue of him meditating and as it were pondering an oracle. They claim to have seen trees there a kind that exists nowhere else on earth, and are called Geryones' trees. They are two

> in number, and grow from the grave that holds Geryones. Each combines the natures of a pine and a fir and drips *blood*, as we are told the Heliad poplar drips *gold*. The island on which the *sanctuary* stands is as large as the temple itself, and is in *no way like a rock*, but resembles a *polished platform*.

The ancient writer mentions two Athenian heroic names. The first is that of Themistocles, the known victor of Salamis naval battle against Persian fleet. This name connects Gadeira with the classical period of Athens. The second, however, is the name of Menestheus, a hero of the Trojan war! This piece of information links, at least in time, Gadeira with 12th century BC from an Athenian perspective. This reminds of Plato, who associates Athenians of 12th century BC with a military operation somewhere west of Gibraltar, in which they were involved. Philostratus describes a specific tree, characterised by the author as a pine tree, existing in the region, which *weeps over* Geryones' tomb shedding *red tears*. This tree is unique from of an observer's standpoint coming from the Mediterranean Sea. It is located at Erytheia, west of Gibraltar, a region near Gadeira. This tree does not grow within the Mediterranean and consequently it was unique for prehistoric Greeks. It is called a Dragon tree and produces red resin! This is the source of the *blood* which the legend describes. The legend connects these trees with Geryones, because two of them grew on the top of his *tomb*. Philostratus further describes the environment as an island, saying that the areas of the island and the temple were the same. He also explains that Geryones' *tomb* had a top, where a certain *valvis*, resembling a polished platform, existed. Since Erytheia is a land resembling a *tomb* with a circular flat top, it is likely to be elevated. Its polished platform *valvis* is in ancient Greece always a point, a line or circle imposing limits on the ground. It has been used by the athletes. A circular *valvis* can be imagined, if one bears in mind how and where modern athletes throw the discus. The geomorphology of Erytheia its likely elevated topography and its flatness implying circular limits reflect an environmental *tomb* and a temple simultaneously. In other words the flat environment described by Philostratus looks like a circular perivolos, as Achaean tombs normally exhibit. The gently hilly-like structure of the island (Erytheia) could not have been considered as a *tomb* by ancient Greek observers, if it had not been slightly elevated having a flat top. Thanks to Philostratus, we understand that the initial observers interpreted an earthly natural phenomenon as a *shouting giant*. The reason was that they saw it as a circular *mouth* giving the impression of somebody *shouting*. We say *shouting*, since Geryones' name itself means shouting. In Figure 4 a flat topped *Geryones* in Andalusia's red soil is shown, in other words in the legendary *Eryhtheia's* (*red land*) environment. In Figure 5 another flat topped *Geryones* from Armenia is shown.

They are both craters of diapyric origin. They both resemble what Philostratus described 2000 years ago. Figures 4 and 5 both exhibit a flat polished *valvis* on their tops. But how can one interpret Geryones' three heads? The only way to interpret this, as an earthly natural phenomenon, is to imagine three heads, one within the other. In other words the *shouting* Geryones exhibited a concentric three-ringed circular crater with a central flat surface on it. It is located somewhere in Erytheia (the red land) close to Gadeira. In that region (Gadeira), thanks to the Atlantic Ocean's climate, appears the first sign of the distribution of the Dragon tree with the *red*

**Fig. 4** A diapyrogenic Geryones from Spain in Andalusia with a flat *valvis* on its top. (Photo after Montexano)

**Fig. 5** A diapyrogenic Geryones from Armenia with circular flat *valvis* on its top.

*tears*. Today's observers can find one at Gibraltar and at Cadiz's Medical School along the coast of West Africa, in the Canaries in the Atlantic Ocean and elsewhere, but not within the Mediterranean Sea. In fact the same thing occurred in ancient times. It was and continues to be, a characteristic unique tree of the Atlantic Ocean. Figures 6, 7 and 8 illustrate a Dragon tree at Cadiz, red *tears* of a *weeping* Dragon

**Fig. 6** A Dragon Dracaena tree in the gardens of the Medical school Cadiz is shown. It can reach a height of 19 m. It looks as if it is *supporting* the celestial sphere like *Atlas or Heracles*.

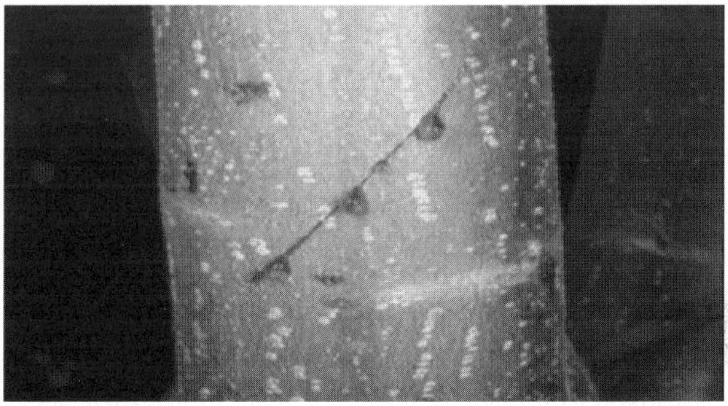

**Fig. 7** Red tears for *Geryones* "killed" by *Heracles*.

**Fig. 8** Red resin from the tree Dragon Dracaena.

tree, over Geryones' death and red resin from another type of Dragon tree respectively.

The Atlantic Ocean Dragon tree is at first a short tree and then it grows very slowly. However, it can reach heights up to 19 m. It looks like a *giant holding the sky with his arms extended upwards.*

But can we find out more about Erytheia? Was it an island or a land and where was its exact location? The positive answer comes from Stesichoros' 7th century BC fragmented passages. The first (Fragmenta 7.1–7.3):

σχεδόν ἀντιπέρας κλεινᾶς 'Ερυθείας Ταρτηςσοῦ ποταμοῦ παρὰ παγάς ἀπείρονας ἀργυρορίζους ἐν κευθμῶνι πέτρας

Almost opposite famous Erytheia ... by the limitless silver-rooted waters of the river Tartessus in the hollow of a rock.

Stesichoros describes Erytheia as lying opposite to a land where the Tartessos River flows, close to some *void* associated with silver mines. It is the first time that a river delta is mentioned to be located in the land opposite Erytheia. But this is Guadalkivir River in southern Spain. In other words Andalusia's region is unavoidably the land described. Stesichoros, however, refers to Geryones in another excerpt as follows (Fragmenta S20, col1.2–S20, col2.3 tit102–103):

{ΔΑΦΝΙΣ} {Γηρυόνης} σχεδόν αντιπέρας κλεινάς 'Ερυθείας Ταρτησσοῦ ποταμοῦ παρά παγάς ⟨τίκτεν⟩ ἀπείρονας ἐν κευθμῶνι πέτρας.

From this particular fragment and from Hesiod's description, we deduce that Geryones was in Erytheia and additionally that was situated between Gadeira and Iberia! However, one cannot be certain whether Erytheia was an island or land, since ancient Greek authors seem not to agree with each other. Erytheia as a word appears in the 7th century BC. The concepts of peninsula and island had not been differentiated yet, therefore any solution is acceptable. If Erytheia was an island or part of Andalusia's coast taken as an island and not visible today, then geoarchaeology is the only means to answer the missing reality. Figure 9 illustrates a map of the distribution of past land slides and mud craters in the Atlantic Ocean [41]. As the reader can see the marks of mud craters exist in detail right in the bay of Cadiz. They are shown in Figure 10 [33].

Since no words of differentiated meanings of islands and peninsula existed before 5th century BC in alphabetic Greek, we can deduce that either geological reality could be meaningful. However, at some stage, it is important to know, the dates of any possible landslides and the scenario under which they were formed in Cadiz's bay and in Andalusia's coast.

But in another fragment Stesichoros continues stating the following (Fragmenta S20, col2.1–col2.6):

διὰ] κ[ύ]μαθ' ἁλός βαθέας ἀφίκοντο θ]εῶν περιλαλλέ[α ν]ᾶσον τ]όθι 'Εσπερίδες π[αγχρ]ύσεα δώμα]τ' ἔχοντι˙ ...] [.]ασς [...]και ...] λυκ [...]λατ[

... over the waves of the deep brine they came to the beautiful island of the gods, where the Hesperides have their homes of solid gold; ... (buds)

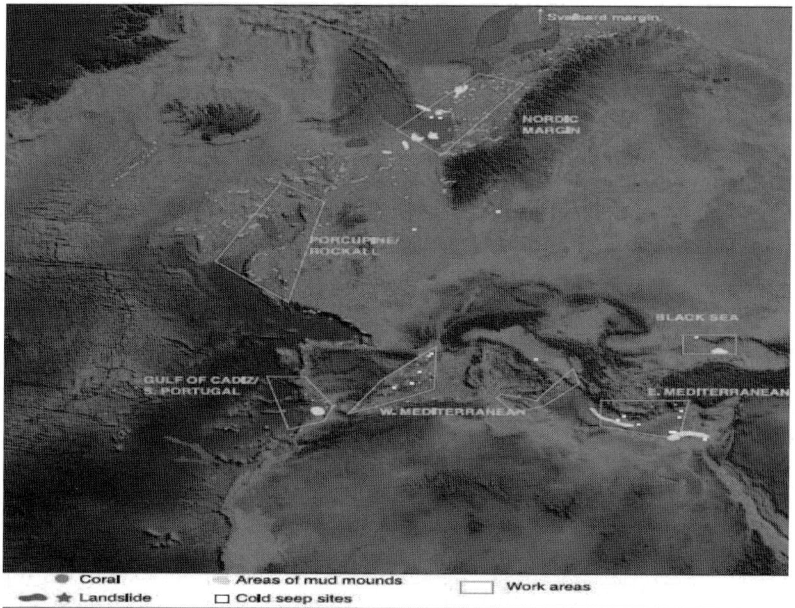


**Fig. 9** Giant land slides (red) and diapyrogenic craters (yellow) (after van Rooij [33]). Reprinted by permission.

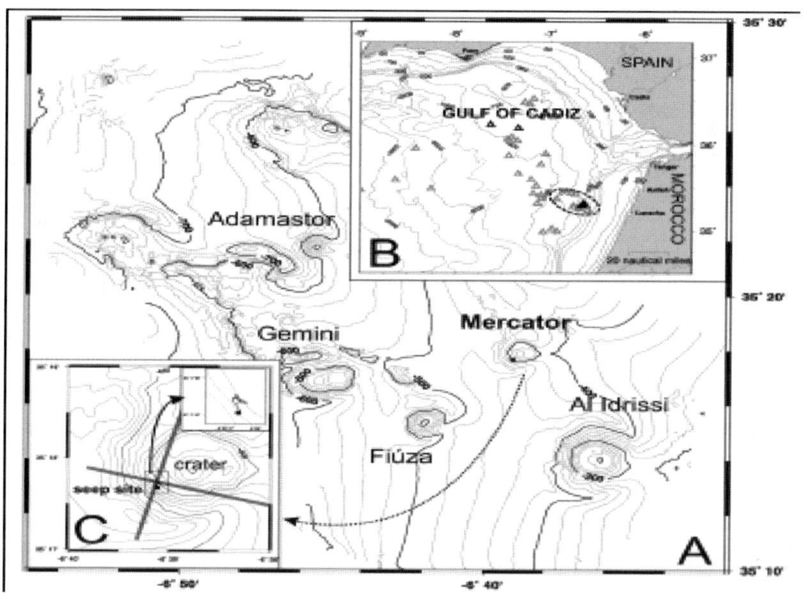


**Fig. 10** Giant diapyrogenic craters at Cadiz's bottom in a bathymetric presentation (after [41]). Reprinted by permission.

Stesichoros describes the wonderful island and relates the Hesperides (Pleiades) to it. But the surprise is that Hesperides is mentioned two centuries before Hellanicus' time. In Hellanicus 5th century BC, Poseidon mates with Kelaeno (*one of the Hesperides*) and their offspring is placed on the islands of the blessed, somewhere west of Gibraltar. In the legend of the Hesperides, also called Atlantides, there is an obvious astronomical layer apart from the earthly. The first does not contradict the second. The astronomical layer reflects knowledge of Pleiades' constellation associated possibly with calendrical systems. But the case of Hesperid Kelaeno illustrates another earthly layer as a piece of additional information in connection to a particular region west of Gibraltar. This information is Poseidon's erotic act on Earth, as conceived by the Greeks, somewhere west of Gibraltar. Versions of the *Earth* having a circular chasm in a marine environment are Calirrhoe in Hesiod, Nereid in Pindar, and Cleito in Plato. These ladies had an erotic contact with Poseidon in an exotic environment for the prehistoric Greeks. The various versions of the legend are multiple attempts of prehistoric Greeks in the region of Gadeira-Andalousia to interpret circular chasms, once visible. Any of these circular chasms became once a sacred and religious center of Iberians, long before the advent of Phoenicians. Only in the bay of Cadiz there are four giant diapyric craters [33], therefore, the legends of Geryones and the like for this region are not too far from existing geological reality.

But let us go deeper in time and reach Homer's period. Do we have a symbolic presentation of a sacred circularity anywhere? The answer is positive. The following passages are very revealing (Od 13.146–13.152):

> ἔπειτα Ποσειδάων ἐνοσίχθων· αἶψα κ' ἐγών ἕρχαιμι, κελαινεφές, ὡς αγορεύεις ἀλλά σὸν αἰεὶ θυμόν ὀπίζομαι ἠδ' ἀλεείνω. νῦν αὖ Φαιήκων ἐθέλω περικαλλέα νῆα ἐκ πομπῆς ἀνιοῦσαν ἐν ἠεροειδέϊ πόντῳ ῥαῖσαι, ἵν' ἤδη σχῶνται, ἀπολλήξωσι δὲ πομπῆς ἀνθρώτων, μέγα δὲ σφιν ὄρος πόλει ἀμφικαλύψαι.
>
> Then Poseidon the earth-shaker, answered him: Quickly should I have close as you say, god of the dark clouds, but always I dread and avoid your wrath. But now I am minded to smite the beautiful ship of the Phaeacians, as she comes back from her convoy on the misty deep, that hereafter they may desist and cease from giving conveyance to men, and to hide their city be hind a huge *encircling* mountain.

In the above text Poseidon suggests to Zeus what he is going to do against the Phaeacians describing his threat into two steps. First he means to smite the Phaeacian boat as it approaches the coast of their land with an earthquake and then go to the second realization of his threat which is the *encircling* of their city by dropping on a huge mountain it. Let us see the realization of his first part of his threat (Od 13.159–13.164):

> αὐτάρ ἐπεί τὸ γ' ἄκουσε Ποσειδάων ἐνοσίχθων, βῆ ῥ' ἴμεν ἐς Σχερίην, ὅθι Φαίηκες γεγάασι. ἔνθ' ἔμεν'· ἡ δὲ μάλα σχεδόν ἤλυθε ποντοπόρος νηῦς ῥίμφα διωκομένη. τῆς δὲ σχεδόν ἦλθ' ἐνοσίχθων, ὅς μιν λᾶαν ἔθηκε

καὶ ἐρρίζωσεν ἔνερθε χειρί καταπρηνεῖ ἐλάσαςἐ.

Now when the Poseidon the earth-shaker heard this he went to Scheria where the Phaeacians dwell, and there he waited. And she drew close to shore, the seafaring ship speeding swiftly on her way. Then near her came the Earth-shaker and turned her to stone and rooted her far beneath by a blow of the flat of his hand and then he was gone.

After the realization of the first part of his threat, Phaeacians were very much scared. One can see what they wanted to do to avoid the imminent catastrophe in the following passage (Od 13.178–13.184):

τὰ δὲ δὴ πάντα τελεῖται, ἀλλ' ἄγεθ', ὡς ἄν ἐγώ εἴπω, πειθώμεθα πάντες· πομπῆς μέν παύεσθε βροτῶν, ὅτε κέν τις ἵκηται ἡμέτερον προτί ἄστυ· Ποσειδάωνι δὲ ταύρους δώδεκα κεκριμένους ἱερεύσομεν, αικ' ἐλεήση μηδ' ἥμιν περίμηκες ὄρος πόλει ἀμφικαλύψη. ὥς ἔφαθ' οι δ' ἔδδεισαν ἑτοιμάσσαντο δὲ ταύρους.

... and now all this is being brought to pass. But now come, as I bid let us all obey. Cease to give convoy to mortals, when anyone comes to our city, and let us sacrifice to Poseidon twelve choice bulls, in hope that he may take pity, and not hide our city behind a huge *encircling* mountain.

Homer keeps repeating the term *encircling mountain* in a marine seismogenous region. This term requires explanation. In order to for us to understand Poseidon's very strange threat we should look at Figure 2. Did Poseidon in the seismogenous marine environment of Nisyros really produce a hole in it by throwing a huge rock or was the pre-existing crater interpreted by the Greeks as done by Poseidon in the past? Certainly the second case is the correct one. In the light of this question, we should see Poseidon's threat against the Phaeacians as a pre-existing crater in a seismogenous environment. It has been interpreted as Poseidon's act of the past. Since the ongoing seismicity in the area was growing, the interpreters of the circular chasm as Poseidon's first act gave the impression to Phaeacians that he will repeat its act, since he was the Earth-shaker. Their fear was based on their interpretation of a pre-existing crater in the vicinity of their seismogenous land, as a result of Poseidon's action. Therefore, the loss of their boat in the sea, due to an earthquake, increased the observing city's people anxiety of a possible repetition of Poseidon's worst act since the god of the sea had sent his message. The frightened citizens of Scheria, who lived in or close to a nature-made crater believed that Poseidon would throw a huge mountain at them as he did in the past producing the crater.

The late Ulf Richter suggested an alternative physical origin of the capital of Atlantis shown in Figure 11. It is located in Mauritania and is a diapyrogenic multi-ringed crater. It is an alternative to the impactogenic crater since it can produce concentric circles, warm water and red, white and black stones.

In Figure 12 the Platonic multi-ringed feature exhibits warm water and red, white and black building stones. The land is depicted with orange (non-dark) whereas the water is shown with blue (dark) colour. Such a feature can be caused by volcanogenic, or impactogenic or diapryrogenic mechanisms.

**Fig. 11** An example of diapyrogenic crater from Mauritania [19]. Reprinted by permission.

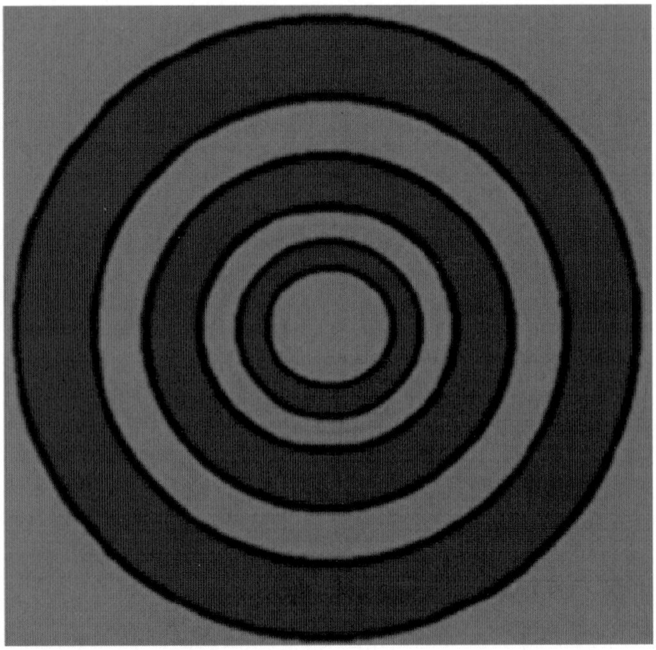

**Fig. 12** The Platonic multi-ringed feature exhibits warm water and red, black and white building stones. The land is depicted with orange colour, whereas the water is shown with blue. Such a feature can be caused by volcanogenic, impactogenic and diapyrogenic mechanisms.

In Figure 13 the result of the arithmetic simulation in the Platonic environment is shown, (after [37]). The authors allowed a projectile to fall on the coastal environment of Atlantis as described by Plato.

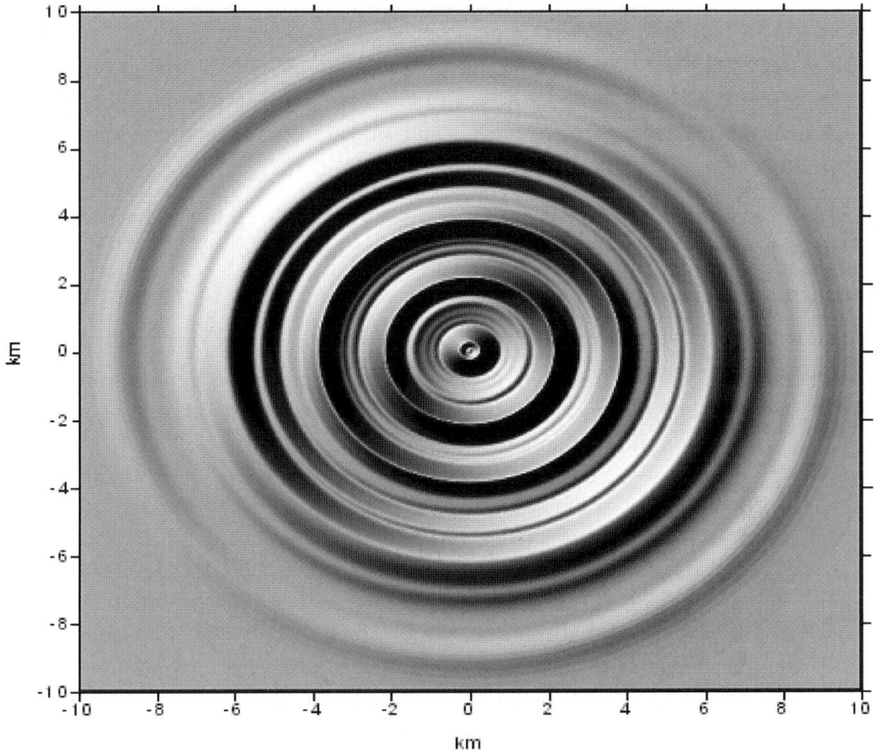


**Fig. 13** The result of the numerical simulation in the Platonic environment (after Tsikalas et al. [36]). The multi-ringed feature has no volcanogenic warm water and no volcanogenic red, black and white building stones. In fact they are impactogenic.

The above-mentioned arithmetic simulation produced warm water and can produce red, white and black impactogenic stones depending on the pre-existing geological setting.

## 4 Nature of the Island

Plato describes Atlantis' nature, particular position, orientation and with unusual details its capital, its flora, fauna and resources. Let us first study the geography implied (Tim 24.e.5–24.e.6):

νῆσον γὰρ πρὸ τοῦ στόματος εἶχεν ὃ καλεῖτε, ὥς φατε, ὑμεῖς ʽΗρακλέους στήλας.

island in front of the mouth which you call Heracles pillars

One may recollect that we are dealing with events of 12th century BC [27], described with phraseology of 4th century BC. The word *island* being unclear, both in Egyptian and in alphabetic Greek up to the 6th century BC (included). It certainly generated ambiguity if the *nēsos* of that remote period was really an island of the classical period. But let us go further and analyze the following passage (Criti 114.b.1–114.b.4):

> λῆξιν δὲ ἄκρας τῆς νήσου πρὸς 'Ηρακλείων στηλῶν εἰληχότι ἐπὶ τὸ τῆς Γαδειρικῆς νῦν χώρας κατ' ἐκεῖνον τὸν τόπον ὀνομαζομένης,
>
> the *extremity* of the *island near* the pillars of Heracles up to the part of the country now called Gadeirus . . .

Without comprehending it, Gill [7] made a very interesting and important observation: He says, the expression in Greek επί το της meaning in English *up to the* is *odd*: "it looks as though Plato intended to say 'facing that part of land'", in other words the peninsula of Gadeira. Gill finds it *odd* because he has in his mind the *giant nēsos* in the middle of the Atlantic Ocean. The possibility that does not cross his mind is that the two concepts of *nēsos* and *peninsula* were not differentiated in Homer's time but only some decades before Plato's time in Herodotus text. Vidal-Naquet [39] was a victim of the same fact. Had he understood this detail, he would have immediately realized that the only way to face the peninsula of Gadeiriki is to be on the giant *island* of Celto-Iberia, which of course is in the Atlantic Ocean. That simple! Plato described Atlantis having placed the observer sitting on the land of Iberia! The confusion of 2600 years is now over. He used the word Atlantis, in a confusing way, for all the three elements of his report, nesos, horseshoe basin and the concentric rings. The latter is *missing* only for the time being . . .

## 5 Orientation of Atlantis

Plato also gives the orientation of Atlantis by saying the following (Criti 118b.1–118b.2):

> ὁ δὲ τόπος οὗτος ὅλης τῆς νήσου πρὸς νότον ἐτέτραπτο, ἀπὸ τῶν ἄρκτων κατάβορρος
>
> and this region, all along the island, faced towards the south and was sheltered from the *northern blasts*.

This description describes Atlantis' orientation and implies the time when this observation was made at Atlantis, because it involves the celestial *bears*. In the past, the latter offered a general indication of the celestial North. We already know that Plato's information in *Timaeos* and *Critias* originates from late 12th century BC and thus has also some Achaean component in addition to the initial Egyptian one. The celestial *bears*, which were not known to the Egyptians as *bears*, remind us of Homer indicating the North who indeed describes North with the *bears*.

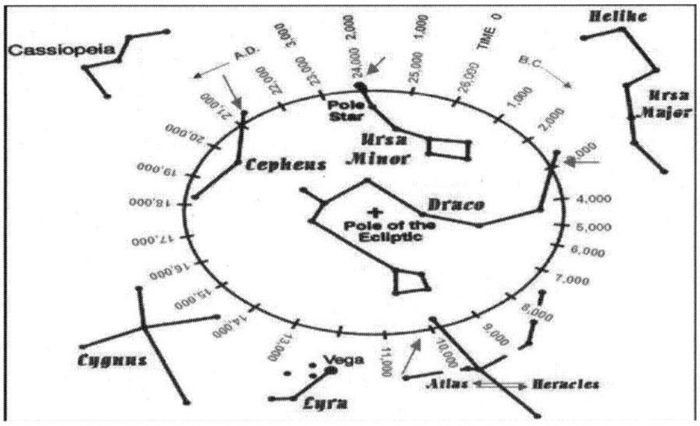


**Fig. 14** A transtellar compass and clock show eight constellations depicted in Greek mythology. Earth's axis of rotation intersects a particular star because it changes position in space in a 26000-year time span due to the metaptosis phenomenon, caused by the gravity attraction of the sun and the moon on Earth. The arrows show four *exact* celestial north indexes.

If the Platonic information was much older, in the range of thousands of years, as he mentions in his text, then the celestial *bears* would not be used. Other constellations offering the general or the exact North had to be used. For instance, Dragon constellation offers a star as an exact North. It was known to some cultures round 2700 BC. The initial *transmitter* in Plato's text does not use it. In Figure 14, one can see time spans in which the exact North was available. In a circle of roughly 26000 years, various constellations offered a star, through which the axis of the earth passed once.

Consequently the *bears* point to an observer of Atlantis much earlier than Dragon's time (2700 yr BC) and certainly much earlier than Heracles' time, which is 10000 yr BC. It is a further proof that the text mentioned time spans in *moon years* and therefore the time of the events is not 9600 BC in *solar years*. Otherwise, the North would be a star in *Heracles* constellation and not the constellation of the *bears*. But there is a further implication, i.e. that the celestial *bears* cannot be seen 25° south of the equator. In other words, the southern hemisphere cannot be the land of Atlantis.

We propose that Solon had information both from Sais in Egypt and from sources going back to the end of the Bronze Age from his Athenian heritage.

## 6 Geomorphology

Foliot [6] and Papamarinopoulos [28] suggested that Andalusia's basin engulfed by the Sierra Morena and Sierra Nevada mountainous range exhibit a striking similarity to Atlantis' geomorphology in the Platonic description. Keeping in mind the previous analysis about nesos we read the following (Criti 118.a.2–118.a.6):

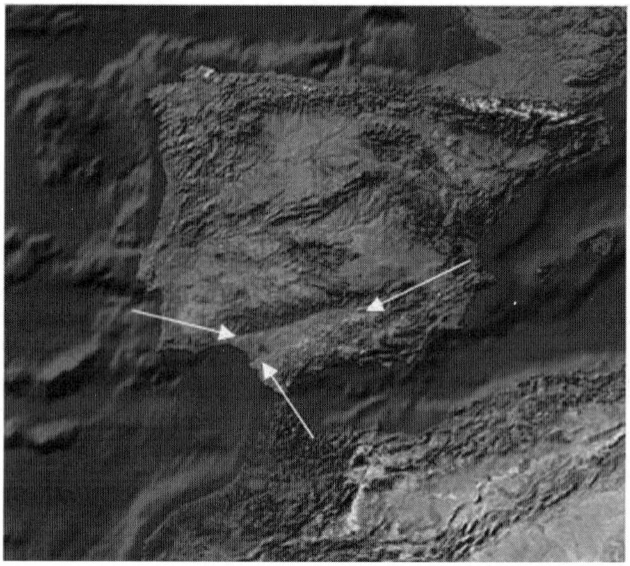

**Fig. 15** Plato described Celto-Iberia as the giant Atlantis *island* and in exceptional detail Andalusia's basin which looks like a horse shoe ending in the Atlantic Ocean. The meaning of the island became what we know today, when Herodotus in the 5th century BC added the word peninsula. Andalusia's basin is absent in Herodotus map in Figure 16. Plato, through unknown sources, recorded prehistoric Greeks' technology in compiling crude geographical notes before Herodotus' era in the pre-cartographic period. Reprinted by permission.

> πρῶτον μὲν οὖν ὁ τόπος ἅπας ἐλέγετο σφόδρα τε ὑψηλός καὶ ἀπότομος ἐκ θαλάττης, τὸ δὲ περί τὴν πόλιν πᾶν πεδίον, ἐκείνην μὲν περιέχον, αὐτό δὲ κύκλῳ περιεχόμενον ὄρεσιν μέχρι πρός τήν θάλατταν καθειμένοις, λεῖον καί ὀμαλές, πρόμηκες δὲ πᾶν,
>
> but the part about the city was all a smooth plain, enclosing it round about, and being itself encircled by mountains which stretched as far as to the sea; and this plain had a level surface and was parallelogram as a whole in shape.

Our analysis leads us to Celto-Iberia, sub-chapter *The nature of the island*. Now, thanks to Plato's text, as above, we conclude that the only place in Iberia, matching the Platonic description is that of Figure 15. Additionally we demonstrate that Herodotus' map, shown in Figure 16, illustrates nothing of the horse-shoe shape Andalusia's flat and elongated rectangular valley reaching the Atlantic Ocean!

In Figure 15, one can see Andalusia's geomorphology and recognize Plato's text fully. Plato himself did not realize that he was describing Iberia. Of course he copied neither Herodotus nor Hecataeus because *they both say nothing* about Andalusia's basin. The richness of metal and wood supplies characterizes Andalusia. Moreover the rich flora, which is a combination of West Mediterranean and Atlantic Ocean sub-tropic microclimate, allowed olive trees to coexist with palm and Dragon trees in the past. The latter is described by Philostratus in the 1st century AD, [26], but

**Fig. 16** In Herodotus' map the horse-shoe shaped Andalusia's flat and orthogonal basin engulfed by Sierra Nevada and Sierra Morena is missing. It is described only by Plato in the giant *Atlantis* land. Reprinted by permission.

it is confirmed even today due to the existence of a Dragon's tree in the gardens of the medical school of Cadiz and of the Gibraltar's land as well. But Plato's description of another tree producing special fruit, which one could eat, drink and use as unguent, can be either an olive or a palm tree. Both interpretations are valid because both trees coexist in Andalusia's environment.

## 7 Mechanism of Catastrophe

If Atlantis' sacred circularity was located in Andalusia's coast, then it would have been destroyed by an earthquake, tsunami and possible land slide [28]. The sacred multi-ringed, nature-made feature was not Tartessos city but it was related to the Tartessian culture. Figure 17 shows the stratigraphic column produced by drillings within the bottom of the Atlantic Ocean. Many turbidity layers have been recognized. At the top the reader can see that produced by the 1755 AD Lisbon earthquake. Below that there is another turbidity layer of 200 BC. All turbidity layers may have as a cause the Goringe Bank seismogenous fault shown in Figures 17 and 18. In the latter it is depicted with a white arrow. In Figure 19 Gutscher's [9] arithmetic simulation is shown. The tsunami hit with a velocity of some hundreds kilometers/hour both coasts of Europe and Africa. The red arrow shows the third turbidity layer which is predicted by the author to be at the end of the Bronze Age. Lisbon's catastrophe of 1755 AD is shown in the etching of Figure 20. In that catastrophe 60000 people died in few minutes.

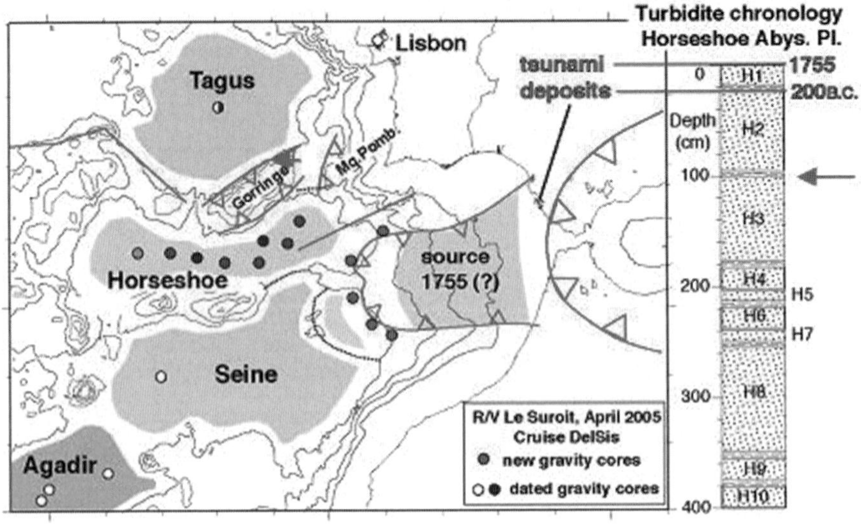


**Fig. 17** In the stratigraphic column produced by Lebreiro et al (1997), a late Bronze Age turbidity event is predicted by the author as responsible for Atlantis' multi-ringed sacred circularity catastrophe. It is marked by an arrow. Reprinted by permission.

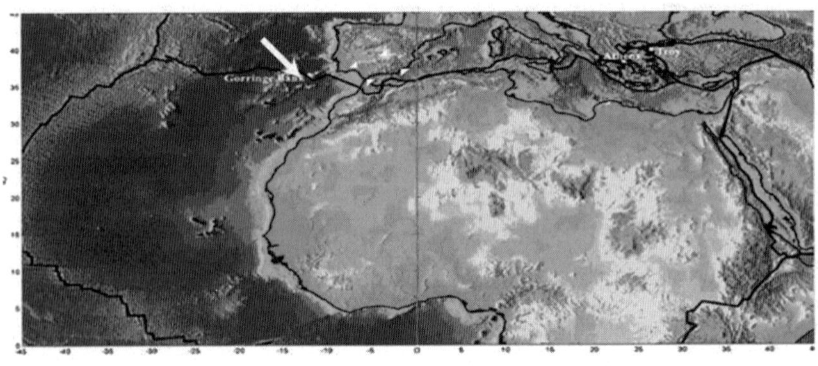

**Fig. 18** The seismogenous fault Goringe Bank is responsible for generating earthquakes above 8R. The white arrow shows the lethal hidden danger in the Atlantic Ocean's bottom. Reprinted by permission.

# 8 Atlantis' Flora and Fauna

In Critias, Plato describes Atlantis' flora as giant trees in the following way (Criti 117.b.5–117.b.7):

> τό δέ ἀπορρέον ἦγον ἐπί τό τοῦ Ποσειδῶνος ἄλσος, δένδρα παντοδαπά κάλλος ὕψος τε δαιμόνιον ὑπ' ἀρετῆς τῆς γῆς ἔχοντα,

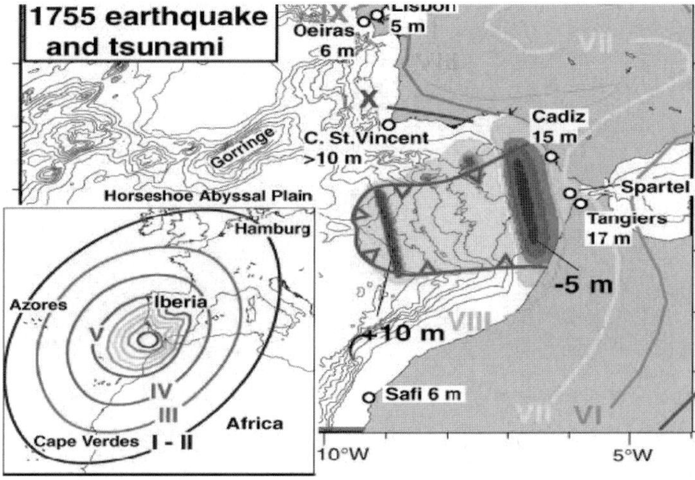


**Fig. 19** The simulated Lisbon's tsunami of 1755 AD. The tsunami moves with some hundreds of kilometers per hour speed ready to hit the coast of both continents [9]. Reprinted by permission.

**Fig. 20** 1755 AD Lisbon's tsunami in a characteristic etching. 60000 people died in few minutes. This simulated tsunami is presented in Figure 19.

> And the outflowing water they conducted to the sacred grove of Poseidon, which contained trees of all kinds that were of marvellous beauty and height because of the richness of the soil;

Trees of a *demonic height* and beautiful could be either very old Dragon or relatively young palm trees or both. They both suit with Andalusia's Atlantic Ocean's and West Mediterranean environment. Dracaena trees can reach a height of 19 m as they

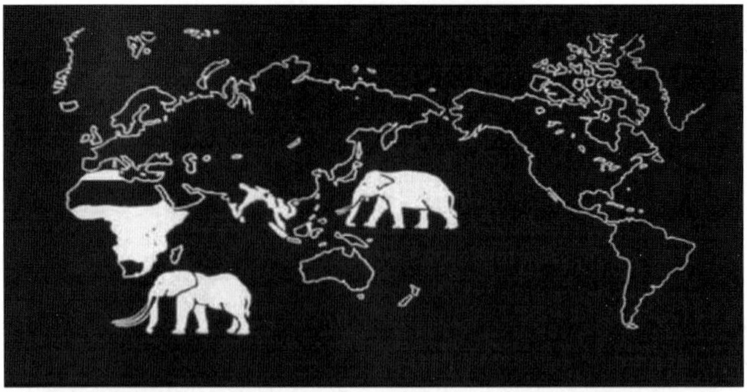

**Fig. 21** The distribution of elephants in Africa before the Roman period.

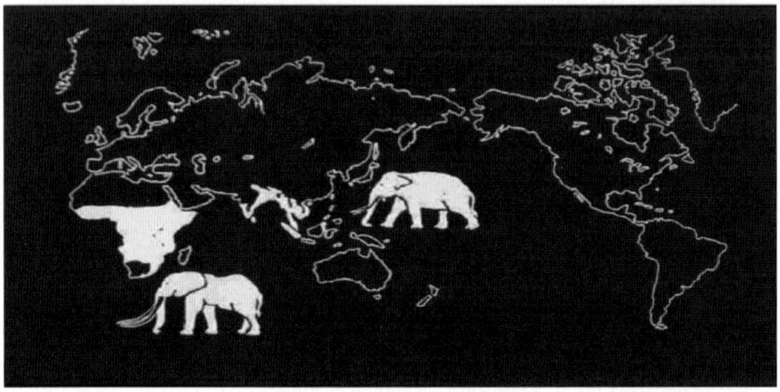

**Fig. 22** The distribution of elephants in Africa after the Roman period.

grow slowly. Plato also describes a particular fruit which one can eat, drink or use it as unguent in Criti 115.b.1–115.b.3. This fruit can be either an olive or coco nut. They can both grow in the Western Mediterranean due to the mixture of the Atlantic Ocean to the Mediterranean climate at Andalusia's basin. Plato describes elephants, bulls and horses. We know that, in prehistory, it was usual to transport animals by boat at relatively small sea distances. Due to the small distance between Africa and Iberia, it is expected that various animals could have been transported both ways. Strabo (Geog 17.4.18–17.4.25) mentions that the Romans were collecting elephants from the basins close to the region of the Atlas Mountains and transported them to Rome. In Figures 21 and 22 we demonstrate the distribution of the elephants before and after the Roman period in North West Africa around Atlas Mountains.

Plato describes horses and bulls in Atlantis. It is known that horses and bulls had existed there since palaeolithic time. Papamarinopoulos and Cosegian (2007) have demonstrated that the famous scene of the bulls with the *Atlantes* in *blue* has nothing to do with the acrobatic leaping over the bull manifested by the prehistoric Cretans in Crete and Egypt. In fact, they demonstrated that the particular ritual capture of the bull was carried out in Iberia.

# 9 The Size of Atlantis

Plato describing Atlantis' size says (Tim24.e.6–24.e.7):

> ἡ δέ νῆσις ἅμα Λιβύης ἦν καί 'Ασίας μείζων
>
> the *island* which was larger than Libya and Asia together.

If Solon's informants were Egyptianised Greeks, who remained in Sais since 12th century BC, after the Trojan War, then the comparison of the areas of Atlantis and the sum of the areas of Libya and Asia would be of great interest. We need to define the concepts of Libya for the Greeks. In Figure 23 Plato's world map is shown. In that Herodotus' map is emplaced. Asia for the Greeks was only Anatolia and not the present day continent. Libya was all Africa known at that period of the 6th

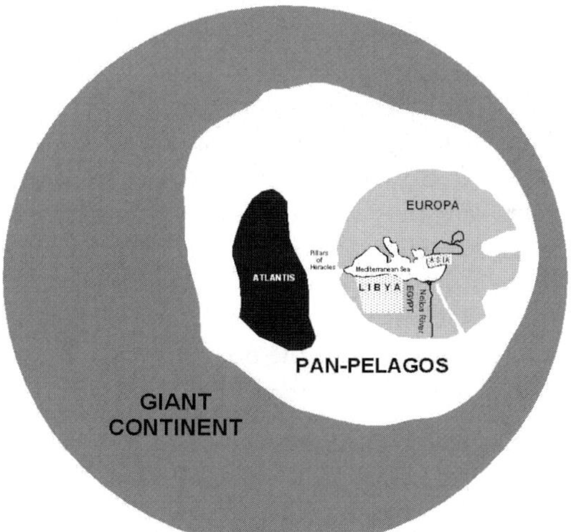


**Fig. 23** Plato's world map is shown with *the giant island of Atlantis* in the middle of the Atlantic Ocean. The world map was plotted from the cartography perspective of the Greeks. Atlantis' giant island produced by those who did not take into account the word's *island* evolution in the Greek language up to the 6th century B.C. and the fact that in Egyptian hieroglyphics did not have the meaning of the island at least before or within the 6th century B.C. Atlantis' zones of influence Libya and Italy are shaded. Libya in accordance with 6th and 5th centuries ancient Greeks is different from the larger land, then known to the Greeks, which much later called Africa. Thomson [35] quotes Euthemenes of Masalia an explorer of the 6th century B.C. who describes most likely Senegal river with crocodiles in a land southwards of Libya. Herodotus describes Libya as three zones. Southwards of these zones he distinctively mentions a huge desert land without giving a name for it.

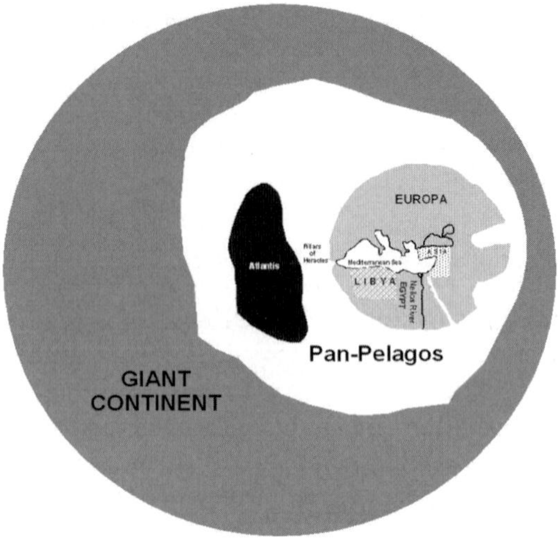


**Fig. 24** This map is as in Figure 23 with Asia as *understood* from the Egyptian standpoint, not of cartography of course, but for locating hostile peoples. This region is larger than Asia of the Greeks of 6th and 5th century B.C. Asiatic peoples were located in Asia Minor and Middle East. Although the Egyptians did not produce maps, they had a sense of distance due to the battles they got involved in with the lands to which they had to go.

century BC and not all of the African continent minus Egypt which is part of Africa. In Figure 23 Atlantis' island, as it was traditionally conceived both by experts and atlantologists, is depicted. Celto-Iberia, Libya and Asia are shaded.

If Solon's informants were only Egyptians, they would have had little interest in comparing areas of countries. However, they would have been interested in comparing war capacities of the *Atlantes* with particular traditional enemies, Libyans and Asians. In Figure 24 we demonstrate the distribution of Egypt's enemies in 12th century BC. This map is different from the previous one only in the distribution of the Asians located both in Anatolia and the Middle East.

## 10 Atlantes

Let us examine the *war between the Achaeans and the 'Atlantes'* as Plato described it in Timaeos and Critias. Plato describes the war of the *Atlantes* as follows (Criti 109.c.3–109.c.6):

> Πάντων δὴ πρῶτον υνησθῶμεν ὅτι τό κεφάλαιον ἦν ἐνακισχίλια ἔτη, ἀφ' οὗ γεγονώς ἐμηνύθη πόλεμος τοῖς θ' ὑπὲρ 'Ηρακλείας στήλας ἔξω κατοικοῦσιν καὶ τοῖς ἐντὸς πᾶσιν·

Now first of all we must recall the fact that 9000 is the sum of years since

> the war occurred, as it is recorded, between the dwellers beyond the pillars of Heracles and all that dwelt within them.

We have already explained that Plato's thousands of years are in fact moon years, because the Egyptian priesthood used three different moon calendars, although other types of calendars were used as well. Therefore, the possible times of the war are within end of 13th or beginning of 12th century BC, if the time spans are divided by the number 12.37 which is the number of full moons in a year. These time spans are modifications of the Platonic 9000 and 8000 years mentioned in *Timaeos* and *Critias*. Plato allows us to understand that Solon's initial informants had the strong impression that all attackers entered the Mediterranean basin through the Atlantic Ocean. Today we know that there had been two wars against Egypt and that some of the *Atlantes* originated from Anatolia.

Others originated from Italy and Libya and some others possibly from Central Europe. Plato adds another point of interest in connection with Atlantis' influence (Tim 25.a.8–25.b.5):

> Πρὸς τούτοις ἔτι τῶν ἐντὸς τῇδε Λιβύη μὲν ἦρχον μέχρι πρός Αἴγυπτον, τῆς δέ Ευρώπης μέχρι Τυρρηνίας. αὕτη δή πᾶσα συναθροιοθεῖσα εἰς ἐν ή δύναμις τόν τέπαρ' ὑμῖν καί τόν παρ' ἡυῖν καί τόν ἐντός τοῦ στόματος πάντα τόπον υιᾷ ποτέ επεχείρησεν όρυῆ δουλοῦσθαι.

> And, moreover, of the lands here within the Straits they ruled over Libya as far as Egypt, and over Europe as far as Tuscany. So this host, being all gathered together, made an attempt one time to enslave by one single onslaught both your country and ours and the whole of the territory within the Straits.

But what is the archaeological opinion about the events and war activities in the specified period toward the beginning of the 12th century BC? O'Connor [24], for instance, discusses the event of 1208 BC, when an aggressive coalition attempted to conquer Egypt in Pharaoh's Menenptah 5th year. Egyptian sources describe the names of the participating coalition being Sherden, Tursha and Shekelesh. These names have been identified as Sardinians, Etruscans and Sicilians, all living in the Atlantis' zone of influence, which is Tuscany, in other words Italy. All these people fought together with Libyans against Egypt. We have a *double correspondence* with the Platonic text, since these two zones are mentioned in *Critias*. But there are also *differences*. The Egyptian sources mentioned two more people the Lukki and Ekwesh The first disagreement between archaeology and Plato's account lies with the Lukki, which are people living in Lykia in Anatolia. Anatolia was not mentioned as Atlantis' zone of influence. The second possible disagreement in accordance with O'Connor is major, because Ekwesh is the name of the Achaean Greeks! However, Iakovides [10] and Sandars [34] fully disagree with O'Connor. Sandars [34] in particular wonders how the Ekwesh could be Achaeans, when they appeared circumcised, according to Egyptian sources. It is a well known fact that Europeans living in Asia or in Europe never practiced circumcision – from prehistory to the historic period. The same scientist says that there is not any linguistic relation between the words Ekwesh and Achaeans or even between Ekwesh and Ahiwaya. The latter was the name by which the Hittites called the Achaeans. Plato continues to describe how the Athenians managed to face the *Atlantes* alone. He describes it like this (Tim 25.c.1–25.c.6):

> τὰ μέν τῶν 'Ελλήνων ἡγουμένη, τά δ' αυτὴ μονωθεῖσα ἐξ ἀνάγκης τῶϝ ἄλλων ἀποστάντων, ἐπί τούς ἐσχάτους ἀφικομένη κινδύους, κρατήσασα μέν τῶν ἐπιόντων τρόπαιον ἔστησεν, τούς δὲ μήπω δεδουλωμένους διεκώλυσεν δουλωθῆναι, τοὺς δ' ἄλλους, ὅσοι κατοικοῦμεν ἐντός ὅρων 'Ηρακλείων ἀφθόνως ἅπαντας ἠλευθέρωσεν.
>
> And acting partly as leader of the Greeks (*meaning the city of Athens*) and partly standing alone by itself when deserted by all others, after encountering the deadliest perils, it defeated the invaders and reared a trophy; where it saved from slavery such as who dwell within the bounds of Heracles it ungrudgingly set free.

Herodotus visited Egypt earlier than Solon, but learnt nothing about the two assaults of the Sea Peoples against East Mediterranean. From Plato's text, one gets the conclusion that Solon's informants had at least mentioned only one to him and associated it with Athenians. Was it possible that Plato had recorded the second assault of the Sea Peoples? The answer is positive, because there were no Achaean mercenaries in Egypt before 1208 BC. There is no textual or any archaeological evidence supporting the case. Homer says that Achaeans arrived to Egypt after the Trojan War, in other words after 1184 BC. Consequently there was no chance to have in Egypt a city such as Sais associated with Achaeans earlier than 1208 BC. Let us examine this second assault, which occurred in Pharaoh Rameses III 8th year. Wachsmann [40], who studied the Egyptian sources, mentioned the names Peleset, Sikala, Sheklesh, Denyen and Weshesh. Sandars [34] explained in her study that the Deneyn were not the Achaeans known by a similar name, but some peoples living close to Syria in Adana and they had been known to the Egyptians since 14th century BC from the Amarna tablets. The Egyptians called their land Danuna, whereas they called Tanaja the country of the Achaeans, which is of course Peloponnesus [1]. We, therefore, conclude that the Achaeans did not contribute to either against or in favour of the Egyptians in the 1176 BC event. But the Sea Peoples lost the battle. This information is not mentioned in Plato's documents.

Consequently, how and where did the Greeks manage to defeat the Atlantes, since the Atlantes lost the war in Egypt and the Greeks did not encounter them in Greece or in Egypt? Marinatos [18] interpreted Plato's Atlantes as a Sea Peoples' remote echo. Iakovides (2007) suggested that the Sea Peoples destroyed the prehistoric ports in Greece and Middle East and indirectly ruined the economy of the Achaeans. This military action is in fact a kind of easy looting and not a war strategy, nevertheless it was effective. From this looting of the ports, many countries suffered in 1176 BC. One of them was Greece. This reminds of the Platonic text about common suffering of Egyptians, Greeks and other peoples within the Mediterranean due to pressures by the Atlantes. Although the military victory of the Egyptians is clear the Platonic text presents the Greeks as victorious but not in Egypt! Assuming correctness of information, a possible place where the Greeks could have been victorious is Atlantis itself. Our scenario requires an Achaean leader of the Greeks of Athenian origin to be in Egypt with his companions after the Trojan War. Foliot [6] says that as we know from Homer, Menelaus, a warrior of the Trojan War, came to Egypt accidentally due to a storm that occurred in the Aegean Sea. It is possible that part of the victorious Achaeans of Athenian origin in Troy came also to Egypt after the Trojan war of 1184 BC and resided there at Sais. Zangger [42] reminds us that Thucydides recounted how the survivors of the Trojan War spread all around the central and eastern Mediterranean. We could add that some of the Mycenaeans went to the West and even passed through the Straits into the Atlantic Ocean as Homer says and Tziropoulou [38] proved in her text. When the Sea Peoples moved against the Eastern Mediterranean in either 1176 or 1177 BC, Athenian warriors, who were already at Sais, decided to go against them in their own strategic place at Atlantis. This means that Solon was talking to Egyptians and possibly Egyptianised Greeks.

The latter had extracted this information from their own archives as a result of a report of some survivors, who had returned to Egypt 600 years before Solon's visit and kept a record of what had happened during the war with the Atlantes in Atlantis. This scenario may explain the friendliness between Gadeirians and Athenians as well as between Saitians and Athenians. We would like to make clear that Sais was the city where the story began in the 6th century BC and Gadeira the city very close to Andalusia's coast where the story had ended. There, the assumed religious circular center, once was flourishing. It *vanished* at the end of the Bronze Age. Between the two localities, the connecting element paradoxically are the Athenians as Plato and Philostratus independently state! For such a scenario to be proved, detailed studies are required both in Sais and the coast of Andalusia. There possibly exists, the multi-ringed nature-made religious center, with all possible proofs, which will either support or refute the present scenario. Plato, however, describes the victory of Athenians, who built a victorious monument in Atlantis in the following way (Tim 25.b.7–25.c.4):

> Πάντων γάρ προστᾶσα υεψυχία καί τέχναις ὅσαι κατά πόλεμον, τά μέν τῶν 'Ελλήνων ἡγουμένη, τά δ' αὐτή μονωθεῖσα ἐξ ἀνάγκης τῶν ἄλλων ἀποστάντων, ἐπί τούς ἐσχάτους ἀφικομένη κινδύνους, κρατήσασα μέν τῶν ἐπιόντων τρόπαιον ἔστησεν.
>
> For if stood pre-eminent above all in gallancy and all warlike arts, and acting partly as leader of the Greeks, and partly standing alone by itself when deserted by all others, after encountering the deadliest perils, it defeated the invaders and reared a trophy.

In other words, Plato describes Mycenaean Athenians as leaders of the Greeks against Atlantis, waging a campaign, in which they were victorious. Similarly, Mycenaean Peloponnesians being leaders of the Greeks, campaigned against Troy. If one requests further sources on what Plato describes, nothing will be found among either historians' texts or any archaeological findings in Mediterranean countries. It is therefore likely to think that Plato imagined this campaign and the victory of his prehistoric compatriots. However, one knows that Plato in *Timaeos* and *Critias* described events, for which the historian's silence was characteristic and yet archaeology has proved his writings one by one.

But was it ever possible that Mycenaean Athenians could have been leaders of the Greeks? From Mycenaean tradition, we know King Eurystheus of Tiryns who was killed in a campaign against Mycenaean Athenians. Eurystheus was the king who had demanded of Heracles to perform his twelve labors. The Mycenaean conflict, between Peloponnesians and Athenians, just described, is a remote echo of the struggle between tribes of Mycenaeans to impose their will on each other. We can not prove that Mycenaean Athenians were ever leaders of the Greeks, but what Plato described seems not to be impossible within the evolution of the Mycenaean world. But Plato does not only describe the victory of the Greeks against the Atlantes, but also a double catastrophe in the following way (Tim 25.c.6–25.d.3):

> ὑστέρῳ δέ χρόνῳ σεισμῶν καί κατακλυσμῶν γενομένων, μιᾶς ἡμέρας καί νυκτός χαλεπῆς ἐπελθούσης, τό τε παρ' ὑμῖν μάχιμον πᾶν ἀθρόον ἔδυ κατὰ γῆς, ἥ τε 'Ατλαντίς νῆσος ὡσαύτως κατὰ τῆς θαλάττης δῦσα ἠφανίσθη·
>
> But at later time there occurred portentous earthquakes and floods, and one grievous day and night befell them, when the whole body of your warriors was swallowed up by the earth, and the island of Atlantis in like manner was swallowed up by the sea and vanished.

In other words, after the victory of the Mycenaean Athenians against the Atlantes at a region in the Atlantic Ocean there was an unknown time span not defined by Plato. After that there was a catastrophe caused by earthquakes and floods, in which Greeks and Atlantes got lost. Let us examine how Foliot [6] evaluates these particular Platonic events. She judges that the catastrophe was in two different places and not in one, as Plato implies. She explains that due to heavy rain and an earthquake, there was a partial landslide at the Acropolis along with the Athenian warriors, whereby most of them were killed. We do not have any doubt about the severe earthquake. It has been proved and its consequence (broken pots) has precisely been dated as being in early 12th century BC. For the rest we have no findings yet. The catastrophe in Andalusia's coast in 12th century BC has not yet been documented. Plato uses the same word for the giant island of Atlantis, which turns to be, following his text strictly, Celto-Iberonesos. He also uses the same word for the horse-shoe shaped flat basin that reaches the Atlantic Ocean and fits geomorphologically with Andalusia. He also uses the same word even for the famous concentric circles formation with warm water springs and multi-colour building rocks. Plato knew, that when he was at the age of 52, Helike vanished in the Corinthian Gulf after an earthquake and a tsunami, but Peloponnese remained intact. Plato knew what a volcanic catastrophe was, because he witnessed himself an explosion of Etna during his years in Italy.

Therefore, he knew that a giant mass cannot easily disappear, even if it was connected with either catastrophe mentioned earlier. However, the use of Atlantis for everything in his report induced unwanted confusion which lasted for centuries. But let us go further. Foliot [6] believes she has the answer of the war of Atlantes against Athenians in Greece and she uses the Hellenic mythology to explain it. She mentions Athena's victory against Poseidon in Attica and Hera's victory over the same god in Argos. She identifies Atlantes with Poseidon and Greeks with Athena and Hera respectively. But she fails in this interpretation, because the Atlantes, in other words the Sea Peoples, as Marinatos [18] first proposed, did not fight against the Greeks in Greece in the way they did in Egypt. No traces of war activity were ever found in Greece of 12th century BC. On the contrary, we have a very good record of what Ramses the III achieved in Egypt against the Atlantes. Plato's view presents a major difference. Let us further examine Foliot's [6] view. She reminds us, as Homer says, that the Achaeans reached Egypt after the Trojan War, where they were warmly received by the Egyptians.

It is very possible that the Egyptians who *understood* the pressure of the Sea Peoples and the Libyans in 1208 BC, received the victors of the Trojan War as mercenaries in order to increase their military capacity against their hostile neighbors. In accordance with Foliot's [6] scenario, Mycenaean Athenians settled in attractive Sais and organized a type of city. In other words, several years before the second assault of 1176 BC of the Sea Peoples against Egypt, the Mycenaeans warriors had already been there. It is reminded that Mycenaean Athenians with Menestheus as well as their leader had been in Troy along with other Mycenaeans just a few years before. Also the fact is reminded that Menestheus, a hero of Trojan War, was honoured in Gadeira by the *Hellenised* Gadeirians as Philostratus has recorded in his text. Foliot's scenario explains the friendliness of the Saitians to Solon the Athenian of 6th century BC and the similarities between Egyptian goddess Neith and Athena [6].

In Plato's text it appears that the discussion between Solon and Saitian priests had no communication problems. It appears that the *priests* knew particular details of the Acropolis of Athens of the 12th century BC. If the people who discussed with Solon were only Egyptians, then clear difficulties would have been to converse. However, the text does not mention any such difficulty. We find very strange that *Egyptian priests* had interest for architectural details on the Acropolis of Athens and its eventful history in 12th century BC. But since no war between Atlantes and Mycenaeans ever existed within Greece and/or Egypt, then is it possible that such a battle took place in Atlantis? And how the information of the catastrophe reached Egypt? Our scenario assumes a contingent of Mycenaean Greeks under Athenian leadership to travel from Egypt to Andalusia, a place already known to Mycenaean traders, and get engaged in a war with the Atlantes. After the natural disaster in Andalusia, some of the survivors came back to Sais and kept a record of what had happened. The archives re-emerged in the 6th century BC and were communicated to Solon, a *compatriot*, so to speak, of theirs.

The victory did not survive in the tradition of the Athenians in the Greek mainland, not only because there was not any eloquent poet, such as Homer, to write about it, but also because the informants resided only in Sais. Another Achaean victory, which did not interest any poet was that of the first fall of Troy 30 years before the legendary Trojan War. In the Eastern part of Aphaea's temple (490–480 BC) on the Greek island of Aegina details of that first war against Troy are given with Ajax's father, whereas in its Western part Ajax himself is shown [12]. Yet some tradition remained even in the absence of an epic poem. Although Plato's story is differentiated with respect to the present day archaeological knowledge, we propose that his differentiation signals a parallel history within the turbulent 12th century BC, so far unknown to scientists. It is not necessarily diametrically opposed to the existing archaeological knowledge in Egypt of the war of 1176 BC. However, it contains exaggerations, mistakes and anachronisms which can be removed. The origin of the Peoples of the Sea remains enigmatic and the whole issue is still open. Betancourt (2000) for instance, does not accept the simplistic view that Sea Peoples were Aegean Sea Peoples, who left the Aegean Sea due to hardships and, en route to settle in Palestine, fought with the Egyptians and then *became* the Philistines of Middle East. Certainly their name is not included in the list of the names of attackers of Egypt in 1176 BC. Wachsmann [40] mentions the possibility that among Sea Peoples were warriors from central Europe! He says that

> the bird-boat motif is not a symbolic device common to Mycenaean cultural realm. Notable in this regard, therefore, is a bird-boat-like depiction painted on the shard of Late Helladic IIIC crater shard found in Tiryns.

Wachsmann [40] continues saying

> The manner in which the bird-heads are positioned on the stem and stern of the ship form a sea going *bird-boat*. As this motif is foreign to the Mycenaean world, we must conclude that the specific ship used by the Medinet Habu artists as a prototype for their five depictions of a Sea Peoples' ship, was manned by a crew that held religious beliefs consistent with those of the Urnfield culture. Or, to put it plainly, the crew was likely to have been composed of Urnfield people.

In Table 1 the military activities of the Achaeans and of the Atlantes are presented against time. We cannot know whether there was any military activity between Achaeans and Atlantes in Andalusia or even whether there was any multi-ringed feature which was destroyed in its coast in the Platonic way. However, we know that some of Plato's passages, for which no other independent sources exist, have been confirmed so far one by one by archaeology. Künhe [14] concludes that Atlantis is in the same place as the author of this study suggests. He believes that Plato knew the Medinet Habu archives and used them the way *he liked*. Künhe's [14] view does not, however, explain the friendliness of the Saitian priests to Solon and the Athenians in general. Neither does he explain the Gadeirian's friendliness toward the Athenians and nor the honoring by them of the Athenian hero Menestheus. He does not take into account that Gadeira was part of Atlantis and of course does not connect it with the Athenians at all.

The same researcher does not take into account Homer, who described Achaeans' arrival to Egypt after the Trojan War. Finally he equates Tartessos with the concentric formation. However, Tartessos, existed up to the Roman period and *vanished smoothly* due to the accumulations of continuously added sediments by Guadalquivir river.

**Table 1** Achaean and Atlantian military operations in 12th century BC.

| | |
|---|---|
| 1224: | First fall of Troy by Ajax's father and his companions |
| 1208: | First assault of the Sea Peoples (*Atlantes*) against Egypt |
| 1195: | Beginning of the Trojan War |
| 1184: | Second fall of Troy by Ajax and his companions |
| 1184+: | Achaean's return to Greece |
| | General instability is exhibited in Greece. *Agamemnon's murder* |
| 1184+: | Achaean's arrival in Egypt |
| 1176: | Second assault of the Sea Peoples (*Atlantes*) against Egypt |
| | Zangger [42] proposes 1177 BC as the year of the conflict |
| 1177: | Odysseus' return to Cephalonia from Atlantis [27, 38] |
| 1177+: | Atlantis' possible catastrophe |

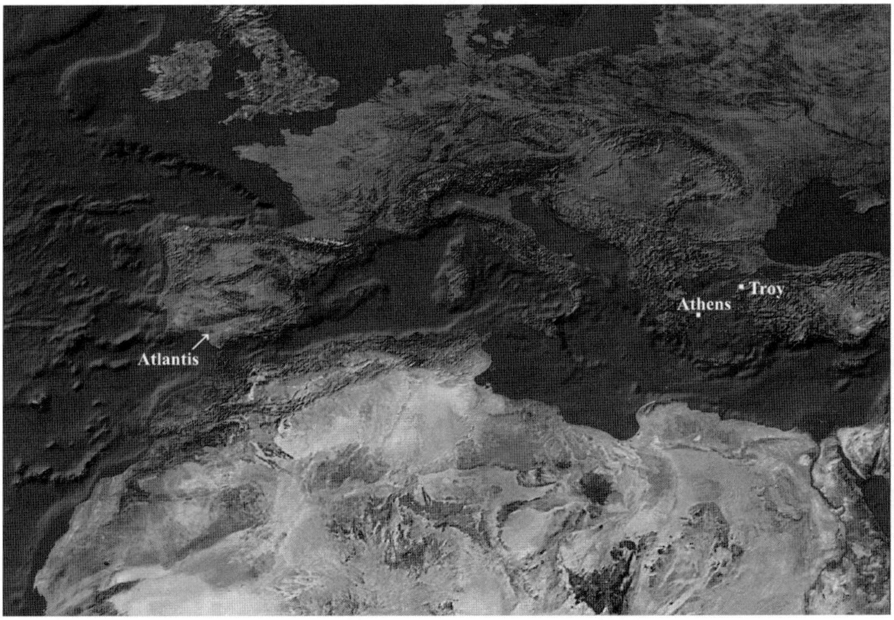


**Fig. 25** Triple seismic catastrophe: Troy (described only by Homer), Athens (described only by Plato) and Atlantis (described only by Plato). May be that of Atlantis corresponds to the one in 12th century BC.

Odysseus' return from Atlantis in 1177 BC [27], in the way Homer described, indicate Achaean knowledge of the region. However, it illustrates that the sacred coastal cyclic feature was still functioning in spite of the beginning of the seismicity and before its complete annihilation. This means that, after Odysseus' return, the Athenians got in conflict with the Atlantes. Since there is no tradition in Athens as a memory of that victory, we can either accept that Plato invented the war, the victory and the catastrophe in Atlantis or that a contingent of Achaeans with Athenians from Egypt as leaders did a military operation in Atlantis. The remaining survivors returned to Egypt and kept a record of the events. These *particular archives* remained in Sais for 600 years and were made known to Solon thanks to his Athenian origin in the 6th century BC. If the year of the conflict in Egypt was 1177 BC, then Achaean's assault against Atlantis could be a parallel event to the War in Egypt. If, on the other hand, the conflict was in 1176 BC, then it could be a target of obliterating a remaining operational site of the Atlantes.

Figure 25 illustrates a triple catastrophe in 12th century BC. The first is in Troy. It was mentioned only by Homer. The second is in Athens. It was mentioned only by Plato. The third one is in Atlantis. It was mentioned only by Plato and is under investigation.

## 11 The Giant Western Continent

In Figure 26 Plato's world map is again presented, in combination with the said Herodotus' map without the major misunderstanding of the giant island in the middle of the Atlantic Ocean, which has been an anathema for the experts. We should explain whether ancient prehistoric societies knew about the existence of a giant continent west of Gibraltar in any possible way. In fact the giant mass is a crude presentation of the two Americas and Antarctica together. Although Balanova's [2] initial results with *artificially mummified bodies from Egypt* were criticized by Buck-

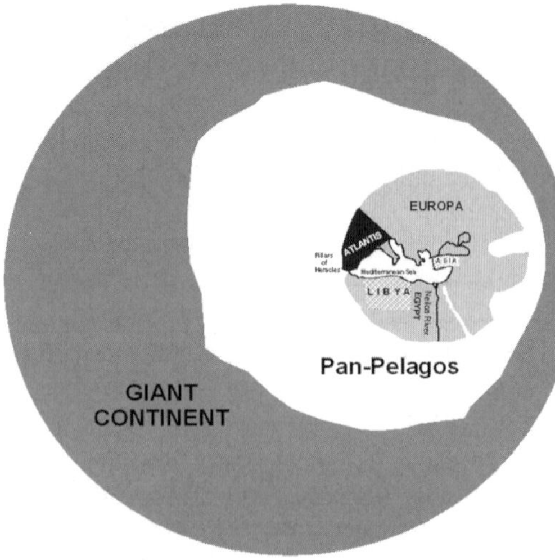


**Fig. 26** The giant peninsula with its zones of influence in Italy and Libya is indeed much larger than Libya and Asia taken as sum. The case of the *Atlantes episode* was not known to historian Herodotus' but it was known to non-historian Solon's and Plato's informants, who visited Egypt earlier and later respectively than the historian. Similarly the Western giant continent, separated by the pan-pelagos, was not known to Herodotus' but it was known to the non-historian Plato's informants and was proved as such. Reprinted by permission.

land and Panagiotakopoulou [5], Balanova's studies [2] on *naturally mummified bodies from Egypt* demonstrated that they contained nicotine, cotinine and cocaine. The existence of cotinine means that, when these people were alive, they were using nicotine which had been metabolized into cotinine. Similarly the existence of cocaine means that, when they were alive, they used cocaine. Consequently, there is no possibility for any contamination of any kind. Although one cannot exclude the possibility that the wild Nicotiana Tombacum plant existed outside South America, it is very difficult to accept that the Erythroxylon Coca plant grows somewhere else outside South America. The time span of the dated naturally mummified Egyptian bodies is between 1070 BC and 395 AD. This unexpected information indicates knowledge of the giant westerly land by Solon's and/or Plato's informants.

## 12 Conclusions

The following conclusions can be drawn:

1. Several ancient authors describe in a symbolic way, through their survived passages, a sacred Poseidonian *circularity* placed west of the Pillars of Heracles.
2. This *circularity* was conceived by the visiting prehistoric Greeks as Poseidon's work, who appeared in the ancient Greek mythology as a *builder*, *lover*, *avenger* or *warrior* in the habit of producing earthquakes and *concentric circles on Earth*.
3. This *circularity* was placed on *Erytheia* (red land). It was either an *island* between Cadiz and Andalusia's coast or a part of the latter. It was mentioned by Hesiod in the 7th century BC. The word could not have the same meaning as in the 5th century BC, when Herodotus added the word *peninsula*.
4. This *circularity* could be either of impactitic or diapyric origin. They can both explain the concentric circles, the warm water and the red, white and black stones. In that environment, no magmatic volganogenic craters exist.
5. It is noted that the Euro-African environment in Morocco and Iberia exhibit marine and land small and giant diapyrogenic craters.
6. This *circularity* has vanished due to successive earthquakes and floods, as Plato described it in *Critias* in a *single day and night*, due to the activity of seismogenous Goringe Bank fault placed in the Atlantic Ocean's bottom, since earthquakes from that fault produced tsunamis. Both could have been destructive for the *circularity*. Such a tsunami, originating from Goringe Bank, hit Cadiz's bay with a 15 m high sea wave in 1755 AD [9]. If the *circularity* had been in the 12th century BC in Andalusia's palaeocoast, it would have been instantly destroyed with all people on it.
7. Plato is proved to be correct both in Eastern Mediterranean, describing architectural details of the Athens' Acropolis 800 years before his century, and in West Mediterranean describing Andalusia's horseshoe geomorphological shape. Both sets of information could not have been known to him, since no historians had any mention of them. Herodotus had described Iberia's map and

had mentioned the Pyrenees, but he has completely missed Andalusia's basin and its surrounding mountains.

8. The word *nēsos had the meaning of peninsula or promontory or an island (with today's meaning) in prehistoric times as well but in 5th century BC. Herodotus added the word [chersonesos* meaning *peninsula* [16]. Therefore it is more than likely that Celto-Iberia was a giant *n=esos (island) in the eyes of prehistoric Greeks*. An example in prehistoric times for the Achaean observers is Peloponnesus, which was never an *island* (*nēsos*) in the way Herodotus described the island's concept in the 5th century BC. We have reached Andalusia's coast based *only* and *strictly* on Plato's text. However, we have reached the same region through the study of other ancient texts.
9. Out of Plato's text emerges a new window for the 12th century BC focusing on the West and is different than Homer's of the same century, focused on the East Mediterranean.
10. Vidal-Naquet's [39] theory of imaginary Athens and imaginary anti-Athens (Atlantis) has collapsed, since prehistoric Athens was real and Atlantis' geomorphology is a real geological reality. They both were unknown to historians.
11. We do not know whether the *sacred circularity* ever existed or if and when it was destroyed.
12. The proposed theory predicts a turbidity layer in the Atlantic Ocean at the end of the Bronze Age. Then Plato's credibility will be tested once more. It also predicts the location of the submerged *circularity* in Andalusia's palaeocoast. Its catastrophe occurred *after* Athens' seismic catastrophe *but within* 12th century BC.
13. A 2600-year old gross misunderstanding is removed. The misunderstanding was the *island* of *Atlantis*. It was not Plato's imagination after all. It was based on the misconception of island's entity, as having in the past the same meaning as in the 5th century BC [16]. This misunderstanding hit the mind of *experts* against Plato and fired the mind of Atlantologists of all epochs in Plato's favour. It resulted in the edition of thousands of books.
14. Plato describes Andalusia's topography and geomorphology, whereas Homer did the same for Troad. Kraft et al. [13] who spent 20 years in the Troad, state the following:

    > Homer's description of the localities, their events and their topography are reflected in our geological reconstructions. Everything shows that Iliad was not a myth, but they all are proved by our palaeogeographic reconstructions of the region.

    Kraft, who is a geologist, does not know the difference between a *genuine myth* and a *fabricated myth*.
15. Atlantis' size, in other words Celto-Iberia's size, is indeed larger than Libya and Asia together but it is much larger, if the zones of its influence, Libya and Italy, are taken into consideration. For the Greeks, Libya was the known Africa up to Herodotus time in terms of area minus Egypt. Asia was only Anatolia.

16. Taking into consideration Hellenic mythology and an ostrich's shell found in tombs of the Achaean period in Mycenae [34], it is shown, that Libya was known to the prehistoric Greeks, independently of Egypt. Similarly taking into consideration Hellenic mythology and the amber jewelry originating from the Baltic and found in tombs of the Achaean period in Mycenae, it is indicated that Europe's western coast was known. Moran [23] proved that the Baltic amber was manufactured into jewelry in *Britain*. In addition the discovery of a Linear B tablet in Munich, Germany and the discovery of a Mycenaean shard in Guadalquivir's estuary indicate again some knowledge of the European western coast by the prehistoric Greek mariners. In the pre-cartographic period, before Herodotus world map, some general geographical notes were possibly kept from these areas, as a result of trade between Achaeans and Libyans and West and/or central Europeans.
17. Without being a historian, Plato uses information originating possibly from Egypt, Greece and Italy. This information is not only confined to the *Egyptian* priests or even possibly to the Egyptianised Greeks. He seems to describe the second assault of the Sea Peoples in 1176 BC, in which all participants from different nations were called *Atlantes* by him. Zangger [42] proposes 1177 BC as the year of conflict between Sea Peoples and Egyptians. For Plato all originated west of the Gibraltar straits. Today we know that some originated from Anatolia. However, he correctly pinpointed Atlantis' two zones of influence, Italy and Libya. He was also correct that some of the *Atlantes* originated from outside Gibraltar, straits since some of them used boats with the Urnfield's culture symbol, which was two birds at either edge of the boat. The Platonic text *ignores* Ramses' III Egyptian victory against *Atlantes* in Egypt and *seems* to know a Greek victory against the inhabitants of Atlantis. The same text describes a lethal end both for victors and defeated by earthquake and flood. Is it possible that a parallel story was developed in Egypt with the massive part of the *Atlantes* and another one with the remaining *Atlantes* in their religious center? The researcher of the future will face three possibilities: Plato either imagined Atlantes' assault and the war between Achaeans and Atlantes at Atlantis, or knew the Egyptian archives and changed them in order to praise the Greeks or he described a parallel truth unknown to historians which requires scientific attention and evidence.
18. There are strong indications that the giant continent (the Americas) west of Atlantis was known to Solon's and/or Plato's informants before 6th century BC, but not to historians' informants, who produced the world maps before Plato's in 4th century BC.
19. Some of Plato's passages, which have been confirmed scientifically, contained unique information not mentioned by ancient historians. The author of this paper has proved that Plato described the following: (1) Prehistoric Athens' Acropolis with its architectural details of 12th century BC. (2) The seismic catastrophe of it in 12th century BC (LH IIIB) period. (3) The end of Achaean culture and the survival of the Greek language from 12th century BC to the

classical period of Greece. (4) The Sea Peoples' second assault in East Mediterranean in 12th century BC. (5) The West Atlantic continent's existence.

20. In spite of some obvious errors, anachronisms or exaggerations, which non-historian Plato made in *Timaeos* and *Critias*, it is very difficult for us to accept that the second part of his story *about Atlantis* is fully imaginary, since he could not judge even the content of the first part of his story *about Athens*, if it was imaginary, and of course he could not judge the existence of the American and Antarctic *continent assumed as one piece*, if it was real, but he mentioned all three. The historians' silence on all of these issues is not an argument against Plato, since they fully failed to produce anything at least about the Athenian prehistoric Acropolis of early 12th century BC.

## Acknowledgements

Sincere thanks are due to Konstantina Vlachou, research student of Athens University for her assistance in the processing of the text and to Georges Seitz, Egyptologist, for his invaluable advice.

## Notes

The author has done his utmost to trace the original copyrightholders of figures and illustrations in order to obtain official permission to reprint them in the present publication. In cases where this search was unsuccessful the author would be pleased to hear from copyright holders so that correct procedures can be followed.

## References


1. Banou, H. (2000). *Amenhotep III and His Relation with the Aegean. The Official Trip. Crete–Egypt. Three Thousand Years Cultural Connections*, Editions Lousy Bratzioti, 479 pp.
2. Balanova, S. (1997). *Die Geschichte der Tabakpflanze vor Columbus außerrhalb Amerikas sowie das Rauchen im Spiegel der Zeiten*. Innovations-Verlag, 112 pp.
3. Betancourt, P.P. (2000). The Aegean and the origin of the Sea Peoples, in *The Sea Peoples and Their World: A Reassessment*, E.D. Oren (Ed.), University of Pennsylvania Museum, 360 pp.
4. Broneer, O. (1948). What happened at Athens, *American Journal of Archaeology* **52**, 111–124.
5. Buckland, P.C. and Panagiotakopoulou, E. (2001). Rameses II and the tobacco beetle, *Antiquity* **75**, 549–556.
6. Foliot, K.A. (1984). *Atlantis Revisited. A Documented Aaccount of the Fabled "Island" as It Was Long Ago and It Is Today*, Information Printing, Eynsham, Oxford, 129 pp.
7. Gill, C. (1980) *Plato: The Atlantis Story*, Bristol Classical Press, 93 pp.
8. Graves, R. (1961). *The White Godess*, first edition. Faber and Faber Ltd., 511 pp.
9. Gutscher, M.-A. (2005). Destruction of Atlantis by great earthquake and tsunami? A geological analysis of the Spartel Bank hypothesis, *Geology* **33**(8), 685–688.

10. Iakovides, S. (2007). Personal communication in Mycenae.
11. Jordan, P. (2001). *The Atlantis Syndrome*, Sutton Publishing, 308 pp.
12. Kakridis, I.Th. (1986). *Hellenic Mythology*, Volume 4, Ekthotiki Athenon.
13. Kraft, J.C., Rapp, G., Kayan, I. and Luce, J. (2003). Harbor areas at ancient Troy: Sedimentology and geomorphology complement Homer's Iliad, *Geology* **34**(2), 163–166.
14. Künhe, R.W. (2008). Did Ulysses travel to Atlantis?, in *Proceedings of the Symposium on Science and Technology in Homeric Epics*, Ancient Olympia, 2006, S.A. Paipetis (Ed.), Springer, Dordrecht, this volume.
15. Lebreiro, S.M., McCave, I.N. and Weaver, P. (1997). Late Quaternary turbitide emplacement on the Horseshoe abyssal plain (Iberian margin), *Journal of Sedimentary Research* **67**, 856–870.
16. Le Noan, G. (2005). *The Ithaka of the Sunset*, Editions Tremen, "Collection Commentaires", 126 pp.
17. Luce, J.V. (1978). The literary perspective, in *Atlantis Fact or Fiction*, E. Ramage (Ed.), Indiana University Press, 210 pp.
18. Marinatos, S. (1950). Peri ton thrylon tis Atlantithos, *Cretika Chronika* **4**, 195–213.
19. Matton, G., Jebrak, M. and Lee, J.K.M. (2005). Resolving the Richat enigma: Doming and hydrothermal karstification above an alkaline complex, *Geology* **35**(8), 665–668.
20. Martin de la Cruz, J.C. (1990). Die erste mykenische Keramik von der Iberischen Halbinsel, *PZ* **65**(1), 49–52.
21. Mommden, H. et al. (1990). Eine mykenische Scherbe in Spanien. Bestätigung ihrer Herkunft mit der neutronenaktivienrung Ananalyse (NAA), *PZ* **65**, 59.
22. Montaxano, D.G. (www.antiguo.com)
23. Moran, J. (2003). Personal communication in Tiryns.
24. O'Connor, D.(2000). The Sea Peoples and the Egyptian sources, in *The Sea Peoples and Their World: A Reassessment*, E.D. Oren (Ed.), University of Pennsylvania Museum, 360 pp.
25. Papamarinopoulos, S.P. (2003). Was Plato's Atlantis the island of Santorini?, in *Proceedings of the International Symposium on Extraordinary Machines and Structures in Antiquity*, Ancient Olympia, 19–24 August 2001, S.A. Paipetis (Ed.), Peri Tecnon.
26. Papamarinopoulos, S.P., Drivaliari, N. and Coseyan, Ch. (2007). Red tears in the Atlantic Ocean, in *Proceedings of the International Conference on The Atlantis Hypothesis: Searching for a Lost Land*, Melos Island, 11–13 July 2005 (in print).
27. Papamarinopoulos, S.P. (2007). Plato and the seismic catastrophe in the 12th century BC, Athens, in *Proceedings of the International Conference on The Atlantis Hypothesis: Searching for a Lost Land*, Melos Island, 11–13 July 2005 (in print).
28. Papamrinopoulos, S.P. (2007). A Bronze Age catastrophe in the Atlantic Ocean?, in *Proceedings of the International Conference on The Atlantis Hypothesis: Searching for a Lost Land*, Melos Island, 11–13 July 2005 (in print).
29. Papamarinopoulos, S.P. and Cosegian, Ch. (2007). The ritual of the bull in 'Atlantis' and its basic parallel in Iberia, in *Proceedings of the International Conference on The Atlantis Hypothesis: Searching for a Lost Land*, Melos Island, 11–13 July 2005 (in print).
30. Papamarinopoulos, S.P. (2008). The memory of a comet during the Trojan war, in *Proceedings of the Symposium on Science and Technology in Homeric Epics*, Ancient Olympia, 2006, S.A. Paipetis (Ed.), Springer, Dordrecht, this volume.
31. Podzuweit, Ch. (1990). Bemerkungen zur mykenischen Keramik von Llanete de los Moros, Montoro, Prov. Córdoba, *PZ* **65**(1), 53–58.
32. Richter, U. (2007). Plato's Atlantis was in a River Delta, in *Proceedings of the International Conference on The Atlantis Hypothesis: Searching for a Lost Land*, Melos Island, 11–13 July 2005 (in print). Strabo Geog 17.3.4.18–17.3.4.25.
33. Van Rooij, D. (2005). *EOS* **86**(49), 509–511.
34. Sandars (1978).
35. Thomson, J.O. (1948). *History of Ancient Cartography*, first edition. Cambridge University Press, 427 pp.

36. Tsikalas, F., Papamarinopoulos, S.P. and Shuvalov, V.V. (2007). The origin of the multi-ringed concentric morphology of Atlantis capital and its relation to the Platonic scripts, in *Proceedings of the International Conference on The Atlantis Hypothesis: Searching for a Lost Land*, Melos Island, 11–13 July 2005 (in print).
37. Tsikalas, F., Shuvalov, V.V. and Papamarinopoulos, S.P. (2007). A new geophysical interpretation of the Platonic multi-ringed concentric morphology of Atlantis capital based on numerical simulations, in *Proceedings of the International Conference on The Atlantis Hypothesis: Searching for a Lost Land*, Melos Island, 11–13 July 2005 (in print).
38. Tziropoulou, A. (2008). Homer and the so called Homeric problems, in *Proceedings of the Symposium on Science and Technology in Homeric Epics*, Ancient Olympia, 2006, S.A. Paipetis (Ed.), Springer, Dordrecht, this volume.
39. Vidal-Naquet, P. (2005). *L'Atlantide. Petite histoire d'un mythe platonicien*, Les Belles Lettres, 198 pp.
40. Wachsmann S.(2000). To the sea of the Philistines, in *The Sea Peoples and Their World: A Reassessment*, E.D. Oren (Ed.), University of Pennsylvania Museum, 360 pp.
41. Weaver, P.P.E. (2005). Hotspot ecosystem research on the margins of European seas, *EOS* **86**(24), 226–227.
42. Zangger, E. (1195). Who were the Sea Peoples?, *Aramco World* **46**(3), 21–31.
43. Zhirov, N. (1970). *Atlantis. Atlantology: Basic Problems*, Progress Publishers, Moscow, 437 pp.

# Did Ulysses Travel to Atlantis?

Rainer W. Kühne

*Braunschweig, Germany*

**Abstract.** Good fiction imitates facts. Plato declared that his Atlantis tale is philosophical fiction invented to describe his fictitious ideal state in the case of war. I suggest that Plato used three historical elements for this tale. (i) Greek tradition on Mycenaean Athens for the description of ancient Athens, (ii) Egyptian records on the wars of the Sea Peoples for the description of the war of the Atlanteans, and (iii) oral tradition from Syracuse about Tartessos for the description of the city and geography of Atlantis.

## 1 Odyssey

Homer's Odyssey is poetry. It is often regarded as pure fiction. This view is supported by Ulysses' adventures with giants like the Cyclops, goddesses like Athena and Circe, nymphs like Calypso, beasts like Scylla, and demons like the sirens.

On the other hand, Schliemann interpreted Scylla and Charybdis as the Strait of Messina. He interpreted Ogygia, the island of Calypso, as the east coast of Sicily, and Scheria, the country of the Phaeaceans, as Corfu [1].

Various authors have pointed out the similarities between Homer's Scheria and Plato's Atlantis [2, 3]. However, these similarities are not very specific.

1. The capitals of both Scheria (Od. 5.401) and Atlantis (Crit. 118a) lay near a steep coast.
2. The capital of Scheria was surrounded by a high wall (Od. 6.262). The capital of Atlantis was surrounded by several walls (Crit. 116a).
3. There were two harbours which allowed only a narrow entrance to the capital of Scheria (Od. 6.263–264). The capital of Atlantis was surrounded by three harbours (Crit. 117d).
4. In the harbours of both Scheria (Od. 6.264–265) and Atlantis (Crit. 116b) there were many ships.
5. In the capital of Scheria there was a realm of Poseidon (Od. 6.266) and the palace consisted of gold, silver, copper and electron (Od. 7.82–91). In the capital of

*S.A. Paipetis (ed.), Science and Technology in Homeric Epics,* 509–514.

Atlantis there was a grove of Poseidon (Crit. 117b) and the temple of Poseidon consisted of gold, silver, ivory and orichalcum (Crit. 116d).
6. In Scheria one could harvest fruit twice a year (Od. 7.117–119). In Atlantis one could harvest twice a year (Crit. 118a).
7. In Scheria there were two springs (Od. 7.129) and warm baths (Od. 8.249). In Atlantis there were two springs, one cold and one hot (Crit. 113e, 117a), and warm baths (Crit. 117b).
8. In Scheria reigned 13 kings simultaneously (Od. 8.390), who were descendants of Poseidon (Od. 13.130). In Atlantis reigned 10 kings simultaneously (Crit. 113e–114a, 116e, 119c), who were descendants of Poseidon (Crit. 113e–114c).
9. Both the kings of Scheria (Od. 8.431) and Atlantis (Crit. 120a) used gold goblets for drinking during ceremonies.

## 2 Fictional Elements

Plato's Atlantis tale (Timaios 20d–27a, Critias 106a–121c) is philosophical fiction invented to describe Plato's fictitious ideal state in the case of war (Tim. 19b–20c). Plato invented the Atlantis tale after he failed twice in educating the tyrants of Syracuse to behave like philosopher kings [4, 5]. The Atlantis tale includes several elements which are based on neither Egyptian nor Greek historical records, but result from Plato's philosophical views:

1. The Phaethon myth has an astronomical explanation (Tim. 22c–d).
2. Several floods (Tim. 23b) and cataclysms (Tim. 22c, 23a–b) have occurred during human history.
3. The Greek city of Athens is older than the Egyptian city of Sais (Tim. 23e).
4. The ancient Athens was reigned like Plato's ideal state (Tim. 24a–b, 25e).
5. The ancient Athens has won a world war (Tim. 24e, 25c, Crit. 108e).
6. A continent exists beyond the Atlantic Sea (Tim. 25a).
7. The war was a joint venture of the Hellenes (Tim. 25b–c) which occurred long before the happenings of the Greek myths (Tim. 22a–c). By contrast, the first joint venture of the Hellenes was the Trojan war (Thucydides 1, 3).
8. There is a mud sea in front of the Strait of Gibraltar (Tim. 25d, Crit. 108e–109a).
9. The then acropolis of Athens was very large (Crit. 112a). By contrast, Thucydides noted that all castles of the Mycenaean time appear small for his contemporaries (1, 10).
10. The Atlanteans had triremes (Crit. 117d). By contrast, the first triremes were used by the Corintheans in the 8th century BC (Thucydides 1, 13).

## 3 Ancient Athens

Plato's description of the ancient Athens resembles the Bronze Age Athens during the period Late Helladic IIIB around 1200 BC. This suggests that Plato has used either Greek tradition or own investigations for the description of ancient Athens. These similarities include the following claims:

1. The accomodations of the warriors were in the northern part of the Athenian acropolis (Crit. 112b). In fact, this was the location of their accomodations from the 15th century until around 1200 BC [6].
2. A spring on the Athenian acropolis has been destroyed by earthquakes just after the war against the Atlanteans (Crit. 112c–d). In fact, Broneer excavated this fountain which has existed for only 25 years and has been destroyed by an earthquake around 1200 BC [7].
3. The Greeks had knowledge of writing until this earthquake (Tim. 23c). Indeed, the Mycenaean Linear B was written in Greek language [8] and its knowledge became lost in Greece around 1200 BC.

## 4 Sea Peoples

Around 1200 BC the countries in Asia Minor and the Near East were destroyed by a war [9]. The attackers were the Sea Peoples described by Pharaoh Ramses III [10]. The similarities between the Sea Peoples and the Atlanteans [11–13] suggest that Plato has used Egyptian records for the description of the Atlantean war. Quotations of the temple inscriptions are given in the combination of plate number and line number:

1. The Atlanteans fought against Europe and Asia (Tim. 24e) and "every country within the mouth", i.e. against the Eastern Mediterranean countries (Tim. 25b). The Sea Peoples destroyed Hatti in Anatolia, Qode and Qarkemish in northern Syria, Arzawa in southwest Anatolia, and Alasia on Cyprus (Plate 46.16–17) and fought against Egypt.
2. The Atlanteans lived in an isle (Tim. 24e, 25a, 25d, Crit. 113c) and reigned over several other islands (Tim. 25a). Also the Sea Peoples came from islands (Pl. 37.8–9, 42.3, 46.16).
3. The Atlanteans reigned in Africa from the pillars of Heracles (Gibraltar) to the frontiers of Egypt (Tim. 25a–b). The war of the Sea Peoples against Egypt occurred simultaneously with the war of the Libyan Meshwesh. According to Ramses' report they appeared to be allies.
4. Atlantis consisted of ten countries (Crit. 113e–114a, 119b). According to the Karnak inscription written under pharaoh Merenptah around 1200 BC, the Sea Peoples consisted of the Ekwesh, Teresh, Lukka, Sherden, and Shekelesh. According to Ramses III their confederation consisted of the union of the countries of the Peleset, Theker, Shekelesh, Denen, and Weshesh (Pl. 46).

5. In case of war the Atlanteans had more than one million soldiers (Crit. 119a–b). Ramses III claimed to have beaten hundreds of thousands of enemies (Pl. 18.16, 19.4–5, 27.63, 32.10, 79.7, 80.36, 80.44, 101.21, 121c.7). Occationally, he spoke of millions (Pl. 27.64, 46.4, 46.6, 79.7, 101.21) and myriads (Pl. 27.64) of enemies numerous like locusts (Pl. 18.16, 80.36) or grasshoppers (Pl. 27.63).
6. The Atlanteans had 1200 warships (Crit. 119b). The ships of the Sea Peoples entered deep into the delta of the Nile (Pl. 42.5) and destroyed the Asian Arzawa, the Cypric Alasia, and the near-eastern Ugarit and Amurru.
7. The Atlanteans had chariots pulled by horses (Crit. 119a). The Meshwesh had horses (Pl. 75.37) and carts (Pl. 18.16, 75.27) which, however, were pulled by oxen (figures to Pl. 32–34).
8. The Atlantean kings reigned for several generations (Crit. 120d–e) and after this they lost their good attitudes (Crit. 121a-b). Ramses III wrote about the Sea Peoples that they had spent a long time, a short moment was before them, then they entered the evil period (Pl. 80.16–17).
9. During a day and a night Atlantis sank by a earthquake into the sea (Tim. 25c–d). Ramses III wrote that he let the Sea Peoples see the majesty and force of (the God of water) Nun when he breaks out and lays their towns and villages under a surge of water (Pl. 102.21), moreover the mountains were in travail (Pl. 19.11).

## 5 Tartessos

Plato described the place of the Atlantean capital. The capital (Crit. 115c) was on a to-all-sides flat hill which was 50 stades (9 kilometres) distant from the sea and lay at the edge of a plain (Crit. 113c). This plain was rectangular (Crit. 118c), smooth and even. The plain lay on the southern part of the isle (Crit. 118a–b), in its middle (Crit. 113c). The plain was surrounded by mountains which reached to the sea (Crit. 118a). Apart from this, the country was very high and had a steep coast (Crit. 118a).

The isle of Atlantis was divided under the ten sons of Poseidon (Crit. 113e). The first born, Atlas, obtained the largest and best territory, namely the region around the capital (Crit. 114a). The second born, Gadeiros, obtained the part at the most distant edge which reached from the pillars of Heracles (Gibraltar) to the Gadeirean country (the region around Cadiz) (Crit. 114b).

The first born, Atlas, obtained the largest and best part. Therefore one can assume that the later born sons obtained smaller and smaller parts. According to this, the second born son, Gadeiros, obtained the second largest part of the "isle of Atlantis". This part included the coastal region of Spain from Cadiz to Gibraltar. Here, the term "isle" should be rather understood as "coast" or "region".

The part of the country belonging to Gadeiros was only a coastal region 100 kilometres long. The parts of the later born sons were probably even smaller. Thus, the part of the country belonging to Atlas cannot have been very far from Cadiz.

In fact, near Cadiz there is a rectangular (Crit. 118c), smooth and even plain which lies at a south coast (Crit. 118a–b). It is the plain south-west of Seville through which the Guadalquivir flows. Was here the capital of Atlantis?

The geography of Atlantis resembles that of the Iron Age Tartessos [13–18]. This suggests that Plato has heard about Tartessos from Greek sailors during his visits in Syracuse. There are the following similarities between Tartessos and Atlantis:

1. Tartessos lay in the mouth of the Guadalquivir river (Strabon III 140; Pausanias VI 19,3). Atlantis lay in front of the pillars of Heracles, i.e. Gibraltar (Tim. 24e, Crit. 108e) and at a south coast (Crit. 118b).
2. The plain of Tartessos included a system of channels (Strabon III 143). Plato reported the same about the plain of Atlantis (Crit. 118d–e).
3. Tartessos was surrounded by the Sierra Morena and the Sierra Nevada. The plain of Atlantis was surrounded by high mountains (Crit. 118a-b).
4. Tartessos was rich in metals (Herodotus IV 152, Diodorus V 35,4). Plato reported the same about Atlantis (Crit. 114e).
5. According to archaeological evidence and Strabon (III 139), the Tartessians had knowledge of writing. Plato wrote the same about the Atlanteans (Crit. 119c–d).

## 6 Conclusion

Plato's Atlantis tale (Timaios 20d–27a, Critias 106a–121c) is philosophical fiction invented to describe Plato's fictitious ideal state in the case of war (Tim. 19b–20c). Plato invented the Atlantis tale after he failed twice in educating the tyrants of Syracuse to behave like philosopher kings [4, 5]. However, the Atlantis tale appears to include three historical elements [13]:

1. Plato's description of ancient Athens resembles Mycenaean Athens of the period LH IIIB around 1200 BC [6, 7]. Plato appears to have used either Greek tradition or own investigations.
2. Plato's description of the Atlantean war resembles the war of the Sea Peoples around 1200 BC [11–13]. Plato appears to have used the temple inscriptions of Medinet Habu written under pharaoh Ramses III.
3. Plato's description of the city and geography of Atlantis resembles that of iron age Tartessos [13–18]. Plato appears to have heard about Tartessos during his visits in Syracuse.

## 7 Outlook

Preliminary evidence for Atlantis in the Guadalquivir valley has been presented by Werner Wickboldt at the conference "The Atlantis Hypothesis: Searching for a Lost

Land", Milos Island, Greece, 11–13 July 2005. He presented a satellite photo of the Donana National Park which showed the following structures.

1. A rectangular structure of size 230 metres times 140 metres which could be a remnant of the "silver temple of Poseidon" (Crit. 116c–d). The geographical coordinates of this structure are 36°57′25″ +/- 6″N and 6°22′58″ +/- 8″W.
2. A rectangular structure of size 280 metres times 240 metres located 500 metres south-west of the first structure. It could be a remnant of the "golden temple of Poseidon and Cleito" (Crit. 116c).
3. These two structures are surrounded by parts of concentric rings. They could be remnants of the concentric rings of earth and water (Crit. 115e–116a).
4. A chain of lakes located north-west of these structures which could be a remnant of the channel which ranged from the capital of Atlantis to the sea (Crit. 115d).

## References


1. Schliemann, H., *Ithaka, der Peloponnes und Troja*, Giesecke & Devrient, Leipzig (1869).
2. Kluge, F., De Platonis Critia, *Rheinisches Museum für Philologie* **75** (1910) 283.
3. Leaf, W., *Homer and History*, Macmillan, London (1915).
4. Ramage, E.S. (Ed.), *Atlantis – Fact or Fiction?*, Indiana University Press (1978).
5. Forsyth, P.Y., *Atlantis – The Making of Myth*, McGill-Queen's University Press (1980).
6. Broneer, O., What happened at Athens, *American Journal of Archaeology* **52** (1948) 111–124.
7. Broneer, O., A Mycenaean fountain on the Athenian Acropolis, *Hesperia* **8** (1939) 317–429.
8. Ventris, M. and J. Chadwick, J., Evidence for Greek dialect in the Mycenaean archives, *Journal of Hellenic Studies* **73** (1953) 86–103.
9. Drews, R., *The End of the Bronze Age*, Princeton University Press (1993).
10. Edgerton, W.F. and J.A. Wilson, *Historical Records of Ramses III*, University of Chicago Press (1936).
11. Marinatos, S., Peri ton Thrulon tes Atlantidos, *Kretica Chronika* **4** (1950) 195–213.
12. Görgemanns, H., Wahrheit und Fiktion in Platons Atlantis-Erzählung, *Hermes* **128** (2000) 405–419.
13. Kühne, R.W., A location for Atlantis?, *Antiquity* **78** (2004) 300; see also http://antiquity.ac.uk/ProjGall/kuhne.
14. Jessen, O., Tartessos-Atlantis, *Zeitschrift der Gesellschaft für Erdkunde* **6** (1925) 184.
15. Hennig, R., Das Rätsel der Atlantis, *Meereskunde* **14** (1925) 1–29.
16. Hennig, R., Zum Verständnis des Begriffs "Säulen" in der antiken Geographie, *Petermanns geographische Mitteilungen* **73** (1927) 80–87.
17. Schulten, A., Tartessos und Atlantis, *Petermanns geographische Mitteilungen* **73** (1927) 284–288.
18. Schulten, A., Atlantis, *Rheinisches Museum für Philologie* **88** (1939) 326–346.

# Homer's Reference to Writing in Proitos' Era

Efthymia Polygiannaki

*Athens, Greece*

**Abstract.** In Book VI of the Iliad Homer refers to a letter written by Proitos, king of Tiryns, and addressed to his father-in-law, Iobates, king of Lycia, requesting him to kill Bellerophontes. Proitos entrusts the closed, folded letter to Bellerophontes himself, so that he would show it to Iobates. But in which script was that message written by Proitos? If we accept the genealogies of the historic tradition, then Proitos must have lived in the Mycenaean era and so he wrote in linear B. If, however, we connect him with the Hellenikon or the Ligourio pyramid, accepting the tradition mentioned by Pausanias (Corint. 25,7), then, according to the dating of these pyramids through the optical thermoluminescence method, Proitos wrote in hieroglyphics (to better determine the relative time-span, excavation data from Argolid and other places are co-evaluated). These hieroglyphics are Cretan hieroglyphics, from which the linear scripts A, B and the Cypriot syllabic script originate.

## 1 Introduction

In Book VI of the Iliad [1], Homer mentions a letter sent by the king of Tiryns, Proitos, to the king of Lycia, Iobates. In which kind of script did Proitos write in that prehistoric era, when the alphabet had not yet been invented?

In order to draw some conclusions on this very interesting matter, one should of course determine with the maximum possible accuracy the time in which Proitos lived and then examine which kind of script was in use at that period in Greece. To travel so far back in time, as to be able to lean over the king's shoulder as he was writing and take a look at his "signs", we have no other guide but archaeology, science and tradition.

## 2 When Did Proitos Live?

At the feet of the Pontino hill, in Argolis, the prehistoric settlement of Lerna has been discovered. Its prime starts at the beginning of the 3rd millennium BC and culminates during the protohelladic period (2500–2100 BC), when a double storey

*S.A. Paipetis (ed.), Science and Technology in Homeric Epics,* 515–523.

palace and a defensive enclosure were built. Around 2100 BC, however, this great settlement is ruined and the stately palace is burned [2, 3, 8]. The conclusion at which specialists have arrived, is that Lerna was destroyed by a group of Danaoi, who founded the settlement that followed, establishing themselves at the place. At the same time, destructions and other changes have been observed also in other sites of Argolis [2].

In fact, according to the tradition, Danaus came in Argolis claiming the throne as descendant of Io, the Princess of Argos, who had fled in the land of Nile in fear of her parents, since she was pregnant while still unmarried. At the end, Danaus became king and master of the Argeian land, whose inhabitants he called Danaoi [5]. (The stories of Danaus, the Danaids, Danae, etc. [5, 14] are well known.)

Abas, the grandson of Danaus, had two twin sons, Proitos and Acrisius, from his daughter Ypermestra. The twins, they say, used to argue, even in their mother's womb. When they grew up, they opened war for power – a war in which newly-invented shields were used. Taking the upper hand, Acrisius chased Proitos away. Proitos fled to Lycia, to Iobates, whose daughter Anteia or Stheneboea he married. Iobates helped Proitos return to his country followed by an army of Lycians who helped him to take Tiryns and build fortifications, Cyclopean walls. The two brothers divided the Argeian land, with Acrisius becoming King of Argos and Proitos King of Tiryns [4, 5, 14]. Excavations in Tiryns have shown that during the Protohelladic period (2500–2100 BC) there was a flourishing settlement with a big circular mansion, home of a king or a castle. However, during the Mesohelladic period (circa 2100–1600 BC) [8], the look of the acropolis changes: upon the circular edifice the older mansion is built, as well as a defensive enclosure whose entrance is in the future site of the propylon (during Mycenaean times, the enclosure was greatly expanded in two phases [3, 8]). Since specialists attribute this to the arrival and settlement of new populations and since the tradition is explicit about Proitos taking Tiryns and becoming its king, one has serious reasons to believe that it was him who built the old mansion and the fortification. By combining the aforementioned, we can thus arrive at the conclusion that Proitos, as third descendant of Danaus, must have lived in the beginning of 2nd millennium BC.

This conclusion is further supported by an important archaeological finding, e.g. the two pyramids of Argolis, one of which must be the one that Pausanias noticed in the area. As the highly dependable traveler states, "In the right, as one goes from Argos to the land of Epidaurus, there is a structure very much like a pyramid, bearing, in relief, images of shields, like the Argolic shields. At this place, they say, Proitos had fought against Acrisius for the Kingdom, a battle whose outcome was even, since neither of them could achieve a clear victory. They say that this was the first combat fought with shields, for them and their soldiers. For the dead of both parts, a common grave was erected in the place, since they all were compatriots and relatives".

Two pyramids are preserved in Argolis to this date. One of them is at the modern village of Hellenikon and the other at Ligourio. The pyramid of Hellenikon is fairly well conserved, while that of Ligourio is ruined; only its foundations are visible today, as well as one corner which is preserved up to a certain height, providing evidence of the form of the structure. The dimensions of the base of both pyramids are roughly equal (about 14 × 12.5 meters). Since both constructions appear to be similar, the form of Ligourio pyramid can be reconstructed according to Hellenikon one, which is a rectangular, rather truncated pyramid, void on the inside, with a gate and a corridor leading to the interior [8].

It is not clear which of the two pyramids was the one noticed by Pausanias, there are arguments for both cases. However, this is immaterial. For the present investigation, what matters is the dating of these structures, i.e. whether ca. 2000 BC, considered as Proitos' era, lies inside the time interval of dating for either the one or the other pyramid.

As a matter of fact, on February 9th, 1995, the results of dating by optical thermoluminescence were announced at the Academy of Athens. The measurements were carried out by Dr. Ioannes Liritzis, inventor of this new method (then researcher at the Academy of Athens and today Professor of Archaeometry at the Aegean University) at the Faculty of Physics of Edinburgh University, with permission and equipment offered by Professor R.B. Galloway [11, 12]. According to these measurements, Hellenikon pyramid was dated at 2720 BC (± 580 years), and Ligourio pyramid at 2100 BC (± 610 years). Consequently, the beginning of the 2nd millennium BC, a time at which the other evidence converges as the time in which possibly Proitos lived, is inside the dating time interval of the pyramids. This is important, not only because it strongly supports previous conclusions, but also because a totally different dating would greatly shake its correctness. Consequently, chronological correlations point at the beginning of 2nd millennium BC as Proitos' era.

## 3 The Letter and Its Script

According to the tradition first transmitted to us by Homer [1, 4, 5], Bellerophontes from Efyra had become a stronger king than Proitos. So Proitos chased him away from the Argeian land, after a slander by his wife: Anteia had fallen passionately in love with Bellerophontes and secretly proposed to him to get together. When the sensible Bellerophontes refused, she asked her husband to chase him away or kill him, since allegedly he had tried to rape her. Proitos was angry, but, not wanting to kill him, he wrote on a folding tablet (ἐν πίνακι πτυκτῷ), a sinister message for his extermination and dispatched him to Lycia asking him to take the letter to his father in law, Iobates, so that he might find death from him. Iobates had Bellerophontes as a guest for nine days and, on the morning of the tenth day, he asked to see the writing he brought from his son-in-law. As soon as he saw the deadly message, he ordered him to fulfill various dangerous missions. Bellerophontes, however, succeeded in all of them and then Iobates kept him there, giving him his daughter as wife and half of his kingdom.

To start with, it is worthwhile to consider the form of Proitos' letter, which is described by Homer as πίναξ πτυκτός (folded table). One is really amazed by the accuracy and the historic consistency of the poems, because one such tablet was in fact found between the merchandise of a Mycenaean ship of the 14th century BC, shipwrecked a few miles west of Chelidonia Cape in Asia Minor! It is a wooden diptych, with ivory hinges and dug-in internal surfaces, which would be covered with wax, topped by red sandaraque. On the wooden margin, three carved signs, resembling characters of linear script B have been preserved [6]. Undoubtedly, this unique evidence offers ground and authenticity to Proitos' letter and helps us imagine it as something similar.

Now, as far as the system of writing is concerned, we must consider, right from the beginning, as absolutely logical that Proitos wrote in a script of his time and place. The period in which, as deduced, Proitos must have lived, coincides with the end of Prepalatial and the beginning of Protopalatial period in Crete, during which Minoan script, in its phase of syllabic hieroglyphic system, was in use. It is true that the erection of the first palaces, defining the beginning of Protopalatial era, was placed at around 1900 BC. However, the fact that the stromatography of the foundations included a "tea-pot" of a former period, (EMIII-MMIA), proves that the old palace of Malia was built between 2100–1900 BC at the latest (MMIA) [7]. For Cnossos, shards of the period 2100–1900 BC found under the floors and the foundations of the first palace prove the same [3]. Consequently, the beginning of the Protopalatial age of Crete essentially begins earlier than 1900 BC and coincides with Proitos' time.

One of the most important characteristics of the great Minoan Civilization is this script. Because all the evolutionary stages of this script were found in Crete, the island is considered to be the place at which they were created and evolved. As a matter of fact, the unprecedented systematic growth of trade (already from the Prepalatial Period) as well as the unprecedented for all contemporary civilizations political and social organization and the impressively pre-designed technical works of the Protopalatial Period, could never have existed without use of a developed writing system and, of course, of an arithmetic system. This writing, Minoan, Cretan or Aegean, as it came to be called, appears in Crete in the following phases, which constitute separate systems: the ideographic, the syllabic hieroglyphic phase and the linear scripts.

In ideography, each ideogram representing beings, plants, objects etc., stands for a word-idea. In the syllabic hieroglyphic script each hieroglyphic symbol – which gradually evolves to Protolinear – does not denote an idea anymore but a syllable. The big leap from the semantic to the phonetic script has already been accomplished! Subsequently, the need for fast writing creates the linear scripts. Linear script A and linear script B (called also Mycenaean) and the Cypriot linear script are subtractive linear renditions of the hieroglyphic symbols, to which by now they remotely resemble [2, 6, 9, 10, 13]. It is worthwhile noting that from these linear scripts the Alphabet originates – i.e. the corner stone of modern civilization – in which a character denotes not a syllable but a pure sound. Among the systems of the Minoan scripts, linear script B and the latter Cypriot script have been fully deciphered, and found to be Greek. Scarce material and other obstacles stand in the way of the full deciphering of the syllabic hieroglyphic script and of the linear script A, although there is serious evidence that they are also Greek scripts [13].

By the vigour of the Minoan civilization, Minoan writing had a wide dissemination in the larger Aegean area and in the whole Mediterranean, where Minoan people conducted their trade. Crete's contacts with the mainland of Greece and with the Aegean are demonstrated by the deep influence of the Minoan civilization in this area – due not only to commerce but also to ethnic bonds [13]. It is characteristic that the Mycenaean civilization, which succeeds Minoan Civilization, is called Cretomycenaean; its script is one of the linear systems we have referred to, i.e. linear script B, and it is not accidental that Greek tradition and mythology are so closely connected to Minoan Crete (Daedalus, Theseus, the Minotaur, Zeus etc.). Between Crete and the nearby Peloponnese, relationships develop from very early times. The excavations show that, in Argolis, Minoan ceramics are introduced in relatively large quantities, while Minoan influence is greatly felt in Lerna in the period 2100–1900 BC (MMI) [2, 3], i.e. in Proitos' times. This means that King Proitos had close relationships with Minoan Crete and accordingly we can consider as certain that he knew its script which, at the time, according to the evidence, was at the phase of the syllabic, hieroglyphic system.

As to this date, no hieroglyphic script in the Greek mainland has been found. This, however, does not necessarily mean that it did not exist. Plutarch, in his biography of Agesilaos, provides a relevant historic evidence: when this Spartan king came to Boetia with his army (394 BC), a strange grave was discovered in Aliartos. Everybody though that this was the grave of Alcmene, Hercules' mother. In the grave, a copper plate with a lot of symbols of an unknown script was found. Being curious to know what was written on the plate, and since the letters resembled Egyptian hieroglyphics, they turned to the King of Egypt to have the text read. The King assigned the task to a young priest by the name of Honoufis who, after profound study and research in old books, presumably willing to give an answer, stated that the plate belonged to the time of Proteus and generally encouraged the Greeks to cast dissensions aside and devote themselves to Letters and Philosophy, as well as to establish Games to honor the Muses. It is obvious that Honoufis did not recognize the script – so the script was not Egyptian. His answer does not even sound truthfull, since this text not only had nothing to do with the buried person – as it should – but neither would such an advise to the Greeks be of any use resting in a dead person's grave. Consequently, since the only hieroglyphic symbols known to exist in the area are those of the Minoan script, we can assume that Agesilaos in fact found himself in front of a text of Minoan hieroglyphic script, like the script in which Proitos must have written to Iobates.

Another approach, however, places Proitos in the Mycenaean period, in which case he must have undoubtedly written in linear script B. This approach is based on the Argeian genealogy, according to which the ancestors of the protagonists of the Trojan wars (that is the Atreides family), as well as Tlepolemus (son of Herakles) are separated from Proitos and Acrisius by 7 or 8 generations [4]. In this way, one can deduce that Proitos lived somewhere around 1500 BC, which means that he wrote in a linear script, since the hieroglyphic script ceased to be used after 1600

BC. Although in Tiryns, Proitos' city, we have the presence of Linear A, Linear B is more likely as Proitos' script because, as we know, this is the system in which Mycenaeans wrote. At this point, it is important to stress that script systems A and B have so many characters in common that essentially we are talking about the same script. Differences – rather of regional character – could be compared to those observed among the numerous Hellenic alphabets (Ionic, Chalcidean, Cretan, Corinthian etc.). The possibility that Proitos lived around 1500 BC would fit with the tradition, according to which the first walls, which Proitos built in Tiryns, were Cyclopean, as these walls, according to the conventional dating, are placed in Mycenaean times. So which is the truth? Either conventional dating (regarding Trojan War, the first building phase for the walls, etc.) must in fact be pushed back by 250–300 years, in which case everything falls into agreement, or there could be some confusion in the genealogy. Perhaps new scientific dating methods in the future, as well as new findings will solve the riddle.

## 4 Conclusion

From what we have mentioned so far, one can deduce that Proitos lived around 2000–1950 BC and wrote in the syllabic hieroglyphic script of Crete, without excluding the case that he lived during the Mycenaean times and wrote in linear B. The answer to our initial question cannot be categorical, however, it is very exact as far as the family to which these systems belong is concerned, since, as undoubtedly and generally accepted linear B, as well as the other linear systems, have a common ancestor, the syllabic hieroglyphic script, which in turn has been derived from ideography. This is confirmed not only by the schematic evolution of each symbol, but also by its phonetic value, as we know it after its deciphering by Ventris [9, 10].

My relevant research and studies [13] have shown that the syllable represented by each character of linear B, is the first syllable of the word corresponding to the image of the ancestral pictorial symbol, as in Table 1.

As I found out, Professor Günter Neumann had come to the same conclusion for the well-known ideogram of linear B for figs, which in Crete were called "νικύλεα" [23]. The same character, in linear B, denotes the syllable "ni" (νι-κύλεα).

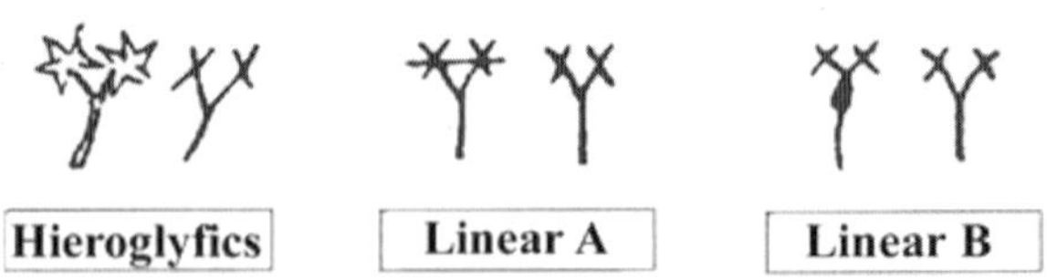

**Table 1**

| Hieroglyfics | Linear A | Linear B | Phonetic value | Word |
|---|---|---|---|---|
|  | 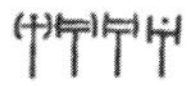 |  | a | α - ξίνη (axe) |
|  |  |  | ka (hα) | hά - ρμο (wheel) |
| | 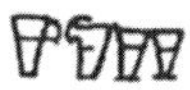 |  | ki | κι - σσύβιον (cup) |
|  |  |  | qe (πε) | πέ - λτη (shield) |
| 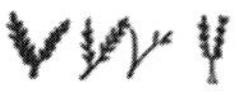 |  |  | qo ($κ^{w}$o) | κό - τινος (olive tree) |
|  | 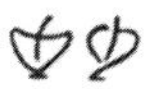 |  | tu (θυ) | θύ - ρσος (ivy) |
|  |  |  | rya | ρύα - ξ (stream) |
|  | 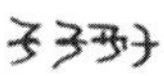 |  | ku (γυ) | γυ – ψ (vulture) |
|  |  |  | ko (γο) | γό - μφος (peg) |

(We know the grammar of linear B, in which there is only one series for the patalal sounds of κ, γ, χ, for the dental τ, θ, etc., while the "q" series has either labial or labial-palatal pronunciation.)

And even Chadwick himself, for the ideogram for wool which, as a syllabic character in linear B, has the value of "ma" (μα-λλός) [10].

Hieroglyfics

Linear B

We are essentially talking about a Greek mother-tongue and her children. It is essential here, to note that it is a genuine Aegean script, representing flora and fauna of the Aegean, its ships and everyday-life objects of the Cycladic-Minoan people. This is the script in which Proitos wrote!

## *Derivation of the Alphabet from the Minoan Syllabic Scripts*

My research has arrived at the conclusion that, in the transition from syllabic script to the alphabet, the acrophonic system was applied. The symbol now keeps the first sound of the syllable which it represented in the syllabic scripts, as its phonetic value [13].

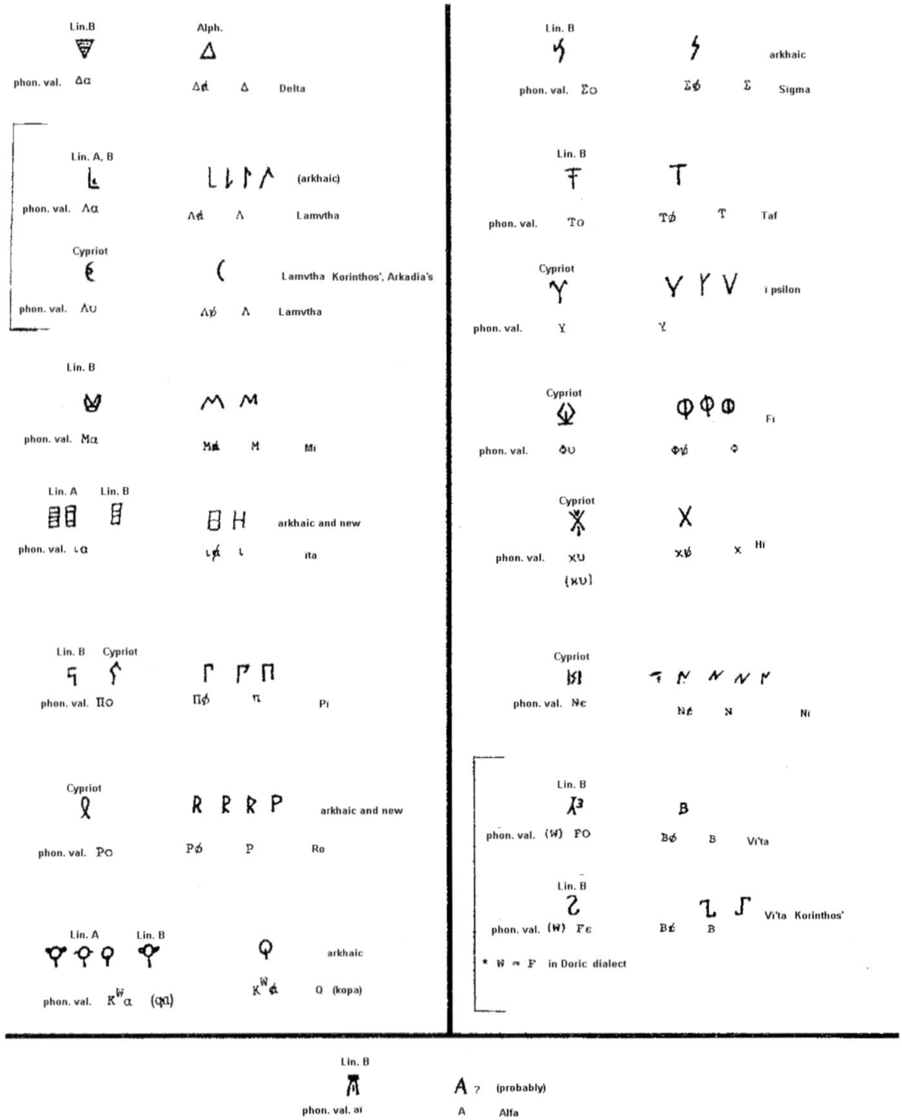


From: Efi Polygiannaki, *The Phaistos Disk Speaks Greek*, 3rd Greek edition, Athens, 2000.

## References

1. Homer, *Iliad.*
2. *History of the Greek Nation*, Ekdotike Athinon, Vol. A.
3. Treuil. R., Darcque, P., Poursat, J. and Touchait, G., *Les civilizations égéenes du Néolitique et de l'Age de Bronze*, Kardamitsa Editions, 1996.
4. Athan, Stageiritis, *Ogygia or Archaeologia*, Vol. D, Nea Thesis Editions.
5. Apollodorus, *Library.*
6. Ruipérez, M.S. and Melena , J.L., *Los Griegos micènicos*, Kardamitsa Editions, 1996.
7. St. Alexiou, *Proceedings of the 4th International Cretological Convention*, Heraklion, 1976, Athens, 1980–1981.
8. Pausanias, *Description of Greece. Corinth, Laconia and Comments* (N. Papahatzis) Ekdotike Athinon.
9. Hooker, J.T., *Linear B, An Introduction*, Educational Institute of the National Bank.
10. Chadwick , J., *Linear B and Related Scripts*, Papadima Editions, 1992.
11. Announcement of the Member of the Academy of Athens, Pericles Theoharis.
12. Liritzis, I., *The Mystery of the Greek Pyramid-Like Structures*, Kardamitsa Editions.
13. Polygiannaki, E., *The Phaistos Disk Speaks in Greek*, 3rd ed., Georgiadis Editions, 2000.
14. Kérenyi, K., *The Mythology of the Greek*, Estia Editions.
15. Plutarch, *The Life of Agesilaos.*

# Linguistic Science and Script Technology: The Homeric Evidence

A. Teffeteller

*Linguistics, Concordia University, Montreal, Canada*

Linguists recognize that language plays many roles: it is, among other things, a system of communication, a social institution, and a medium of ethnic and cultural identity. These fundamental uses of language have been recognized since antiquity but it is only recently that the sociolinguistic issues have been afforded serious concern and subjected to the detailed analysis that has marked the more formal aspects of the scientific study of human language. But in the Homeric epics we see already an interest in these concerns in the awareness of multilingualism, of the problems in communication that can result from the lack of bilingualism where it is needed, and of ethnic identification along linguistic lines.

When Odysseus in disguise spins one of his tall tales of past life and lineage, he mentions in a famous passage the island of Crete, set, he says, in the midst of the wine-dark sea, with many men in it, their numbers past counting, and 90 cities. "They have not", he says, "all the same speech; their tongues are mixed. There dwell Achaeans, there great-hearted True-Cretans, there Cydonians, and Dorians in their three divisions, and noble Pelasgians" (Od. 19.172–77). Homer here describes, as Russo notes, "a society of mixed, international composition, the only such settlement known in early Greece, whether we assume the description to be valid for the Bronze Age, for Homer's time, or for some period in between" [28, p. 83]. In the narrative period of the *Iliad*, the Achaeans would be the Mycenaeans who dominated Crete in the late Bronze Age and whose leader, Idomeneus, plays such a prominent part in the war against Troy. The name "True-Cretans" ("Eteocretans") clearly designates the original inhabitants of the island; the Cydonians also are said by Strabo (10.4.6) to be autochthonous. And the reference to Dorians "in their three divisions" indicates that already by the time this passage was composed the Dorians had penetrated as far south as Crete and that they were already socially and ethnically delineated in the three tribes known to us from the classical period. Strabo cites the lost historian Staphylos in assigning the Dorians to the eastern part of the island,

*S.A. Paipetis (ed.), Science and Technology in Homeric Epics,* 525–530.

the Eteocretans to the south, and the Cydonians to the west, in the region of modern Khania.

In the first millennium BC Crete was dominated by Dorians, although non-Greek languages continued in use throughout the period (see, e.g., [2, 9, 37]; for an historical sketch of Greek and the dialects, see [22, 23, 31]). From the second millennium we have evidence of non-Greek languages in the written records which are not yet fully understood (see most recently [10], with references to earlier literature; see also [5, 6, 8, 9, 14, 17, 25, 38, 39]); the evidence of place names also indicates the presence of a non-Greek language prior to the appearance of Greek in Crete and on the mainland in the mid-millennium. While lexical items with no Greek etymology such as *laburinthos* "labyrinth", *erebinthos* "chick-pea", *kuparissos* "cypress", *kolossos* "statue", *apene* "cart", *sagene* "drag-net" could be simple borrowings from contact with other languages in the process of trade, place names on Greek soil indicate the early presence of speakers of the languages in the areas so designated, for instance, *Korinthos*, *Zakunthos*, *Parnassos*, *Halikarnassos*, *Athenai*, *Mukenai* [3, 7, 10].

The pre-alphabetic scripts on Crete include the hieroglyphic script, used on sealstones dating as early as 1900 BC, the Linear A script, in wide-spread use on the island from about 1800 to 1450 (and in use also in Minoan settlements in the Cyclades), and Linear B, used to write the Greek language from as early as the 16th century [4, 5, 10, 25, 34]. Finally, the famous Phaistos Disk, an isolated document using a pictographic (perhaps syllabic) script, has resisted all attempts at decipherment [8, 10]. Linear A and the Cretan hieroglyphic script hold out more hope but have not yet been adequately deciphered. In none of these cases is the language recorded on these documents known, although the suggestion of Luwian, an Anatolian language, has recently proved attractive to a number of researchers as the language represented by Linear A (see [10] and references therein).

In the various languages spoken on Crete one might not unreasonably expect to find widespread bilingualism, even extensive multilingualism. At Troy, on the contrary, the lack of bilingual competence presents a problem for the Trojan allies. Iris, in the form of his brother Polites, cautions Hector to ensure that each linguistic group has a commander who speaks the language, for "there are many allies about the great city of Priam and language differs from language among the scattered nations" (Il. 2.802–6). The effect of the sound of many unknown tongues on an uncomprehending hearer is well expressed in the simile likening the clamour of the disparate Trojan armies to the ceaseless bleating of sheep (Il. 4.433–8), while the (unilingual) Greeks march silently and inexorably into battle. We are reminded that Homer characterizes the Carians as *barbarophonoi* (Il. 2.867) and the non-Greek Sintians on Lemnos as "of wild speech" (*agriophonoi*, Od. 8.294).

The Trojan catalogue in Book 2 of the *Iliad* lists the Trojan contingents and those of their allies, providing geographical and topographical details which allow us to locate with greater or lesser precision most of the contingents named (see [15, pp. 248–263]). As Kirk observes, "The Trojan catalogue starts logically enough from the main city involved; it then passes to other contingents from the Troad, then radiates outwards in four different directions to take in the allies. … Thus the poetical tradition remained conscious that the whole Trojan force consisted of three elements: (i) the Troes proper, from the city of Ilios and perhaps a few other places in the vicinity; (ii) the Dardanoi or Dardanioi, from the foothills of Ida; (iii) the allies, *epikouroi*, from farther afield [15, pp. 250–253]. Not all the language alignments, however, are immediately obvious from these geographical indications. Thus, while we may reasonably assume that the Lycians (the Lukioi) spoke Lycian, and the Carians (the *barbarophonoi* Kares) Carian, what are we to suppose of the Pelasgoi, or the Paiones who come "from afar", or the Halizones from "distant Alube, the birthplace of silver"?

The linguistic situation in Anatolia in the second and early first millennia BC is preserved for us to some extent by the written records of the periods, primarily in the second millennium by the multilingual records of the Hittite empire and in the early first millennium by local inscriptions both in monumental scripts and in local script and dialect. Since the Homeric poems may – and are generally assumed to – draw on the historical circumstances of various periods from the Mycenaean down to the Iron Age, we may expect to find a range of circumstances reflected in these texts, from the mid-second millennium down to the early first. In Anatolia we are able to specify the following languages in these periods: first, the most copiously attested language of the region, Hittite, the official language of the Hittite Empire, in records from c. 1900 to c. 1100 BC; and Luwian, a language closely related to Hittite, spoken in adjoining regions south and west of Hittite lands, areas which were at different times under the control of, in liege to, or in conflict with the Hittites. Hittite is attested in the cuneiform script which the Hittites borrowed from Akkadian, and which had been in turn borrowed at an earlier date from the Sumerians. Luwian is attested in two dialects, "Cuneiform" Luwian, recorded in glosses and brief passages written in the cuneiform script of Hittite texts, and "Hieroglyphic" Luwian (also referred to in earlier scholarly publications as "Hieroglyphic Hittite"), recorded in texts written in a hieroglyphic script on seals and in monumental inscriptions on rock in formal and sacral contexts (for example, the Sacred Pool inscription at the Hittite capital of Hattusa, discovered only in 1988). Hieroglyphic Luwian is in fact the only Anatolian dialect that is attested in both the 2nd and the 1st millennia BC. The script in which it is written is native to the region and was developed by the 14th century BC, although in the early period it was largely limited (as far as our evidence indicates) to the writing of proper names on seals. The monumental use of Hieroglyphic Luwian increased markedly with the fall of the Hittite Empire shortly after 1200 BC, with the majority of inscriptions dating from the 9th to the 7th centuries. (For Anatolian languages, see [13, 18, 21].)

It is looking increasingly likely that the language of the Trojans may have been Luwian (see [10] with references therein, and [16, 36]). There is general agreement now that Troy is to be identified with an Anatolian city known in the Hittite records as Wilusa, a name remarkably similar to *Wilion*, the early form of *Ilion*, a Greek name for Troy. The Hittite king Muwatalli signed a treaty with a ruler of Wilusa known as Alaksandu, whose name almost exactly matches the "Greek" name of the Trojan prince Paris, Aleksandros (see [36] and [1, pp. 82–88]).

Other languages of Anatolia include:

- Palaic, in the region of Pala in north-central Anatolia (classical Paphlagonia), attested only in a few ritual and mythological texts from the Old Hittite period.
- Lycian, on the southwest coast of Anatolia, found on coins and in tomb inscriptions from the 5th and 4th centuries BC, in a script derived from the Greek alphabet. We have two extensive texts, the Xanthos Stele and the Letoon Trilingual, written in Lycian, Greek, and Aramaic. Lycian is widely regarded as a descendant of a dialect closely related to but not identical with attested Luwian.
- Lydian, an Anatolian language spoken on the west-central coast in the 1st millennium, attested in inscriptions on stone mostly from the 5th and 4th centuries, in an alphabet similar to the Greek and in the Old Phrygian alphabet as well.
- Carian, spoken in the far southwest of Anatolia, north and west of Lycia. A few inscriptions have been found in this area and more in Egypt, where epitaphs and graffiti were left by Carian mercenaries from the 7th century to the 4th.
- Pisidic, known from a few alphabetic inscriptions dating from perhaps the 3rd or 2nd centuries BC in the region east of Caria and north of Lycia.
- Sidetic, attested in a handful of inscriptions from the 3rd century BC on the Pamphylian coast. (On these languages, see further [11, pp. 171–176] with references therein, [19,21]; see also [29].)

These and other languages may have been among the many tongues spoken by the "scatttered nations" of the Trojan allies.

Finally, we will look briefly at the evidence for script technology in Homer, who gives us our first reference to writing in European literature, and it is a sinister enterprise, incomprehensible "baleful signs" that are intended to bring about the death, unawares, of their bearer, the hero Bellerophon (Il. 6.152–202).

The reference to these "baleful signs" seems to indicate that the poet of this passage was not closely acquainted with writing, but whether this is indeed the case or whether he was deliberately archaizing we are not able to determine. In any case it is widely assumed that in the Mycenaean period knowledge of writing was largely restricted to the scribal class and that a general acquaintance with writing on the part of the populace arose some time after our first attestations of alphabetic writing in Greece around the middle of the 8th century BC. (It is notable that when the Greek champions draw lots for the privilege of fighting Hector in single combat the mark made by Aias is recognizable only by him: Il. 7.170–192; clearly, this is not "writing" in the sense of a medium of general communication.) The international aspect of the Homeric episode is important as well; as Anna Morpurgo Davies so insightfully comments, "Anyone interested in the Mediterranean area between the second and the first millennium BC is necessarily concerned with the relationship (differences and similarities) between the various types of writing available in what obviously was a melting pot of both languages and scripts" [20, p. 54] (see also [12,26]).

Bellerophon, falsely accused of attempted rape, is sent off to Lycia with a folded tablet on which are written "baleful signs, many and life-destroying" (*semata lugra … thumophthora polla*, Il. 6.168–9), which he is to give to the Lycian king who is thereby instructed to kill him. (Presumably the tablet is sealed or else Bellerophon is unable to read the script or the language written on it.) Of course in true folktale fashion, Bellerophon triumphs in all the trials intended to bring about his death and ends by marrying the king's daughter. But our concern here is with the "folded tablet" on which the signs were written (*en pinaki ptuktoi*).

At various periods in the Greek world letters were written on metal, on wood coated with wax, on earthenware fragments, on vellum (animal skin), and especially on papyrus (see [33] with references therein). We have no letters from the Mycenaean period as we do in the Near East in the second millennium, with the possible exception of the Hittite tablet KUB 26.91, which has recently been reinterpreted as a letter from a Mycenaean king, perhaps the king of Thebes, to the Great King of the Hittites [32]. Diplomatic letters exchanged between Near Eastern kingdoms in the second millennium were usually written in Akkadian, the *lingua franca* of the period. So the *semata lugra* written on the tablet given to Bellerophron could have been Akkadian, Hittite, or Luwian (Cuneiform or Hieroglyphic), Linear A, Linear B, Syllabic Cypriot, Alphabetic Greek, or Lycian (written in a Greek-style alphabet or possibly in cuneiform or hieroglyphics).

As for the material of Bellerophon's letter, the term "folded tablet" seems to indicate a wooden tablet of the type known from the second millennium, which consisted of two "pages" (hence the term "diptych"), each recessed to hold a layer of wax, and joined by a hinge (see [15, pp. 181–182], with references therein). The Ulu Burun shipwreck, off the south coast of Turkey, excavated from 1984 to 1994 and dated to the late 14th century BC, has yielded just such a tablet, made of boxwood with an ivory hinge [27, 30, 35].

The Homeric epics thus contain rich material in the areas of linguistic science and script technology, evidence that is all the more compelling for being so seamlessly interwoven into the dramatic narrative. Once again Homer proves an invaluable source for much that is of interest in our contemporary world.

## References


1. Beckman, G., 1996. Hittite diplomatic texts. In *Writings from the Ancient World*, Vol. 7, H.A. Hoffner, Jr. (Ed.), Atlanta.
2. Brown, R.A., 1985. *Evidence for Pre-Greek Speech on Crete from Greek Alphabetic Sources*. Amsterdam.
3. Carruba, O., 1995. L'arrivo dei greci, le migrazioni indoeuropee e il "ritorno" degli Eraclidi. *Athenaeum* 83:5–44.
4. Chadwick, J., 1987. *Linear B and Related Scripts. Reading the Past*. Berkeley.
5. Chadwick, J., 2003a. Pre-alphabetic scripts (Greece). In *OCD 2003*, p. 1243.
6. Chadwick, J., 2003b. Pre-Greek languages. In *OCD 2003*, pp. 1243–1244.
7. Drews, R., 1997. PIE speakers and PA speakers. *JIES* 25:153–177.
8. Duhoux, Y., 1977. *Le disque de Phaestos: Archéologie, épigraphie, édition critique, index*. Louvain.
9. Duhoux, Y., 1982. *L'Étéocrétois: les texts – la langue*. Amsterdam.
10. Finkelberg, M., 2005. *Greeks and Pre-Greeks: Aegean Prehistory and Greek Heroic Tradition*. Cambridge.
11. Fortson, B.W. IV, 2004. *Indo-European Language and Culture*. Oxford.
12. Hawkins, D., 1979. The origin and dissemination of writing in Western Asia. In *The Origins of Civilization. Wolfson College Lectures 1978*, P.R.S. Moorey (Ed.), Oxford, pp. 128–168.

13. Hawkins, D., 2003. Scripts and texts. In *The Luwians*, H.C. Melchert (Ed.), Leiden, pp. 128–169.
14. Hutchinson, R.W., 1962. *Prehistoric Crete*. London.
15. Kirk, G.S., 1985. *The Iliad: A Commentary. Volume I: Books 1–4*. Cambridge.
16. Latacz, J., 2004. *Troy and Homer*. Translated by Kevin Windle and Rosh Ireland. Oxford.
17. Lochner-Hüttenbach, 1960. *Die Pelasger*. Vienna.
18. Melchert, H.C., 2003. Language. In *The Luwians*, H.C. Melchert (Ed.), Leiden, pp. 170–210.
19. Melchert, H.C (Ed.), 2003. *The Luwians*. Leiden.
20. Morpurgo Davies, A., 1986. Forms of writing in the ancient Mediterranean world. In *The Written Word: Literacy in Transition. Wolfson College Lectures 1985*, G. Baumann (Ed.), Oxford, pp. 51–77.
21. Morpurgo Davies, A., 2003a. Anatolian languages. In *OCD 2003*, pp. 81–82.
22. Morpurgo Davies, A., 2003b. Dialects, Greek (prehistory). In *OCD 2003*, pp. 461–462.
23. Morpurgo Davies, A., 2003c. Greek language. In *OCD 2003*, pp. 653–656.
24. *OCD: Oxford Classical Dictionary 2003*, 3rd ed., S. Hornblower and A. Spawforth (Eds.), rev. Oxford.
25. Packard, D.W., 1974. *Minoan Linear A*. Berkeley.
26. Palaima, T., 2005. The Triple Invention of Writing in Cyprus and Written Sources for Cypriote History. Annual Lecture in Memory of Constantine Leventis, 6 November 2004. Nicosia.
27. Payton, R., 1991. The Ulu Burun writing-board set. *Anatolian Studies* 41:99–106.
28. Russo, J., 1992. *A Commentary on Homer's Odyssey, Volume III*, Books XVII–XXIV. J. Russo, M. Fernández-Galiano, A. Heubeck (Eds.). Oxford.
29. Steiner, G., 1990. The immigration of the first Indo-Europeans into Anatolia reconsidered, *Journal of Indo-European Studies* 18:185–214.
30. Symington, D., 1991. Late Bronze Age writing-boards and their uses: Textual evidence from Anatolia and Syria. *Anatolian Studies* 41:111–123.
31. Teffeteller, A., 2006. Greek, ancient. In *Encyclopedia of Language and Linguistics*, 2nd ed., K. Brown (Ed.). Oxford, pp. 149–151.
32. Teffeteller, A. (Ed.), forthcoming. *Mycenaeans and Anatolians in the Late Bronze Age: The Ahhiyawa Question* (http://modlang-hale.concordia.ca/ahhiyawa.html).
33. Trapp, M.B.., 2003. Letters, Greek. In *OCD 2003*.
34. Ventris, M. and Chadwick, J., 1973. *Documents in Mycenaean Greek*. 2nd ed. Cambridge.
35. Warnock, P. and Pendleton, M., 1991. The wood of the Ulu Burun diptych. *Anatolian Studies* 41:107–110.
36. Watkins, C., 1986. The language of the Trojans. In *Troy and the Trojan War*, A Symposium held at Bryn Mawr College, October 1985, M. Mellink (Ed.), Bryn Mawr, PA, pp. 45–62.
37. Willetts, R.F., 1965. *Ancient Crete*. London.
38. Van Windekens, A.J., 1952. *Le Pélasgique*. Louvain.
39. Van Windekens, A.J., 1960. *Études pélasgiques*. Louvain.

# The Miraculous Homeric Metre

S.A. Paipetis

*University of Patras, Greece*

**Abstract.** Meditation practices, used by many religious and esoteric systems, mainly in the Far East, but also in Christianity, are in fact methods of coordinating heart and respiration rates. This explains their several favourable effects on human body and has nothing to do with divine or metaphysical influences. The amazing thing, proved by recent scientific research, is that the same effect can be achieved by reciting the Homeric Epics with their proper metre: The wondrous dactylic hexameter is the base, which in fact maintains its benevolent effect, even with translation to other languages, such as English or German.

## 1 Introduction

Besides the stunning knowledge contained in the Homeric Epics, it is the structure of the poetic word used, which leads to a phenomenally paradoxical question: How can Homer help your heart?

To this question a phenomenally even more paradoxical answer is given by several scientists in Europe and the USA. *By reciting the Odyssey*! This is not a witticism. Recitation of the Homeric Epics in their proper metre causes synchronization of heart and respiration rates in the body! Equally benevolent effect appears to have Christian prayer with the use of rosaries as well as the various mantras used in yoga, both Hindu and Buddhist.

In other words, the Homeric Epics can be used in the same way that the sacred scriptures of various religions and esoteric traditions are used for meditation.

## 2 Meditation

By this term one means either individual worship action or a mental exercise, consisting of many different techniques for concentration, contemplation and subtraction, supposedly leading to a higher spiritual realization or bodily relaxation.

*S.A. Paipetis (ed.), Science and Technology in Homeric Epics,* 531–536.

**Fig. 1** Left: The Hindu mantra AUM[2] (Sanskr.) Right: Tibetan OM.

Exercising meditation is a most ancient and universal practice in many different ambiences. It may serve *hesychastic* (soothing, calming) purposes as with certain mystics in isolation, being considered as a method of rehabilitation and enrichment of every-day life, which is the case with numerous religious and secular bodies and individuals. Also, in the form of concentration, in view of an extreme effort or trial, such as before a tough game, a performance or examinations. In any case, according to recent medical and psychological studies, meditation techniques for trained persons are substantially efficient in controlling heart and respiration rates and, to various degrees, in controlling the symptoms of such syndromes as migraine headaches, high blood pressure, haemophilia, etc.[1]

Several great religions have developed their own meditation schools, such as Hindu yoga, Tibetan and Japanese Zen, etc. Respective techniques include *spoken* (legomena), *shown* (deiknymena) and *acted upon* (drōmena).

As spoken, besides chanting and music, special syllables, words or full phrases are used, which oriental religions call *mantra* (in Sanskrit "mental tool"), islam *dhikr*, while in Christianity the role is taken up by repeated phrases, such as "Kyrie eleison" or "Ave Maria", etc. During the first stages of meditation, the novice learns how to control, in fact, to lower his/her brain activity and to be able concentrate on respiration rate or to use as mantra typical words that create no associations possibly causing thought trains. On the contrary, at advanced contemplation levels, meditation aims at a straightforward communication with the divine, practically evolving into prayer, with full phrases repeatedly uttered. In Hindu meditation AUM is the supreme mantra, the creative sound of the Universe, whose three letters express respectively birth, conservation and destruction, as expressed by the Hindu divine triad (Brahma, Vishnu, Siva). In Tibetan Buddhism the word has the form OM and is contained in the equally important mantra "Om Mani Padme Hum" (Figure 1).

With *shown* objects meditation is focusing attention on a picture, such as a flower or mountain. In many traditions may acquire typical forms, as in Tibetan Tantric

[1] The author, out of personal experience, can confirm that through meditation periodic migraine headaches can be alleviated or even disappear, and also that a strong back-ache was cured instantly, apparently after successful relaxing of the respective nerve.

[2] The mantra in Sanskrit appears on a PC screen by clicking the reverse slash \ and converting it into Microsoft *Wingdings* font.

**Fig. 2** A Buddhist mandala. In practice, mandala has become a generic term for any plan, chart or geometric pattern that represents the cosmos metaphysically or symbolically, a microcosm of the Universe from the human perspective.

Buddhism,[3] where a *mandala* (in Sanskr. *circle*) is considered as a concentration point of universal forces, which a human may contact through meditation (Figure 2).

For meditation, many traditions are using objects or mechanical devices, such as rosary and prayer wheel. Finally, *drōmena* are various motions or gestures, walking etc. synchronized with recitation of a mantra.

[3] Tantras (in Sanskrit "loom"), any of the numerous scriptures dealing with esoteric practices of certain Hindu, Buddhism or Jaina cults. Buddhist Tantras date back to the 7th century or even earlier. Tathāgataguhyaka is an early and extreme work. Tantras have been translated into Tibetan and Chinese from 9th century onwards. Only some texts in these languages are extant only, while the Sanskrit originals have been lost. An important text among Buddhist Tantras is Kālacakra-tantra.

Decreased brain activity can be obtained by exclusion of thought trains and focusing on a mantra and, as mentioned, may have benevolent effects on the human body. The weight of human brain is equal only to 2% of that of the body, however, it absorbs 20% of blood quantity and 20% of the oxygen it carries.

By EEG it is proved that during sleep the brain keeps on thinking, especially while dreaming. Decrease of thinking activity, which can be achieved through systematic meditation, causes brain energy consumption to decrease, i.e. less blood circulation, lower metabolism rate and general relaxation and soothing of the body. Similar is the change of the brain waves, whereby *alpha waves* (frequency 10 cycles per second) prevail with respective decrease of the irregular, resembling to noise, *beta* waves. With deep meditation very low frequency *theta* waves appear. Finally, results similar to those of meditation can be obtained by using *Biofeedback*.[4]

The effect of meditation on the cardio-vascular function is now under scientific investigation and its favourable effects have been confirmed. Research continues, in fact, encouraged by Dalai Lama himself.[5]

## 3 The Homeric Metre

According to Francois Haas,[6] the results concern reciting metrical Homeric poetry and its effects on human physiology: It is certain that all of the internal rhythms can be modified by external stimulations.

In a more recent study, researchers from Austria, Germany and Switzerland investigated 20 healthy individuals, men and women, of average age 43, who repeated excerpts from a German translation of the Odyssey, which somebody had been reading them. Their heart and lungs were mechanically interconnected and their behaviour was monitored. The respective findings were recently published [1].

In the German edition of the Odyssey the complex rhythmic verse, the dactylic hexameter, is maintained. Here, the verse consist of six parts, consisting of a *long*

---

[4] Biofeedback: Information instantly supplied to a person in relation to his/her own physiological parameters, e.g. blood pressure, heart rate, body temperature, brain wave of muscle tension. This information, in the form of an electronic signal, is returned to the person through a measuring element or a light or sound indication. In this way the autonomous neural system is "bridged" with the thought, so that the individual, with proper exercise, is able to control the involuntary functions of the body at will, for example, to decrease the symptom of an ailing, such as pain or muscle tension, but also migraine headaches, colitis, blood hypertension, nervous ticks, as well as frequency and intensity of epileptic fits; Through biofeedback of brain waves the brain functions are enhanced. In particular, it generates all tranquillizing and holistic effects of meditation, while exercising with theta waves leads to improved attention focusing and control of mental hindrances and stress.

[5] Leader of the ruling class Dge-lugs-pa (of the Yellow Hat) of Tibetan Buddhism and religious as well as political leader of Tibet until 1959, when independence of Tibet was abolished by Communist China. Present 14th Dalai Lama Bstan-'dzin-rgya-mtsho is Head of a Government in exile situated in Dharmsala, India at the Himalayans. He is a Nobel Prize for Peace Laureate, thanks to his non-violent struggle for the independence of his country.

[6] Research Director, Cardiovascular Rehabilitation, Medical School, New York University, USA.

*syllable* followed by a *long syllable* and a *short syllable* or of two *short syllables*. A detailed account follows.

While reading the verses, the patients, their respiration rate was decelerating, while heart and respiration rates were synchronizing more and more. These rhythms were fully de-synchronized when the pantients started breathing normally without reading, i.e. they return to their usual every day situation.

Concerning the above-mentioned studies, i.e. of the effect of Christian prayer by the use of rosaries or the utterance of Hindu or Buddhist mantra, it was found that, in that way, respiration rate may drop to 6 per minute, helping thus the heart to function more efficiently. Also, under the circumstances, oxygen content in blood reaches saturation point, e.g. an optimum condition.

In fact, scientist have been wondering if rosaries have been so popular, because people feel better by using it and more perceptive of religious messages.

One of the scientists, Dirk Cysarz, Chairman of the Department of Medical Theory and Complementary Medicine of Witten University, Herdecke, Germany, suggested that low respiration rate is associated with low blood pressure. In addition, other investigations prove that low respiration rate leads to better lung function. It is also noted that classical poetry must be properly pronounced in order to be able to affect human body, i.e. mumbling is not helpful, but every syllable of the semi-verse must by carefully pronounced and every semi-verse to be followed by a relaxed respiration, in which case the heart rate can be influenced.

Francois Haas adds that these phenomena are fully rational, without any trace of mystical dimension, since all organs within a living body are very closely interconnected. He emphasizes the existence of marches accompanying rhythmic walking or rhythmic singing of South America natives, when rowing. In essence, all of these are techniques synchronizing the respective functions.

## 4 The Dactylic Hexameter

The dactylic hexameter is a form of *metric* poetry or a rhythmic formation. Traditionally, it is related to classical poetry, mainly with the Homeric Epics, but also with Latin ones, such as Virgil's Aeneid. Dactyl (δάκτυλος) is a system consisting of three syllables: The first one is *long* and the others are *short*. Accordingly, the ideal verse of dactylic hexameter consists of six metres or feet, each one of which is dactylic. However, as a rule, the last foot of the verse is not pure dactylic, but rather a two-syllable *spondee* (σπονδείος) or *trochaic*, i.e. the syllable before the last is always long, while the last one is either long or short (such a syllable with optional accentuation is called an *anapaestic*). In essence, it is difficult to arrange words according to this metre. So, poets often replace these dactyls with spondees, which are feet with two long syllables. Traditionally, the fifth foot of a verse is pure dactylic. About one out of twenty Homeric verses has a spondee at the fifth foot. Such a verse is called *spondaic*. For example, separation of verses into feet and semi-verses in the first seven verses of the *Iliad* is as follows:

μῆνιν ἄ/ειδε, θε/ά, ‖ Πηληιά/δεω Ἀχι/λῆος
οὐλομέ/νην, ἣ / μυρί᾿ ‖ Ἀ/χαιοῖς /ἄλγε᾿ ἔθ/ηκεν,
πολλάς /δ᾿ ἰφθί/μους ‖ ψυ/χάς Ἄι/δι προί/αψεν
ἡρώ/ων, αὐ/τούς δὲ ‖ ἑ/λώρια / τεῦχε κύ/νεσσιν
οἰω/νοίσί τε /δαῖτα ‖ Δι/ὸς δ᾿ ἐτε/λείετο / βουλή,
ἐξ οὗ / δὴ τὰ π/ρῶτα ‖ διαστή/την ἐρί/σαντε
Ἀτρεί/δης τε ἄ/ναξ ‖ ἀν/δρῶν καὶ / δῖος Ἀ/χιλλεὺς

# 5 Conclusion

Starting from the above and, in particular, from the last example, one may discover that reading the ancient Greek verse is an acoustic issue. People who can read dactylic hexameters according to grammar, cannot necessarily recite it loud and with proper rhythm, therefore, the musicality of poetry is lost. Reading with sensitivity and satisfaction requires effort and exercise, while the initial rules that must be adhered to are few and simple. Certainly, with the motive of a favourable influence on physical health, it is worth trying!

# References

1. Cysarz, D. et al., Oscillations of heart rate and respiration synchronize during poetry recitation, *American Journal of Physiology – Heart and Circulatory Physiology* **287**, 2004, 2 (online edition).